IET BUILT ENVIRONMENT SERIES 01

Handbook of Ventilation Technology for the Built Environment

Handbook of Ventilation Technology for the Built Environment

Design, control and testing

Edited by
Shi-Jie Cao and Zhuangbo Feng

The Institution of Engineering and Technology

Published by The Institution of Engineering and Technology, London, United Kingdom

The Institution of Engineering and Technology is registered as a Charity in England & Wales (no. 211014) and Scotland (no. SC038698).

First published 2021

The Institution of Engineering and Technology
Michael Faraday House
Six Hills Way, Stevenage
Herts, SG1 2AY, United Kingdom

www.theiet.org

British Library Cataloguing in Publication Data
A catalogue record for this product is available from the British Library

ISBN 978-1-83953-147-7 (hardback)
ISBN 978-1-83953-148-4 (PDF)

Typeset in India by MPS Limited
Printed in the UK by CPI Group (UK) Ltd, Croydon

Contents

About the editors

Shi-Jie Cao is a full professor at Southeast University, China; a visiting professor at the University of Surrey; and the President of International Society of Built Environment (ISBE). Dr Cao's research interests are sustainable built environment design, building ventilation and air pollution control. He is now serving as the deputy editor of Indoor and Built Environment, an editorial board member of Sustainable Cities and Society, Safety Science and Buildings.

Zhuangbo Feng is an associate professor of School of Architecture, Southeast University. His research areas cover building environment and air quality in hot and humid areas, air purification, indoor air cleaning technology, ventilation. Dr Feng is now serving as an assistant editor of Indoor and Build Environment, and a guest editor of Sustainable Cities and Society.

Chapter 1
Introduction

Shi-Jie Cao[1] and Zhuangbo Feng[1]

1.1 What is building ventilation?

Building ventilation system is introducing fresh air to indoor environment or processing the indoor air to remove air pollutant and maintain indoor air quality (IAQ) at a satisfactory level. A conventional ventilation system typically consists of air distribution components, air cleaning components, a duct network, and fluid machinery. Building ventilation has been widely adopted to remove airborne pollutant/waste heat and improve indoor air quality (Cao, 2019; Deng *et al.*, 2018). Building ventilation is proven to be effective to control disease transmission and ensure indoor safety, such as airborne precautions of SARS-CoV-2 aerosol (Zhao *et al.*, 2020). Besides, ventilation can be used in building cooling/heating, improving thermal comfort level (Liu *et al.*, 2014) as well as significantly reduces energy consumption (Ren and Cao, 2019; Wang *et al.*, 2019).

Building ventilation types can be classified into natural ventilation (NV), mechanical ventilation (MV) and hybrid ventilation (HV) (Sha and Qi, 2020). NV and MV are, respectively, based on natural and mechanical power. HV combines the two basic ventilation types together.

A typical MV system consists of indoor air-distribution components, air-cleaning components, a duct network and fluid machinery (Zhao and Liu, 2020). For indoor air-distribution components, architecture design, space pattern, layout of diffuser and ventilation rate determine indoor airflow pattern and ventilation performance. For air-cleaning components, different pollutant removal devices are used to remove/disinfect particulate matter, gaseous pollutants (e.g. volatile organic compounds (VOC), ozone) and biological aerosols. For duct networks, ducts, connectors and dampers are utilized. Air dampers are selected for air balance and airflow adjustment. For fluid machinery, two essential factors (airflow rate and pressure head) are determined based on the hydraulic calculation of duct network.

NV utilizes buoyant force induced by indoor–outdoor air temperature difference and wind pressure difference to introduce outdoor air into buildings (Yuan *et al.*, 2018). No mechanical devices (e.g. fan, air filter, duct and value) are adopted

[1]School of Architecture, Southeast University, Nanjing, China

in NV. The influential factors of NV performance include building orientation, space layout, forms of door/window forms. Different from MV, NV could save a lot of energy.

1.2 Limitations and developments

For traditional building MV engineering, three stages should be considered: (1) design stage, (2) construction stage and (3) operation stage. In design stage, engineering designers utilize standards/specifications to quantitatively design ventilation system, including duct network, indoor airflow pattern and power device. A large number of semi-empirical–theoretical formulas are directly adopted by designers. The construction process of ventilation system is based on design drawings. In operation stage, the so-called TBA (testing, balancing and adjusting) method is widely used after construction. Before the actual operation of ventilation system, engineers conduct TBA to satisfy user needs. The TBA process includes damper adjustment, airflow rate testing, duct network balancing, air-purification testing, indoor air-velocity measuring and others. Up to now, the effect of TBA process strongly depends on experiences of engineers. After TBA process, users can adjust speed (e.g. high/medium/low) of fan installed in rooms to satisfy individual needs.

However, the traditional design/operation methods have some limitations. For design stage, the empirical–theoretical formulas could not accurately predict complex airflow patterns and pollutant removal efficiency in ventilation system (Chen, 2009), such as large space building, rooms with complex geometries, new ventilation patterns (e.g. displacement ventilation + radiation cooling) and ventilation devices (e.g. air filters) (Feng *et al.*, 2014). Therefore, some advanced simulation tools are proposed for improving prediction accuracy. Building ventilation performance is basically dependent on the airflow and heat/mass transfer in indoor environment/duct devices, which is governed by the Navier–Stokes equation. Computational fluid dynamics (CFD) become more and more popular in performance predictions of ventilation system. Gao *et al.* (2015) utilized the CFD method to accurately predict the pressure loss of a novel low-resistance duct tee. Tong *et al.* (2017) adopted CFD simulation to investigate and estimate NV potential for high-rise buildings considering boundary layer meteorology. However, a direct use of CFD in ventilation design has two obvious limitations: (1) computing cost and (2) optimal design. When the target simulation domain is large and turbulent flow pattern is complex, CFD simulation requires a large amount of computing meshes, resulting in very long computing time and high computing cost. In order to overcome the disadvantages of CFD, many newly proposed fast models begin to draw more and more attentions, including low-dimensional linear ventilation model, fast fluid dynamic, zonal model, multi-zone model, CFD with advanced turbulence model, CFD with coarse mesh and others (Feng *et al.*, 2019). With the rapid development of artificial intelligence and big data technology, CFD+AI can significantly improve the computing efficiency of traditional CFD method in ventilation design (Cao and Ren, 2018). The traditional optimal design of ventilation system by CFD uses a trial-and-error process. The trial-and-error process is robust and easy to use. However, the design process can take days and the obtained air distributions may not be optimal. The CFD-based adjoint methods have been developed for indoor environment optimization (Liu *et al.*, 2015), which can be utilized to automatically find the optimal design values. One of the aims of this book is to introduce and summarize the newly developed prediction models of ventilation system, including theory, methodology and case application.

For operation stage, traditional user experience based TBA could not satisfy the dynamic need of indoor environment, resulting in uncomfortable/unhealthy environment and extra energy consumption. In order to overcome the limitations, an "online monitoring and intelligent control" system has been proposed, which draws more and more attention including three key components: online monitoring, "faster-than-real-time" prediction, and optimal evaluation and control (Cao and Ren, 2018). The "online monitoring and intelligent control" system automatically adjusts building ventilation system based on limited monitored data1 of indoor environment (e.g. temperature, humidity and airborne pollutant). First, indoor environment data are monitored by "online monitoring" component. Then "faster-than-real-time" predictions are repeated, and fast evaluations are conducted to obtain optimal values for ventilation system operation/adjustment. This intelligent system can satisfy the dynamic characteristics/need of indoor environment, creating safe, healthy and comfortable indoor environments for occupants in a long operation period. The bottleneck issues are how to achieve "faster-than-real-time" prediction of indoor environment and how to combine limited monitoring and optimal air-supply parameter evaluation/control. The fast models developed for building ventilation design could provide essential reference for "faster-than-real-time" predictions in intelligent control system. This book will introduce cutting edge research achievements about ventilation monitoring and control (Figure 1.1).

For NV design and control, it is quite necessary to couple outdoor and indoor environments. However, traditional semi-empirical formulas, which assumed

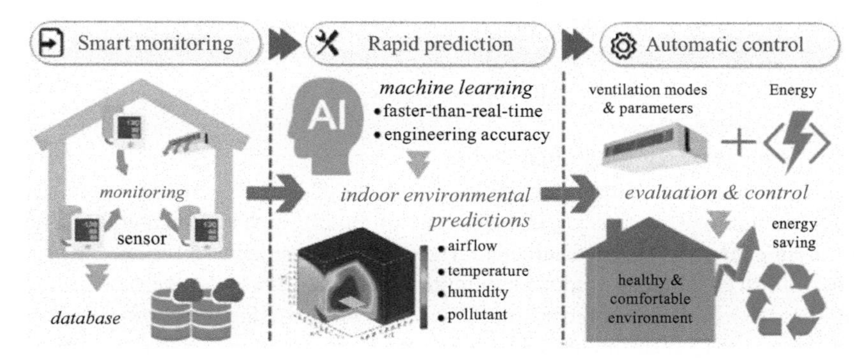

Figure 1.1 The flow-chart and key components of "online monitoring and intelligent control" system

constant outdoor climate factors, could not be directly applied in accurate design or operation control. In order to overcome such disadvantages, CFD-based simulations have been adopted to design NV (Zhou *et al.*, 2014). The influences of atmospheric environment on NV are considered in numerical design. It is quite challenging to develop design strategies to couple urban and indoor environments. Systematic methodology has also been proposed to comprehensively evaluate the NV potential. This book will present the available methodology and research achievement of urban environment simulations and coupling of indoor/outdoor predictions.

Recently, some advanced ventilation technologies have been newly proposed and effectively improve building ventilation performance, such as electrostatic-assisted air-purification system, displacement ventilation and radiation cooling system, vortex ring ventilation in industrial buildings, ventilation technology for high-rise buildings. The newly developed ventilation technologies will be summarized in this book.

One of the designs proposed for building ventilation is to dilute/remove air-borne pollutant and ensure occupant health. The widely used quantitative design strategy of building ventilation is based on pollutant concentration. However, human health (e.g. allergic diseases and respiratory infections) is not only related with indoor air pollution but also influenced by indoor climatic factors (such as air temperature, relative humidity, air velocity and turbulence characteristics). This book also introduces the most cutting edge researches focusing on the complex relationship between building ventilation and human health issue.

The global outbreaks of corona-virus disease 2019 (COVID-19), induced by severe acute respiratory syndrome corona-virus 2 (SARS-CoV-2), have yet caused serious threats to human health and economy development. Aerosol transmission of pathogens has been proven as an important pathway. Different from common air pollutant types (e.g. particulate matter, carbon dioxide), airborne SARS-CoV-2 aerosols have the following characteristics: stronger toxicity, respiratory emission and source mobility. It is quite challenging to control SARS-CoV-2 aerosols due to its complexity. To effectively control the transmission of SARS-CoV-2 aerosols, some advanced ventilation and purification/disinfection technologies have been proposed, such as occupant based ventilation control, UV+Filter system, electrostatic disinfector (ESD), etc. (Wang *et al.*, 2021; Feng *et al.*, 2021a, 2021b). The advantages and disadvantages of the COVID-19 related technologies will be introduced in this book.

1.3 The structure of this book

This book summarizes and presents the state-of-the-art methods for building ventilation design/control. The cutting edge research achievements are also introduced, including fast ventilation predictions, online monitoring and intelligent control, coupled simulation of urban/indoor simulation, CFD-based adjoint design strategy and others. In total, 17 chapters are included in this book. Chapter 1 introduces the overview of this book. Chapters 2 and 3 introduce the basic concept and knowledge of building ventilation: "ventilation definition and requirements" and "ventilation type."

Chapters 4–7 describe basic components and patterns of building ventilation: "duct network/fluid machinery," "air-cleaning technologies," "air distribution in MV system" and "air distribution in NV system." Besides, Chapter 8 introduces the advanced building ventilation system with a function of heating/cooling. Chapters 9–11 describe prediction models of different ventilation types: "NV," "MV," "fast prediction model." Chapter 12 describes the cutting edge building ventilation control system: "online monitoring and intelligent control" system. Then four special topics (Chapters 13–16) are described and analyzed: "ventilation and health," "ventilation in industry buildings," "ventilation and fire safety for high-rise buildings" and "urban ventilation and design." Finally, conclusions and future perspectives are presented in Chapter 17.

References

Cao, S-J. Challenges of using CFD simulation for the design and online control of ventilation systems. *Indoor and Built Environment*, 2019, 28, 3–6.

Cao, S-J. and Ren, C. Ventilation control strategy using low-dimensional linear ventilation models and artificial neural network. *Building and Environment*, 2018, 144, 316–333.

Chen, Q. Ventilation performance prediction for buildings: A method overview and recent applications. *Building and Environment*, 2009, 44, 848–858.

Deng, H., Feng, Z., and Cao, S-J. Influence of air change rates on indoor CO_2 stratification in terms of Richardson number and vorticity. *Building and Environment*, 2018, 129, 74–84.

Feng, Z., Cao, S-J., and Haghighat, F. Removal of SARS-CoV-2 using UV+Filter in built environment. *Sustainable Cities and Society*, 2021a, 74, 103226.

Feng, Z., Cao, S-J., Kumar, P., and Haghighat, F. Indoor airborne disinfection with electrostatic disinfector (ESD): Numerical simulations of ESD performance and reduction of computing time. *Building and Environment*, 2021b, 200, 107956.

Feng, Z., Long, Z., and Chen, Q. Assessment of various CFD models for predicting airflow and pressure drop through pleated filter system. *Building and Environment*, 2014, 75, 132–141.

Feng, Z., Yu, C., and Cao, S-J. Fast prediction for indoor environment: Models assessment. *Indoor and Built Environment*, 2019, 28, 727–730.

Gao, R., Zhang, H., Li, A., Liu, K., Yu, S., and Deng, B. A novel low-resistance duct tee emulating a river course. *Energy and Buildings*, 2015, 91, 91–100.

Liu, W., Huangfu, H., Xiong, J., and Deng, Q. Feedback effect of human physical and psychological adaption on time period of thermal adaption in naturally ventilated building. *Building and Environment*, 2014, 76, 1–9.

Liu, W., Zhang, T., Xue, Y., *et al.* State-of-the-art methods for inverse design of an enclosed environment. *Building and Environment*, 2015, 91, 91–100.

Ren, C. and Cao, S-J. Development and application of linear ventilation and temperature models for indoor environmental prediction and HVAC systems control. *Sustainable Cities and Society*, 2019, 51, 101673.

Sha, H. and Qi, D. A review of high-rise ventilation for energy efficiency and safety. *Sustainable Cities and Society*, 2020, 54, 101971.

Tong, Z., Chen, Y., and Malkawi, A. Estimating natural ventilation potential for high-rise buildings considering boundary layer meteorology. *Applied Energy*, 2017, 193, 276–286.

Wang, J., Huang, J., Feng, Z., Cao, S. J., and Haghighat, F. Occupant-density-detection based energy efficient ventilation system: Prevention of infection transmission. *Energy and Buildings*, 2021, 240, 110883.

Wang, H., Olesen, B., and Kazanci, O. Using thermostats for indoor climate control in offices: The effect on thermal comfort and heating/cooling energy use. *Energy and Buildings*, 2019, 188–189, 71–83.

Yuan, S., Vallianos, C., Athienitis, A., and Rao, J. A study of hybrid ventilation in an institutional building for predictive control. *Building and Environment*, 2018, 128, 1–11.

Zhao, L. and Liu, J. Air purifiers: A supplementary measure to remove airborne SARS-CoV-2. *Building and Environment*, 2020, 170, 106600.

Zhao, B., Liu, Y., and Chen, C. Air purifiers: A supplementary measure to remove airborne SARS-CoV-2. *Building and Environment*, 2020, 177, 106918.

Zhou, C., Wang, Z., Chen, Q., Jiang, Y., and Pei, J. Design optimization and field demonstration of natural ventilation for high-rise residential buildings. *Energy and Buildings*, 2014, 82, 457–465.

Chapter 2
Ventilation definition and requirements

Li Bai[1]

2.1 Concept of building ventilation

2.1.1 Definition of building ventilation

A ventilation system is designed to improve indoor polluted air by introducing fresh air or by processing the indoor air to maintain the indoor air quality (IAQ) and thermal comfort at a satisfactory level that meets the standards of residential or industrial use. A conventional ventilation system typically consists of air distribution components, air cleaning components, a duct network, and fluid machinery.

Before the development of modern ventilation systems, the IAQ was controlled by natural ventilation, which is a passive ventilation method in architectural design. Due to the limitation of science and technology, people had relatively few options to adjust the indoor environment. Windows were opened to bring in fresh air for cooling and air clearing because stoves and radiators were the main sources of heat in cold weather. Open stoves produce smoke and dust, reducing the IAQ. In addition, the carbon monoxide (CO) produced by unburned combustion poses serious threats to people's health and lives.

Due to the limits of natural ventilation to adjust the indoor environment, mechanical ventilation was invented and became the main ventilation method. By the 1950s, the supply of cheap energy and the widespread use of air-conditioning had a profound impact on the type and design of buildings. The ability to control indoor temperature and humidity by mechanical means eliminated the constraints architects faced regarding the type and design of buildings and window openings. Architects and engineers were no longer limited to traditional buildings that relied on natural ventilation. Curtain wall windows became common to allow for more daylight indoors. However, large-area curtain wall windows increase the sun's heat gain, placing a higher load on the air-conditioning system.

The energy crisis and environmental degradation have also prompted people to reevaluate existing ventilation methods. People realized the need to save energy and began to focus on reducing global energy consumption by reducing the energy consumption of heating, cooling, and ventilation in buildings. The proposed

[1]School of Municipal and Environmental Engineering, Jilin Jianzhu University, Changchun, China

solutions have focused primarily on improving the insulation level of the building enclosure and reducing air infiltration by sealing the building. In other words, the main goal is to reduce heat loss by enhancing the building insulation and reducing unwanted natural ventilation. However, the development of energy-saving buildings significantly increases the airtightness of buildings, and the amount of air entering buildings is minimized to reduce energy consumption. Coupled with the use of various new construction and decoration materials, indoor pollutants cannot be removed rapidly, and the resulting health problems have attracted increased attention. Therefore, the demand for "healthy houses" has increased. The health problems of housing have intensified in the last few decades. Since the sick building syndrome (SBS) was officially defined by the World Health Organization (WHO) in the 1970s, new ventilation standards were established, or existing ones were revised to deal with deteriorating IAQ. Currently, ventilation occupies an important position in the building design process because people have high standards of IAQ and comfort. However, most ventilation standards focus only on the minimum IAQ and thermal comfort. With an increasing focus on green methods regarding ventilation design (MacNaughton *et al.*, 2016; Juan *et al.*, 2017), the energy efficiency of ventilation systems has to be addressed. The designers and operators should be familiar with the ventilation and energy efficiency requirements to achieve satisfactory operation of ventilation systems, which can pose great challenges to the design and control. Novel ventilation systems, coupled with intelligent control, are necessary to balance IAQ and energy efficiency (Ren and Cao, 2020). Besides, novel ventilation components and advanced simulation methods also provide solutions to the trade-off between high air quality and energy efficiency. This chapter is a summary of the remaining chapters in this book.

2.2 Purpose of building ventilation

Building ventilation is necessary for the health of inhabitants for the following reasons: (a) to bring fresh air into the building from the outside to improve the oxygen content of the air; (b) to remove excess heat and moisture from the air to ensure the thermal comfort of inhabitants; and (c) to remove pollutants from the air emitted by human beings and construction materials. We spend more than 90% of our time indoors, including at home, in the office, at school, and other public facilities. Human beings need more than 10,000 L of fresh air every day. Aside from ensuring the basic physiological needs of the human body, building ventilation plays an important role in indoor thermal comfort (ITC) and IAQ. The quality of the indoor environment is paramount for people's health and well-being. The types of indoor pollutants include biological pollutants (human, pets, and plants), particulate matter (outdoor, tobacco), and chemical pollutants (building materials, cosmetics). Because a person's sensory organs cannot detect harmful substances, their harm to the human body is concealed and latent. Hence, we should recognize the importance of building ventilation and our motivation to create a comfortable, healthy, and safe indoor environment.

2.2.1 Types of ventilation systems

The objective of building ventilation is to recirculate the indoor air and exchange indoor air and outdoor air, which requires driving forces. There are two categories of ventilation systems, i.e., natural ventilation systems driven by thermal or wind pressure and mechanical ventilation systems. Natural ventilation is also called passive ventilation in contrast to active ventilation, which relies on mechanical equipment to power fans.

2.2.1.1 Natural ventilation systems

Natural ventilation uses thermal pressure caused by the air density difference between indoors and outdoors or wind pressure to introduce outdoor air for ventilation. No mechanical power is required.

The physical mechanism of natural ventilation depends on the pressure difference at the opening of the building enclosure. The pressure difference is caused by the following factors:

1. the influence of wind;
2. the temperature difference between the air inlet and outlet (gravity acts on density);
3. a combination of both.

According to the physical mechanism, natural ventilation can be divided into "wind-induced" and "buoyancy-induced" ventilation. The wind produces pressure around the building, causing wind-induced ventilation (see Figure 2.1). The pressure difference drives the air into the periphery of the windward side (positive pressure zone) of the building, and the air exits the building through the opening on the leeward side (negative pressure zone). The pressure effect of the wind on a building is primarily determined by the shape of the building, the wind direction, the wind speed, and the surrounding environment. These factors affect the pressure coefficient. In addition to the pressure coefficient, the average pressure difference

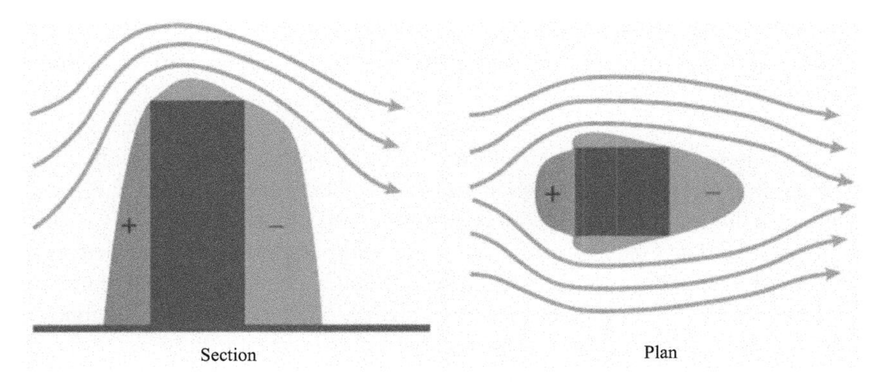

Figure 2.1 Effect of the wind on the pressure distribution around buildings (Wood, 2013)

on the building enclosure also depends on the average wind speed at the height of the upwind building. The indoor air density is a function of atmospheric pressure, temperature, and humidity.

Buoyancy can create ventilation (also known as the "chimney effect") due to density differences caused by temperature and height changes inside and outside the building or between certain areas. The pressure difference produced by buoyancy mainly depends on the height of the chimney (the height difference between the intake and exhaust ports) and the air density difference, which is a function of the temperature and moisture content in the air. The outdoor temperature has to be lower than the indoor temperature to achieve buoyancy-induced ventilation in the absence of wind. When the indoor air temperature exceeds the outdoor temperature, negative pressure occurs in the lower part of the building, and the air is sucked in through the openings on the periphery of the building. (If the outdoor temperature is equal to the indoor temperature, buoyancy is still generated due to the indoor load.) As the air passes through the building, it is heated by the internal load and people in the building. The density difference caused by the temperature difference between the indoor and outdoor causes different pressure gradients in the building. The overpressure area on the top of the building vents the air out of the building's opening (because the airflows from the high-pressure area to the low-pressure area). However, at a certain height of the building, the indoor pressure is equal to the outdoor pressure. This level is called the "neutral level" or "neutral stress level" (see Figure 2.2).

A temperature difference between the air inlet and the outlet is required for buoyancy-induced ventilation. It should be noted that when the outdoor air temperature is significantly higher than the internal temperature of the building, a "reverse stacking effect" will occur. Under this condition, air can enter at the upper areas of the building and exhaust from the low areas; this reverse stacking effect is difficult to control. Finally, it should be noted that the two driving forces of natural ventilation (wind and buoyancy) can occur separately but typically co-occur. In calm weather with no wind, thermal buoyancy is usually the dominant driving force, whereas, in windy conditions, the pressure difference generated by the wind is usually the main driving force.

According to the airflow mode, natural ventilation systems can be divided into the following:

1. One-sided ventilation. The fresh air enters the room through an opening and then exits on the same side. If the depth of the room is up to 2.5 times its height, this strategy can effectively ventilate the space (Figure 2.3). The driving force of one-sided ventilation is the temperature difference between the low-level air inlet and the high-level air outlet, which causes air movement. When the ventilation openings are located at different heights, the buoyancy effect can also induce one-sided ventilation.
2. Cross ventilation. Cross ventilation occurs due to the pressure difference between the openings on both sides (the air moves from the windward side to the leeward side). It relies on the airflow between the two sides of the building enclosure.

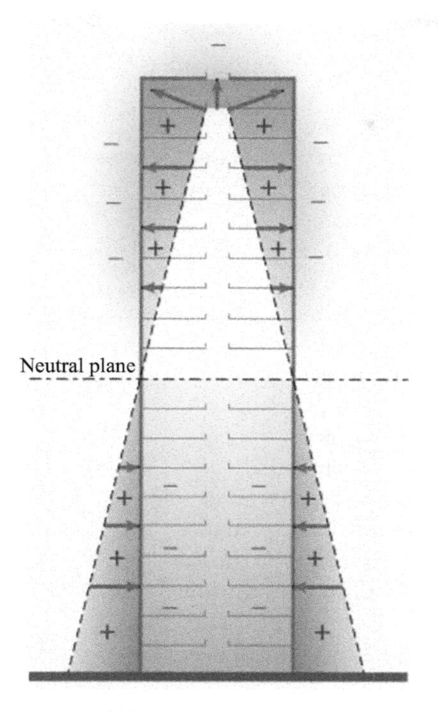

Figure 2.2 Thermal buoyancy diagram of high-rise buildings with different pressure gradients inside the building caused by indoor–outdoor temperature differences (Wood, 2013)

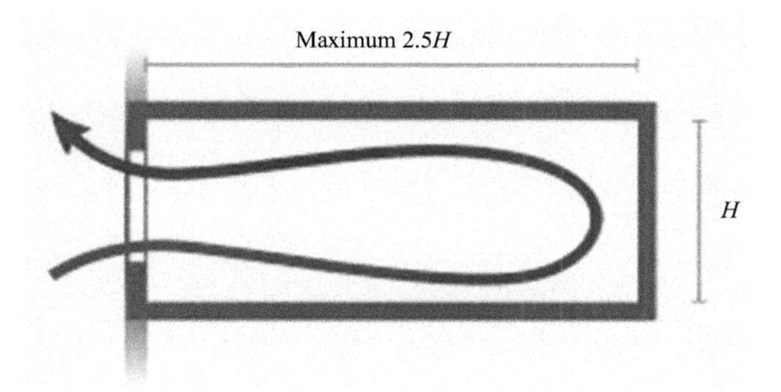

Figure 2.3 One-sided ventilation (Wood, 2013)

The depth of the room must not exceed five times its height (Figure 2.4) to achieve effective ventilation. The buoyancy effect can also induce convective ventilation when there is a high open space (such as an atrium).

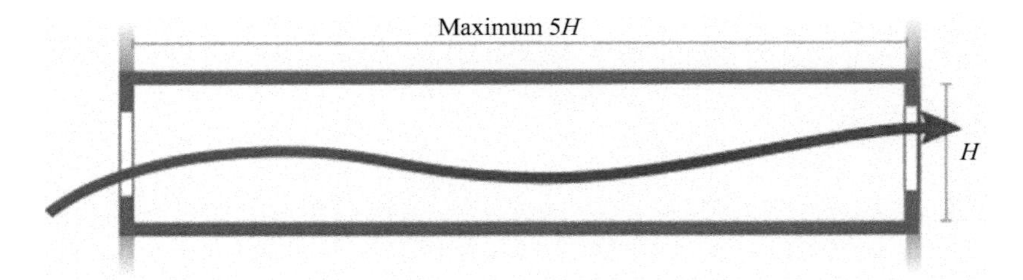

Figure 2.4 Cross ventilation (Wood, 2013)

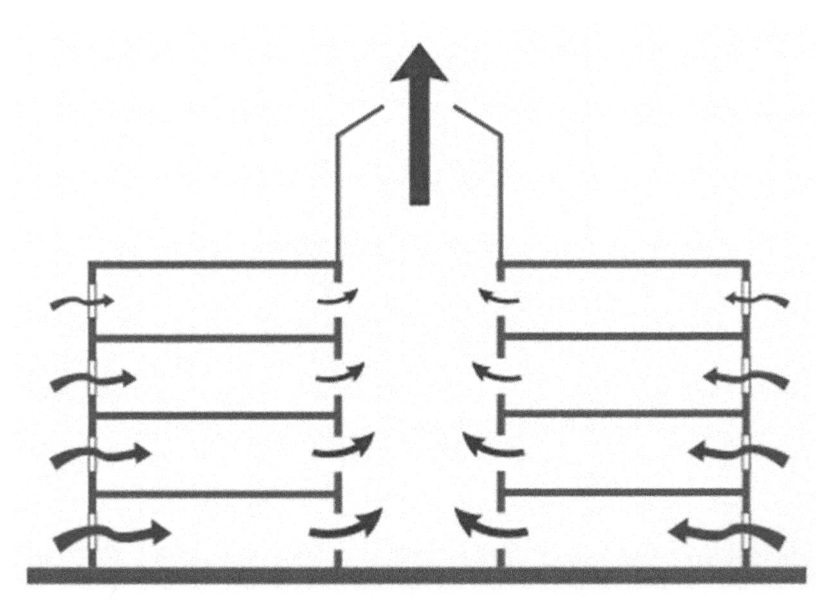

Figure 2.5 Stacked ventilation (Wood, 2013)

3. Stacked ventilation means that the fresh air enters the building at a low level and is discharged at a high level due to the temperature, density, and pressure differences between the inside and outside of the building or between different areas in the building. Stacked ventilation is often used in buildings with atriums and chimneys or elevated structures (Figure 2.5).

Local and climatic conditions influence the ability to rely solely on natural ventilation for cooling. For example, the use of natural ventilation may not be suitable for buildings located in areas with extreme weather (extremely cold, hot, and/or humid), or in places with high noise and pollution. In this case, a suitable strategy for mixed ventilation (MV) should be used, such as the combination of natural ventilation (using windows and/or vents) and mechanical systems. Both systems can be used simultaneously, or the systems can be converted. Mixed-mode buildings usually contain complex building management systems and control strategies that allow overlap/alternation between natural ventilation and mechanical cooling systems.

2.2.1.2 Pros and cons of natural ventilation

Natural ventilation is an important passive green building technology that can reduce the energy consumption of air-conditioning and mechanical ventilation. The effectiveness of natural ventilation depends largely on the climate, building type, function, design, control strategy, and other aspects. Studies have shown that natural ventilation can significantly reduce energy costs (Cardinale *et al.*, 2003). Under certain conditions of wind pressure or thermal pressure, natural ventilation can provide high ventilation rates, improving occupant comfort and IAQ and reducing the energy consumption of air-conditioning and mechanical ventilation. Apart from the large ventilation rates, natural ventilation is widely applicable and satisfies the IAQ requirements. Thus, it is an economic ventilation method suitable for residential buildings, office buildings, industrial buildings, and high-temperature workshops.

However, a strong and stable wind/thermal pressure depends on multiple factors (building structure, prevailing wind direction, and climatic conditions) (van Moeseke *et al.*, 2005). Wind and thermal pressure differ at different times of the day. Designers should consider these factors to ensure the effectiveness of natural ventilation systems, which may pose challenges to the design and operation of ventilation systems. Low-energy mechanical ventilation systems combined with efficient heat recovery and effective energy storage have higher annual comfort ratings than natural ventilation strategies and have relatively low-energy consumption (Braham, 2000). Therefore, natural ventilation systems are usually used for auxiliary purposes.

2.2.1.3 Mechanical ventilation systems

The type of mechanical ventilation system depends on the design requirements of the indoor built environment. Mechanical ventilation systems can be categorized based on the air supply method into MV (van Hooff and Blocken, 2020), diffuse ceiling ventilation (Lestinen *et al.*, 2018), displacement ventilation (DV) (Schmeling and Bosbach, 2020), underfloor air distribution (UFAD) (Fathollahzadeh *et al.*, 2015), wall-attached ventilation (Cho *et al.*, 2008), stratum ventilation (Cheng and Lin, 2016), personalized ventilation (PV) (Zhai and Metzger, 2019), and other systems (Yang, 2019) (Figure 2.6). Here we briefly introduce the most common ventilation systems (MV, DV, UFAD, and PV). See Figure 2.6 for the specific mechanical system legend. The other ventilation systems will be discussed in Chapter 3.

Mixed ventilation

MV systems mix the air inside the occupied zone with an external air supply at a high velocity (momentum). Due to the high velocity rate of the jet stream, the air inside the occupied zone is entrained into the air jet and mixed. The mixing of the external and indoor air results in a uniform indoor thermal environment and air quality. MV systems result in a comfortable indoor thermal environment.

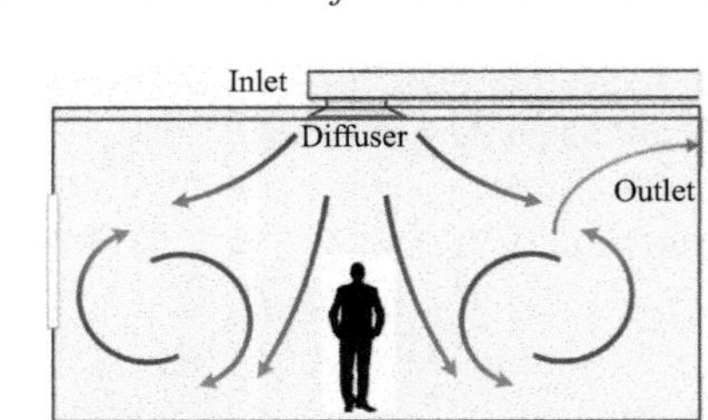

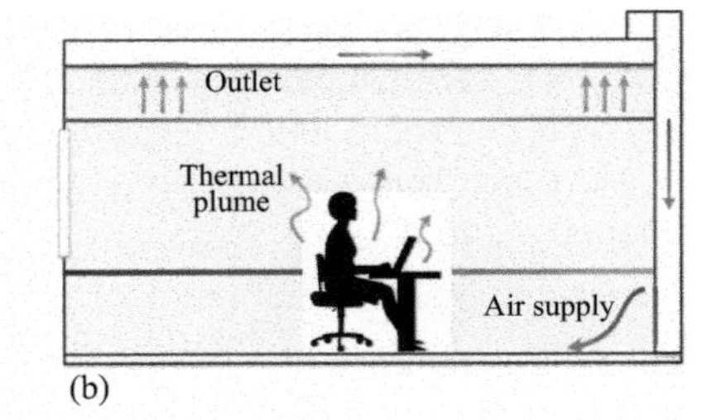

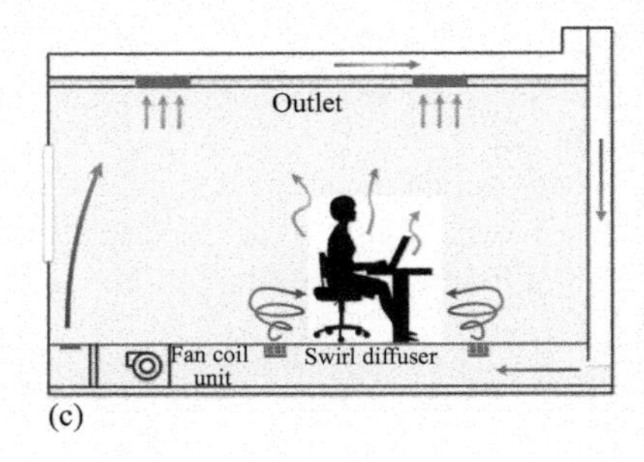

Figure 2.6 Conventional mechanical ventilation systems: (a) mixed ventilation system, (b) displacement ventilation system, and (c) underfloor air distribution system

In summer, the air supplied by the air-handling units is too cold to bring into the building. Thus, the mixing of external and indoor air creates a comfortable thermal environment. In winter, the fresh cold air entering the room is premixed with the warmer room air to achieve a comfortable temperature. The polluted air is diluted efficiently by mixing it with high-velocity fresh air. Compared with the conventional active air-conditioning system, it is possible to save nearly half of the energy consumption of the equipment by using the hybrid mode to maintain the indoor environmental conditions (Ezzeldin and Rees, 2013). However, if the outdoor air is polluted, this type of ventilation system brings contaminants inside the building (Melikov *et al.*, 2003).

Displacement ventilation

DV systems displace the indoor air by supplying external air through a stratified flow without mixing. The most common airflow patterns of DV are thermal plume and piston flow. When a significant temperature difference exists between the indoor air and supply air, the colder fresh air replaces the warmer indoor air, or the warmer air replaces the colder air by buoyancy as the main driving force of the process, i.e., the thermal plume. When there are no or negligible temperature differences in the room, piston flow DV may be used. The objective of this type of DV is to keep the fresh and polluted air separated. Temperature and contaminants have different distributions. The temperature gradient is linear, whereas contaminants are distributed in layers. The air quality of DV is better than that of MV; however, it is more difficult to achieve thermal comfort using DV due to the temperature gradient. In a room with DV, nearly half (49%) of the residents were not satisfied with the local discomfort caused by ventilation and vertical temperature difference (Melikov *et al.*, 2005).

Underfloor air distribution

As a new method to provide air-conditioning and ventilation for commercial buildings and data centers, UFAD has attracted the attention of researchers and designers because of its potential advantages, such as good ITC and IAQ, layout flexibility, life cycle cost reduction, and energy-saving (Zhang *et al.*, 2014). UFAD systems deliver conditioned air into the occupied zone using an underfloor supply system located between the concrete slab and the floor system. The air enters through ground level diffusers, and the mixed polluted air is exhausted from top outlets (American Society of Heating, Refrigerating and Air-Conditioning Engineers (ASHRAE), 2013). Unlike in DV systems, there is no layered contaminant distribution since fans drive the diffusers. Thus, the mixing of the indoor air and fresh air results in lower pollutant removal efficiency than in DV systems but in higher efficiency than in MV systems. The diffusers in the UFAD system provide localized air distribution (with individual control), resulting in excellent thermal comfort.

Personalized ventilation

PV systems aim to provide high IAQ directly to the breathing zone of each occupant based on individual needs and preferences. PV systems were first implemented in aircraft cabins (Gao and Niu, 2008). In recent years, PV systems have been used as a viable ventilation concept for indoor built environments (Gao and Niu, 2004). PV can improve the occupants' comfort, decrease SBS symptoms, and reduce the risk of contagion transmission between occupants. However, the design (air distribution and control) and type of occupant activity (occupancy rate and density) have to be carefully considered to ensure high efficiency.

2.2.1.4 Pros and cons of mechanical ventilation

The ventilation rate of mechanical ventilation systems can be adjusted throughout the year and is not affected by the outside climate. The air brought into the room can also be preprocessed (heated or cooled, humidified, or dehumidified) according to requirements. Thus, mechanical ventilation is highly personalized and caters to the high demand for air quality in specialized enclosures.

However, high-energy inputs are required for buildings with high ventilation rate requirements. An example of this is health-care facilities. The WHO recommends a minimum ventilation rate of 288 m^3/h per person to minimize the infection risk in health-care facilities (Atkinson, 2009). Most mechanical systems do not satisfy this requirement by diluting the indoor air. In addition, the mixing of indoor air and fresh air at a relatively low ventilation rate may lead to the spread of pollutants to the remainder of the building for certain airflow patterns.

2.2.2 Indoor thermal comfort

Among the five factors contributing to the thermal comfort of human beings, three are environmental factors (air temperature and radiant temperature, humidity, and airflow speed), and two are personal factors (physical activity and clothing). The five factors affect the heat balance of the human body, which is defined as follows:

$$S = M - W - R - C - K - E$$

where S is the heat storage in the body, W/m^2 (same unit for the other variables); M is the metabolic rate, W is the mechanical work, R is the heat exchange by radiation, C is the heat exchange by convection, K is the heat exchange by conduction, and E is the evaporative heat loss.

When the value of S is positive, a person feels warm, and when it is negative, a person feels cold. When $S = 0$, the gain and loss of heat are in balance. More details on the calculation of these terms will not be discussed here, and readers can refer to the American Society of Heating, Refrigeration and Air-Conditioning Engineers (ASHRAE) handbook (ASHRAE, 1981).

However, the gain and loss of heat can not be directly expressed by the cold and hot feeling of the human body. Therefore, thermal comfort is introduced to describe people's perception of the cold and hot state of the surrounding environment. Due to individual variations, it is difficult to determine whether a certain condition represents high or low thermal comfort. Thus, many indices have been developed to quantify thermal comfort, such as the predicted mean vote (PMV) (Fanger and Toftum, 2002) and the predicted percentage of dissatisfied (PPD) (Gil-Lopez *et al.*, 2013). The PMV represents the average vote of the sensation of a large group of people exposed to the thermal conditions of interest. The scale is $++$ hot, $+2$ warm, $+1$ slightly warm, 0 neutral, -1 slightly cool, -2 cool, and -3 cold. The PPD is related to the PMV and is based upon the individual variation in response to a given set of conditions. A value of PMV $= 0$ is neutral and reflects the comfort conditions with an associated PPD of 5%. A PMV of $+1$ or -1 provides a PPD of around 25%.

Ventilation should keep the indoor air temperature and humidity at comfortable levels while meeting indoor pollutant standards. The comfort of a person depends on the degree of activity. In addition to reducing the air supply parameters, the cooling effect of ventilation on a human body can be achieved by increasing the air supply speed because it increases the sweat evaporation rate of the skin. However, a very high air velocity will affect thermal comfort. Figure 2.7 shows the comfort zones for different intensities of human activity based on the ISO 7730 standard. The red area is the optimum working temperature in winter. The higher the intensity of human activity, the lower the temperature required to be in a comfortable state. Therefore, designers have to plan for different indoor temperature and humidity conditions according to different building functions to meet people's different production and living needs. Figure 2.8 shows the relationship between the comfort zone and the temperature, air velocity, and turbulence (ISO 7730).

Ventilation standards have changed over time. People have recognized that the ventilation standards have lagged behind the requirements of the sustainable

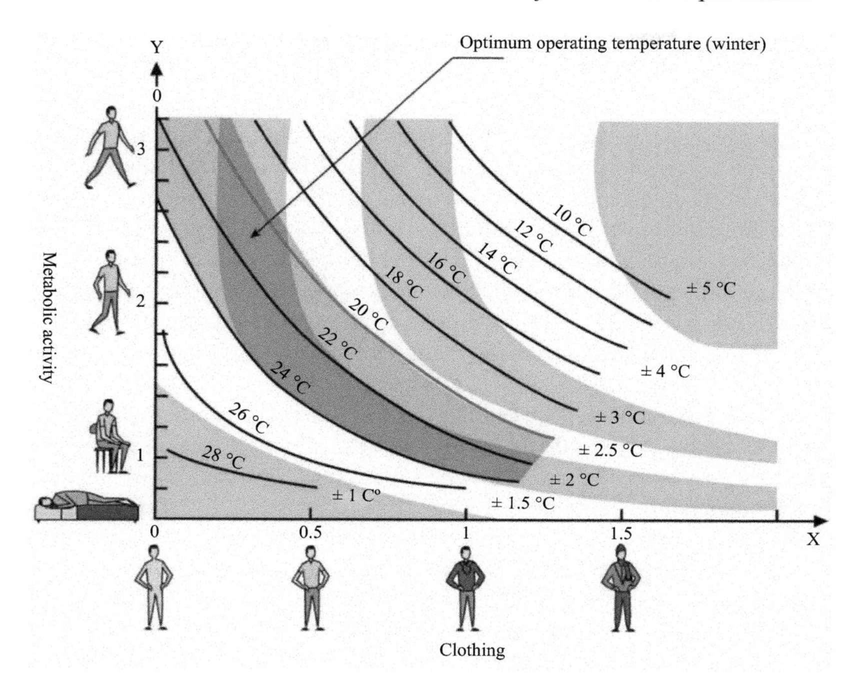

Figure 2.7 Changes in the comfort zone for different intensities of human activity (Bernard, 2016)

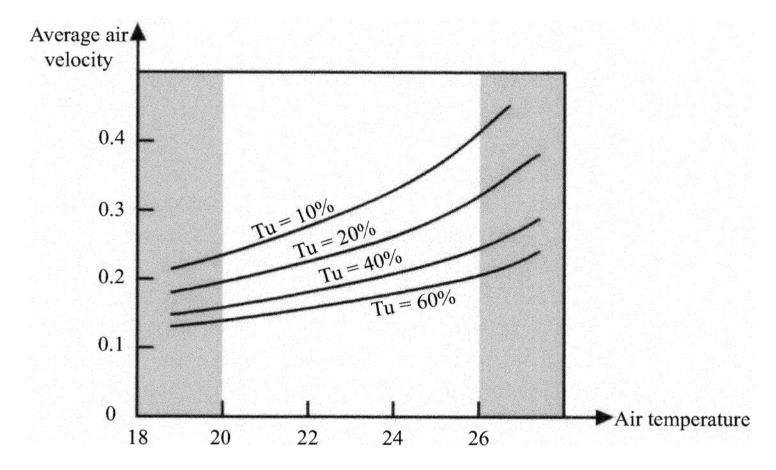

Figure 2.8 ISO 7730 relationship between the comfort zone and the temperature, air velocity, and turbulence (Bernard, 2016)

development of the building environment. Natural ventilation or multiple ventilation technologies should be used to achieve better climate control in buildings. The parameters of the traditional thermal comfort standards ignore cultural, social, and traditional factors and do not consider the complexity of the interaction between humans and the environment. Humans can actively change their behavior or gradually adapt to the environment. New research has shown that the thermal balance model cannot explain the comfort of natural ventilation. Human comfort also changes with external climatic conditions, which is because the feeling of comfort is subjective and objective adaptive. European countries have proposed adaptability standards for natural ventilation buildings in response to this situation. There is no fixed comfort temperature value in the adaptability standard; instead, a comfort temperature range with seasonal variation is used. The acceptable indoor temperature range is related to the monthly average outdoor temperature. The wide range of the comfort standard allows designers to design and operate buildings to optimize thermal comfort and achieve energy-savings.

For naturally ventilated buildings, a localized and more environmentally responsible method should be used for comfort evaluation instead of the conventional PMV method for air-conditioned buildings. Psychological adaptability, behavioral adaptability, and physiological thermal adaptability are all necessary for evaluating the thermal comfort of naturally ventilated buildings. In the 1990s, the Passive Cooling Project (PASCOOL) research group composed of members from various European countries conducted a systematic study on natural ventilation methods. Many experimental studies were conducted, and the members of the PASCOOL analyzed various methods of natural ventilation: one-sided ventilation, cross-flow ventilation, airflow prediction, and night ventilation. The results provided significant research progress in this field. However, the models used in most studies, such as the multi-area network model, are not sufficiently accurate. The model regards indoor air distribution as a homogeneous field, making it impossible to predict the velocity and temperature of the airflow in sufficient detail.

Among the relevant standards, the NF EN ISO 7730 (ISO, 2005) standard specifies the following: (a) the diffusion of heated or cooled fresh air without airflow or discomfort; (b) the ventilation equipment must be installed in a location that does not generate damp heat (cold airflow) and acoustics (equipment noise, external noise propagation). The NF EN ISO 7730 standard also defines control areas related to the room's operating temperature, air humidity, and occupants' clothing.

In 2013, the ASHRAE 62-2013 Standard 3 proposed a new definition of acceptable IAQ: air in which there are no known contaminants at harmful concentrations as determined by cognizant authorities and with which a substantial majority (80% or more) of the people exposed do not express dissatisfaction (WHO, 2010). The first part of this definition means that the allowable concentration of known pollutants should be used as an objective indicator; the latter part means using human perception as a subjective evaluation index. The air quality should meet the objective and subjective evaluation indicators. For example, if the known pollutants do not exceed the allowable concentration but more than 20% of the people in the environment are not satisfied with the air quality, then the air quality of the environment is unacceptable. The subjective and objective evaluation of air quality reflects the current stricter requirements for air quality.

France stipulates that the main pollutants in residential buildings are those related to metabolic activities and human activities (humidity and CO). Therefore, the flow rate is established according to the size of the residence. In June 2010, the standard NF ISO 16814 "Building Environment Design-Indoor Air Quality-Human Occupational Indoor Air Quality Expression Method" (BN ISO, 2008) helped designers define IAQ standards for building design. The standard requires that all pollutants have to be considered according to the building type, occupied area, and use. Pollution must be reduced or captured at the source and diluted by ventilation.

2.2.3 Indoor air quality

The WHO defines "Health" as a person's physical, mental, and social adaptations in good condition, not just the absence of disease. Accordingly, a "healthy building" must ensure that the occupants are in a state of "physical and psychological well-being." The IAQ, thermal environment, acoustic environment, light environment, and other aspects of a "healthy house" should meet the requirements. Maintaining good IAQ is the basic requirement of a "healthy building." The most basic and effective means to maintain good IAQ is ventilation. Ventilation brings in fresh air (fresh air volume) and dilutes and eliminates various indoor pollutants. Therefore, building ventilation has the following functions:

1. Meet the needs of fresh air.
2. Eliminate indoor pollutants or odor.
3. Ensure the thermal comfort of the inhabitants.

Hazardous substances emitted from buildings, construction materials, and indoor equipment or resulting from indoor human activities, such as combustion of using fuel for cooking or heating, lead to a broad range of adverse health problems and may even be fatal. Indoor air pollutants include fine and ultrafine particulate matter ($PM_{2.5}$, PM_{10}) (Wu *et al.*, 2019), formaldehyde (Yu and Kim, 2010), volatile organic compounds (VOCs) (Yu and Kim, 2010), airborne fungi (Yassin and Almouqatea, 2010), CO (Zhang *et al.*, 2020), nitrogen oxide (NO_x) (Karakitsios *et al.*, 2015; Wang *et al.*, 2019), ozone (O_3) (Gall and Rim, 2018), and radon (Pacheco-Torgal, 2012). Table 2.1 lists six typical indoor air contaminants (CO, NO_2, formaldehyde, VOCs (benzene and naphthalene)) and their threshold limits. Short- and long-term exposure limits are given. Note that for some pollutants, there are no safe levels of exposure, either short- or long term. Exposure to high levels over a short period or to low concentrations over a long period will pose serious health risks to occupants or death. Strict standards and guidelines are imperative for people's environmental health. Apart from the already-mentioned indoor air contaminants, carbon dioxide (CO_2) should also be considered since the CO_2 concentrations have been used for decades to characterize building ventilation and IAQ (Persily and de Jonge, 2017).

Many diseases are related to the office environment, such as SBS and building-related symptoms (BRSs). SBS describes an increasingly common pattern of symptoms found among employees in modern office buildings. The core symptoms include

Table 2.1 Threshold limits of typical indoor air contaminants according to the WHO IAQ guidelines (WHO, 2010)

Agent	TLV-TWA*		TLV-STEL†		Critical outcome(s)
	Level	**Time-weighted average**	**Level**	**Exposure time**	
$PM_{2.5}$	10–40 μg/m^3	1 year	–	–	Risk of cardiovascular disease and death
CO	7 mg/m^3	24 h	100 mg/m^3	15 min	Ischemic heart disease
NO_2	40 μg/m^3	1 year	200 μg/m^3	1 h	Respiratory infection
Formaldehyde	–	–	0.1 mg/m^3	30 min	Sensory irritation
Benzene	No safe threshold	–	No safe threshold	–	Acute myeloid leukemia
Naphthalene	0.01 mg/m^3	1 year	–	–	Respiratory tract lesions leading to inflammation and malignancy
Radon	100 bq/m^3	1 year	–	–	Lung cancer

No safe threshold means no safe level of exposure is recommended.

*TLV-TWA, threshold limit value–time weighted average.

†TLV-STEL, threshold limit value–short-term exposure limit.

drowsiness, mucous membrane irritation, headache, eye irritation, and dry skin (Lyles *et al.*, 1991). BRSs also refer to a set of symptoms that are associated with the office environment but diminish away from the workplace (Buchanan *et al.*, 2008). SBS and BRS are not related to a certain cause, but there are still some common differences between them. SBS and BRS are affected by different ventilation volumes, and the detection mechanism of two IAQ diseases is very different. SBS is more complex than BRS. SBS is more dependent on the quality of air filter than BRS, and SBS can be used as an index to classify environmental symptoms (Alsumaiti, 2013).

According to the WHO (WHO, 2016), 91% of the world's population live in places where indoor air contaminants exceed the WHO guideline limits. In view of the seriousness of indoor air pollution and SBS and BRS caused by air pollution, building ventilation plays a more important role in people's health.

The diffusion of pollutants is mainly due to pollution sources, transmission routes, and pollutant receptors. Among them, the most effective way is to cut off the pollution source. However, due to the complexity of indoor environment and the diversity of pollution sources, besides controlling the emission of these pollutants from the source, building ventilation is also an effective method to remove pollutants and dilute indoor air pollutants. Table 2.1 shows the threshold limits of typical indoor air pollutants in accordance with the WHO IAQ guidelines.

Some of the mentioned harmful substances, such as formaldehyde, can persist indoors for several years. Therefore, people will be affected by these harmful substances for a long time. Because the human body's sensory organs cannot detect the existence of these harmful substances, their harm to the human body is concealed and latent. Major diseases may occur within 10 years or more, or congenital diseases of newborn infants caused by genetic mutations and contamination of pregnant women may develop. Children and the elderly are highly vulnerable because of their weaker resistance.

The most straightforward and effective method to reduce indoor air pollution and improve IAQ is to increase the airflow velocity of the indoor air and the ventilation air volume and accelerate the discharge of indoor polluted air and the inflow of fresh outdoor air.

It is observed in Figure 2.9 that the indoor pollution level decreases exponentially with an increase in the natural ventilation air volume. The airflow speed inside the building should not be excessively high because this will transport the deposited bacteria and dust back inside and adversely affect the thermal comfort level of the occupants (such as the sensation of blowing and the temperature difference caused by uneven airflow). Therefore, it is necessary to find the best balance between IAQ, ventilation efficiency, energy use, and thermal comfort.

2.2.4 *Indoor environmental quality (IEQ) and productivity*

The indoor thermal environment, IAQ, luminous environment, and the acoustic environment are indices that describe the IEQ. Investigations have shown that indoor environmental conditions may influence the performance, productivity, health, and well-being of office occupants (Lan *et al.*, 2011). Residents consider that the main

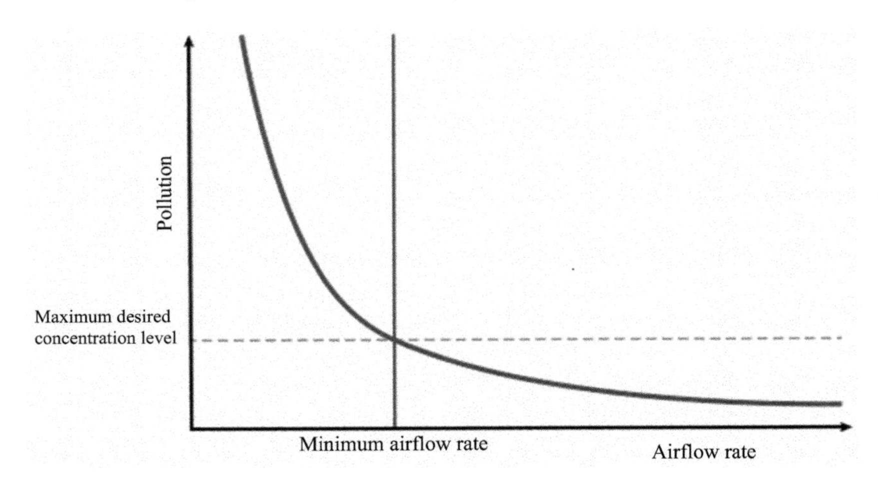

Figure 2.9 Relationship between indoor air quality and air velocity under natural ventilation (Salib, 2012)

indoor environmental parameters (visual, acoustic and thermal environment, as well as air quality) are the most important parameters determining comfort (Frontczak *et al.*, 2012). Thermal discomfort and low IAQ may have negative impacts on work performance. Many studies have tried to quantify the effect of the IEQ on work performance (Elnaklah *et al.*, 2020). It is important to improve the awareness of IAQ by assessing the health effects and consider worker complaints, increases in the amount of sick leave, absenteeism, and reduced productivity. Zhang *et al.* (2016) evaluated the comfort and severity of acute health symptoms of 25 subjects exposed to different levels of carbon dioxide (CO_2) and biofilters. During exposure to 3,000 ppm CO_2, heart rate decreased less when CO_2 was added or when ventilation was restricted compared to changes under reference conditions, and DBP and salivary amylase levels increased significantly after exposure to biofilters (Zhang *et al.*, 2016).

Research conducted in the United Kingdom indicated that a healthy office environment could increase productivity by up to 20% (Clements-Croome, 2015). Using the thermal gap between rooms to control the room temperature of office and rest room at the same time can reduce the time needed to resume work after rest, so as to improve the intelligence concentration (Ueda *et al.*, 2019). Hence, it is essential to create a comfortable thermal environment and guarantee adequate ventilation to achieve optimal work performance and reduce health risks if people show adverse health symptoms.

2.3 Requirements of building ventilation

Sustainable development, energy-savings, IEQ, and thermal comfort are crucial design standards (Bai *et al.*, 2015). In engineering applications, the removal efficiency of indoor air contaminants depends on the ventilation rate, airflow pattern, and airflow direction (Qian and Zheng, 2018). The solution to remove indoor air contaminants through the ventilation system is to find the optimal ventilation rate and airflow pattern in the ventilation design.

2.3.1 Ventilation efficiency

Ventilation efficiency is defined as the percentage of outdoor air provided by the heating, ventilation, and air-conditioning (HVAC) system. It is important in diagnosing IAQ problems and building design (Rask *et al.*, 2020). Variations in the efficiency with which the outdoor air is distributed to the occupants under different ventilation system airflows and temperatures shall be permitted as an optional basis of dynamic reset (Frontczak *et al.*, 2012).

The definition of ventilation efficiency in a steady state is based on two characteristics of the ventilation system:

1. The relative ventilation efficiency, which indicates the variability in the system's ventilation ability between different parts of a room.
2. The absolute ventilation efficiency, which is the ability of the ventilation system to reduce the pollutant concentration in relation to the feasible theoretical maximum.

The relative ventilation efficiency, ε_j^r, at a given point, j, in the room may be defined as follows using the values of the steady-state condition (Sandberg and Environment, 1981):

$$\varepsilon_j^r = \frac{C_f^s - C_t}{C_j^s - C_t} \tag{2.1}$$

where C_f^s is the concentration in the exhaust air terminal, C_j^s is the concentration at point j, and C_t is the concentration in the supply air terminal.

The relative ventilation efficiency is a measure of dispersion and does not take into account the absolute concentration levels or concentration changes from the initial concentration level. An overall measure of the relative ventilation efficiency is provided by substituting the mean concentration in the room $\overline{C^s}$ in the following equation for the local contamination concentration, C_j^s:

$$\overline{\varepsilon}^r = \frac{C_f^s - C_t}{\overline{C}^s - C_t} \tag{2.2}$$

The ventilation system is designed by considering ventilation efficiency.

However, in the face of emergencies, such as outbreak of epidemic, biochemical terrorist attacks, to correctly evaluate the ventilation performance at any time, it is necessary to comprehensively consider the spatial and temporal distribution of indoor personnel and pollutants. The temporal and spatial distribution characteristics of indoor personnel and pollutant concentration should be considered, and the ventilation efficiency should be evaluated from different angles.

It should be noted that the goal of building ventilation should be based on people's feeling and comfort at any time. It is impractical to carry out architectural ventilation design without the research object. It is easy to enter the misunderstanding of building ventilation design simply chasing ventilation efficiency. The designer should fully reflect the satisfaction degree of ventilation to the needs of residents.

2.3.2 Ventilation parameters (ventilation rate, airflow distribution, and airflow direction)

2.3.2.1 Ventilation rate

The most common index to represent the amount of outside air moved into a building is the air change rate per hour (ACH). The ACH is a measure of the frequency of the air changed within a defined space:

$$ACH = \frac{V}{Q} \tag{2.3}$$

where V represents the volume of the room (m^3); Q represents the airflow rate (m^3/h); ACH is the air change rate per hour ($1h^{-1}$).

The ACH refers to the total air supply (including fresh air and returned air); however, the minimum change rate of the fresh air shall be specified for certain places. The required ACH of a ventilated enclosure can vary widely depending upon the use of the building. According to the requirements for personnel, production process, and other functions, the design principles of the ACH are as follows: (a) the ACH should not be less than the minimum fresh air volume required for personnel in accordance with health standards or documentation; (b) supplement the air consumed by indoor combustion and local exhaust volume; and (c) ensure positive pressure of the room. Table 2.2 lists the minimum ventilation rates in occupied zones, as recommended by the ANSI/ASHRAE Standard 62.1-2013.

$$R_{combined} = R_p + R_a \cdot occupant\ density.$$

A higher ventilation rate indicates a higher dilution capacity to reduce the concentration of air contaminants. Figure 2.10 displays the CO_2 concentration for different ACHs. As the ACH of the ventilated enclosure increases from 4 to 12 with a step of 2, the CO_2 concentration decreases significantly. The zone of high concentration is near the CO_2 source (A).

If all other factors remain the same, increasing the ventilation rate is not the best option since higher ventilation rates result in higher energy costs for mechanical ventilation. We will discuss the balance between energy costs and ventilation efficiency in detail in Section 2.3.2.

2.3.2.2 Airflow pattern

The air supply methods of the ventilation system determine the airflow pattern, e.g., the up-supply and down-return, up-supply and up-return, and down-supply and up-return. The type and size of vents (Cao *et al.*, 2019), the parameters of the supplied

Table 2.2 Guidelines for minimum ventilation rates in breathing zones (ANSI/ ASHRAE, 2013)

Occupancy category	**People outdoor air rate R_p (L/s per person)**	**Area outdoor air rate R_a (L/s m^2)**	**Occupant density (per 100 m^2)**	**Combined outdoor air rate (L/s per person)**
Classrooms (ages 5–8)	5	0.6	25	7.4
Classrooms (ages 9 plus)	5	0.6	35	6.7
Computer lab	5	0.6	25	7.4
Music/theatre/dance	5	0.3	35	5.9
Bedroom/living room	2.5	0.3	10	5.5
Transportation waiting	3.8	0.3	100	4.1
Museums/galleries	3.8	0.3	40	4.6
Courtrooms	2.5	0.3	70	2.9
Conference	2.5	0.3	50	3.1
Office space	2.5	0.3	5	8.5
Supermarket	3.8	0.3	8	7.6
Libraries	2.5	0.6	10	8.5
Hotel lobbies	3.8	0.3	30	4.8
Restaurant	3.8	0.9	70	5.1

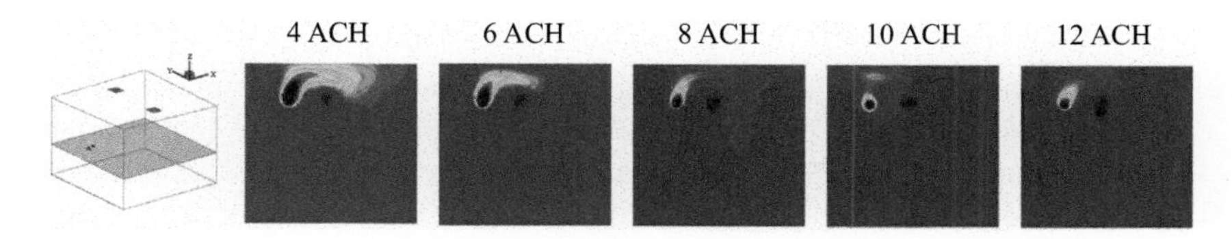

Figure 2.10 CO_2 concentration in the up-supply and up-return ventilation mode for different ACHs (4, 6, 8, 10, 12) (Cao and Ren, 2018)

air, and the interior of the building also influence the airflow pattern since turbulence and eddies occur when moving air comes in contact with the human body and surfaces. Thus, the same air supply method may lead to different airflow patterns.

The investigations of airflow patterns are typically case dependent. There is no point in investigating the airflow pattern without considering practical pollution diffusion and the occupants' thermal comfort. Thus, the purpose is to find a method to remove pollutants rapidly so that they are not circulated locally by eddies, and fresh air can enter the occupied zone. Figure 2.11 displays the CO_2 concentrations for different airflow patterns. Ventilation mode 2 results in the minimum pollutant diffusion. In ventilation mode 3, the pollutants are dispersed locally, but the area of dispersion is large.

Among different air supply modes and airflow patterns, vertical unidirectional flow is the most recommended ventilation mode in several guidelines since it

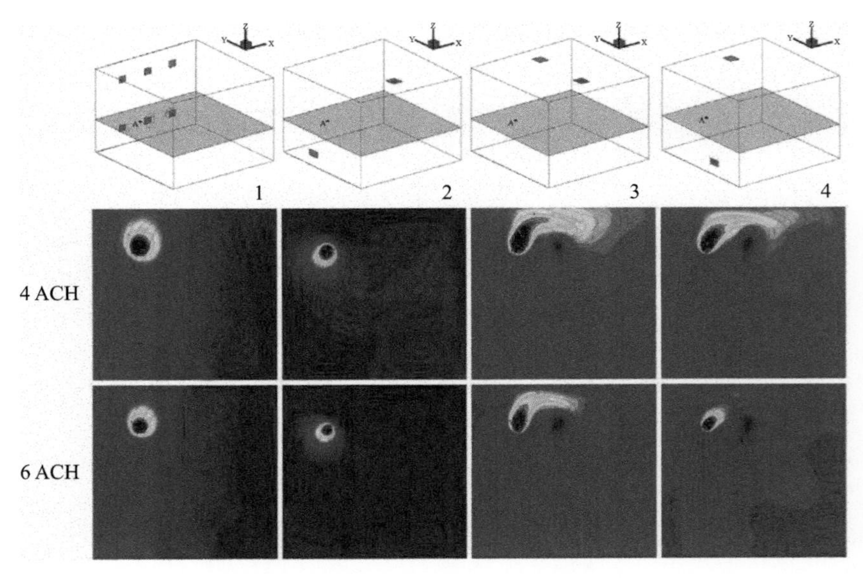

Figure 2.11 *CO_2 concentrations for different airflow patterns; ACH = 4 and 6: (1) up-supply and down-return at the sidewalls; (2) down-supply and up-return; (3) up-supply and up-return at ceiling; and (4) up-supply and down-return (Cao and Ren, 2018)*

efficiently removes indoor air contaminants. This mode is widely applied in operating rooms and isolation wards to reduce the patients' exposure to fatal germs (Whitcomb and Clapper, 1966). The "laminar" streams of vertical unidirectional flow minimize the effect of eddies on the airflow. However, it is impractical to implement vertical unidirectional airflow in large spaces since the uniformity of the air velocity decreases with an increase in the room width (Cheng *et al.*, 1999).

2.3.2.3 Airflow direction

Unlike the airflow pattern, the airflow direction should be considered on multiple levels in a ventilation system. When different cleanliness levels are needed, or the cleanliness requirement is high in buffer zones, the airflow direction must be controlled to prevent the transport of particle-laden aerosols between rooms with different functions. This type of directional flow is used in facilities requiring high cleanliness, such as biosafety laboratories, multilevel clean rooms in the electronic industry, and airborne infection isolation rooms.

The airflow direction between rooms is achieved by a pressure difference. A positive pressure difference prevents the intrusion of contaminants from the surrounding environment, whereas a negative pressure difference prevents the dispersion of contaminants to the surrounding environment. The airflow direction should be from the clean zone to the polluted zone to ensure that air with a high concentration of pollutants does not spread to low-pollution areas, as shown in Figure 2.12.

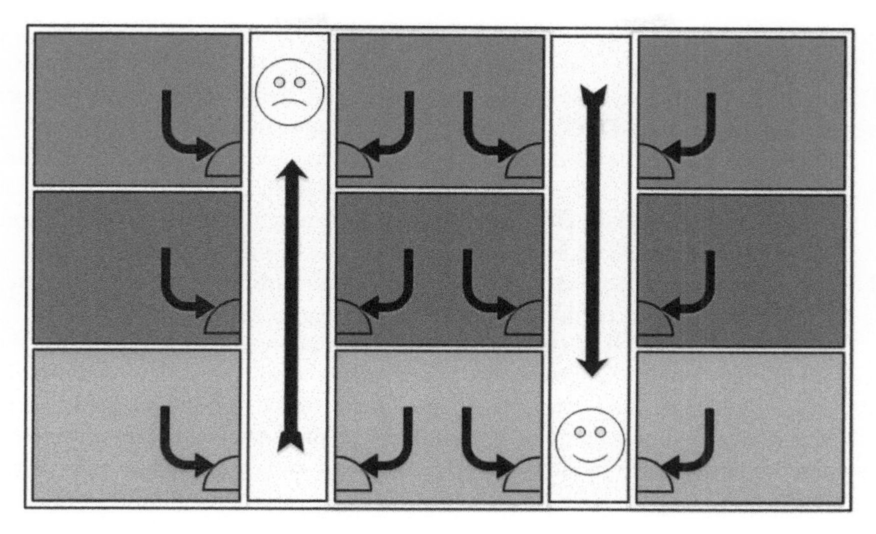

Figure 2.12 Recommended airflow direction in a ventilation system for multiple levels of cleanliness

2.3.3 Balance between ventilation efficiency and energy efficiency

As mentioned in Section 2.2, the IEQ may suffer due to suppression of indoor pollutants and a bad thermal environment because of limited ventilation and infiltration, adversely affecting the occupants' well-being and productivity. Thus, the ventilation rate may be increased to counteract the negative effects and meet the IEQ standards, increasing energy consumption. Except when high ventilation rates are required, mechanical ventilation systems are usually operated at the recommended ventilation rates based on the occupant level multiplied by the ventilation requirement per person, as shown in Table 2.2 (Cao and Ren, 2018). For example, ventilation systems are typically operated based on designed occupant levels instead of actual levels obtained from online monitoring, often resulting in increased energy consumption. Hence, there is an optimum balance between ventilation efficiency, IAQ, energy efficiency, and thermal comfort.

A variable air volume (VAV) system was proposed to deal with the energy consumption problem in mechanical ventilation systems and changes in the load due to the constant flow of cold/hot medium. The need for cold/heat transfer is adjusted by changing the air volume to meet the needs. Significant amounts of energy can be saved by adjusting the ventilation rates using dampers or stopping and starting fans instead of maintaining the design flow rate all the time.

Although a VAV system balances heating and cooling requirements, it is not the optimum solution to remove indoor air contaminants. One feasible solution is to develop demand-controlled ventilation (DCV) systems, an energy-saving strategy that controls the air change rate according to the occupancy and indoor pollution load. For most DCV systems, the CO_2 concentration monitored by sensors is used to control the air supply rate. The CO_2 concentration is maintained below the allowable maximum (i.e., 1,000 ppm) by adjusting the fresh air intake rate. If there is low or no occupancy, DECV systems can save large amounts of energy.

However, DCV systems do not work well when different pollutants are produced at different rates. Therefore, a combination of different ventilation modes is required in different scenarios. A novel ventilation strategy (multimode ventilation, MMV) was developed to address different scenarios to improve ITC. The results showed that MMV performed better than systems with single-airflow patterns because less energy was required for cooling (Shao *et al.*, 2017).

The objective of DCV systems was to provide ventilation based on the demand. These systems represent the first attempt in advanced ventilation technologies. In addition, the continuous improvement of the ventilation mode, technologies, including but not limited to ventilation components and control systems, also play important roles in the trade-off between air supply efficiency and energy conservation.

2.3.4 State-of-the-art ventilation technologies

New and emerging technologies provide new opportunities and solutions to the demand for a better indoor environment and energy efficiency; however, there are several challenges (Cao *et al.*, 2020). The trade-off between air supply efficiency and energy conservation, which was not possible in the past, has become possible due to new technologies, including advanced ventilation components and systems (Wang *et al.*, 2019, 2020), optimized system sensing, monitoring and controlling technologies (Cao *et al.*, 2020; Jeon *et al.*, 2018; Ottosen and Kumar, 2019), self-adjusted MMV systems using artificial intelligence (Shao *et al.*, 2017; Belmans *et al.*, 2019), and rapid prediction of the indoor environment using computational fluid dynamics (Shao *et al.*, 2019; Feng *et al.*, 2019).

2.3.4.1 Novel ventilation components and systems

Conventional ventilation systems do not always operate efficiently due to time-varying indoor occupancy and heat sources. The airflow pattern should match the real-time conditions of the heat source and the contaminants. Novel ventilation components, such as an adjustable fan network (Wang *et al.*, 2020) consisting of multiple small and adjustable axial fans, can change the airflow pattern according to the demand. As a novel ventilation system, radiant cooling and ventilation coupled systems provide temperature- and humidity-independent control in residential buildings (Wang *et al.*, 2019).

2.3.4.2 Optimized system sensing and monitoring and controlling methods

As a result of the rapid development of the Internet of Things (Jeon *et al.*, 2018), monitoring and low-cost sensor techniques (Ottosen and Kumar, 2019) have refined the monitoring of indoor conditions to meet personalized demand. Advanced DCV systems (Taal and Itard, 2020) that provide high ventilation efficiency and personal needs have become possible, such as CO_2 monitoring-based ventilation systems (Chan *et al.*, 2020).

2.3.4.3 Artificial intelligence for self-adjusted MMV systems

Conventional MMV systems required a set of air supply inlets or outlets, which is not conducive to easy installation, operation, and maintenance. With the help of artificial intelligence, self-adjusting or small axial fans can be used to reduce the number of air supply inlets or outlets.

2.3.4.4 Fast prediction and online control of ventilation systems

There is a need to couple modeling and monitoring of the indoor environment due to nonuniform characteristics for safe, healthy, and energy-efficient building environments. Numerous rapid prediction models have been proposed recently, but these models are limited by the prediction accuracy and speed and may not meet the engineering requirements (Cao, 2019).

2.4 Modern and future ventilation

2.4.1 Influence of building property on ventilation efficiency

The comprehensive utilization of energy must be considered in practical design. It is necessary to understand the interaction between the building structure and the ventilation equipment to achieve maximum energy-savings. The primary building characteristics influencing the ventilation efficiency are as follows:

1. Building location. The location of the building affects the outdoor airflow, and the "street canyon effect" is a common phenomenon. Buildings on both sides of city streets, especially tall buildings, form a street canyon, causing increases in the wind speed, strong local strong, and complex airflow phenomena, such as vortices and variable up-and-down airflow.
2. Orientation. Although the orientation of the building does not directly affect the pollutant emission, it influences the indoor cooling and heating load. The effect of the orientation on the maximum temperature and heat inside the building varies for different cases (Rawat *et al.*, 2017).
3. Structure. Due to lighting considerations, the depth of the building shall not exceed 15 m. The depth limit of buildings with a single natural ventilation system is typically 6 m. The use of atrium ventilation can enhance the effect of natural ventilation. The buoyancy effect of natural ventilation is improved in high-rise buildings due to the thermal pressure effect. If this phenomenon is utilized, energy-savings can be achieved without affecting comfort.
4. Building enclosure. The effective thermal performance and air tightness of building envelope are the basis for selecting and operating HVAC system to

control indoor conditions (Pape-Salmon *et al.*, 2014). Enclosures that effectively address thermal bridging and air leakage problems can significantly improve load predictability and reduce building HVAC system peaks. Therefore, the optimization of building envelope structure can not only resist the wind, wind-blown particles, but also make the building free from the influence of cold and hot, and stabilize the internal thermal environment. Optimizing the performance of building envelope can improve the performance of ventilation system and energy-saving effect.

5. Airtightness. Regardless of whether natural ventilation or mechanical ventilation is used, a well-sealed building is required for optimum indoor airflow. The level of airtightness affects the building's cooling and heating load. Therefore, an airtightness inspection of the building must be carried out to detect air leakages.
6. Sun exposure. The radiation heat gain of buildings is an important aspect for calculating the building cooling load. The use of trees and other shade structures can significantly reduce the thermal radiation of the sun. Commonly used vertical shading methods include external shading (high effectiveness), window shades, and internal shading (low effectiveness). Horizontal sunshade structures are mainly used to reduce the heat gain of the sun at higher altitude angles, such as the south-facing wall in summer. Vertical sunshade structures are used to reduce the sun's heat gain at a lower angle, such as east- and west-facing walls. External shading can be combined with ventilation systems to optimize the ventilation design and reduce glare by controlling the opening angle of windows or curtain walls.
7. Window type. The outdoor air temperature is moderate in the transitional seasons, and natural ventilation can be used by opening windows. In severe cold winter and hot summer, the infiltration air volume should be reduced as much as possible. However, the window area should not be too large because large windows have a high solar heat gain. Thus, ventilation efficiency can be improved by the appropriate design of the size and position of the windows. The glass type should be appropriate to balance the solar heat gain, ventilation efficiency, and lighting needs. The window structure also affects the heating and cooling load of the building. Common window types include single-, double-, and triple-pane windows with coatings or air gaps. The ideal glass window transmits long-wave radiation and reflects short-wave radiation. Choosing a window with a low-emissivity coating is equivalent to a triple-pane window with air gaps. The window frame and its details should also be considered.

However, it is impossible to design energy-efficient buildings using only one approach, such as lowering the heat transfer rate of the exterior walls, roofs, and windows or increasing the thickness of thermal insulation (Mahto *et al.*, 2015). In addition to an appropriate building design, it is also necessary to consider the ventilation design of buildings in conjunction with the relevant policies and energy-saving requirements.

2.4.2 Urban ventilation

Architecture, city, and all kinds of human living space (built environment) are the result of the interaction between human and natural environment to a certain extent. On the one hand, human beings obtain a suitable living environment by transforming the natural environment; on the other hand, the natural environment reacts on human society and affects the inherent attributes of public life and human culture. Similar to the influences of the indoor environment, the microclimate of the built environment is crucial because it should provide a platform for healthy and comfortable outdoor activities for urban residents. However, our demand for a better life has accelerated social development and urbanization, changing the balance between the natural environment and the built environment. As a result, environmental deterioration and associated social problems have occurred, running counter to the intention of pursuing a high-quality life. Therefore, the boundary between the built environment and natural environment is crucial to the sustainable development of cities.

The microclimate of the built environment is a crucial aspect in terms of safety and comfort. As important elements of the urban microclimate, the urban thermal environment and wind environment are vital for the diffusion of air pollutants, the mitigation of the heat island effect, and the perception of the thermal/wind comfort in open spaces (Yang and Fu, 2020; Li and Chen, 2020). In the process of urbanization, urban high-density buildings make the land cover change rapidly, which increases the surface roughness, leads to the reduction of urban ventilation, and then affects the urban wind environment and thermal environment (Chen *et al.*, 2016). The thermal environment and wind environment in urban areas are the result of various external factors (e.g., local climate, landform, and hydrology) and internal factors (e.g., urban function, population density, surface patterns, and urban geometries). However, few guidelines or standards have considered the urban wind/thermal environment, and there are far fewer studies on the urban microenvironment than on the rate of urbanization and construction. Future studies should focus on the urban thermal environment and wind environment for urban microclimate optimization and urban planning.

Therefore, designers should consider the urban thermal/wind environment to achieve sustainable development and create a comfortable urban space and balance the needs of people and nature.

2.4.3 Future ventilation

The development of building ventilation is always based on the dual requirements of improving energy efficiency and environmental quality, which is affected by scientific development and people's thermal comfort requirements.

In terms of efforts to improve energy utilization, machinery, construction, sensing, control, computer, micro electromechanical, and other related technologies are gradually mature. Future energy systems will also use forecast meteorological data, such as from weather forecasts, to predict future energy use and costs, so as to reduce energy consumption (Angerame and Harroun, 2014).

Overall building energy simulation will also be widely used to evaluate building energy consumption, so that people can determine that building ventilation is economically feasible. With the building energy consumption moving toward zero energy consumption, the energy consumption will be greatly reduced as the window-to-wall ratio, shading, thermal quality, and ventilation are the main parameters for evaluation. And the low-exergy system creates more flexibility and new possibilities for the design of high-performance buildings (Meggers *et al.*, 2012). By reducing the use of energy, the demand for primary energy can be reduced, which provides a number of heating and cooling methods for buildings with medium supply temperatures and utilizes more valuable heat pumps without energy sources.

In terms of thermal comfort, people also began to have a deeper understanding of the concept of "comfort." People's thermal comfort because of the timescale of climate change may affect the thermal performance of buildings and HVAC systems (Wilde *et al.*, 2010). Therefore, it is very important to implement adaptive thermal comfort model. It is true that the design of building ventilation system has a relatively complete specification and standard, but looking forward to the development of building ventilation, the balance between energy consumption and comfort will certainly achieve a better balance. In the future development, it is necessary to develop a behavior model suitable for domestic residential culture. Besides opening and closing windows, building energy consumption can be predicted more accurately if the behavior model of heating and cooling equipment can be established (Mun *et al.*, 2015).

Looking at the development history of building ventilation, ventilation and building thermal process have changed from separation to combination; the main ventilation mode has changed from natural ventilation to mechanical ventilation; the ventilation object has changed from the whole building space to the breathing area; the ventilation control has changed from simple start-up and stop operation of the whole system to local or overall "on-demand" control according to the needs of control objectives. Accordingly, the main form of building envelope changes from "transparent" type to "closed" type. At the same time, what the development of science and technology brought is people's demand for a more comfortable and cleaner indoor environment.

Looking forward to the future development of building ventilation, the building ventilation will not be ventilation alone but also should be integrated with the whole process of architectural design and construction, e.g., the close combination of building ventilation and modern buildings and the mutual integration of digital information system in building ventilation design, simulation, and prediction makes building ventilation system and micro environment indicators sensitive response and can make real-time adjustment. More perfect standards and regulations are hoped to be formulated to achieve the goal of higher level of energy efficiency and comfortableness.

References

Alsumaiti, A. *The Effect of Indoor Air Quality on Occupants' Health and Performance in Office Buildings in Dubai*. The British University in Dubai. 2013.

American Society of Heating, Refrigerating and Air-Conditioning Engineers (ASHRAE). *ANSI/ASHRAE Standard 170-2013*. Atlanta: ASHRAE. 2013.

Angerame, CJ and Harroun, DJ. *Predictive Alert System for Building Energy Management*. 2014. United States Patent Application 20140350738.

Atkinson, J. World Health Organization. *Natural Ventilation for Infection Control in Health-Care Settings*. Geneva: World Health Organization. 2009, 106 p.

ANSI/ASHRAE. *Ventilation for Acceptable Indoor Air Quality*. 2013, https://www.ashrae.org/File%20Library/Technical%20Resources/Standards%20and%20Guidelines/Standards%20Addenda/62_1_2013_p_20150707.pdf.

ASHRAE. Handbook—1981 Fundamentals. Building Services Engineering Research and Technology. 1981;2(4):193-193. doi:10.1177/014362448100200409.

Bai, G; Gong, G; Yu, CW; and Zhen, O. A combined, large, multi-faceted bulbous facade glazed curtain with open atrium as a natural ventilation solution for an energy efficient sustainable office building in Southern China. *Indoor and Built Environment*. 2015, 24(6), 813–32.

Belmans, B; Aerts, D; Verbeke, S; Audenaert, A; and Descamps, F. Set-up and evaluation of a virtual test bed for simulating and comparing single- and mixed-mode ventilation strategies. *Building and Environment*. 2019, 151, 97–111.

Braham, GD. Mechanical ventilation and fabric thermal storage. *Indoor and Built Environment*. 2000, 9(2), 102–10.

Buchanan, ISH; Mendell, MJ; Mirer, AG; and Apte, MG. Air Filter Materials, Outdoor Ozone and Building-Related Symptoms in the BASE Study. *Indoor Air*. 2008, 18(2), 144–55.

Cai, H; Long, WD; Cheng, BY; and Ma, XB. Efficiency factor indices for evaluating ventilation effectiveness based on spatial and temporal distribution of occupants and contaminants. *Journal of Harbin Institute of Technology*. 2007, 14, 213–6.

Cao, SJ and Ren, C. Ventilation control strategy using low-dimensional linear ventilation models and artificial neural network. *Building and Environment*. 2018, 144, 316–33.

Cao, SJ. Challenges of using CFD simulation for the design and online control of ventilation systems. *Indoor and Built Environment*. 2019, 28(1), 3–6.

Cao, SJ; Deng, HY; Zhou, XQ; and Deng, YL. Ventilation inlets design based on ventilation performance assessment using a dimensionless time scale. *Indoor and Built Environment*. 2019, 28(8), 1049–63.

Cao, SJ; Ding, J; and Ren, C. Sensor deployment strategy using cluster analysis of fuzzy C-means algorithm: Towards online control of indoor environment's safety and health. *Sustainable Cities and Society*. 2020, 59, 102190.

Cao, SJ; Yu, CW; and Luo, XL. New and emerging building ventilation technologies. *Indoor and Built Environment*. 2020, 29(4), 483–4.

Cardinale, N; Micucci, M; and Ruggiero, F. Analysis of energy saving using natural ventilation in a traditional Italian building. *Energy and Buildings*. 2003, 35(2), 153–9.

Chan, WR; Li, XW; Singer, BC; *et al.* Ventilation rates in California classrooms: Why many recent HVAC retrofits are not delivering sufficient ventilation. *Building and Environment*. 2020, 167, 106426.

Chen, YC; Fröhlich, D; Yao, CK; Matzarakis, A; and Lin, TP; editors. The urban roughness estimation based on digital building model for urban wind and thermal environment application coupled with thermal image. 2016 ICEO&SI Conference. 2016.

Cheng, M; Liu, GR; Lam, KY; Cai, WJ; and Lee, EL. Approaches for improving airflow uniformity in unidirectional flow cleanrooms. *Building and Environment*. 1999, 34(3), 275–84.

Cheng, Y and Lin, Z. Experimental investigation into the interaction between the human body and room airflow and its effect on thermal comfort under stratum ventilation. *Indoor Air*. 2016, 26(2), 274–85.

Cho, YJ; Awbi, HB; and Karimipanah, T. Theoretical and experimental investigation of wall confluent jets ventilation and comparison with wall displacement ventilation. *Building and Environment*. 2008, 43(6), 1091–100.

Clements-Croome, D. Creative and productive workplaces: A review. *Intelligent Buildings International*. 2015, 7(4), 164–83.

Bernard AM, Leprince Valérie. *Ventilation Mécanique Contrôlée Dans le Résidentiel*. 2015.

Elnaklah, R; Fosas, D; and Natarajan, S. Indoor environment quality and work performance in "green" office buildings in the Middle East. *Building Simulation*. 2020, 13, 1043–62.

Ezzeldin, S and Rees, SJ. The potential for office buildings with mixed-mode ventilation and low energy cooling systems in arid climates. *Energy and Buildings*. 2013, 65(4), 368–81.

Fanger, PO and Toftum, J. Extension of the PMV model to non-air-conditioned buildings in warm climates. *Energy and Buildings*. 2002, 34(6), 533–6.

Fathollahzadeh, MH; Heidarinejad, G; and Pasdarshahri, H. Prediction of thermal comfort, IAQ, and energy consumption in a dense occupancy environment with the underfloor air distribution system. *Building and Environment*. 2015, 90, 96–104.

Feng, ZB; Yu, CW; and Cao, SJ. Fast prediction for indoor environment: Models assessment. *Indoor and Built Environment*. 2019, 28(6), 727–30.

Frontczak, M; Andersen, RV; and Wargocki, P. Questionnaire survey on factors influencing comfort with indoor environmental quality in Danish housing. *Building and Environment*. 2012, 50(50), 56–64.

Gall, ET and Rim, D. Mass accretion and ozone reactivity of idealized indoor surfaces in mechanically or naturally ventilated indoor environments. *Building and Environment*. 2018, 138, 89–97.

Gao, NP and Niu, JL. Personalized ventilation for commercial aircraft cabins. *Journal of Aircraft*. 2008, 45(2), 508–12.

Gao, NP and Niu, JL. CFD study on micro-environment around human body and personalized ventilation. *Building and Environment*. 2004, 39(7), 795–805.

Gil-Lopez, T; Castejon-Navas, J; Galvez-Huerta, MA; and Gomez-Garcia, V. Predicted percentage of dissatisfied and air age relationship in ventilation systems: Application to a laboratory. *HVAC&R Research*. 2013, 19(1), 76–86.

BS ISO 16814:2008. *Building Environment Design. Indoor Air Quality. Methods of Expressing the Quality of Indoor Air for Human Occupancy*. 2009. Retrieved from https://doi.org/10.3403/30191905.

ISO/TC 159. Ergonomics, Ergonomics of the physical environment Subcommittee SC 5, and International Organization for Standardization. *Ergonomics of the Thermal Environment: Analytical Determination and Interpretation of Thermal Comfort Using Calculation of the PMV and PPD Indices and Local Thermal Comfort Criteria = Ergonomie Des Ambiances Thermiques: Détermination Analytique et Interp*. 2005.

Jeon, Y; Cho, C; Seo, J; *et al.* IoT-based occupancy detection system in indoor residential environments. *Building and Environment*. 2018, 132, 181–204.

Juan, YH; Yang, AS; Wen, CY; Lee, YT; and Wang, PC. Optimization procedures for enhancement of city breathability using arcade design in a realistic high-rise urban area. *Building and Environment*. 2017, 121, 247–61.

Karakitsios, S; Asikainen, A; Garden, C; *et al.* Integrated exposure for risk assessment in indoor environments based on a review of concentration data on airborne chemical pollutants in domestic environments in Europe. *Indoor and Built Environment*. 2015, 24(8), 1110–46.

Lan, L; Wargocki, P; and Lian, ZW. Quantitative measurement of productivity loss due to thermal discomfort. *Energy and Buildings*. 2011, 43(5), 1057–62.

Lestinen, S; Kilpelainen, S; Kosonen, R; Jokisalo, J; and Koskela, H. Experimental study on airflow characteristics with asymmetrical heat load distribution and low-momentum diffuse ceiling ventilation. *Building and Environment*. 2018, 134, 168–80.

Li, Y and Chen, L. The relationship between the internal condition and the ventilation environment in a modular building. *Indoor and Built Environment*. 2020, 30.

Lyles, WB; Greve, KW; Bauer, RM; *et al.* Sick building syndrome. *Southern Medical Journal*. 1991, 84(1), 65–71, 76.

MacNaughton, P; Spengler, J; Vallarino, J; Santanam, S; Satish, U; and Allen, J. Environmental perceptions and health before and after relocation to a green building. *Building and Environment*. 2016, 104, 138–44.

Mahto, S; Bartaria, VN; and Agarwal, RK. Energy efficient building structure. *Engineering Research and Management (IJERM)*. 2015, 02(04), 106–108.

Meggers, F; Ritter, V; Goffin, P; Baetschmann, M; and Leibundgut, H. Low exergy building systems implementation. *Energy*. 2012, 41(1), 48–55.

Melikov, A; Pitchurov, G; Naydenov, K; and Langkilde, G. Field study on occupant comfort and the office thermal environment in rooms with displacement ventilation. *Indoor Air*. 2005, 15(3), 205–14.

Melikov, AK; Cermak, R; Kovar, O; *et al.* Impact of airflow interaction on inhaled air quality and transport of contaminants in rooms with personalized and total volume ventilation. 7th International Conference on Healthy Buildings. Singapore: National University of Singapore, Department of Buildings. 2003.

Mun, SH; Bae, WB; and Huh, JH. Effects on energy saving and thermal comfort improvement by window opening behavior. *Journal of the Architectural Institute of Korea Planning & Design*. 2015, 31(4), 213–20.

Ottosen, TB and Kumar, P. Outlier detection and gap filling methodologies for low-cost air quality measurements. *Environmental Science-Processes & Impacts*. 2019, 21(4), 701–13.

Pacheco-Torgal, F. Indoor radon: An overview on a perennial problem. *Building and Environment*. 2012, 58, 270–7.

Pape-Salmon, A; Ricketts, L; and Hanam, B, Ashrae, Eds. The Measured Impacts of Building Enclosure Renewals on HVAC System Performance and Energy Efficiency. *ASHRAE Winter Conference*; 2014; New York, NY. 2014.

Persily, A and de Jonge, L. Carbon dioxide generation rates for building occupants. *Indoor Air*. 2017, 27(5), 868–79.

Qian, H and Zheng, XH. Ventilation control for airborne transmission of human exhaled bio-aerosols in buildings. *Journal of Thoracic Disease*. 2018, 10, S2295–S304.

Rask, DR; Woods, J; and Sun, J. *Ventilation Efficiency*. 2020.

Rawat, V; Shivankar, S; Das, B; and Sil, BS. Effect of orientation of window on building heat load: Perspective of N-E India. *IOP Conference Series Materials Science and Engineering*. 2017, 225.

Ren, C and Cao, SJ. Implementation and visualization of artificial intelligent ventilation control system using fast prediction models and limited monitoring data. *Sustainable Cities and Society*. 2020, 52, 101860.

Sandberg, M. What is ventilation efficiency?. *Building and Environment*. 1981, 16(2), 123–35.

Schmeling, D and Bosbach, J. Influence of shape and heat release of thermal passenger manikins on the performance of displacement ventilation in a train compartment. *Indoor and Built Environment*. 2020, 29(6), 835–50.

Shao, XL; Wang, KK; and Li, XT. Rapid prediction of the transient effect of the initial contaminant condition using a limited number of sensors. *Indoor and Built Environment*. 2019, 28(3), 322–34.

Shao, XL; Li, XT; Ma, XJ; and Liang, C. Multi-mode ventilation: An efficient ventilation strategy for changeable scenarios and energy saving. *Building and Environment*. 2017, 115, 332–44.

Taal, A. and Itard, L. Fault detection and diagnosis for indoor air quality in DCV systems: Application of 4S3F method and effects of DBN probabilities. *Building and Environment*. 2020, 174, 106632.

Ueda, K; Sugita, K; Shimoda, H; Ishii, H; Obayashi F and Taniguchi, K. *Proposal of the Integrated Thermal Control in Office Environment to Improve Intellectual Productivity*. 2019, 44, 1–10.

Van Hooff, T and Blocken, B. Mixing ventilation driven by two oppositely located supply jets with a time-periodic supply velocity: A numerical analysis using computational fluid dynamics. *Indoor and Built Environment*. 2020, 29(4), 603–20.

Van Moeseke, G; Gratia, E; Reiter, S; and De Herde, A. Wind pressure distribution influence on natural ventilation for different incidences and environment densities. *Energy and Buildings*. 2005, 37(8), 878–89.

Wang, C; Li, F; Sun, ZS; and Song, QJ. A highly active K/Cu-Mn-O catalyst for the removal of nitric oxide in indoor air. *Indoor and Built Environment*. 2019, 28(1), 7–16.

Wang, H; Wang, GJ; and Li, XT. Implementation of demand-oriented ventilation with adjustable fan network. *Indoor and Built Environment*. 2020, 29(4), 621–35.

Wang, Y; Yin, YG; Zhang, XS; and Jin, X. Study of an integrated radiant heating/cooling system with fresh air supply for household utilization. *Building and Environment*. 2019, 165, 106404.

Whitcomb, JG and Clapper, WE. Ultraclean operating room. *American Journal of Surgery*. 1966, 112(5), 681–5.

WHO. *Ambient Air Pollution: A Global Assessment of Exposure and Burden of Disease*. 2016, https://appswhoint/iris/bitstream/handle/10665/250141/9789241511353-engpdf?sequence=1.

WHO. *WHO Guidelines for Indoor Air Quality: Selected Pollutants*. 2010, https://www.euro.who.int/__data/assets/pdf_file/0009/128169/e94535.pdf.

Wilde, PD and Tian, W. The role of adaptive thermal comfort in the prediction of the thermal performance of a modern mixed-mode office building in the UK under climate change. *Journal of Building Performance Simulation*. 2010, 3(2), 87–101.

Wood, A and Salib, R. *Natural Ventilation in High-rise Office Buildings: An Output of the CTBUH Sustainability Working Group: CTBUH Technical Guide*. Routledge, 2013.

Wu, CL; Zhang, HL; Fu, SC; Chan, KC; Qin, DD; and Chao, CYH. Ultrafine particle emissions from a smouldering cigarette in a residence and its associated lung cancer risk. *Indoor and Built Environment*. 2019, 28(10), 1396–405.

Yang, B; Melikov, AK; Kabanshi, A; *et al.* A review of advanced air distribution methods – Theory, practice, limitations and solutions. *Energy and Buildings*. 2019, 202.

Yang, J and Fu, X. Optimization strategy of wind environment in urban central area. *The Centre of City: Wind Environment and Spatial Morphology*. Singapore: Springer Singapore. 2020, 167–86.

Yassin, MF and Almouqatea, S. Assessment of airborne bacteria and fungi in an indoor and outdoor environment. *International Journal of Environmental Science and Technology*. 2010, 7(3), 535–44.

Yu, CWF and Kim, JT. Long-term impact of formaldehyde and VOC emissions from wood-based products on indoor environments; and issues with recycled products. *Indoor and Built Environment*. 2012, 21(1), 137–49.

Yu, CWF and Kim, JT. Building pathology, investigation of sick buildings – VOC emissions. *Indoor and Built Environment*. 2010, 19(1), 30–9.

Zhai, ZQ and Metzger, ID. Insights on critical parameters and conditions for personalized ventilation. *Sustainable Cities and Society*. 2019, 48, 101584.

Zhang, J; Pang, LP; Cao, XD; *et al.* The effects of elevated carbon dioxide concentration and mental workload on task performance in an enclosed environmental chamber. *Building and Environment*. 2020, 178, 106938.

Zhang, K; Zhang, X; Li, S; and Jin, X. Review of underfloor air distribution technology. *Energy and Buildings*. 2014, 85, 180–6.

Zhang, X; Wargocki, P; and Lian, Z. Physiological responses during exposure to carbon dioxide and bioeffluents at levels typically occurring indoors. *Indoor Air*. 2016, 27, 65–77.

Chapter 3
Ventilation systems

Guohui Feng[1]

3.1 Mechanical ventilation

3.1.1 Definition and classification of mechanical ventilation

3.1.1.1 Definition

Depending on the air pressure and air volume provided by the fan, fresh or processed outdoor air can be effectively delivered to any workplace of the building through air supply and exhaust systems; the polluted air in the building can be discharged to the outdoor or be sent to the purification devices before discharging. This type of ventilation is called mechanical ventilation.

3.1.1.2 Classification

Mechanical ventilation can be divided into local ventilation and overall ventilation according to the distribution of hazardous air and the scope of the system. Local ventilation includes local supply air system and local exhaust air system. Overall ventilation comprises a total supply air system and a total exhaust air system. According to whether the fresh air is mixed with pollutants, it can be further divided into displacement ventilation and mixed ventilation.

3.1.2 Requirements for mechanical ventilation

3.1.2.1 Mechanical ventilation

Mechanical ventilation is different from natural ventilation. The airflow is driven by fans and is not limited by natural conditions. It can supply and exhaust air as required to obtain a stable ventilation effect. Mechanical ventilation and natural ventilation are often used together in some situations. Priority should be given to the use of natural air compensation, including the possibility of using clean air from adjacent rooms (Belmans *et al.*, 2019).

Mechanical ventilation, such as operating rooms in hospitals and precision instruments in experimental buildings, can only be used when natural ventilation fails to meet sanitary conditions and production requirements or is technically and economically unreasonable. “Unable to meet the indoor sanitary conditions” means

[1]School of Municipal and Environmental Engineering, Shenyang Jianzhu University, Shenyang, China

that the indoor ambient temperature is too low or the concentration of air hazard exceeds the standard, which affects the work and health of operators; "Production process requirements" means the requirements of the production process for the amount of dust and temperature in the air that permeates the room; "Technical and economic irrationality" means that a large number of radiators are required in order to maintain the thermal balance.

3.1.2.2 Position of mechanical air supply system air intake

In order to ventilate the air into the room from the adverse effects of the outside environment and keep it clean, the provisions of the air inlet are directly arranged in the cleaner outdoor air location (Cao *et al.*, 2020).

In order to prevent the exhaust air (especially the exhaust of the plant that releases harmful substances) from leading to air pollution, it is required that the inlet should be lower than the outlet when there are exhaust outlets at a close distance.

In order to prevent dust and debris near the air inlet from being raised and inhaled by the air supply system, it is stipulated that the lower edge of the air inlet should not be less than 2 m from the outdoor floor, and it is also stipulated that it should be no less than 1 m when it is arranged in the green belt.

The short circuit in inlets and outlets should be avoided. When a skylight, roof ventilator and other exhaust devices are provided on the roof, if an air inlet is set on the roof at the same time, a certain distance should be kept between the air inlet and the roof exhaust device.

3.1.2.3 Demand for fresh air and heat in winter

In the design of buildings with the demand for heat and fresh air, how to take into account that the replenishment of fresh air and heat in winter is in need. Under the condition of a certain amount of exhaust air, in order to maintain the balance of indoor air volume, there are two ways of replenishing air: one is to rely on the natural infiltration of building envelope; the other is to use the supply air system to compensate. No matter which method is adopted, in order to maintain the indoor room temperature to reach the established standard, there is the problem of heat supplement, so as to achieve the thermal balance under the design condition. The main measure to ensure indoor air quality (IAQ) is ventilation, even though the indoor air with very low pollutants.

In buildings with central heating, excessive heating demand due to the use of ventilation will increase the energy consumed by indoor heat dissipation equipment. In practice, when the ventilation system stops operating, the excessive heat provided by the heat dissipation equipment will make the temperature in the building too high. If heat dissipation equipment is set only according to the load of the enclosure structure without considering the load of fresh air, it is difficult to ensure the heating temperature in the building when the ventilation system is in operation (Chan *et al.*, 2020). Therefore, the air volume balance and heat balance shall be calculated when the mechanical air supply system is set up in this code.

3.1.2.4 Air distribution

The concentration of pollutants is often used as the control index for IAQ of civil buildings. If the concentration of one or more air pollutants exceeds the control target, the air is considered unclean. The amount of ventilation required should match the dilution of the indoor pollutant concentration and eliminate excess heat and moisture as needed (Cheng and Lin, 2016).

3.1.2.5 Treatment of air pollutants

In order to ensure human health and meet the cleanliness requirements of certain industrial processes (such as the food industry), the air brought into the room must be treated with different degrees of purification. The air filters commonly used in the ventilation system can be divided into three categories based on the filtering efficiency, i.e., crude, medium and high efficiency (Karakitsios *et al.*, 2015).

3.1.3 Mixed ventilation and displacement ventilation

3.1.3.1 Mixed ventilation

The ventilation form of mixed ventilation is derived from mechanical ventilation based on the dilution method, which is based on the principle of uniform mixing and is used to ensure the air environment of the whole space. The so-called dilution principle is to send some air with low physical quantity content into the target space and fully mix the air with high physical quantity content in the space, so as to ensure the physical quantity content satisfying the requirements of life and technology. This principle is the main basis of traditional building ventilation.

Different forms of ventilation are the products of different stages in the historical evolution of indoor air environment construction methods. The pursuit of a uniform indoor environment gives rise to mixed ventilation, that is, the air is sent into the room in the form of a jet from outside the working area in one or more streams. During the shooting process, a certain amount of indoor air is sucked up. With the diffusion of the supply air, the wind speed and temperature difference will soon attenuate. This type of ventilation is mainly realized by fully mixing the incoming air with the air in the space, so it is called mixed ventilation. This is the most recent fundamental application of the dilution method. In an ideal state, the supplied airflow is fully and uniformly mixed with indoor air, and the concentration of indoor generalized pollutants can be considered to be basically the same regardless of the adjacent area of the tuyere. Set the backflow area near the working area, so as to ensure that the wind speed in the working area is appropriate and the temperature is relatively uniform (Cho *et al.*, 2008).

Most of the traditional ventilation and air conditioning methods adopt mixed ventilation. The treated air is sent into the room at a relatively high speed, which drives the indoor air to be fully mixed with it and makes the temperature of the whole space tend to be uniform and consistent. At the same time, indoor pollutants are "diluted," but the air arriving at the workplace is far less fresh than the air arriving at the outlet. In various indoor air environment construction methods, mixed ventilation has been an extremely common application, such as office, conference room, shopping mall, workshop, stadium and other tall space.

3.1.3.2 Displacement ventilation

Displacement ventilation is to send the treated or untreated air directly into the lower part of the indoor personnel activity area by means of low wind speed, low turbulence and small temperature difference. The air into the room is first evenly distributed on the floor and then flows to the heat source (personnel or equipment) to form a hot gas flow in the form of a hot plume, forming a stagnant layer in the upper space, and discharging waste heat and pollutants from the retention layer out of the outdoor (Melikov *et al.*, 2005).

In the building space, people only stay in the activity area. Take a civil building with a net height of 2.4 m or greater and a factory with a floor height of 5.5 m as an example, and the ratio between the height of the breathing zone and the height of the building space is about 0.46–0.27. Direct delivery of fresh air to the activity area not only meets the indoor hygiene requirements but also ensures good thermal comfort and the effectiveness of ventilation to the maximum extent. The vertical flow pattern of displacement ventilation is based on buoyancy, and indoor pollutants flow upward under the action of thermal buoyancy. As the airflow ascends, it absorbs the surrounding air, increasing the hot plume flow. Under the action of heat, the air in the building becomes stratified (Schmeling and Bosbach, 2020).

When displacement ventilation is in a stable state, the indoor air will form two different areas in the flow state: the upper turbulent mixing zone and the lower one-way flow zone. There is no circulating airflow (close to displacement airflow) in the lower area (people's activity zone), while there is circulating airflow in the upper area (retention zone). The hot and cloudy air in the room is trapped in the upper area, while the cool and clean air in the lower area. The height of the stratified interface of the two areas depends on the air supply volume, heat source characteristics and indoor distribution. In the design of the displacement ventilation system, the layered interface should be controlled above the personnel activity area to ensure the air quality and thermal comfort of the personnel activity area.

Compared with the commonly mixed ventilation, the design of displacement ventilation is required to ensure the minimum mixing degree of airflow in the personnel activity area. The purpose of displacement ventilation is to maintain air quality close to the supplied air condition in the personnel activity area. At the same time, since displacement ventilation is first distributed evenly on the floor and then flows upward, in order to avoid the discomfort caused to human body by the lower air supply, the air outlet speed of the displacement ventilator should not be more than 0.5 m/s.

3.2 Natural ventilation

Natural ventilation is a way of using natural energy without relying on air conditioning equipment to maintain a suitable indoor environment. The principle of natural ventilation is to use the thermal pressure caused by the difference in indoor and outdoor temperature or the wind pressure caused by outdoor wind force to achieve ventilation. It is a manageable, organized way of total ventilation and can be used to dilute the concentration of hazardous materials in the work area (Gall and Rim, 2018).

Natural ventilation can provide plenty of outdoor fresh air, improve indoor comfort and reduce the cooling load of the building. In Japan and some European countries, natural ventilation is preferred by architects and designers. It is widely used in many residential and nonresidential buildings (such as industrial plants and stadiums) (Cardinale *et al.*, 2003). A large number of dwellings in many places are also adapted to local conditions, combined with different climates, creating a variety of natural ventilation, which solves the problem of ventilation and cooling in summer well. For example, Osaka Gymnasium in Japan, Nanning Gymnasium in Guangxi, China, Frankfurt Commercial Bank Headquarters building and Bavaria House in Germany are all the examples of using natural ventilation to cool down and improve the indoor environment (Bai *et al.*, 2015).

3.2.1 The form of natural ventilation and its advantages and disadvantages

3.2.1.1 Cross ventilation

Generally speaking, cross ventilation mainly refers to the entrance and exit of a room opposite. Natural wind can enter directly from the entrance and pass through the whole room to the exit. If there is a partition between the entrance and exit, this kind of wind will be blocked and the ventilation effect will be greatly reduced. In general, the distance between the inlet and outlet should be 2.5–5 times the room height.

3.2.1.2 The single-sided ventilation

When the entrance and exit of natural ventilation are on the same side of the building, this ventilation mode is called single-sided ventilation, which usually has three conditions: (1) mainly relying on the turbulent pulsation of the air to carry out indoor and outdoor air exchange; (2) mainly relying on hot pressure or wind pressure to carry out indoor and outdoor air exchange; (3) the draught caused by the gap between the indoor parts.

3.2.1.3 Passive air well ventilation

Passive air well systems have been widely used in many buildings in Scandinavia, usually to remove wet air from more humid rooms and also to improve IAQ. The airflow through the chimney is driven by both heat pressure and wind pressure.

3.2.1.4 The atrium ventilation

The atrium is a common building component in some modern office buildings, which can be used as an air shaft to realize natural ventilation. Airflow distribution in the midcourt is generally complex, and computational fluid dynamics technology can be used to predict airflow flow.

3.2.1.5 The advantages of natural ventilation

Natural ventilation is suitable for many types of buildings in a temperate climate, which is more economical than other mechanical ventilation systems. If the number of openings is sufficient and the position is appropriate, the airflow rate will be relatively large. Natural ventilation does not need special machine room and the corresponding maintenance.

3.2.1.6 Disadvantages of natural ventilation

Natural ventilation is difficult to control, so it may lead to IAQ less than expected requirements and excessive heat loss.

In the large and deep multiroom building, the natural wind ensures the sufficient fresh air intake and balanced distribution of fresh air. However, natural ventilation is not applicable in areas with serious noise pollution. Some natural ventilation designs may bring safety hazards, and prevention measures should be considered in advance. Moreover, natural ventilation usually requires the occupant to adjust the air outlet to meet the demand. Current natural ventilation rarely filters and purifies the imported air. Besides, air ducts of natural ventilation require large spaces and are often limited by architectural forms.

3.2.2 Principles of natural ventilation design

3.2.2.1 General rules

Full and reasonable use of natural ventilation is an economic and energy-saving measure. In general industrial buildings, natural ventilation should be widely used to improve the working conditions of the working area of the organization of natural ventilation. Mechanical ventilation should be considered only when natural ventilation cannot meet the requirements.

Natural ventilation measures (such as the selection of skylight, hood and intake and exhaust window) should be based on the process requirements and emission characteristics of harmful equipment, combined with local specific conditions of the building design.

In the calculation of natural ventilation, the air volume required to enter can ensure the process requirements and sanitary conditions in the workshop and compensate for the process of local ventilation of the air volume.

3.2.2.2 General layout of the plant

When deciding the orientation of the general chart of the workshop, the vertical axis of the workshop should be arranged as far as possible into the east, west, to avoid a large area of windows and walls affected by the sun, especially in hot areas.

Generally, the main air inlet surface of the plant should be at an angle of 60°–90° with the upwind direction in summer (at least no less than 45°) and should be considered at the same time to avoid the problem of western sun exposure. In the hot area of the cold processing plant, generally one should avoid the west mainly.

3.2.2.3 Layout of process equipment

When the natural ventilation is designed based on hot pressing plant with a skylight, the heat dissipation equipment should be arranged in the position below the skylight.

When the heat dissipation equipment is installed in a multistory building, under the condition that the technology allows, it should be arranged in the building on the top floor. Measures should be taken to prevent hot air from affecting the upper layer if it must be placed on other layers.

If the heat source is arranged on one side of the plant near the outer wall, the heat source should be arranged on both sides of the air inlet hole as far as possible, as shown in Figure 3.1.

Heat source with large heat dissipation (such as the heating furnace and hot material) should be arranged outside the plant and the leeward of the dominant wind direction in summer. At the same time, measures should be taken to effectively insulate the heat source in the workshop.

3.2.2.4 Arrangement and selection of air inlet and outlet

1. The main incoming wind surface of the workshop should be arranged according to the building form, the most favorable orientation in summer and generally can be arranged in the windward side of the dominant wind direction during the summer day. When arranging air inlet holes, even if the area of air inlet holes on one side of the outer wall can meet the requirements, an appropriate number of air inlet holes should be arranged on the other side of the outer wall.
2. When the cold and hot cross adjacent and the use of cold cross skylight into the air, should avoid the air into the pollution by exhaust air. Generally, the side of the exhaust skylight is not suitable for air intake.
3. Multi-span hot workshop can use its adjacent cold span skylight or external wall hole into the air. When the adjacent span is used to enter the air, the concentration of harmful gas or dust in the air should be less than 30% of its maximum allowable concentration.
4. The elevation of natural ventilation of air inlet can be selected according to the following conditions:
   (i) The height of the air inlet in summer from its lower edge to the indoor floor should be 0.3–1.2 m. When the air inlet is high, the effect of air inlet efficiency reduction should be considered.

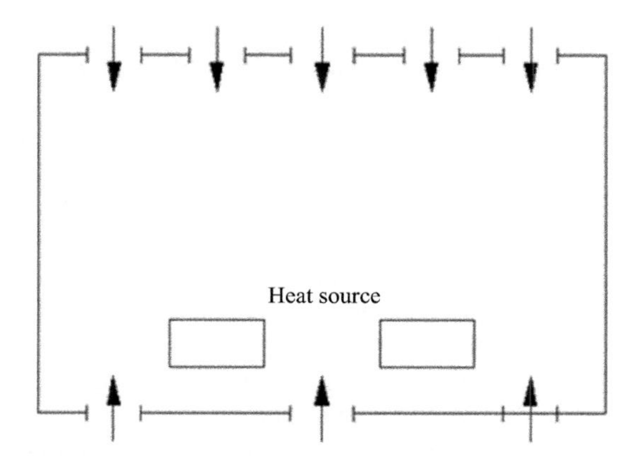

Figure 3.1 Heat source layout

   (ii) In the central heating area, the lower edge of the air inlet in winter should not be lower than 4 m. If it is lower than 4 m, measures should be taken to prevent the cold wind from blowing toward the working point.

5. When the heat source is close to the side of the production plant and there is no working point between the outside wall and the heat source, the air inlet on the side of the outside wall should be arranged as far as possible at the discontinuity of the heat source, as shown in Figure 3.1.
6. The air inlet with natural ventilation in summer could use doors, holes, flat windows or vertical rotating window panels, etc.
7. The production plant using skylight for ventilation shall adopt the windproof skylight when it conforms to one of the following conditions:
   (i) Where the average temperature of the hottest month in a year is higher than or equal to 28 °C and indoor heat dissipation load is greater than 23 W/m^2 at most time;
   (ii) In other areas, indoor heat loss is greater than 35 W/m^2;
   (iii) Reverse irrigation is not allowed;
   (iv) For the adjacent skylight of the multi-span production plant or the skylight adjacent to the building on both sides and in the negative pressure zone, the skylight without windscreen can be regarded as the windshield skylight.

8. The production workshops and auxiliary buildings that utilize skylights to exhaust the wind are provided with no sheltered skylights when they conform to any of the following conditions:
   (i) When the skylight can stabilize the air exhaust.
   (ii) When the average outdoor wind speed is less than or equal to 1 m/s in summer.

3.2.2.5 Principles of civil building design

1. Considering the ventilation factor, the windward side of the building is blocked, and the wind and the windward side of the building generate the maximum wind pressure. The windward side of the building should generally be at an angle of 70°–90° with the dominant wind direction in summer and should not be less than 45°. In the interior design, the ventilation can be north–south or south–north by east preference.
2. When the design of the buildings is arranged in a north–south direction, the shielding of the buildings in the front will reduce the wind pressure in the rear, which will affect the ventilation effect of the buildings in the rear. By increasing the distance between buildings, the buildings in the back row can avoid the vortex area, which will be conducive to the formation of wind pressure ventilation, and the buildings in the back row can also obtain greater wind pressure to achieve better ventilation.
3. In order to better realize the natural ventilation, the architectural complex layout should use the determinant and the free form, in which the determinant

and the wrong column and oblique column layout are better so that most of the rooms can get good natural ventilation and sunshine.

(i) After the height of the building is <24 m, the maximum continuous expansion surface $\ngtr 80$ m.
(ii) After the height of the building is 24–60 m, the maximum continuous expansion surface $\ngtr 70$ m.
(iii) When the building height is >60 m, the projection of its maximum continuous expansion surface is 60 m.

4. In summer, in order to avoid blocking the wind from the higher buildings, the ventilation vents are usually designed on the north side of the buildings so that the natural wind can enter the residential buildings smoothly, and at the same time, the entry of winter wind is blocked, so as to enhance the natural ventilation effect of the residential areas.

3.2.3 Selection of natural ventilation equipment

3.2.3.1 Air intake device

The air inlet device can be the opposite window, the push–pull window, the top turning window, the middle turning window, the air inlet shutter and so on. A push–pull window has beautiful shapes and good sealing with high durability, but the window–wall ratio is generally around 50%.

In hot summer and cold winter and hot summer and warm winter areas, the adoption of air inlet shutter is in the majority. This kind of window is convenient to open, durable and beautiful, and the opening angle can be realized by remote control. Other forms of windowing need to be agreed with the architectural profession.

In the cold area and severe cold area, because the cold wind permeates a lot in winter, fixed louvers outside are commonly set.

3.2.3.2 Air exhaust device

The exhaust device mainly has the skylight and the roof ventilator. The skylight is a common exhaust device. In regions of hot summer and cold winter, indoor heat loss was greater than 23 W/m^2. Indoor heat loss in other areas is greater than 35 W/m^2. In summer, the average outdoor wind speed is greater than 1 m/s. When airflow is not allowed to reverse irrigation, the wind shelter skylight should be adopted. For high-temperature industrial buildings that do not need to adjust the opening angle of skylights, skylights without window sashes should be adopted, but rainproof measures should be taken.

In actual use, the skylight's resistance coefficient is large, the flow coefficient is small, its opening and closing are very tedious, and its glass is easy to damage. As a result, the desired effect is often not achieved.

Roof ventilator is a new type of natural ventilation device, which is made of steel and is composed of colored steel plates (or glass fiber reinforced plastics). It has the advantages of simple structure, light weight and good ventilation effect without electricity. The shape of this ventilator is shown in Figure 3.2. The

Figure 3.2 New roof natural ventilator (Pictures from the Internet)

equipment has small local resistance and has been mass-produced by the factory. It is especially suitable for tall industrial buildings.

3.3 Air infiltration

Air penetration refers to the phenomenon that air flows from high pressure to low pressure through the envelope structure when there is a pressure difference between the air inside and outside the building.

Air penetration includes the penetration of air through the solid material of the enclosure structure and of the gaps between doors and windows. The airtightness of the building is the decisive factor for air infiltration/exfiltration of the building's facade. Studies have shown that air infiltration through the building envelope has a significant impact on the building's HVAC system, heat and humidity performance of building materials, IAQ, building energy consumption, thermal comfort, and noise and fire performance.

In addition, air penetration helps to spread airborne pollutants (such as fungal spores, gaseous pollutants and fine particles) from the outside to the indoor environment. In severe cases, it can cause human diseases such as asthma and cancer. Therefore, accurately predicting the air permeability of a building is the basis for finding the coupling relationship between air, building energy consumption and IAQ.

Table 3.1 Test instrument

Equipment name	Performance parameter	Quantity	Remarks
Telaire 7001 portable CO_2 tester	Temperature: 20–60 °C CO_2 concentration: 0–10,000 ppm	12	Produced by GE to measure CO_2 concentration in real time
Ruler	Ordinary 2-m-long tape measure	1	Used to measure the height of the sampling point
Swing fan	Ordinary swing fan	2	For uniform indoor gas distribution
CO_2 gas tank	Conventional gas tank, 4-L capacity	1	Used to generate CO_2 tracer gas
CO_2 pressure reducer	Pressure control 0–25 MPa Flow control 25 L/min	1	Pressure reducing device for generating gas

According to the national testing standard GB/T18204.1-2013 "Public Place Hygiene Inspection Method Part 1: Physical Factors," the test of the number of permeable air changes in residential rooms adopts the CO_2 concentration attenuation method with CO_2 as the tracer gas, and the average method is used for analysis. Test results calculate the number of permeable air changes in a single room of the residence. The equipment used for the test is shown in Table 3.1.

The test process is as follows:

1. Close the windows of the test room.
2. Release carbon dioxide into the bedroom and turn on the fan to mix the tracer gas with indoor air.
3. When the carbon dioxide concentration in the test room reaches 4,000 ppm, stop the injection of carbon dioxide and turn off the two mixing fans after 5 min.
4. Open the window (inner door) to allow outdoor air to enter the bedroom through natural ventilation (open window area is 30%, this step is not performed in the air permeability test).
5. Continuously test the concentration of carbon dioxide at two measuring points.
6. Complete the test when the carbon dioxide concentration decays to the background concentration level.
7. Repeat the test twice to ensure data quality.
8. The number of air changes N is calculated by the following formula:

$$N = \frac{\ln(C_1 - C_0) - \ln(C_\tau - C_0)}{\tau} \tag{3.1}$$

where N is the number of air changes, the ratio of the total amount of indoor air entering from outside to the total amount of indoor air per unit time; C_0 is the environment background concentration of tracer gas, mg/m^3; C_1 is the concentration of tracer gas at the beginning of the measurement, mg/m^3; C_τ is the tracer gas concentration at time t, mg/m^3; τ is measurement time, h.

3.4 Hybrid ventilation

As the name implies, hybrid ventilation is a form of ventilation that combines natural ventilation with mechanical ventilation. In order to save energy, save investment and avoid noise interference, natural ventilation should be adopted as much as possible. When natural ventilation is difficult to ensure hygienic requirements, mechanical ventilation or the combination of mechanical ventilation and natural ventilation can be used (van Hooff and Blocken, 2020). Taking the workshop can be considered an example for the mixed ventilation design scheme.

3.4.1 Natural air supply and natural exhaust

The aim of eliminating indoor excessive heat and moisture excess by natural ventilation should be closely combined with the workshop's technological layout, building type and general plane position. The following should be considered in the design:

1. Lower the height of the approach wind window from the ground as far as possible, generally no higher than 1.2 m.
2. The main air inlet surface of the workshop should be perpendicular to the dominant wind direction in summer as much as possible, and the angle should not be less than 45°. Meanwhile, large areas of windows and walls should not block the sun.

The pros and cons are as follows:

1. Make full use of natural energy with low operating cost and less investment.
2. The large area of inlet and exhaust air leads to a large amount of cold air infiltration, which increases the heating load in winter.
3. It can only be used in heating plants.

3.4.2 Natural air supply and mechanical exhaust

The system is a negative pressure ventilation system. For the high-temperature workshop, the hot air is removed by the power of the roof ventilator, and the natural air is brought in from the air inlet device at the bottom.

The pros and cons are as follows:

1. The ventilation system is simple, and it can run when the fan and the air inlet window are opened.
2. The ventilation effect will not be affected by outdoor wind speed and direction.
3. Due to negative pressure ventilation, there is a large amount of outdoor air infiltration, which is easy to infiltrate into the workshop with other harmful gases.
4. It does not apply to rooms that release and emit dust or steam, gas with a density greater than air, without simultaneously exothermic heat.

3.4.3 Mechanical air supply and natural exhaust

The system is positive pressure ventilation. According to the requirements of different temperatures, humidity and cleanliness of the workshop, the outdoor air is filtered, cooled (or heated) and directly sent to the working area of the workshop. The hot air after eliminating excessive heat and moisture excess is discharged outside through the exhaust device on the top of the workshop.

The pros and cons are as follows:

1. The inlet air can be pretreated, which can not only meet the requirements of ventilation and cooling in summer but also provide heating in winter and meet the clean requirements of some workshops so that the inlet air quality is good.
2. Positive pressure ventilation reduces the infiltration of cold air and heating load.
3. Convenient operation and management.
4. The equipment air duct and pipeline cover a large area, and the construction investment and operation cost are higher.

3.4.4 Mechanical air supply and mechanical exhaust

The system is fully mechanical ventilated. Compared with the other three schemes, it can improve the air intake and exhaust capacity, eliminate the hot and humid air in the room more effectively and improve the space air diffusion. It is an ideal way of ventilation that can be designed according to the requirements of indoor sanitation, environmental protection and technological conditions. However, the construction investment and operation maintenance costs are relatively high, and the operation management personnel also need higher requirements.

3.4.5 Comparison of several ventilation schemes (Table 3.2)

Table 3.2 Hybrid ventilation scheme

Item	Ventilation type			
	Natural air supply, natural exhaust	**Natural air supply, mechanical exhaust**	**Mechanical air supply, natural exhaust**	**Mechanical air supply, mechanical exhaust**
The initial investment	B	B	B	C
Operating costs	A	B	B	C
Operation management	A	B	B	C
Running effect	D	C	B	A

Note: A—best; B—good; C—poor; D—worst.

3.5 Air supply principles

3.5.1 Indoor air quality standards

People who have been living and working in modern buildings for a long time show more and more serious pathological reactions, so the concepts of sick building and sick building syndrome (SBS) were proposed.

According to the WHO (World Health Organization), SBS is a condition caused by the use of buildings, including red eyes, runny nose, sore throat, drowsiness, headache, nausea, dizziness and itchy skin.

The investigation shows that people spend 80% of their time indoors. Therefore, the problem of SBS is mainly caused by poor IAQ. Therefore, IAQ has become a hot topic in the field of the built environment.

1. In ASHRAE 62.1-2016, the concepts of acceptable IAQ and acceptable perceived IAQ were first proposed.
2. In ASHRAE Standard 62.1-2016, acceptable IAQ is defined as follows: The vast majority of people ($\geq$80%) in air-conditioned spaces are not dissatisfied with indoor air, and no known pollutant in the air has reached the concentration determined by a recognized authority that may pose a serious health threat to the human body.
3. The acceptable perceived IAQ is defined as follows: The vast majority of people ($\geq$80%) in air-conditioned spaces do not express dissatisfaction because of the smell.
4. The acceptable perceived IAQ is a necessary condition for meeting the standard definition of "acceptable IAQ," but it is not a sufficient condition. Some pollutants do not produce odor and irritation, such as radon and carbon monoxide. Because there is no smell and irritation to people, it will not be felt by people, but it is very harmful to people. Therefore, only the perceived IAQ is not enough, and acceptable IAQ must be introduced at the same time. Under the condition of acceptable IAQ, not only do people feel comfortable, but there is also no peculiar smell in the environment, and pollutants do not reach harmful levels.
5. In order to maintain a good IAQ, the concentration of indoor gas and particulate pollutants must be lower than acceptable levels. Indoor pollutants include carbon dioxide, carbon monoxide, other gases, vapors, radioactive substances, microorganisms, viruses, pathogens and suspended particulate matter.

3.5.2 Evaluation method of IAQ

At present, the comprehensive evaluation method, which combines subjective evaluation and objective evaluation, is widely used.

1. Subjective evaluation: Use human sensory organs to describe and evaluate IAQ. The commonly used method is to determine whether the IAQ is satisfied or not based on the feedback through questionnaire surveys.
2. Objective evaluation: Through quantitative monitoring, based on the directly measured concentrations of various pollutants, the satisfaction of IAQ can be assessed objectively. Because there are too many pollutants involved, it is impossible to measure all of them. Therefore, some representative pollutants are usually selected as evaluation indicators.

The measured data needs to be sorted, analyzed and summarized into an index. The measured pollutant concentration C_i (average value) is used as an objective evaluation index, and the reciprocal of the standard value S_i of the pollutant (using the standard value in current standards and specifications) is used as the weighting coefficient. Dimensionless C_i/S_i as the distribution index of pollutants can be used to reflect the difference in pollution degree among various indoor pollutants. Usually, the superposition value of each C_i/S_i is called the arithmetic superposition index p, which is

$$p = \sum \frac{C_i}{S_i} \tag{3.2}$$

The average pollution level of indoor pollutants is reflected by the arithmetic average index Q:

$$Q = \frac{1}{n} \sum \frac{C_i}{S_i} \tag{3.3}$$

The comprehensive pollution degree of pollutants is reflected by the comprehensive index I:

$$I = \sqrt{\frac{1}{n} \left(\sum_{i=1}^{n} \frac{C_i}{S_i} \right) \cdot \max \cdot \left| \frac{C_1}{S_1} \cdot \frac{C_2}{S_2} \cdots \frac{S_n}{S_n} \right|} \tag{3.4}$$

It is generally believed that, when the composite index $I<0.50$, it belongs to a clean environment and the maximum acceptance rate of indoor personnel can be obtained; when $I>1.00$, it belongs to a polluted environment. The grade index evaluation of IAQ and its relationship with the composite index and the corresponding indoor environment characteristics are shown in Table 3.3.

3.6 Ventilation technologies

The primary air system mainly brings outdoor fresh air into the room organizationally through the fan that forced air supply and exhaust, so as to eliminate the dirty air in the room, ensure the IAQ, improve the living environment and protect human health. Primary air system is not only widely used in residential design but also has many types and styles. At present, the residential mechanical primary air system can be divided into two categories according to whether there are pipes or not. According to the form of air distribution, it can be divided into two categories: one-way primary air system and bidirectional primary air system. According to the form of heat recovery, it can be divided into total heat recovery primary air system and sensible heat recovery primary air system.

The pipeless primary air system is mainly divided into a wall-mounted fresh air handling unit and cabinet fresh air handling unit, as shown in Figure 3.3. The fresh air is sent to the room by installing a wall-mounted fresh air handling unit on the outside wall or placing a cabinet fresh air handling unit indoors. Each room can be installed separately wall-mounted fresh air handling unit or cabinet fresh air handling unit; each room is independent and noninterference. This type of primary air system has the following advantages: (1) no pipe design, simple selection; (2) small size,

Table 3.3 Indoor air quality grade indicators and corresponding environmental characteristics

Composite index	Indoor air quality rating	Rating comments	Corresponding environmental characteristics
$I \leq 0.50$	I	Clean	Suitable for human life
$I = 0.50–0.99$	II	Uncontaminated	The pollutants of various environmental elements are not exceeded, and human life is normal
$I = 1.00–1.49$	III	Lightly polluted	At least one environmental element has a pollutant exceeding the standard, and generally, no acute or chronic poisoning occurs except for sensitive persons
$I = 1.49–1.99$	IV	Polluted	Generally, there are 2–3 environmental elements, the pollutants of which exceed the standard. People's health is obviously affected. Sensitive people are seriously affected
$I \geq 2.00$	V	Heavily polluted	Generally, there are 3–4 environmental elements, the pollutants of which exceed the standard. The health of the population suffers severely, and sensitive people may die

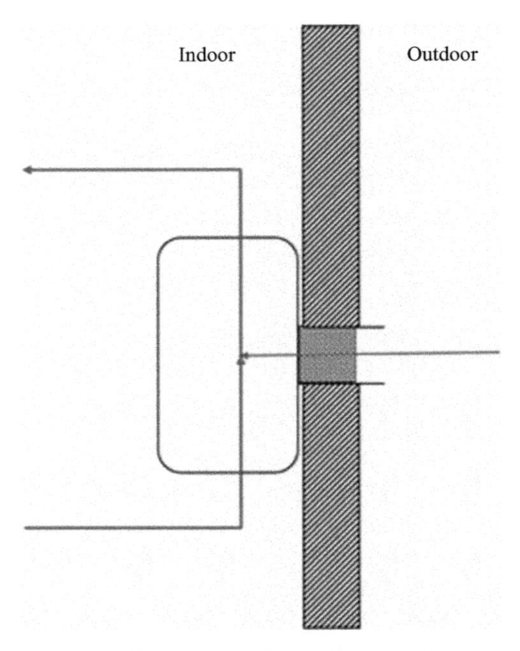

Figure 3.3 Wall-mounted fresh air handling unit

easy to install, operate and maintain; (3) low power, low cost and operating expenses. Its disadvantages are also prominent: (1) it cannot scientifically define the air inlet and exhaust path, which will produce a corner of ventilation; (2) the air intake and exit are close to each other, which is easy to cause fresh air pollution; (3) low filtering level and limited filtering effect; (4) lack of heat exchange function will cause indoor heat loss. Wall-mounted fresh air handling or cabinet fresh air handling unit is suitable for the residential building where IAQ needs to be improved but is limited by existing decoration conditions.

Unidirectional flow primary air system is mainly divided into unidirectional flow negative pressure primary air system and unidirectional flow positive pressure primary air system. As shown in Figure 3.4, the unidirectional flow machine fresh air system, characterized by natural air supply and the form of mechanical exhaust, the exhaust fan is generally installed in the indoor suspended ceiling, each room installed separately. Negative pressure is formed in the room during exhaust, which will drain the polluted air to the outside, while fresh air will enter the room through the natural air intake or windows installed on the outside wall. The advantages of this system form are as follows: (1) only the exhaust outlet has power, which relatively reduces the use cost and saves the electric energy; (2) the air intake is not powered, and no fan or pipe is needed, thus avoiding secondary pollution of air supply through the pipe; (3) the natural air intake is equipped with primary filtration device, which is easy to replace. Disadvantages are as follows: (1) ordinary unidirectional negative pressure primary air system does not have heat exchange function, which will cause indoor heat loss; (2) low filtering level of the natural air intake, unable to filter PM2.5 effectively; (3) if the house is not airtight, it will

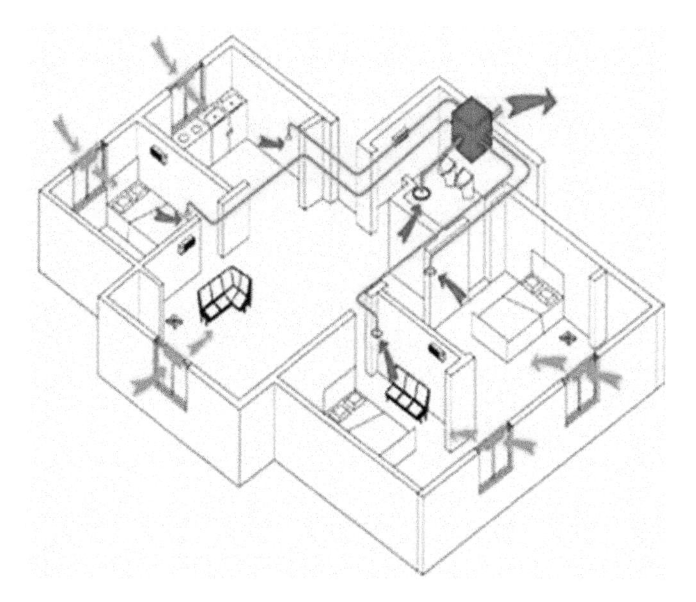

Figure 3.4 Unidirectional flow negative pressure fresh air system (Sun, 2020)

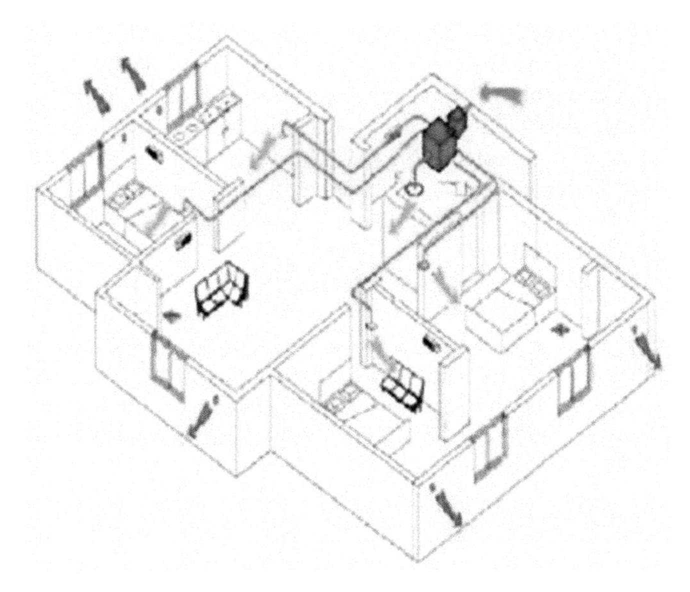

Figure 3.5 Unidirectional flow positive pressure fresh air system (Sun, 2020)

affect the ventilation effect. This system is suitable for where the outdoor air quality is better so as to avoid the secondary pollution of the indoor environment caused by natural air supply.

In Figure 3.5, unidirectional flow positive pressure primary air system, characterized by mechanical air supply and the form of natural exhaust, blower installed on interior suspended ceiling, general each room individually installed air supply outlet are shown. Positive pressure is formed in the room during air supply, fresh air is sent to all rooms after being processed by fresh air handle unit, and indoor polluted air is discharged to the outside from the exhaust outlets or windows and doors of the external wall installation. The advantages of this system form are as follows: (1) only the air intake needs fans and pipes, so the operation cost is relatively low; (2) the supply fan is installed with a filter device, which can effectively filter outdoor pollutants. The disadvantages are as follows: (1) there are pipes for air supply, and a long time without operating will cause secondary pollution; (2) lack of heat exchange function will cause indoor heat loss; (3) positive pressure air supply will reduce body comfort; (4) high position of air intake and exhaust, if the room span is not large enough, it will cause short circuit of fresh air, which cannot guarantee full circulation.

As shown in Figure 3.6, the bidirectional flow primary air system is in the form of mechanical air supply and mechanical air exhaust, mainly composed of a heat exchange device, air duct, exhaust outlet, air supply outlet and other accessories. In the process of indoor and outdoor air circulation exchange and IAQ assurance, the exhaust channel discharges the indoor dirty air from the outdoor and the intake channel introduces the outdoor fresh air through the working operation of the heat

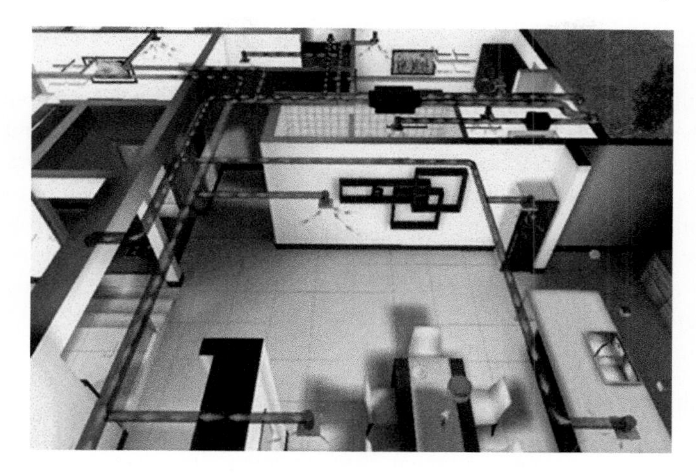

Figure 3.6 Bidirectional flow fresh air system (Sun, 2020)

exchange fresh air handle unit. At the same time, the air intake and exhaust carry out sufficient heat exchange to ensure the IAQ and reduce the cooling load of indoor air conditioning, thus ensuring the comfort of the indoor environment. Advantages of this system form the following: (1) both intake and exhaust air are provided with fans and pipes, which can realize fixed-point air supply and exhaust; (2) it has a heat exchange function, which can reduce indoor heat loss; (3) it has certain filtering function, which can filter PM2.5. The disadvantages are as follows: (1) the fan is large and occupies the suspended ceiling space, so it can only be installed during decoration; (2) pipeline air supply is easy to cause secondary pollution, which requires ceiling decoration and increases decoration cost; (3) the air intake and exhaust require electricity, which consumes a lot of power and costs a lot; (4) the filter device needs to be cleaned and replaced regularly, which may cause inconvenience.

The main difference between the fresh air system of the total heat recovery machine and the sensible heat recovery machine is whether humidity exchange takes place. The total heat recovery primary air system can exchange both temperature and humidity. The heat exchange device is mainly specially processed paper material, which cannot easily produce condensed water. The fresh air system of sensible heat recovery machinery can only carry out temperature exchange. The heat exchange device is usually an aluminum plate or steel plate. In the process of heat exchange, if the temperature difference between indoor and outdoor is too large and reaches the dew point of the air, the condensed water will be produced. In summer, the condensed water will block the air passage, which will seriously reduce the heat exchange rate of equipment and reduce the energy-saving effect. In winter, condensed water frosting can clog air passages and affect the heat exchange of equipment.

The primary air system has been widely used in many countries. In Occident, the proportion of the primary air system installed in buildings is nearly 100%. For example, in Denmark and Poland, residential buildings must be installed with the primary air system as the main ventilation mode. In Japan, the primary air system has become a standard for buildings. It has been installed in both family houses and public buildings and has become a standard to measure the quality of buildings. In China, the prevalence rate of the primary air system is still low, but with the development of the social economy and people living standard rise, more and more people begin to pay attention to the quality of life, especially when the housing decoration and outdoor pollution problems affect residents' health, so primary air system gradually becomes a big choice of improving IAQ. Many property developers have installed and sold primary air systems as part of their homes, making them a selling point.

3.7 Sensor technologies

In an indoor ventilation environment, there are many important indicators, such as temperature, humidity, wind speed, air pressure, PM2.5, CO_2, formaldehyde, VOC and other pollutant concentrations. These indicators cannot be measured without sensors. From the type of test objects, they can be divided into physical factor and chemical factor sensors. Next, the test principle of some sensors is introduced.

3.7.1 Physical factor test sensor

3.7.1.1 Principle of the temperature sensor

1. Resistance sensing, the resistance of metal changes with temperature. For different metals, when the temperature changes by 1°, the resistance value changes differently, and the resistance value can be directly used as the output signal.
2. Thermocouple sensing, the thermocouple consists of two metal wires of different materials, which are welded together at the end. If the ambient temperature of the unheated part is measured, the temperature of the heating point can be accurately known. Because it must have two different materials of the conductor, it is called the thermocouple. Thermocouples made of different materials are used in different temperature ranges, and their sensitivities are also different.

3.7.1.2 Principle of the humidity sensor

1. The double pressure method and double temperature method are based on the thermodynamic P, V, T equilibrium principle, and the balance time is long. The split flow method is based on the accurate mixing of absolute moisture and absolute dry air. Due to the use of modern measurement and control means, these devices can be made quite precise. However, due to the complexity, high cost and time-consuming operation of the equipment, they are mainly used for standard measurement, and the measurement accuracy can reach more than ±2% RH.
2. The saturated salt method of static method is the most common method in humidity measurement, which is simple and easy to operate. However, the

saturated salt method has strict requirements for the balance of liquid and gas phases, and high requirements for the stability of ambient temperature.

3. The dew point method measures the temperature at which the humid air reaches saturation and is a direct result of thermodynamics. It has high accuracy and a wide measurement range. The accuracy of the precision dew point meter for measurement can reach ±0.2 °C or even higher.
4. Wet and dry bulb method, which was invented in the eighteenth century. It has a long history and is widely used. The wet and dry bulb method is an indirect method. It uses the wet and dry bulb equation to convert the humidity value, and the equation is conditional: the wind speed near the wet bulb must reach more than 2.5 m/s. The general wet and dry bulb thermometer simplifies this condition, so its accuracy is only 5%–7% RH.
5. Electronic humidity sensor method, electronic humidity sensor products and humidity measurement belong to the industry rising in the 1990s. In recent years, the research and development field of humidity sensors has made great progress. The humidity sensor is developing rapidly from simple humidity sensor to integrated, intelligent and multiparameter detection, which creates favorable conditions for the development of a new generation of humidity measurement and control system and also improves the humidity measurement technology to a new level.

3.7.1.3 Principle of wind speed sensor

1. The thermal anemometer is a kind of speed measuring instrument that converts the flow rate signal into an electrical signal. A thin metal wire heated by electricity is placed in the airflow. The heat dissipation in the airflow is related to the flow rate, and the heat dissipation causes the temperature change that in turn causes the resistance change. The flow velocity signal is converted into the electrical signal.
2. The working principle of the impeller anemometer and vane wheel probe of anemometer is based on converting the rotation into an electrical signal. After a near induction start, the rotation of the impeller is "counted" and a pulse series is generated, and then the rotation speed value can be obtained through the conversion processing of the detector.

3.7.2 Chemical factor test sensor

3.7.2.1 CO_2

1. Infrared method: The most commonly used method of determination. The working principle is that CO_2 mainly absorbs infrared rays with a wavelength of 4,260 nm. The gas is sent to the test room; one side is irradiated by infrared ray, and the other side is measured by a sensor to measure the attenuation degree of infrared ray received. The attenuation degree is directly proportional to the concentration of CO_2.
2. Mass spectrometer method: The exhaled and inhaled gases are input into the mass spectrometer at a rate of 60 mL/min, the gas molecules are dissociated

and converted into ions under the bombardment of the cathode electron beam. Some positive ions are accelerated and electrostatically focused into electron beams and then enter the test chamber. A strengthened magnetic field is applied in the vertical direction of the ion beam outlet to disperse them into an arc orbit and deposit them on a disk. The orbital radius of each gas ion is directly proportional to its mass charge ratio, thus dispersing in space to form a mass spectrum. Then collect and measure the current carried by different gas ions. The electric flow is directly proportional to the number of ions in the gas. After being processed by a calculator, the value can be reported in 200 μs and the waveform can be displayed.

3. Colorimetric method: The color change of the detector is used to determine $CETCO_2$ and also determine whether the catheter is in the trachea. The color of the detector cannot be restored after contact with the acid substance. It is simple and useful, but its accuracy is not guaranteed.

3.7.2.2 PM2.5

1. In the light scattering test method, when the light irradiates the suspended particles in the air, the scattered light will be generated. Under the condition of certain particle properties, the scattering light intensity of the particles is directly proportional to their mass concentration. By measuring the intensity of scattered light and applying the conversion coefficient of mass concentration, the mass concentration of particulate matter can be obtained.
2. In the micro-oscillation balance method, an oscillating hollow cone tube is used in the mass sensor, and a replaceable filter membrane is installed at the oscillating end. The oscillation frequency depends on the characteristics of the conical tube and its mass. When the sampling airflow passes through the filter membrane, the particles in the filter membrane are deposited on the filter membrane itself, and the mass change of the filter membrane results in the change of the oscillation frequency. The mass of the particles deposited on the filter membrane is calculated by the change of the oscillation frequency, and then the mass concentration of the particles in this period is calculated according to the flow rate, the ambient temperature and the air pressure.
3. In β-ray method, the β-ray instrument is based on the principle of β-ray attenuation. The ambient air is drawn into the sampling tube by the sampling pump and discharged after passing through the filter membrane, and the particles precipitate on the filter membrane. When β-ray passes through the filter membrane with particles deposited, the energy of β-ray will be attenuated, and the concentration of particulate matter can be calculated by measuring the attenuation.

3.7.2.3 Formaldehyde

1. Phenol reagent spectrophotometry: Formaldehyde in the air reacts with phenol reagent to form piperazine, which is oxidized by high iron ion in an acidic solution to form the blue-green compound. Quantitative analysis is carried out according to color depth.

2. Gas chromatography: Formaldehyde in the air is adsorbed on the support coated with 2,4-dinitrophenylhydrazine (2,4-DNPH) 6201 under acidic conditions to form stable formaldehyde hydrazone. After elution with carbon disulfide, it was separated by 0V-column and determined by hydrogen flame ionization detector in order to retain time qualitatively and quantify peak height.
3. Photoelectric method: When formaldehyde gas passes through the detection unit, the color of the paper soaked with chromogenic agent changes from white to yellow due to chemical reaction. The change of reflected light intensity caused by the degree of discoloration is a function of formaldehyde concentration. Formaldehyde concentration is determined according to the change rate of reflected light intensity.
4. Electrochemical sensor method: Formaldehyde gas is passed through the sensor, under the catalysis of electrolyte, formaldehyde molecules on the electrode oxidation–reduction reaction to form electron transfer, under the action of external voltage to form a current proportional to the concentration of formaldehyde.

3.7.2.4 TVOC

The method of thermal desorption/capillary gas chromatography was used to select the suitable adsorbent (Tenax GC or Tenax TA), and a certain volume of air sample was collected by adsorption tube, and the volatile organic compounds in the airflow were retained in the adsorption tube. After sampling, the adsorption tube is heated to desorb volatile organic compounds, and the sample to be measured enters the capillary gas chromatography with the inert carrier gas. The retention time was used for qualitative analysis and peak height or peak area was used for quantification.

References

Bai, G; Gong, G; Yu, CW; and Zhen, O. A combined, large, multi-faceted bulbous facade glazed curtain with open atrium as a natural ventilation solution for an energy efficient sustainable office building in Southern China. *Indoor and Built Environment*. 2015, 24(6), 813–32.

Belmans, B.; Aerts, D.; Verbeke, S.; Audenaert, A.; and Descamps, F. Set-up and evaluation of a virtual test bed for simulating and comparing single- and mixed-mode ventilation strategies. *Building and Environment*. 2019, 151, 97–111.

Cao, SJ; Ding, J; and Ren, C. Sensor deployment strategy using cluster analysis of fuzzy C-means algorithm: Towards online control of indoor environment's safety and health. *Sustainable Cities and Society*. 2020, 59, 102190. 10.1016/j.scs.2020.102190.

Cardinale, N; Micucci, M; and Ruggiero, F. Analysis of energy saving using natural ventilation in a traditional Italian building. *Energy and Buildings*. 2003, 35, 153–9.

Chan, WR; Li, XW; Singer, BC; *et al.* Ventilation rates in California classrooms: Why many recent HVAC retrofits are not delivering sufficient ventilation.

Building and Environment. 2020, 167, 106426. 10.1016/j.buildenv.2019.106426.

Cheng, Y and Lin, Z. Experimental investigation into the interaction between the human body and room airflow and its effect on thermal comfort under stratum ventilation. *Indoor Air*. 2016, 26(2), 274–85.

Cho, YJ; Awbi, HB; and Karimipanah, T. Theoretical and experimental investigation of wall confluent jets ventilation and comparison with wall displacement ventilation. *Building and Environment*. 2008, 43(6), 1091–100.

Gall, ET and Rim, D. Mass accretion and ozone reactivity of idealized indoor surfaces in mechanically or naturally ventilated indoor environments. *Building and Environment*. 2018, 138, 89–97.

Karakitsios, S; Asikainen, A; Garden, C; *et al.* Integrated exposure for risk assessment in indoor environments based on a review of concentration data on airborne chemical pollutants in domestic environments in Europe. *Indoor and Built Environment*. 2015, 24(8), 1110–46.

Melikov, A; Pitchurov, G; Naydenov, K; and Langkilde, GJIA. Field study on occupant comfort and the office thermal environment in rooms with displacement ventilation. *Indoor Air*. 2005, 15(3), 205–14.

Schmeling, D and Bosbach, J. Influence of shape and heat release of thermal passenger manikins on the performance of displacement ventilation in a train compartment. *Indoor and Built Environment*. 2020, 29(6), 835–50.

Sun, W. *Study on Air Quality Monitoring and Ventilation Mode in Mechanically Ventilated Residential Buildings*. Shenyang Jianzhu University. 2020.

van Hooff, T and Blocken, B. Mixing ventilation driven by two oppositely located supply jets with a time-periodic supply velocity: A numerical analysis using computational fluid dynamics. *Indoor and Built Environment*. 2020, 29(4), 603–20.

Chapter 4

Ventilation system: duct network and fluid machinery

Angui Li[1] and Ran Gao[1]

4.1 Basic process of duct design

Duct system design is an indispensable step to realize duct system from concept to practice. If the duct is designed in a building, the design process of the duct system needs to follow the following steps:

First, the material, form and working pressure of the duct system should be confirmed based on production process, production safety and human needs for ventilation and air-conditioning systems.

Second, the orientation and layout of the air duct should be confirmed according to the position of tuyeres.

Third, relevant parameters (such as economic flow velocity and economic friction factor) should be confirmed based on factors of cost, noise and pressure to carry out the hydraulic calculation (Li *et al.*, 1995). Based on the hydraulic calculation, frictional pressure loss ΔP_m and local pressure loss ΔP_j for the air duct need to be confirmed.

Subsequently, air dampers need to be selected for air balance and air adjustment when necessary (Gao *et al.*, 2019).

Finally, in addition to the fan or the air-handling unit, the specification, size, thickness of air ducts, as well as relevant hoisting, sealing and reinforcement, should be selected.

4.2 Basic theory of duct design

The Bernoulli equation proposed in 1726 by the Swiss scientist Daniel Bernoulli is used as the basic theory for duct design. It is an equation expressing the conservation of mechanical energy of a fluid, obtained from the equation of motion (or Euler equation) through streamline integral when an ideal fluid is in steady motion

[1]School of Building Services Science and Engineering, Xi'an University of Architecture and Technology, Xi'an, China

under potential body force. Thus, it is the conservation of a fluid's mechanical energy in nature:

$$P + \frac{1}{2}\rho v^2 + \rho g h = c \tag{4.1}$$

where P is the fluid's pressure intensity (Pa); ρ is the fluid density (kg/m^3); v is the fluid velocity (m/s); g is the gravity acceleration, $g = 9.81$ m/s^2; h is the height (m); c is the constant.

By the Bernoulli equation, (4.1) can be written as

$$P_1 + \frac{1}{2}\rho_1 v_1^2 + \rho_1 g h_1 = P_2 + \frac{1}{2}\rho_2 v_2^2 + \rho_2 g h_2 \tag{4.2}$$

In fact, the previous formula is only adaptable to ideal fluid. Considering the resistance caused by the fluid viscidity, ζ (local resistance coefficient) and λ (friction factor) are introduced, and the previous Bernoulli equation can be represented by

$$P_1 + \frac{1}{2}\rho_1 v_1^2 + \rho_1 g h_1 = P_2 + \frac{1}{2}\rho_2 v_2^2 + \rho_2 g h_2 + \sum \varsigma_i \cdot \frac{1}{2}\rho_i v_i^2 + \sum \frac{\lambda_i l_i}{D_i} \cdot \frac{1}{2}\rho_i v_i^2 \tag{4.3}$$

where l_i is the length of the ith duct section; D_i is the hydraulic diameter of the ith duct section; $\sum \varsigma_i \cdot \frac{1}{2}\rho_i v_i^2$ is the sum of local resistance of duct sections; $\sum \frac{\lambda_i l_i}{D_i} \cdot \frac{1}{2}\rho_i v_i^2$ is the sum of frictional resistance of duct sections (Li *et al.*, 2009; Li *et al.*, 2006).

Except the devices (such as solar walls) driven by hot pressing, the duct system is generally considered isothermal, so the density in the previous formula is considered unchanged, i.e., $\rho_1 = \rho_1 = \rho$. Meanwhile, h_1 and h_2 in the previous formula are a relative height to some datum plane. Because the air density in the low-pressure duct system is basically equal to that in the atmospheric environment, $\rho_1 g h_1$ and $\rho_2 g h_2$ can be ignored, and the previous formula can be expressed as

$$P_1 + \frac{1}{2}\rho v_1^2 = P_2 + \frac{1}{2}\rho v_2^2 + \sum \varsigma \cdot \frac{1}{2}\rho v_i^2 + \sum \frac{\lambda_i l_i}{D_i} \cdot \frac{1}{2}\rho v_i^2 \tag{4.4}$$

The duct system is usually used to bring air from outdoor to indoor, or exhaust air from indoor to outdoor. It is thus an open system. When P_2 is 0, formula (4.4) can be re-simplified as

$$P_1 + \frac{1}{2}\rho v_1^2 = P_2 + \frac{1}{2}\rho v_2^2 + \sum \varsigma_i \cdot \frac{1}{2}\rho v_i^2 + \sum \frac{\lambda_i l_i}{D_i} \cdot \frac{1}{2}\rho v_i^2 \tag{4.5}$$

The previous formula can also be transformed into

$$P_1 = \frac{1}{2}\rho v_2^2 - \frac{1}{2}\rho v_1^2 + \sum \varsigma_i \cdot \frac{1}{2}\rho v_i^2 + \sum \frac{\lambda_i l_i}{D_i} \cdot \frac{1}{2}\rho v_i^2 \tag{4.6}$$

For common duct systems, $1/2\rho v_2^2 - 1/2\rho v_1^2$ can be ignored because of the following three reasons:

1. The value for the differential dynamic pressure of $1/2\rho v_2^2 - 1/2\rho v_1^2$ is one to two orders of magnitude smaller than that of the resistance sum of $\sum \varsigma_i \cdot 1/2\rho v_i^2 + \sum (\lambda_i l_i / D_i) \cdot 1/2\rho v_i^2$.
2. Duct systems often adopt the assumed flow velocity – which is within a certain range and thus has a relatively small differential velocity – for hydraulic calculation.
3. The resistance value of the differential dynamic pressure of $1/2\rho v_2^2 - 1/2\rho v_1^2$ is generally included in local components (such as reducers and tuyeres) with a changing cross-sectional area. In this case, the previous formula can be changed into

$$P_1 = \sum \varsigma_i \cdot \frac{1}{2}\rho v_i^2 + \sum \frac{\lambda_i l_i}{D_i} \cdot \frac{1}{2}\rho v_i^2 \tag{4.7}$$

Moreover, according to the equation of continuity, it can be known that

$$Q_1 = \sum Q_i \tag{4.8}$$

Because these two formulas need to be met at the same time, they can be combined as

$$\begin{cases} P_1 = \sum \varsigma_i \cdot \frac{1}{2}\rho v_i^2 + \sum \frac{\lambda_i l_i}{D_i} \cdot \frac{1}{2}\rho v_i^2 \\ Q_1 = \sum Q_i \end{cases} \tag{4.9}$$

This simultaneous equation is basic for hydraulic calculation. It can be known from (4.7):

1. The role of the fan is to provide static pressure (P_1) and air volume (Q_1) the whole system needs.
2. The air volume Q_1 offered by the fan is used to meet the sum ($\sum Q_i$) of air volume from each branch.
3. The static pressure P_1 offered by the fan is used to tackle the sum $\sum \zeta_i \cdot 1/2\rho v_i{}^2 + \sum (\lambda_i l_i / D_i) \cdot 1/2\rho v_i{}^2$ of frictional resistance and local resistance of each duct section.

During the hydraulic calculation, ζ_i is generally only related to the form of local components, which can be selected from the previous studies (Vedavarz *et al.*, 2007; Gao *et al.*, 2018; Liu *et al.*, 2021). λ_i is related to the *Re* number and the relative roughness of *k*/*D* associated with duct materials. Therefore, λ_i can also be obtained by selecting or calculating duct materials and the *Re* number.

The duct length l_i is obtained from the duct layout, which is a known parameter for hydraulic calculation. Therefore, the confirmed parameters for the actual hydraulic calculation include hydraulic diameter (D_i) of each duct section; based on this, the actual duct size needs to be selected as per duct specification:

1. Air velocity (v_i) of each duct section: It can be given by inverse calculation as per hydraulic diameter (D_i) with the known air volume (Q_i).
2. The sum ($\sum \zeta_i \cdot 1/2\rho v_i^2 + \sum (\lambda_i l_i/D_i) \cdot 1/2\rho v_i^2$) of fractional resistance and local resistance. It is used to select power devices like fans.

Therefore, the content in this chapter mainly includes the following:

1. Basic requirements for duct design: They partly dictate the specific duct materials and duct layout (while determining the length of l_i and the number of ζ_i forms).
2. Hydraulic calculation methods for ducts, which mainly determine the selection and calculation of local resistance coefficient ζ_i and λ_i (Gao *et al.*, 2018).
3. Methods to adjust air volume in the duct system, and the characteristic curve of the fan and duct network.
4. Main forms and connecting methods of the duct system's components, including tees, reducers and elbows (Gao *et al.*, 2018).
5. Confirmation of the specification, reinforcement, sealing and installation of ducts.

4.3 Basic requirements of duct design

4.3.1 Requirements for duct strength

The duct strength is regarded as qualified when there is no crack at seams and permanent deformation and damage on the entire structure after the test pressure remains for 5 min and above. The testing pressure needs to abide by the following rules: the pressure ($P \leq 500$ Pa) for micro- and low-pressure air ducts should be 1.5 times more than the working pressure, the pressure (500 Pa$<P\leq$1,500 Pa) for medium-pressure air ducts should be 1.2 times more than the working pressure and no less than 750 Pa, and the pressure (1,500 Pa$<P\leq$2,500 Pa) for high-pressure air ducts should be 1.2 times more than the working pressure.

It needs to be noticed that among clean air-conditioning systems for duct leakage measurement, those featuring high-pressure air ducts and air cleanliness grade of 1–5 should be inspected as per the standards for high-pressure ducts. Those with a working pressure within 1,500 Pa and air cleanliness grade of 6–9 should be tested as per the standards for medium-pressure air ducts *(GB 50243-2016).*

4.3.2 Requirements for installation

Before installation, the position, elevation, route measurement, location, pay-off and technical review of air ducts need to be completed, while relevant design requirements should be met. For the holes reserved in building structures, their positions should be surely correct and no less than 100 mm of the external size of air ducts.

The air duct system has to be installed after the completion of building envelope for construction area and the cleaning up of installation position and operation site. The purifying air duct system should be installed when there are anti-dust measures or free of dusts indoors and the initial purification can be met, after the completion of the installation ground and wall surface procedures. Moreover, dusts from installation should be cleaned up to meet the requirements of initial purification (Zhang *et al.*, 2008; Gao and Li, 2012).

The connections of air ducts should not be installed in walls or floors. When air ducts are installed along walls or floors, they should keep a distance of no less than 200 mm from walls and more than 150 mm from floors. Meanwhile, no other pipelines get through the air ducts. When assembling air ventilation equipment and components with air ducts, the connecting ducts should not be shortened. Flexible short ducts should be installed at a proper tightness to avoid any twist. Metallic or non-metallic flexible ducts should have a length of within 2 m, without any dead twist or dent.

For air ducts crossing the closed anti-fire or anti-explosion wall slabs, protective steel tubes with a thickness of no less than 1.6 mm should be set. For air ducts and relevant components passing through walls and roofs, rain covers should be set. When the duct is 1.5 m out of the roof, a steel rope or a galvanized inhaul cable for outdoor duct systems required not to be fixed on duct flanges, lightning rod or lightning protection network should be set.

For duct systems installed at the inflammable and explosive environment, reliable anti-static grounding devices should be set in place; for those conveying inflammable and explosive gases, interfaces should be set outside the living areas or other auxiliary production rooms.

4.3.3 Requirements for air leakage rate

The air leakage grade, the permissible leakage rate and the specified static pressure limit should conform to the regulations in Table 4.1.

The air leakage rate should be inspected as per duct system classification after installation, when air leakage grades fail to be given out in the design. The permissible leakage rate for air duct systems should be subject to the regulations in Table 4.2.

Equipment, ducts and fittings need to conduct cold insulation, when cold loss of equipment and ducts caused by cold medium lower than room temperature needs to be reduced, when condensation on the equipment and duct surface caused by cold medium lower than room temperature needs to be prevented, when the temperature rise or gasification of cold medium during production and transmit needs to be mitigated, when the cooling capacity by the failure of cold insulation has an adverse impact on parameters of temperature and humidity in plants.

4.3.4 Requirements for air velocity

The air velocity in ducts has a big impact on the economy of ventilation and air-conditioning systems. A higher air velocity means smaller duct cross sections, less

Table 4.1 Level of air leakage rate and permissible leakage rate for air duct system (BS EN 12237-2003)

Level of air leakage rate	Permissible leakage rate for air system [$m^3/(h\ m^2)$]	Static pressure detection limits (Pa)	
		Positive pressure	Negative pressure
Level A	$0.1056 \times P^{0.65}$	500	500
Level B	$0.0352 \times P^{0.65}$	1,000	750
Level C	$0.0117 \times P^{0.65}$	2,000	750
Level D	$0.0036 \times P^{0.65}$	2,000	750
Level E	$0.0010 \times P^{0.65}$	2,000	750

Note: (1) According to the duct system's five grades classified by usage purpose, the maximum air leakage rate for medium- and high-pressure air ducts should be no more than Level B and no more than Level C, respectively, and that for the ducts with special requirements should be no more than Level D. (2) The air leakage rate of smoke exhaust systems, dust removal systems and low-temperature air supply systems should be no more than Level B. (3) The air leakage rate of 1–5 grade purification air-conditioning systems should be no more than Level C. (4) Level E is only for the ducts with special purpose such as those used in virological laboratories. (5) P (Pa) is the static pressure that imposes inside the duct during testing.

Table 4.2 Permissible leakage rate for air duct system (GB 50738-2011)

Pressure (Pa)	Permissible leakage rate [$m^3/(h\ m^2)$]
Air duct with micro- and low pressure ($P \leq 500$ Pa)	$\leq 0.1056 \times P^{0.65}$
Air duct with medium pressure (500 Pa$<P \leq$1,500 Pa)	$\leq 0.0352 \times P^{0.65}$
Air duct with high pressure (1,500 Pa$<P \leq$2,500 Pa)	$\leq 0.0117 \times P^{0.65}$

Note: (1) The air leakage rate of loading test (loading by insulation material and by 80-kg external force) in labs should conform to the specified values in the table. (2) The air leakage rate of the systems featuring smoke extraction, dust removal and low-temperature air supply should abide by the specified values for medium-pressure systems in the table. (3) The air leakage rate of 1–5 grade purification air-conditioning systems should obey the specified values for high-pressure systems in the table. (4) The air leakage rate of special air duct systems applicable to Grade D or E should be measured by corresponding limits during testing.

material consumption and low construction cost, but it also means bigger resistance and higher power consumption and operation cost. A lower air velocity means smaller resistance and less power consumption, but it also means bigger duct cross sections, high material and construction cost and larger space for ducts.

Therefore, a reasonable air velocity needs to be selected by comprehensive comparison in techniques and economy. For example, the air velocity designed for the ducts of industrial dust removal systems should be no less than the values shown in Table 4.3. The economical flow velocity in ducts for civilian buildings should be confirmed by Table 4.4. The air velocity in ducts of ventilation and

Table 4.3 Recommended air velocity for industrial dust removal duct (m/s) (GB 50019-2015)

Dust property	Vertical duct	Horizontal duct
Wet soil	15	18
Pulverized coal	11	13
Clay	13	16
Dry sand	17	20
Ground dust furnace ash	18	20
Light metal shavings	16–18	18–20
Fibrous dust	14–16	16–18
Heavy metal shavings and wet wood chips	16–19	20–23
Light and dry dust (ground wood dust)	14–16	16–18

Table 4.4 Economic flow rate in air ducts of civil buildings (m/s) (GB 50736-2012)

Category	Location		Air velocity
Mechanical ventilation system	Main pipe	Residential	3.5–4.5 (6.0)
		Public building	5.0–6.5 (8.0)
		Metal and non-metal duct	6–14
		Brick and concrete duct	4–12
	Branch pipe	Residential	3.0 (5.0)
		Public building	3.0–4.5 (6.5)
		Metal and non-metal duct	2–8
		Brick and concrete duct	2–6
	Fresh air inlet	Residential and public buildings	3.5–4.5
		Computer room and warehouse	4.5–5.0
	Fan outlet	Residential and public buildings	5.0–10.5
		Computer room and warehouse	8.0–14.0
Natural ventilation system	Air supply shaft		1.0–1.2
	Horizontal main pipe		0.5–1.0
	Ventilation vertical shaft		0.5–1.0
	Exhaust duct		1.0–1.5
	Inlet shutters		0.5–1.0
	Air outlet		0.5–1.0
	Ground air inlet		0.2–0.5
	Ceiling air inlet		0.5–1.0

Note: The velocity in brackets is the maximum velocity.

air-regulating systems with noise elimination requirements should be selected as per Table 4.5.

During the placement of air ducts, those with noise elimination treatment should avoid rooms with higher noise; those with high noise should keep away

Table 4.5 Air velocity in the duct with noise reduction requirements (m/s) (GB/T 50466-2018)

Allowable indoor noise level dB (*A*)	**Air velocity of main pipe**	**Air velocity of branch duct**
25–35	3–4	≤2
35–50	4–7	2–3
50–65	6–9	3–5
65–85	8–12	5–8

from rooms with lower noise requirements, but when failed, soundproofing treatment should be done. The air velocity in ducts of ventilation and air-regulating systems with noise elimination requirements should be selected as per Table 4.5.

4.3.5 Requirements for thermal and cold insulation

Equipment, ducts and fittings need to conduct thermal insulation when the external surface of equipment and ducts (excluding indoor heating pipeline) has a temperature of higher than 50 °C, when thermal medium has to remain at a certain state or parameter, when thermal loss is huge and uneconomical due to the failure of thermal insulation, when installation or laying is conducted at frozen places, and when the heat generated by the failure of thermal insulation would have an adverse impact on parameters of temperature and humidity in plants.

4.3.6 Anti-fire requirements

The air duct system should be made of non-flammable materials, except that the air ducts contacting corrosive medium and flexible joints can be made of flame retardant (resistant) materials. The air ducts of ventilation air-conditioning systems in plants, single- and multi-storey buildings and large-space buildings (such as stadia, exhibition halls and waiting rooms for airplane, bus and ship) should be made of flame-resistant materials when fire compartments are not stepped over and fire dampers are set at the partition of the room that would be crossed.

Therefore, the ducts for ventilation and air-conditioning systems must adopt non-flammable materials. Except anti-fire ducts whose fire-resistant limit time should conform to the requirements of relevant fire prevention code as well, duct framework and fixation materials and sealing padding should opt for non-flammable materials.

Ducts conveying air with a temperature of higher than 80 °C should adopt safe, reliable prevention measures according to design regulations.

4.4 Calculation of the frictional and local resistance loss of air ducts

There are two kinds of airflow resistance in ducts. One is the frictional resistance loss caused by the viscosity of air itself and the friction between air and duct wall, which is called frictional resistance or frictional resistance. The other is when the air runs through the components and equipment (such as elbows, tees and reducers) in ducts, the velocity and direction of airflow would change and generate vortex, thus causing energy losses, which in this case are called local resistance losses.

4.4.1 Frictional resistance

According to the principles of fluid mechanics, when air is running in ducts with a cross section of unchanged area and shape, the frictional resistance losses can be calculated by the following formula:

$$\Delta P_m = R_m l = \lambda \frac{1}{4R_s} \cdot \frac{v^2 \rho}{2} l(\text{Pa}) \tag{4.10}$$

where λ is the friction factor; v is the average air velocity in ducts (m/s); ρ is the air density (kg/m^3); l is the duct length (m); R_m is the frictional resistance (Pa/m) (or specific frictional resistance) of unit duct length; R_s is the hydraulic radius (m) of ducts:

$$R_s = \frac{f}{P} \tag{4.11}$$

where f is the cross-sectional area (m^2) for the part filled with fluid in ducts; P is the wetted perimeter (m), i.e., the duct perimeter for ventilation and air-conditioning systems.

The friction factor λ can be calculated as per the following formula:

$$\frac{1}{\sqrt{\lambda}} = -2lg\left(\frac{K}{3.7D_e} + \frac{2.51}{Re\sqrt{\lambda}}\right) \tag{4.12}$$

where K is the absolute roughness (m) of the inner duct wall; D_e is the equivalent diameter (m) of ducts; Re is the Reynolds number.

The Moody chart shows the functional relationship between friction factor λ and e/D and Re (Figure 4.1). First, confirm the flow's Reynolds number by checking the corresponding abscissa on the Moody chart; subsequently, check the absolute duct wall roughness e before dividing by the duct diameter D (or equivalent diameter D_e for non-circular ducts) for the value of e/D, which is corresponding to the right ordinate and the central curve on the Moody chart. The Reynolds number Re on the abscissa and e/D on the right ordinate are used to confirm a point – which is corresponding to the friction factor λ on the left ordinate of the Moody chart – on the central curve on the Moody chart. Then the frictional resistance in ducts can be calculated by λ.

We can use an alignment chart to calculate the ventilating duct resistance. With any two being given among four parameters (namely, flow, duct diameter, flow velocity and resistance), we can use this chart to get the other two. The alignment chart is obtained by the value of λ in the transition region, under the pressure of B_0 = 101.3 kPa, temperature of t_0 = 20 °C, air density of ρ_0 = 1.204 kg/m^3, kinematic viscosity of $\nu_0 = 15.06 \times 10^{-6}$ m^2/s, wall roughness of K = 0.15 mm, and

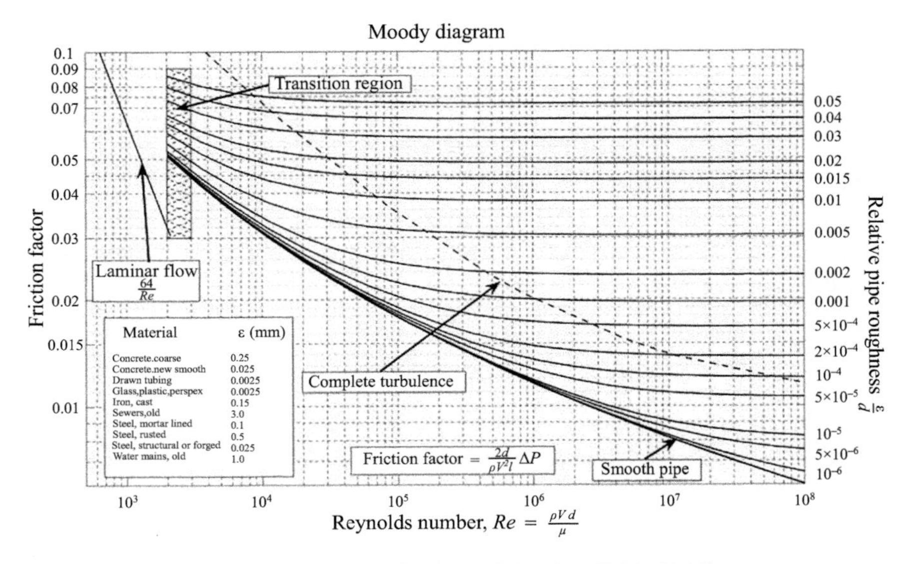

Figure 4.1 Moody chart (Moody, 1944, 1947)

circular ducts. Corrections must be done when the real usage conditions are inconsistent with the previous ones.

1. Correction of air density and viscosity in ducts,

$$R_m = R_{m0}(\rho/\rho_0)^{0.01}(v/v_0)^{0.1}(\text{Pa/m}) \tag{4.13}$$

where R_m is the frictional resistance (Pa/m) of the actual unit length; R_{m0} is the frictional resistance (Pa/m) of unit length on the chart; ρ is the actual air density (kg/m^3); v is the actual air kinematic viscosity (m^2/s).

2. Correction of air temperature and atmosphere pressure,

$$R_m = K_t K_B K_{m0} \tag{4.14}$$

where K_t is the temperature correction coefficient; K_B is the ambient air pressure correction coefficient.

$$K_t = \left(\frac{273 + 20}{273 + t}\right)^{0.825} \tag{4.15}$$

where t is the actual air temperature (°C).

$$K_B = (B/101.3)^{0.9} \tag{4.16}$$

where B is the actual ambient air pressure (kPa).

K_t and K_B can be directly checked in Figure 4.2, from which the temperature correction K_t decreases with temperature rise, and the pressure

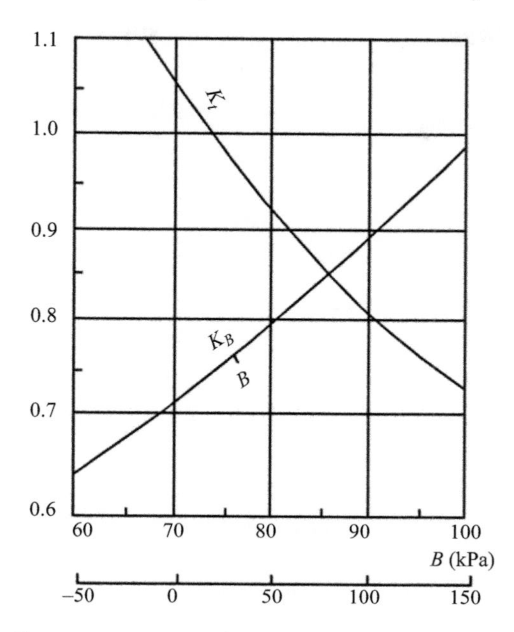

Figure 4.2 Correction curve of temperature and atmospheric pressure

Table 4.6 Roughness of various materials in air duct manufacturing (Sun, 2010)

Duct materials	**Roughness (mm)**	**Duct materials**	**Roughness (mm)**
Steel sheet or galvanized steel sheet	0.15–0.18	Plywood	1.0
Plastic plate	0.01–0.15	Brick masonry	3–6
Slag gypsum sheet	1.0	Concrete	1–3
Slag concrete slab	1.5	Board	0.2–1.0

correction coefficient K_B decreases with the decrease in static pressure of the air in ducts.

3. Correction of duct wall roughness.

In the engineering involving ventilation air-conditioning systems, the ducts are generally made of different materials, with their roughness shown in Table 4.6.

With the duct wall roughness of $K \neq 0.15$ mm, we can first check out R_{m0} on the alignment chart and then conduct correction as per the following formulas:

$$R_m = K_t R_{m0} (\text{Pa/m}) \tag{4.17}$$

$$K_t = (Kv)^{0.25} \tag{4.18}$$

where K_t is the correction coefficient of duct wall roughness; K is the duct wall roughness (mm); v is the airflow velocity (m/a) in ducts.

Table 4.7 Absolute roughness factors (mm) (Lu, 2008)

Duct velocity (m/s)	Absolute roughness factors (mm)				
	0.03	**0.09**	**0.15**	**0.9**	**3.0**
2	0.95	1	1	1.20	1.50
3		0.95		1.25	1.60
4	0.90			1.30	1.70
5–7				1.35	1.80
8–12	0.85			1.40	1.85
13				1.45	1.90
14–16	0.80	0.90			1.95

For a more accurate calculation, the value of K_t can be checked out in Table 4.7.

When calculating the frictional resistance of rectangular ducts via the alignment chart, we should first convert the equivalent diameter, (which is the diameter of circular ducts that have the same unit-length frictional resistance as rectangular ducts), i.e., convert the cross-sectional size of rectangular ducts into corresponding circular duct diameter to obtain the unit-length frictional resistance of rectangular ducts. The equivalent diameter of rectangular ducts includes the flow-equivalent diameter and the flow-velocity-equivalent diameter, as shown next.

The unit-length frictional resistance in a rectangular duct is equivalent to that in a circular duct. The airflow in a rectangular duct is equivalent to that in a circular duct. In that case, the diameter of the circular duct is the flow-equivalent diameter of the rectangular duct, represented by D_L.

According to the previous deduction, the flow-equivalent diameter can be approximately calculated by the following formula:

$$D_L = 1.3\frac{(ab)^{0.625}}{(a+b)^{0.25}} \tag{4.19}$$

When using the equivalent diameter method to calculate rectangular duct resistance, the corresponding relations should be heeded. Under the flow-velocity-equivalent diameter method, the resistance must be calculated by using airflow velocity in rectangular ducts. Under the flow-equivalent diameter method, the resistance must be calculated by using airflow in rectangular ducts.

The air velocity and the unit-length frictional resistance in a circular duct are equivalent to those in a rectangular duct. In this case, the diameter of the circular duct is the flow-velocity-equivalent diameter of the rectangular duct, represented by D_v. According to this definition, the hydraulic radius of circular and rectangular ducts must be equivalent from the following formula:

$$D = \frac{2ab}{a+b} = D_v \tag{4.20}$$

where D_v is the flow-velocity-equivalent diameter of the rectangular duct with a side length of $a \times b$. If the flow velocity in a rectangular duct is the same as that in a circular duct with a diameter of D, their unit-length frictional resistance is also equivalent.

The unit friction pressure loss of a rectangular duct is equivalent by the previous two methods, among which the flow-velocity-equivalent diameter method is much easier.

4.4.2 Local resistance

When air is passing through the components with changing cross sections (such as reducers, tuyeres and dampers), through those that could change flow direction (such as elbows), and through those that could change flow volume (such as tees, crosses, and air supply and exhaust tuyeres at the sides of ducts), the generation of vortexes would lead to local resistance, which could be calculated by the following formula:

$$\Delta P_j = \xi \frac{v^2 \rho}{2} \tag{4.21}$$

where ξ is the local resistance coefficient; v is the average air velocity in ducts (m/s); ρ is the air density (kg/m^3).

The value of the local resistance coefficient (ξ) for the fittings (such as reducers, elbows, dampers and tees) that generate local resistance in ventilation air-conditioning systems is generally confirmed by experiment. In the experiment, the total differential pressure (i.e., the local resistance ΔP of the fittings) before and behind the fittings needs to be first measured before dividing by the dynamic pressure $\rho v^2/2$ corresponded to the velocity v, in order to obtain the value of the local resistance coefficient ξ. Part of the previous process has been sorted out into empirical formula. Some common fittings' local resistance coefficient can thus be checked up in Section 5.2 in ASHRAE Manual 130-2016. When calculating local resistance, the air velocity corresponding to the value of ξ needs to be heeded. Because the air in ventilation and air-conditioning systems generally runs in the self-modelling zone, the local resistance coefficient ξ is only dictated by the shape of fittings, rather than the impact of relative roughness and the Reynolds number.

4.5 Hydraulic calculation method for ducts

The hydraulic calculation for ventilation ducts is conducted only when system and equipment arrangement, duct material, the position for air supply and exhaust ducts, and air volume are all confirmed. It includes verifying calculation and design calculation. The former is to confirm whether the air pressure meets the requirements by calculation with given duct diameter and flow; the latter is to confirm duct diameter and resistance with given air volume. Hydraulic calculation, the basic method for the design of ventilating duct systems, can ensure the design quality of duct network.

The hydraulic calculation methods include the assumed velocity method that is commonly used at present, pressure-loss averaging method and static regain method.

4.5.1 Assumed velocity method

The characteristic of the assumed velocity method is to first select flow velocity as per technical and economical requirements and then confirm the duct's cross-sectional size and resistance according to air volume in ducts.

Its calculation steps and methods are as follows:

1. Draw an axonometric diagram of ventilating or air-conditioning systems, number the duct sections, and then mark their length and air volume to confirm the index circuit. A duct section's length is generally calculated as per two fittings' centreline length, without deducting the length of the fittings (such as tees and elbows).
2. Confirm a reasonable air velocity. The flow velocity by the assumed velocity method needs to be selected by the application places of ventilating systems and the requirements in noise elimination, economy and dust removal (see details in the relevant design manual).
3. Confirm the size of duct sections as per the ducts' air volume and the selected flow velocity to calculate frictional resistance and local resistance.
4. Balance the resistance for parallel ducts.

   To ensure an anticipated air volume at air supply and exhaust tuyeres, the resistance of two parallel branches must be balanced (confirm the diameter of parallel ducts to make their calculated resistance and working power equal by using the pressure-loss averaging method). This is the key to distributing flow as per requirements. In the case of unequal calculated resistance and working power of parallel ducts, the duct network would automatically adjust flow in parallel ducts in practical application until such equality is realized. At this point, the flow in parallel ducts is not the required one.

   As for common ventilating systems and dust removal systems, the resistance difference of two branches should not exceed 15% and 10%, respectively. If violating the previous regulations, the resistance balance needs to be done. Adjusting branch diameter, regulating dampers and increasing air volume can realize hydraulic balance:

   First, adjust branch diameter to alter branch resistance to reach resistance balance:

$$D' = D\left(\frac{\Delta P}{\Delta P'}\right)^{0.225} \text{(mm)} \tag{4.22}$$

where D' is the duct diameter (mm) after adjustment; D is the duct diameter (mm) in the original design; ΔP is the branch resistance (Pa) in the original design; $\Delta P'$ is the required resistance (Pa).

It needs to be pointed out that when using this method, the tee branch diameter has to keep unchanged, and that a section of reducing (extending) duct can be added on the tee branch, to avoid the change of the tee's local resistance.

Second, adjusting dampers is to change the damper opening to adjust duct resistance, which is theoretically an easy, simple method. It must be pointed out that commissioning a multi-branch ventilation air-conditioning system in practical applications is technically a complex work. Only repeated adjustment and test can reach the anticipated flow distribution. In addition, dampers are only adjusted under certain conditions, which are not omnipotent, i.e., they are generally adjusted in the range of unbalance rate.

Last, when the resistance difference of two branch ducts is small (e.g., within 20%), instead of changing duct diameter, the flow of the branch duct with small resistance can be appropriately increased to reach hydraulic resistance balance. The increased air volume can be calculated by the following formula:

$$L' = L\left(\frac{\Delta P'}{\Delta P}\right)^{1/2} \quad \left(\frac{m^3}{h}\right) \tag{4.23}$$

where L' is the duct diameter (m) after adjustment; L is the branch's air volume (m^3/h) in the original design; under this method, the flow and the resistance in the front main pipe would increase correspondingly, so do the fan's air volume and pressure.

5. Calculate the system's total resistance.

The sum of the resistance (including equipment resistance) of tandem ducts in index circuits is the total resistance ΔP of the duct network. According to the hydromechanics theory, the characteristic curve equation of duct network resistance is shown as follows:

$$\Delta P = SL^2 \tag{4.24}$$

where S is the duct network impedance (kg/m^7); L is the total flow (m^3/s) of duct network.

The duct network impedance is related to its geometric size, frictional resistance coefficient, local resistance coefficient and fluid density. When these factors keep unchanged, the duct network impedance S is a constant. According to the calculated total duct network resistance ΔP and the required total air volume L, the duct network impedance can be calculated by the following formula to achieve the characteristic curve of duct network:

$$S = \frac{\Delta P}{L^2} \tag{4.25}$$

The characteristic curve of duct network can also be achieved by first calculating duct impedance rather than duct resistance and total duct network resistance before calculating duct network impedance as per the following impedance relationship between tandem and parallel pipelines:

$$S_i = \frac{8\left(\lambda \frac{l}{d_i} + \sum \zeta_i\right)\rho_i}{\pi^2 d_i^4} \tag{4.26}$$

Parallel pipeline: $S^{-1/2} = \sum S_i^{-1/2}$

Tandem pipeline: $S = \sum S_i$

The previous formula shows that the change of relevant parameters of any duct in the duct network would lead to the alternation of the whole duct network's characteristic curve, thus changing the total flow in the duct network and flow distribution in ducts. This dictates the complexity of duct network adjustment. It can be theoretically further proved that it is unfeasible to adjust flow by completely relying on dampers instead of prioritizing resistance balance during the design of duct network, especially the duct network with many parallel pipelines. After achieving the characteristic curve for the duct network, appropriate power equipment can be selected for the duct network by combining the fan's characteristic curve.

6. The selection of the fan.

The fan type can be selected by the nature of the delivered air and the air volume and resistance of the system. For example, common fans are used to transmit clean air; and anti-explosion fans are adopted to transmit explosive gases or dusts.

Given the safe margin, the fan's air volume and pressure are shown as follows:

$$P_f = K_p \cdot \Delta P(\mathrm{Pa}) \tag{4.27}$$

$$L_f = K_L \cdot \Delta L\left(\frac{m^3}{h}\right) \tag{4.28}$$

where P_f is the fan's air pressure (Pa); L_f is the fan's air volume (m^3/h); K_p is the additional coefficient of air pressure. For ordinary air supply and exhaust systems, $K_p = 1.1-1.15$; for dust removal systems, $K_p = 1.15-1.20$; K_L is the additional coefficient of air volume. For ordinary air supply and exhaust systems, $K_L = 1.1$; for dust removal systems, $K_L = 1.1-1.15$. When the fan is working at the non-standard state, its performance should be converted by the following formula, by which a proper fan can be selected from samples:

$$L_f^{'} = L_f \tag{4.29}$$

$$P_f = P_f^{'}\left(\frac{\rho^{'}}{1.2}\right) \tag{4.30}$$

where L_f is the fan's air volume (m^3/h) under the standard state; $L_f^{'}$ is the fan's air volume (m^3/h) under the non-standard state; P_f is the fan's air pressure (Pa) under the standard state; $P_f^{'}$ is the fan's air pressure (Pa) under the non-standard state; $\rho^{'}$ is the air density (kg/m^3) under the non-standard state.

After the fan selection, the motor power is calculated by the fan's air pressure and volume under the non-standard state, by which a proper motor can be selected from samples:

$$N = \frac{L_f^{\prime} \cdot P_f^{\prime}}{\eta \cdot 3,600 \cdot \eta_m} \cdot K(\mathrm{W}) \tag{4.31}$$

$$N_y = \frac{L'_f \cdot P'_f}{3{,}600} (\mathrm{W}) \tag{4.32}$$

$$\eta = \frac{N_y}{N} \tag{4.33}$$

where N is the motor power W; N_y is the fan's effective power W; η is the full-pressure efficiency. Because of the energy loss during the operation of the fan, the power (N) consumed on the fan shaft should be larger than the effective power N_y; η_m is the fan's mechanism efficiency; K is the motor capacity's safety coefficient.

The fan's performance parameters (or characteristic curve) are usually given out by special environment parameters (such as temperature and atmospheric pressure). For example, the ambient temperature of testing room-temperature fans is 20 °C, 1 atm (or 104 mmH_2O), and that of testing boiler fans is 200 °C, 1 atm (or 104 mmH_2O). The environment temperature, the ambient gas pressure and the transmitted gas density are all different from those under testing conditions when the fan is working, so the fan's motor power needs to be corrected:

$$N_2 = N_1 \frac{\rho_2}{\rho_1} (\mathrm{W}) \tag{4.34}$$

where N_2 is the motor power (W) under standard working conditions; N_1 is the motor power (W) during operation; ρ_1 is the air density (kg/m^3) under standard working conditions; ρ_2 is the air density (kg/m^3) during operation.

When the transmitted gas during operation has a smaller temperature than that under testing conditions, ρ_1 is smaller than ρ_2, and the fan's motor power is relatively small, so that the normal operation requirements cannot be met; on the contrast, the motor power is relatively large, and the energy consumption during operation would increase.

4.5.2 Average pressure-loss method

The characteristic of the average pressure-loss method is to distribute the known total pressure head equally to each duct section by the main pipe length to confirm the duct resistance and then the cross-sectional size as per the air volume in each duct section. This method is more applicable to the situation where the fan's pressure head for the duct system is confirmed, or the resistance balance calculation is conducted for branch pipeline, with its basic steps shown as follows:

1. Draw an axonometric diagram for the duct network to confirm the index circuit.
2. Calculate the pressure loss of the unit duct length for the index circuit.
3. Confirm the diameter of each duct by the flow in each duct and the pressure loss of unit duct length for the index circuit.
4. Confirm the working pressure head for parallel pipelines to calculate the pressure loss of unit duct length.
5. Confirm the diameter of each branch by the flow in each duct and the pressure loss of unit duct length for the branch pipeline.

4.5.3 Static regain method

The characteristic of the static regain method is to reduce flow velocity, overcome resistance in ducts and remain the required static pressure in ducts by changing the duct's cross-sectional size. It is applicable to the design of air supply and exhaust pipelines, with the basic steps shown as in the following:

1. Confirm the outflow velocity at each hole in the pipeline.
2. Calculate the static pressure P_j at each hole.
3. Assume the flowing velocity in the duct at the first hole along the inside airflow (e.g., the air supply velocity is given out by the required range of air distribution) to calculate the inside full pressure Pq_1 and the duct's cross-sectional size at this position.
4. Calculate the resistance ($\Delta P_{1\text{–}2}$) from the first hole to the second one.
5. Calculate the dynamic pressure at the second hole – $Pd_2 = Pq_1 - \Delta P_{1\text{–}2} - P_j$.
6. Calculate the inside flow velocity at the second hole, to confirm the duct's cross-sectional size at this position.
7. Repeat the previous calculation until the duct's cross-sectional size at the last hole is confirmed.

4.6 Air dampers and air-volume adjustment

4.6.1 The operating principles and classification of air dampers

The air damper is the controlling and adjusting mechanism for air transmission and distribution network, with its basic functions of blocking or opening air ventilation pipelines and adjusting or distributing flow in the pipeline.

The check damper, a damper to prevent backflow by utilizing airflow to automatically open and close damper plate, should be set during the parallel installation of two fans or above in ventilation systems. When air is running along the specified direction, the damper plate is opened by the impact of airflow; in the case of backflow, the damper plate is closed because its own weight and the impact of airflow resistance lead to the closing of the sealing surface with damper seat, so that the backflow is prevented. Thus, it has two performances – one is resistance performance when the air is running forward, the other is air leakage performance when the air is running backward.

The installation of the check damper must ensure the rotation flexibility of damper plate and shaft. The usage temperature for rectangular check dampers is generally 10–50 °C, and that for circular and square check dampers is generally 0–95 °C (Figure 4.3).

The air-volume-regulating damper, also known as air regulation gate, is an indispensable fitting in ventilation, air adjustment and purification engineering in civilian buildings and industrial plants. It is generally applied into air-conditioning and ventilation system pipelines, aiming for the adjustment of air volume in ducts and the mixed adjustment of return air and fresh air.

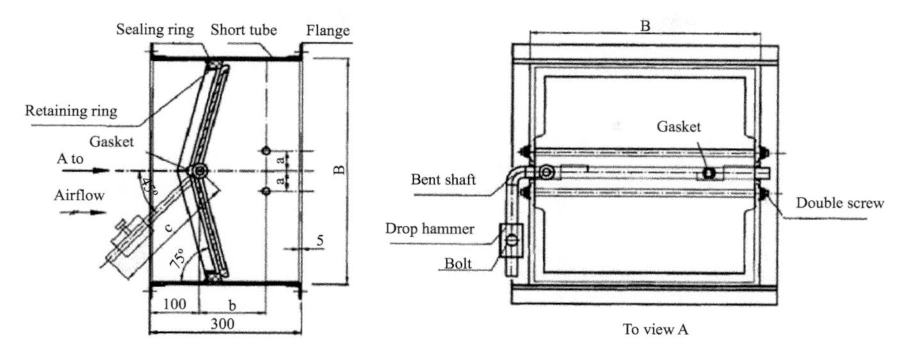

Figure 4.3 Check damper (GJBT 1026-2007)

The basic structure of air-volume regulation damper is a damper plate placed at the duct centre that can spin around a central shaft paralleled to it to change the angle with the duct cross section, thus changing air volume by altering the duct flow-section. A single-damper plate, which is relatively thick for bearing air pressure, needs a large space for spinning, so multiple damper plates are designed to be arranged like window blinds to realize air-volume adjustment when spinning.

Based on the motion pattern of air damper blades, air dampers can be divided into opposed multi-blade dampers, parallel multi-blade dampers, diamond-shaped dampers and butterfly dampers.

The multi-blade damper, the most common damper in ventilation systems, can be applied into almost all ventilations systems thanks to its simple installation, easy operation, great regulation performance and excellent resistance characteristics. By constitution, it includes parallel, opposed and lightweight opposed patterns; by sealing requirements, it has ordinary and closed types; by motion pattern, it can be driven by power, gas or hand. Except gear-driven lightweight opposed dampers, others are all driven by connecting rod.

The multi-blade damper is to realize the air-volume adjustment by regulating the opening of blades. The blades of the parallel multi-blade damper spin in the same direction. The adjacent two blades of the opposed multi-blade damper spin in the opposite direction (Figures 4.4 and 4.5).

As a single-plate air damper, the butterfly damper can open, close and adjust air volume by rotating a switch fitting back and forth at around 90°. Despite its simple structure, convenient to be machined, low cost and easy operation, the butterfly damper, which can be fixed at any angle of within 90°, is only used as switch or for rough adjustment in ventilation and air-conditioning systems due to its worse precision.

Operated by hand, power or zipper, the switch fitting for the butterfly damper is a disc- or square-shaped butterfly plate, rotating around its own axis inside the damper to realize the opening, closing and adjusting purpose. It can be mainly divided into handle circular butterfly damper, zippered circular butterfly damper, zippered circular insulation butterfly damper, handle square butterfly damper,

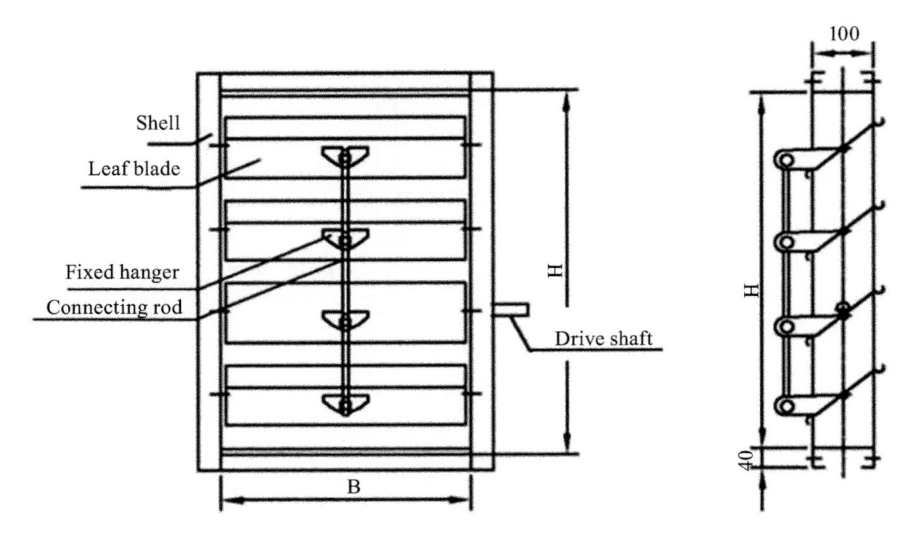

Figure 4.4 Parallel multi-blade damper (GJBT 1026-2007)

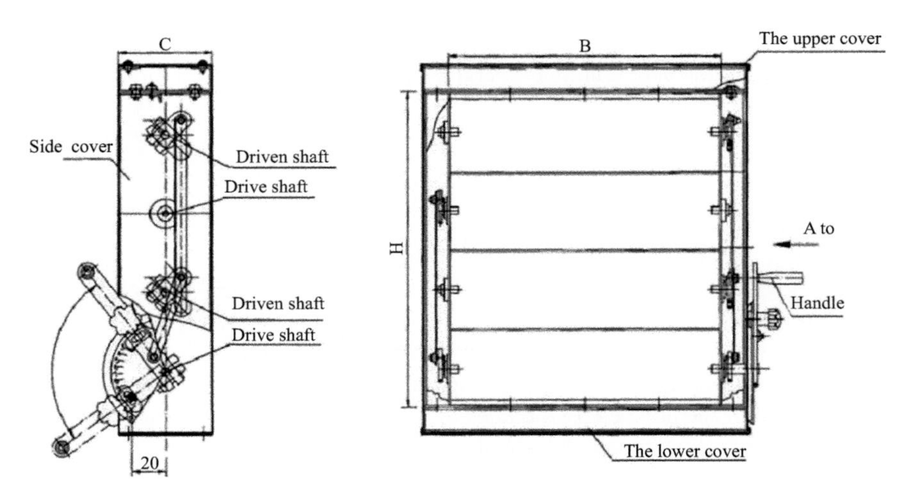

Figure 4.5 Manual opposed multi-blade damper (GJBT 1026-2007)

zippered square butterfly damper, zippered square insulation butterfly damper, handle rectangular butterfly damper, zippered rectangular butterfly damper and zippered rectangular insulation butterfly damper. Figure 4.6 is a diagram of a handle square butterfly damper.

Featuring a simple mechanism, easy adjustment and low cost, the three-way control damper (Figure 4.7) is mainly used for initial air-volume distribution of main duct, to make the air volume in branches meet the design requirements. With

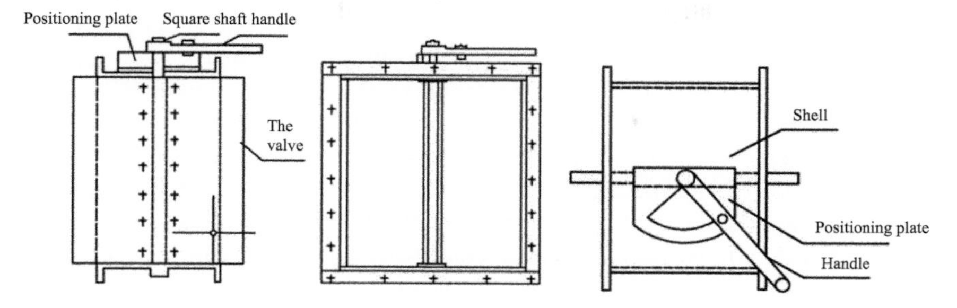

Figure 4.6 Handle square butterfly damper (GJBT 1026-2007)

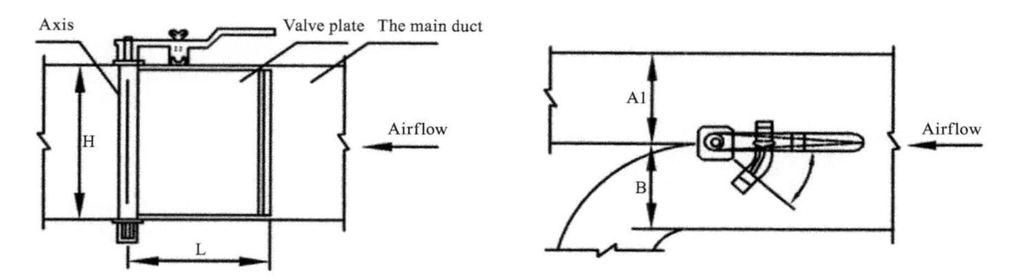

Figure 4.7 Three-way control damper (GJBT 1026-2007)

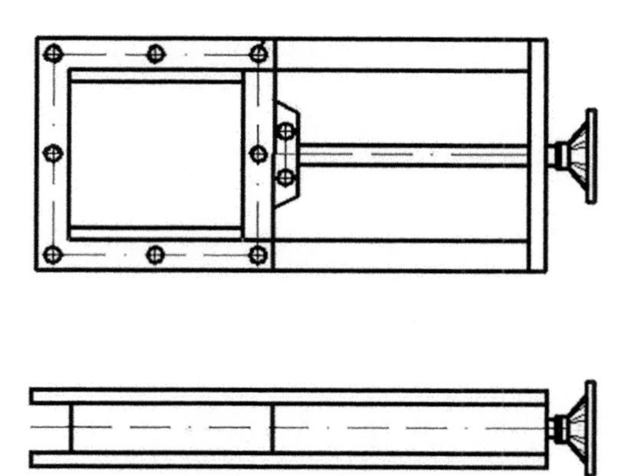

Figure 4.8 Slide damper (GJBT 1026-2007)

the patterns of handle and pull rod, it is generally processed together with tee fittings only for air-volume divergence, rather than convergence. The initial air-volume adjustment can be realized by changing the damper plate position.

The slide damper (Figure 4.8), a damper whose seat and plate closely contact with each other to realize a close sealing, can close and open by the plate's repeated movement along the direction vertical to air duct. It can realize the air-volume adjustment by the depth of inserting the plug board into the duct.

The constant volume damper, a device of mechanically, automatically adjusting air volume, can work without any external force. The control damper plate is borne by the shaft in the damper, and positioned by the power caused by airflow, thus enabling the flow to be kept at a set value in an allowable differential pressure range. The air resistance imposed on the damper plate makes it closed and strengthens the aerated gasbag, which could shake the damping. The calibrated spring piece, which would generate a force against closing, is using a cam to adjust resistance, which could automatically adjust the damper plate angle during the variation of differential pressure, thus keeping the flow at a set value with a minor error.

The constant air-volume damper (Figure 4.9), applicable to the pipeline systems with constant air supply and exhaust rates, is an air-volume-adjusting device unrelated to pressure, which can be customized and installed by a duct diameter while equipped with an electric actuator and an additional sound-absorption shell to reduce radiation and noise.

The fire-resisting damper, which would automatically close when fire breaks out, should be set in place on the air ducts of ventilation and air-conditioning systems to prevent the sweeping of the fire and the toxic high-temperature smoke in the ducts. Its closing is due to the weight of the eccentric damper plate itself, whose heavy end is pulled up by steel wire, on which a fusible alloy piece is installed to keep the damper open. When the airflow temperature in air ducts exceeds the set temperature, however, the alloy piece is fused off and the damper is automatically closed. At the same time, the fire-resisting damper can connect with a signal-warning device to give an alarm.

The fire-resisting damper (Figure 4.10) is used for air ducts of air-conditioning ventilation systems to prevent the sweeping of fire along the ducts. When the air temperature is at 70 or 150 °C, the temperature fuse would automatically close the damper, while releasing electric signals for manual resetting.

The fire damper in smoke-venting system (Figure 4.11), generally in an opening state, is set on the pipeline of mechanical exhaust systems. When the

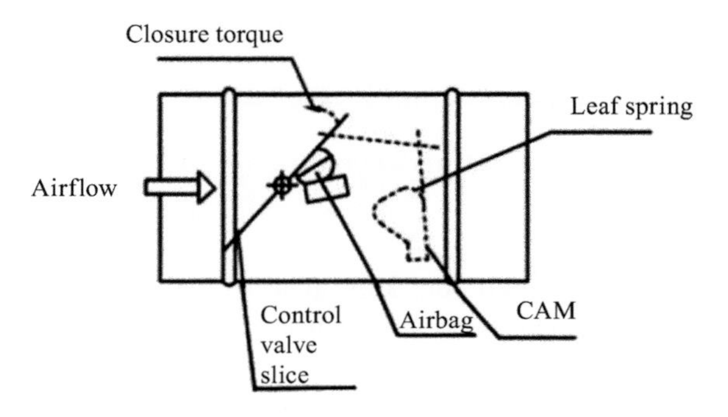

Figure 4.9 Constant volume damper (GJBT 1026-2007)

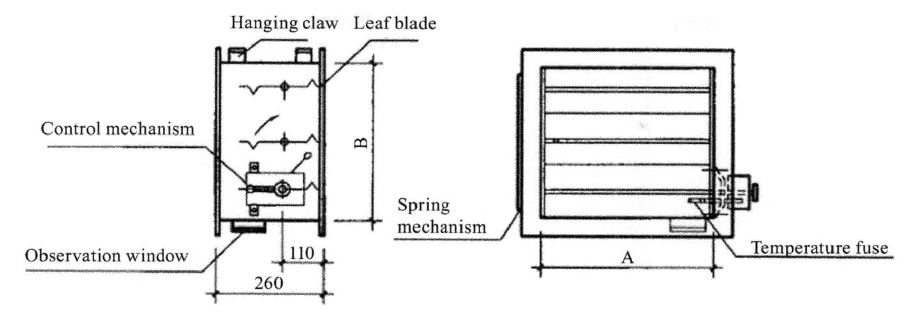

Figure 4.10 Fire-resisting damper (Wang, 2003)

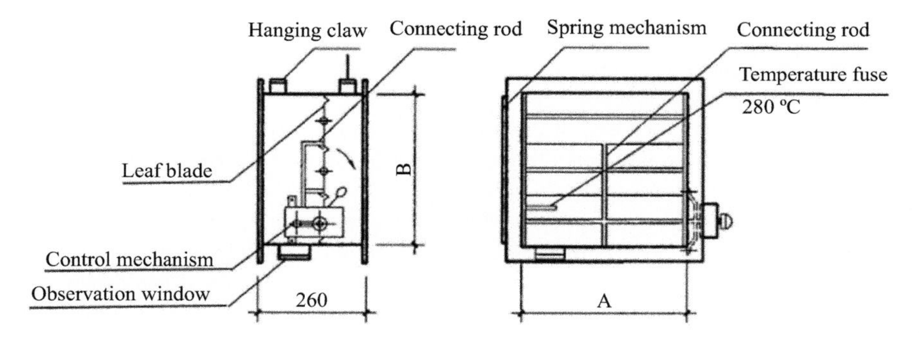

Figure 4.11 Fire damper in smoke-venting system (Wang, 2003)

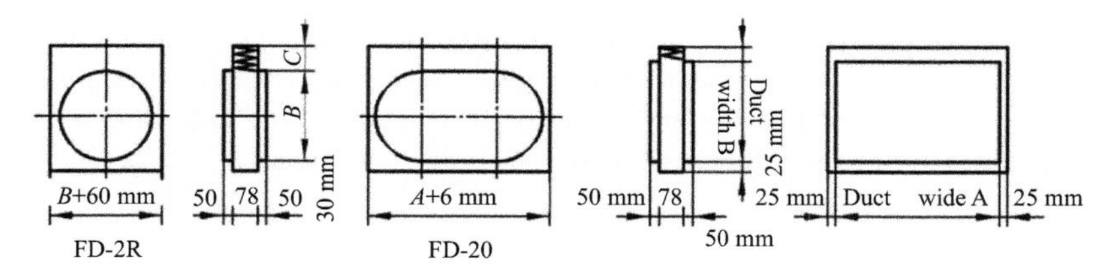

Figure 4.12 Rolling shutter fire damper (Sun, 2010)

temperature in the smoke extraction pipeline reaches 280 °C as fire breaks out, the damper is closed to successfully block smoke and fire, while meeting the requirements of smoke leakage rate and fire integrity in a certain time.

Installed on air-conditioning and ventilation pipelines or in firewalls, the blades of the fire-resisting damper, which are generally rolled up and kept at an opening state, adopt a structure of rolling shutters (Figure 4.12). When the temperature in ducts or in ambient environment exceeds that (70 or 280 °C) for the

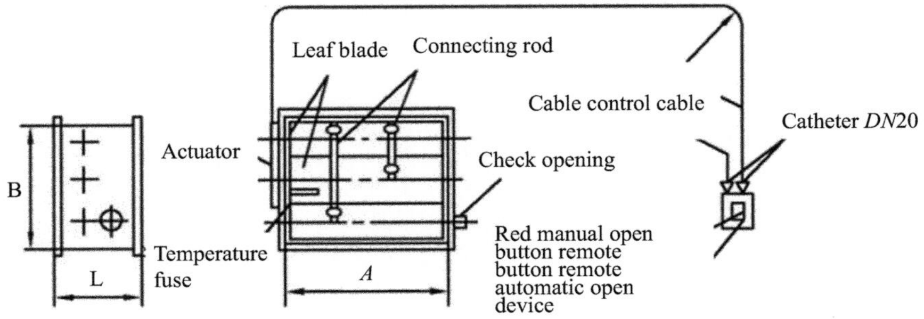

Figure 4.13 Smoke exhaust damper (Wang, 2003)

fire-resisting damper's fusible piece, the blades would rapidly close due to the spring force.

Installed at the branch end of the mechanical exhaust system (Figure 4.13), the smoke exhaust damper, which is often closed while meeting the requirements of air leakage rate, only is opened by hand or power in the case of fire occurrence or smoke extraction and is automatically closed when the temperature reaches 280 °C.

Before selecting and installing air dampers, technical personnel responsible for ventilation and air-conditioning engineering must fully understand the functions, structures and controlling modes of air dampers shown in Table 4.8. Correctly selecting and installing air dampers is important for ventilation systems to operate under the designed working conditions.

4.6.2 Main performance parameters of air dampers

Generally, represented by β, the damper authority is the percentage of the damper pressure loss to the total pressure loss in the regulating branch (including the damper itself) when the regulating damper is fully opened. It is also known as the design damper authority, which can be represented as follows:

$$\beta = \frac{\Delta P_{\min}}{\Delta H} \tag{4.35}$$

where $\Delta P_{\min}$ is the pressure loss under the designed flow when the control damper is completely opened; ΔH is the available pressure head of the circuit where the control damper is located.

Theoretically, the larger its value, the better its performance. This indicates that the damper is able to effectively adjust flow and thus control energy output effectively. However, the pressure drops on the circuit where the control damper is located are generally not constant, and their practical characteristics are different from theoretical ones. When there are no other facilities to ensure the damper authority, to realize a larger damper authority means a larger pressure drop on the electric control damper, which would consume much head of the water pump and

Table 4.8 *Functions, structures and controlling modes of air dampers (GJBT 1026-2007)*

Air duct damper category	Function			Structural style		Control method			
	Switch	Regulate	Seal	Simple leaf	Multi-layered	Manual	Electric	Pneumatic	Self-balancing
Multi-blade damper	○	●	○	○	●	●	●	●	–
Butterfly damper	●	○	○	●	–	●	●	●	–
Constant volume damper	○	●	–	●	–	○	○	○	●
Check damper	●	–	○	–	●	–	–	–	●
Three-way control damper	○	●	–	●	–	●	–	–	–
Closed inclined damper	●	○	●	●	–	●	–	–	–
Excess pressure damper	○	●	○	●	○	●	–	–	●

Note: ● is for recommendation; ○ is for reference; - means unavailable.

thus lead to an uneconomical operation. Therefore, its value is generally taken at around 0.5 (in the case of non-hydraulic balancing damper). The acceptable minimum damper authority by air-conditioning systems must be larger than 0.25 (generally taken as 0.3).

The flow characteristics of dampers refer to the special relations between the relative flow of fluid medium passing through the damper and the relative opening of the damper, i.e.,

$$\frac{Q}{Q_{\max}} = f\left(\frac{l}{l_{\max}}\right) \tag{4.36}$$

where Q is the flow when the damper is at a certain opening; $Q_{\max}$ is the flow when the damper is fully opened; $L_{\max}$ is the running route of the trim when the damper is at a certain opening; $L_{\max}$ is the running route of the trim when the damper is fully opened.

The ratio of the maximum flow to the minimum flow that the damper can control is the adjustable ratio R:

$$R = \frac{Q_{\max}}{Q_{\min}} \tag{4.37}$$

where Q_{min} is the lower limit of the damper's adjustable flow, which is unequal to the leakage rate when the damper is completely closed. In general, the minimum adjustable flow is 2%–4% of the maximum flow, while the leakage rate is only 0.1%–0.01% of the maximum flow.

Generally, the flow can be adjusted by changing the restriction area between the trim and seat. But in fact, when the restriction area is changing, the impact of multiple factors would change the differential pressure before and behind the damper, thus altering flow. Thus, both ideal flow and working flow features constitute flow characteristics. To facilitate analysis and comparison, the differential pressure before and behind the damper is first assumed at a certain value, i.e., first study ideal flow characteristics before discussing the reality, i.e., working flow characteristics.

The flow characteristics obtained under the unchanged differential pressure before and behind the damper are called ideal flow characteristics, or inherent flow characteristics. The typical ideal flow characteristics include linear flow characteristics, equal percentage flow characteristics, quick opening flow characteristics and parabola flow characteristics, as shown in Figure 4.14.

Because of their single structure, air dampers, mostly butterfly dampers and multi-blade dampers, almost feature an equal percentage pattern. From the figure, it is known that the magnification factors (i.e., slope of curve) of dampers with equal percentage flow characteristics would gradually increase with the increase in route. With the same route, the flow would change less in the case of low load (small opening) and vary largely in the case of high load (large opening). Therefore, such a damper works steadily in almost full closing and features sensitivity and efficiency in nearly full opening.

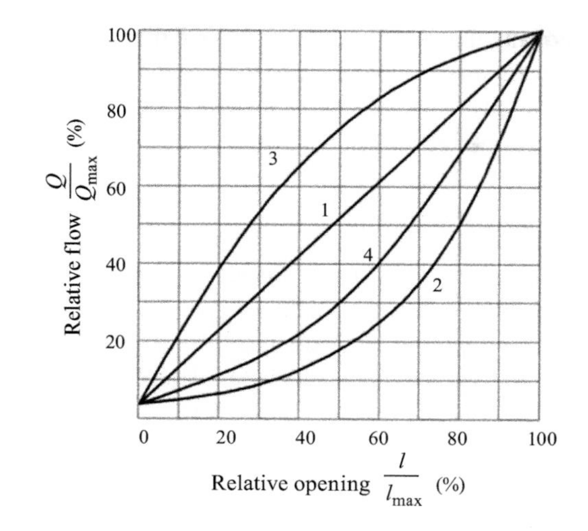

Figure 4.14 Ideal flow characteristics of dampers (R = 30): (1) linear, (2) equal percentage, (3) quick opening, (4) parabola

The equal percentage characteristics of air dampers can be mathematically represented by

$$\frac{d(Q/Q_{max})}{d(l/l_{max})} = k(Q/Q_{max}) \tag{4.38}$$

The air volume of equal percentage air dampers can be calculated as per the following formula:

$$\frac{Q}{Q_{max}} = R\left(\frac{l}{l_{max}} - 1\right) \tag{4.39}$$

The flow coefficient refers to the flow capacity of dampers at a certain opening under specified conditions.

Before implementing legal unit of measurement, the flow capacity represented by C refers to the hourly flow (m^3/h) of fluid passing through the damper when the damper is fully opened, the differential pressure before and behind the damper is 1 kgf/cm^2, and the fluid density is 1 g/cm^3. However, after the implementation of legal unit of measurement, it is changed as the hourly flow (m^3/h) of fluid passing through the damper when the damper is fully opened, the differential pressure before and behind the damper is $\Delta P = 1$ bar (105 Pa), and the fluid density is 1 g/cm^3.

Known as the C_V value by European and American standards, it is defined by a non-international unit system based on the US flow measurement system, proposed by Instrument Society of America, and wildly applied internationally. It refers to the US gallons of water at 40–100 °F (4.44–37.78 °C) running through the damper within 1 min when the pressure drops at 1 psi.

As an internationally universal coefficient defined by International Electrotechnical Commission, the flow capacity K_V refers to the cubic meters (m^3/h) of water at the temperature of 278K–313K (5–40 °C) passing through the damper or other duct components with the pressure drop of 105 Pa (1 bar). For other fluid, it can be corrected by relative density:

$$C = 1.167K_V, K_V = 0.865C_V \tag{4.40}$$

The damper's flow coefficient is an index of measuring its flow capacity. The bigger the flow coefficient, the larger the flow capacity, the smaller the pressure loss when fluid is running through the damper. The flow coefficient is changing with the size, type and structure of dampers. Thus, for dampers with different types and specification, their flow coefficient should be confirmed. Here, C is only introduced.

The volume flow of the gas passing through dampers on the cross section of damper outlet is shown as

$$Q_2 = \frac{F}{\sqrt{\xi}}\sqrt{\frac{2\Delta P}{\rho_2}} \tag{4.41}$$

With the mass flow shown as

$$G = \rho_2 Q_2 = \frac{F}{\sqrt{\xi}}\sqrt{2\rho_2\Delta P} \tag{4.42}$$

where F is the sectional area (cm^2) of the duct connecting the damper; ρ_2 is the air density on the cross section of the damper outlet, 1 kg/m^3 = 10–6 kg/cm^3; ΔP is the damper's differential pressure, 1 Pa = 10^{-2} kg/(s^2 cm).

The following formula can be obtained by modifying the unit in the previous formula:

$$G = \frac{F}{\sqrt{\xi}}\sqrt{\rho_2\Delta P}\sqrt{2\times 10^{-6}\times 10^{-2}}\ \ (\text{kg/s}) = \frac{F}{1.9642\sqrt{\xi}}\sqrt{\rho_2\Delta P}\ \ (\text{kg/h}) \tag{4.43}$$

i.e.,

$$\frac{1.9642G}{\sqrt{\rho_2\Delta P}} = \frac{F}{\sqrt{\xi}} \tag{4.44}$$

As shown earlier, the flow capacity C is

$$C = 5.09\frac{F}{\sqrt{\xi}} = 5.09\times\frac{1.9642G}{\sqrt{\rho_2\Delta P}} = \frac{10G}{\sqrt{\rho_2\Delta P}} \tag{4.45}$$

If ρ_2 (air density, kg/m^3) behind the damper in operation in the previous formula is converted into ρ_N (air density, kg/N m^3) under the standard state (0 °C, 101,324.71 Pa), the following formula can be obtained:

$$\rho_2 = \rho_N\frac{T_N P_2}{T_2 P_N} \tag{4.46}$$

where ρ_N is the density (kg/m^3) ($\rho_N = 1.293$ kg/m^3 under 0 °C and 101,324.71 Pa); T_N is the standard temperature, $T_N = 273$K; P_N is the standard atmospheric pressure, $P_N = 101{,}324.72$ Pa; T_2 is the air temperature behind the damper, $T_2 = (273.15+t)$K (t is the operational temperature (°C)).

A formula can be given by substituting ρ_2 into formula (4.45) as follows:

$$C = \frac{10G}{\sqrt{\rho_N \times \frac{273P_2}{T_2 \times 101{,}324.72} \times \Delta P}} = \frac{193G}{\sqrt{\frac{\rho_N P_2 \Delta P}{T_2}}} = 193G\sqrt{\frac{T_2}{\rho_N P_2 \Delta P}} \tag{4.47}$$

It should be noticed that the previous formula is only applicable to the situation of $P_2 > P_1/2$ (ordinary flow). When the air is in a supercritical state, both air pressure $P_2{}'$ and the air density $\rho_2{}'$ on the cross section at the damper outlet remains unchanged, no matter how smaller the air pressure behind the damper is. When using ρ_N and P_1 to calculate the C value, a formula can be obtained as follows:

$$C = 193G\sqrt{\frac{273.15+t}{\rho_N \times \frac{P_1}{2} \times \frac{P_2}{2}}} = \frac{386G}{P_1}\sqrt{\frac{273.15+t}{\rho_N}} \tag{4.48}$$

The damper's flow capacity is related to the resistance coefficient, the damper's size, pattern and structure, as well as the direction along which the air runs through it.

Take manual opposed multi-blade damper, handle rectangular butterfly damper and handle circular butterfly damper, the formula (4.48) is used to calculate the flow capacity, with the results shown in Tables 4.9–4.11.

The previous tables only provide approximate values. If more precise results are required, the values of the temperature and pressure behind the damper and the differential pressure before and behind the damper can be used for calculation based on the formula (4.47).

Because of various reasons, the air volume needs to be adjusted during the operation of ventilating systems. Changing the fan's operational air volume needs to change its working point. Changing the fan's working point can be realized by changing the characteristic curve of the ventilating duct or the fan.

The damper adjustment aims to change the fan's working point by regulating the damper opening to change the duct network features. As shown in Figure 4.15 (a), the network's characteristic curve is changed to C_1 from C_2, the fan's working point is changed to A_1 from A_2, and the fan's air volume is changed from L_2 to L_1. This method is easy and simple, but part of energy (the wind differential pressure ΔP of two curves) would be consumed on the damper. It is therefore applicable to the systems with small air volume.

Regulation of the rotational velocity of a fan can be seen in Figure 4.15(b). This method is to change its characteristic curve by adjusting the fan's rotational velocity. For example, when the fan's rotational velocity changes from n_2 to n_1, its characteristic curve n_2 is transformed to the curve n_1, its working point to A_1 from A_2 and its air volume to L_1 from L_2.

Table 4.9 Flow coefficient of several typical specifications (B×H) of manual opposed multi-blade dampers

Aperture (°)		**90**	**80**	**70**	**60**	**50**	**40**	**30**	**20**	**15**	**10**
Resistance coefficient ξ		**0.4**	**0.7**	**1.5**	**4**	**9**	**25**	**90**	**400**	**900**	**3,000**
Flow coefficient	160×320	0.41	0.31	0.26	0.13	0.09	0.05	0.03	0.01	0.01	0
	800×320	2.06	1.56	1.3	0.65	0.43	0.26	0.14	0.07	0.04	0.02
	630×630	3.19	2.41	2.02	1.01	0.67	0.4	0.21	0.1	0.07	0.04
	800×630	4.06	3.07	2.57	1.28	0.86	0.51	0.27	0.13	0.09	0.05
	1,250×800	8.05	6.08	5.09	2.55	1.7	1.02	0.54	0.25	0.17	0.09
	1,250×1,000	10.06	7.6	6.36	3.18	2.12	1.27	0.67	0.32	0.21	0.12
	1,600×1,250	16.1	12.17	10.18	5.09	3.39	2.04	1.07	0.51	0.34	0.19

Table 4.10 Flow coefficient of several typical specifications (B×H) of handle rectangular butterfly dampers

Opening angle of blade (°)		**0**	**10**	**20**	**30**	**40**	**50**	**60**	**Closed**
Resistance coefficient of rectangular butterfly damper ξ		**0.04**	**0.33**	**1.2**	**3.3**	**9**	**26**	**70**	**1,665**
Flow coefficient	320×250	2.04	0.71	0.37	0.22	0.14	0.08	0.05	0.01
	400×320	3.26	1.13	0.59	0.36	0.22	0.13	0.08	0.02
	500×250	3.18	1.11	0.58	0.35	0.21	0.12	0.08	0.02
	630×320	5.13	1.79	0.94	0.56	0.34	0.20	0.12	0.03
	800×500	10.18	3.54	1.86	1.12	0.68	0.40	0.24	0.05
	800×630	12.83	4.47	2.34	1.41	0.86	0.50	0.31	0.06

Table 4.11 Flow coefficient of several typical specifications (D) of handle circular butterfly dampers

Opening angle of blade (°)		**0**	**10**	**20**	**30**	**40**	**50**	**60**	**Closed**
Resistance coefficient of circular butterfly damper ξ		**0.2**	**0.52**	**1.5**	**4.5**	**11**	**29**	**108**	**1,620**
Flow coefficient	160	0.23	0.14	0.08	0.05	0.03	0.02	0.01	0.00
	200	0.36	0.22	0.13	0.08	0.05	0.03	0.02	0.00
	320	0.91	0.57	0.33	0.19	0.12	0.08	0.04	0.01
	360	1.16	0.72	0.42	0.24	0.16	0.10	0.05	0.01
	400	1.43	0.89	0.52	0.30	0.19	0.12	0.06	0.02
	630	3.55	2.20	1.29	0.75	0.48	0.29	0.15	0.04

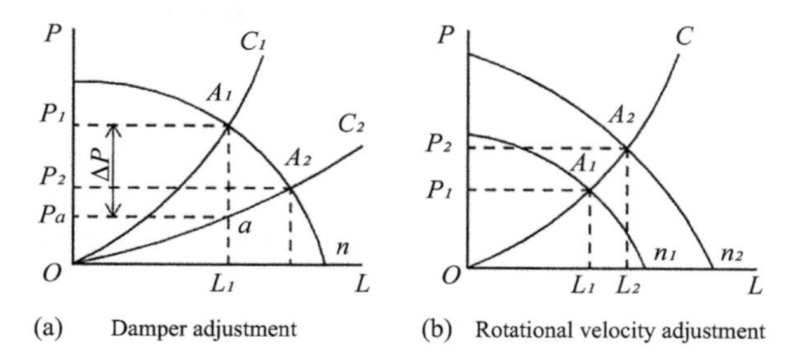

Figure 4.15 Fan adjustment during working

The relations between the fan's rotational velocity and air volume, air pressure and power under similar working conditions (the fan's efficiency keeps unchanged) can be described as the following formula:

$$\frac{L_1}{L_2} = \frac{n_1}{n_2}; \frac{P_1}{P_2} = \left(\frac{n_1}{n_2}\right)^2; \frac{N_1}{N_2} = \left(\frac{n_1}{n_2}\right)^3 \tag{4.49}$$

When $L_1 = 0.8L_2$, $N_1 = 0.512N_2$ and the fan's energy consumption is halved compared with previous energy consumption. At present, the common ways to change rotational velocity include changing the rotational velocity ratio of belt pulley and adopting hydraulic coupler and variable-speed motor.

With the scientific and technological development, the price of frequency converters drops, so the frequency-converting, velocity-adjusting, energy-saving method has been wildly applied. The relations between the AC motor's rotational velocity and the power-supply frequency are shown as follows:

$$n = \frac{60f(1-s)}{M} \tag{4.50}$$

where n is the motor's rotational velocity; f is the motor's power supply frequency; M is the number of pole pairs of the motor; s is the slip ratio.

From the previous formula, the rotational velocity n is proportional to the frequency f, i.e., changing frequency could alter the motor's rotational velocity. When the frequency f is changing in the range of 0–50 Hz, the rotational velocity can be adjusted within a wide range. The frequency variation and velocity adjustment are realized by changing the motor's power frequency. In practical applications, only reducing the frequency would increase the current of the motor winding, especially when the frequency decreases greatly, the motor would be easily burnt out. When the fan has a bigger rotational velocity variation, its efficiency would relatively drop. Therefore, during the operation of changing rotational velocity, the rotational velocity of the fan should be no less than 50%–60% of the designed one.

4.7 Connection of duct fitting and fan

Two kinds of resistance would generate when air is running through air ducts. One is the energy loss along the way (also known as frictional resistance or frictional resistance) due to the viscosity of the air itself and the friction between air and duct wall. The other is the concentrated energy loss (also known as local resistance loss) caused by the variation of the airflow velocity and direction and the resultant vortexes when air is passing through fittings and equipment in air ducts.

4.7.1 Elbow

Try to arrange ducts straight to avoid the emergence of elbows. The curvature radius of circular duct elbows is generally one to twice larger than the duct diameter. When airflow is making a turn passing through an elbow, two vortex zones will generate on the outside and inside of the elbow because of inertia (Figure 4.16).

Additionally, the air velocity in the duct centre is greater than that near the duct wall. This would lead to a larger centrifugal force during the turning, thus creating a torque and causing paired airflow spinning (secondary flow) outwards from the centre until an extended length of 10*d*-15*d* (*d* is the duct diameter) is reached sometimes. Relevant experiment showed that the increase of the elbow's bending radius would weaken the spinning of vortex zones and airflow. However, an overlarge curvature radius would make the elbow occupy more space. The curvature radius (*R*) in air ducts is generally 1*d*-2*d*. The larger the length-to-width ratio of the rectangular elbow cross section, the smaller the resistance (Figures 4.17 and 4.18).

4.7.2 Tee

Tees aim to distribute air or make it converge together. Shunt tees are used for air distribution, confluence tees for air converging. When the air velocity in confluence tees is larger than that in branch ducts, the straight airflow would inject branch airflow, i.e., the straight airflow with a larger velocity would lose energy, while the branch airflow with a small velocity would obtain energy. Thus, the local resistance coefficient in branch ducts would be negative sometimes, but not always

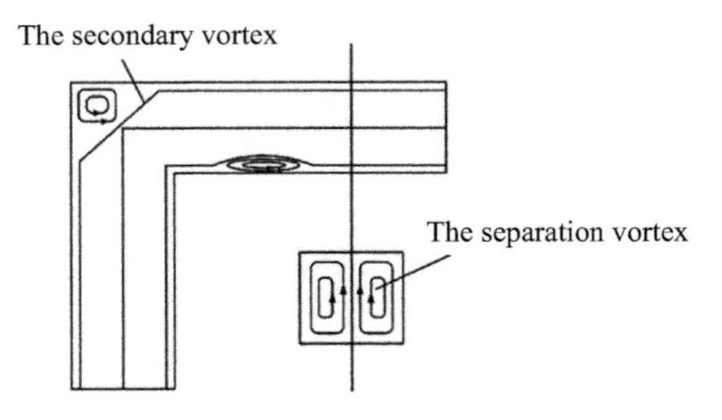

Figure 4.16 Airflow in an elbow

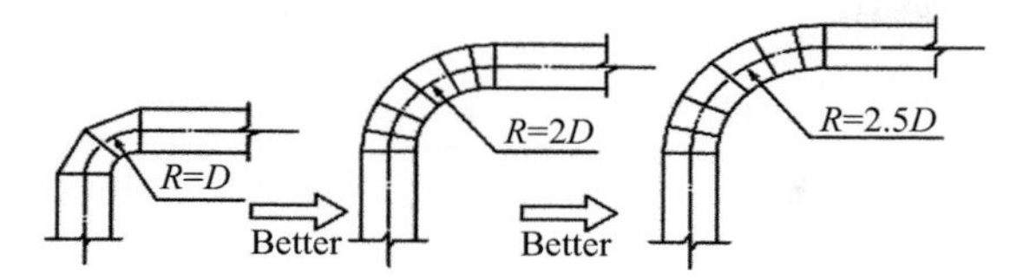

Figure 4.17 Circular duct elbows

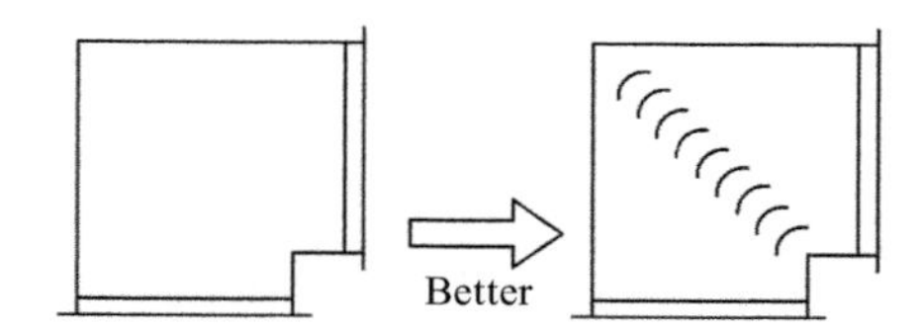

Figure 4.18 A 90° elbow equipped with a turning vane

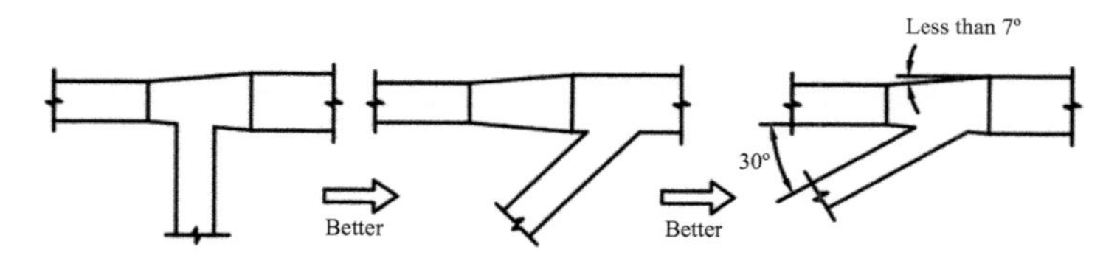

Figure 4.19 Connection between branch and trunk ducts

(Jing *et al.*, 2021). It must be pointed out that to reduce the local resistance of tees, the injection leading to energy loss should be avoided.

To reduce the tee's local resistance, the connection between branch duct and trunk duct (Figure 4.19) – reducing the included angle of both – must be heeded. At the same time, the velocity in the branch and truck ducts should remain the same.

4.7.3 Connection of duct and fan

Local vortex should be prevented at the duct connection. Connection of the fan inlet and fan outlet can be seen in Figures 4.20 and 4.21.

4.8 Material and specification of ventilation ducts

4.8.1 Material of ventilation ducts

The duct material determines the duct thickness, combustion performance, strength and usage range. The common materials for relevant engineering are shown in Tables 4.12–4.16.

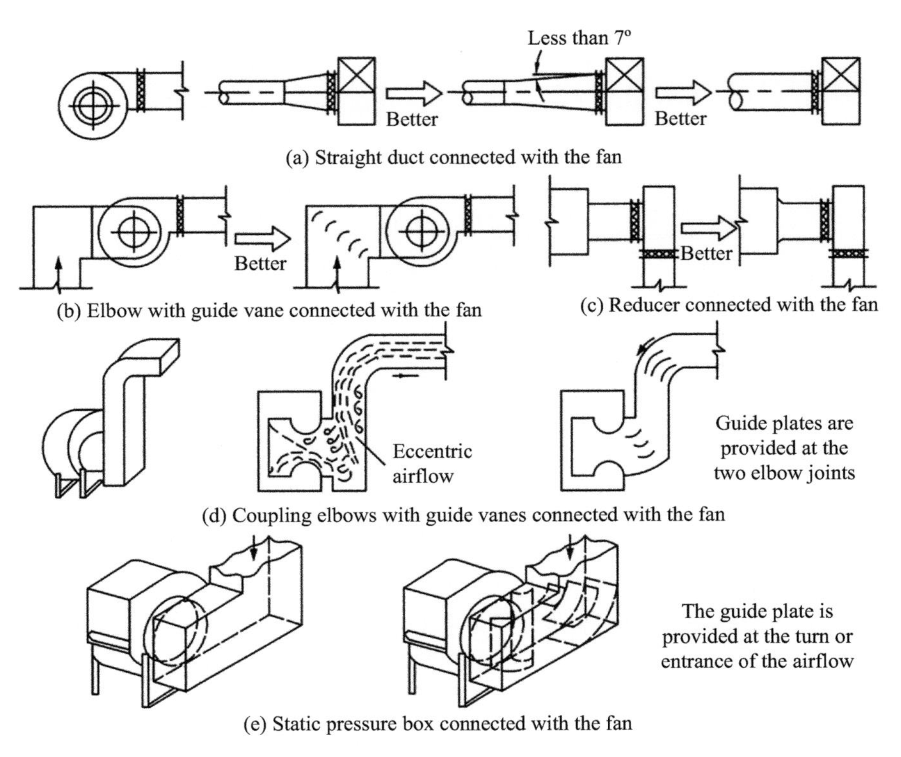

Figure 4.20 Connection of the fan inlet (Lu, 2008)

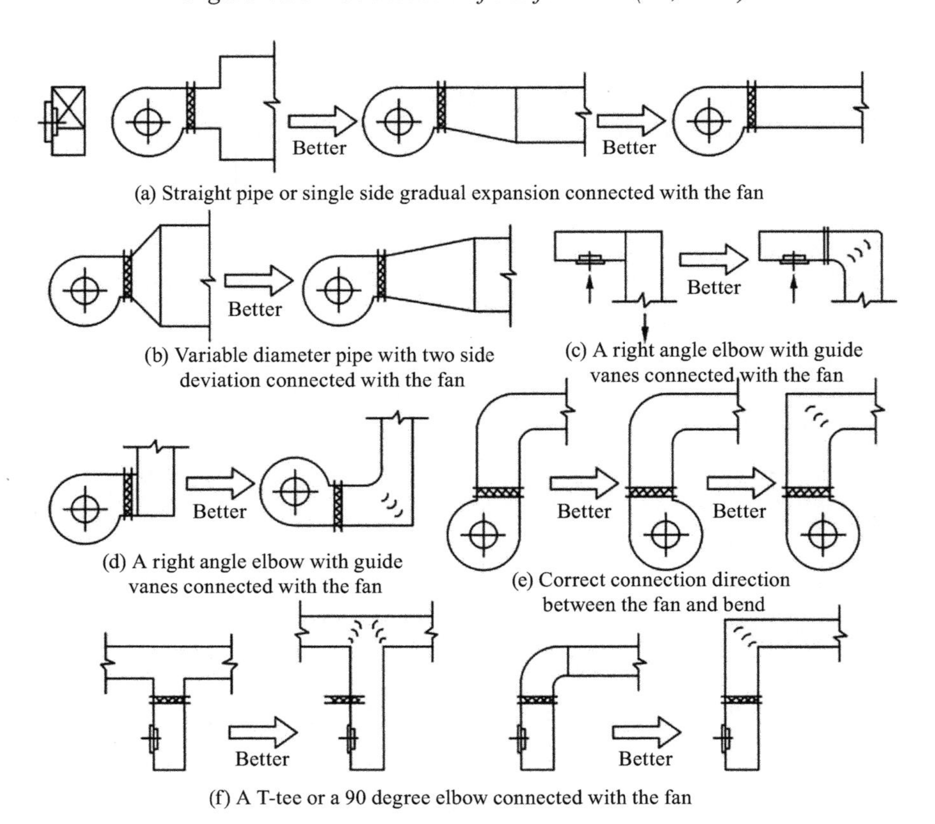

Figure 4.21 Connection of the fan outlet (Lu, 2008)

Table 4.12 List of common metallic ducts in engineering (Lu, 2008)

Type	Insulation material density	Thickness	Combustion performance	Strength	Features	Applicable range
Common steel ducts	≤ 1,700	Table 4.23	Grade A	High	1. There are circular and rectangular cross sections. The rectangular ones should have a length-to-width ratio of no more than 4, up to 10 2. There is a small resistance and excellent strength due to smooth inner walls, with anti-fire and non-flammable features. Ordinary cold-rolled steel sheets generally adopt locking or welding connection. Galvanized steel sheets or colour-coated (plastic composite) steel sheets should adopt locking or riveting connection 3. Albeit low cost, ordinary cold-rolled steel sheets are easy to corrode; galvanized steel sheets have a better anti-corrosion performance but are expensive; plastic composite steel sheets and stainless steel sheets, albeit anti-corrosive and dust-free, are expensive	Air supply and exhaust systems for ventilation, and smoke extraction and dust removal systems
Galvanized steel ducts		Table 4.23				1. Low-, medium- and high-pressure air-conditioning systems, especially air supply and return systems with higher requirements on temperature and humidity 2. Air supply systems in the front of the medium-efficiency filters, and between the medium- and high efficiency filters in clean air-conditioners
Coloured, plastic-coated steel ducts		Table 4.23				Air supply ducts behind high-efficiency filters and smoke removal systems for surface treatment plants
Stainless-steel ducts		Table 4.24				Air supply ducts behind high-efficiency filters of ultra-clean systems and ducts transmitting corrosive gases

(Continues)

Table 4.12 (Continued)

Type	Insulation material density	Thickness	Combustion performance	Strength	Features	Applicable range
					4. Ducts and their fittings by the completely equipped processing and manufacturing devices have a great performance and a long service life	
Spiral circular ducts of galvanized steels		Table 4.21			1. Have merits of galvanized steel sheet ducts 2. Ducts and their fittings by the processing machinery and production line have a good working performance, great strength, high gas tightness, and easy to be assembled 3. The ducts adopt non-flange connection, which is tidy and beautiful, and more applicable to exposed installation 4. Oblate ducts can save more installation space	Low-, medium-, high-pressure air-conditioning systems. Spiral circular ducts also apply to high-speed air-conditioning air supply systems
Spiral oblate ducts of galvanized steels		Table 4.22				
Aluminium alloy ducts		Table 4.26			The thin aluminium sheets adopt locking or riveting connection, while thick aluminium sheets adopt the connection of argon arc welding or gas welding. Aluminium alloy sheets are dust free but expensive	Air supply systems behind high-efficiency filters in ultra-purification air conditioners

Table 4.13 List of common non-metallic ducts in engineering (Lu, 2008)

Type	Insulation material density	Thickness	Combustion performance	Strength	Features	Applicable range
Hydraulic inorganic FRP duct	$\leq$ 1,700	Table 4.30	Grade A	Flexural strength $\geq$ 70	1. The circular or rectangular cross sections 2. Hydraulic inorganic FRP ducts are made by fibreglass mesh and sulphates as gel material; magnesium oxychloride cement ducts are made by modified magnesium oxychloride cement as gel materials and fibreglass mesh 3. Inorganic glass-steel ducts can be divided into integral ordinary type (non-thermal-insulation), integral thermal insulation type (inorganic glass steel on the outside and inside surfaces, and thermal insulation material in the centre), combined type (using clad plates, special gel flanges and strengthening to form ducts) and combined thermal insulation type 4. Rough inner surface leads to a big resistance, a large density and a heavy mass	Low-, medium- and high-pressure air-conditioning systems and smoke control systems; air supply and exhaust systems in humid environment such as basement; ducts conveying corrosive gases
Magnesium oxychloride cement duct	$\leq$ 2,000	Table 4.30		Flexural strength $\geq$ 65		

(Continues)

Table 4.13 (Continued)

Type	Insulation material density	Thickness	Combustion performance	Strength	Features	Applicable range
Hard PVC duct	1,300–1,600	Table 4.27	Grade B1	Tensile strength ≥ 34	1. The circular or rectangular cross sections 2. Duct plates are connected by plastics joining by welding, leading to smooth inner surface, free of ducts and anti-corrosion, but they are expensive and easy to carry static	Air supply ducts behind high-efficiency filters in clean air conditioners; exhaust systems in acid-base treatment plants
Flexible duct					1. Cross sections are circular or rectangular. Woven fibres can be used to send air in rooms, or a slot or circular outlet can be made on the duct surface to supply air 2. Feature light mass, free of noise and of dew formation on the surface, as well as easiness for installation, disassembly and transportation	1. Used for the connection of ventilation air-conditioning ducts and end devices, and for some production plants, as well as public air-conditioning systems whose ducts are allowed to be obviously installed on the ceilings 2. Temporary air-conditioning systems for exhibition halls

Table 4.14 List of common clad plate ducts in engineering (Lu, 2008)

Type	Insulation material density	Thickness	Combustion performance	Strength (MPa)	Features	Applicable range
Phenolic aluminium foil composite board air duct	≥60	≥20	Grade B1	Flexural strength ≥1.05	1. Rectangular cross sections 2. Made by phenolic or polyurethane aluminium foil composite board, with the inside and outside surfaces made of aluminium foil. The inner walls have a medium smoothness and small resistance 3. The joint of duct plates is made by 45° splice or 'H-shaped' reinforcing bar. At the joint, adhesive or aluminium foil tape is used to ensure the stiffness and gas tightness. Such ducts feature thermal insulation performances 4. Belong to Grade B1 representing great fire resistance 5. Feature advanced processing techniques, a light mass and a long service life	Air-conditioning systems with the working pressure of no more than 2,000 Pa or humid environments
Polyurethane aluminium foil composite board air duct	≥45	≥20	Grade B1	Flexural strength ≥1.02		
Glass magnesium composite board air duct	See Table 4.16	See Table 4.17	Grade A	Bending strength ≥65	1. As per duct structure ducts can be divided into integral common ducts, integral insulated ducts and combination insulated ducts	Ducts made by modified magnesium oxychloride cement as the adhesive, alkali-free or medium alkali glass cloth as the reinforcing materials, isolated

(Continues)

Table 4.14 (*Continued*)

Type	Insulation material density	Thickness	Combustion performance	Strength (MPa)	Features	Applicable range
					2. Integral common ducts are non-insulated ducts formed wholly by once and made by glass cloth and magnesium oxychloride cement 3. Integral insulated ducts are insulated ducts formed wholly by once and made by glass cloth and magnesium oxychloride cement as inner and outer surface structure layer, isolated modelling polystyrene foam (EPS) as the central layer 4. First use glass cloth and magnesium oxychloride cement to make a surface course, subsequently use isolated modelling polystyrene foam (EPS) as the core materials to make lightweight insulation sandwich panels, then use the lightweight insulation sandwich panel and special cohesive to make combination insulated ducts	modelling polystyrene foam (EPS) as the core material

Fibreglass composite board air duct	≥70	≥25	Grade B1	–	1. Feature rectangular cross sections 2. Formed under high temperature and pressure by centrifugal glass fibre plates, aluminium foil cloth laid on the outer walls, flame-retardant alkali-free or medium-alkali glass cloth on inner walls, special reinforced framework and Grade A adhesives. The outer surface can be sprayed with coloured airtight glue 3. Have functions of thermal insulation, noise elimination, fire and damp resistance, and feature a light mass and a long service life	Air-conditioning systems with the working pressure of no more than 1,000 Pa

Note: The facing material for ducts made of composite materials must be fire resistant, and the isolation material for the inner layer should adopt non-inflammable, fire-retardant materials harmless for humans.

Table 4.15 Areal density of insulation materials (kg/m²) (JC/T 646-2006)

Integral insulated duct	Nozzle width $(b)\leq 500$	Nozzle width $(b)\leq 1{,}500$	Nozzle width $(b)>1{,}500$
	≤ 10.0	≤ 15.0	≤ 20.0
Combination insulated duct	≤ 9.0		

4.8.2 Duct specification

The specifications of ducts (including circular ducts, rectangular ducts, spiral circular ducts, oblate ducts and metallic circular flexible ducts) are generally selected by the duct's cross-sectional shape.

When there is an enough available space, the mass and perimeter of circular ducts are both smaller than those for rectangular ducts. Due to a great capability of resisting low-frequency noise, circular ducts are more preferable than rectangular or oblate ducts. Rectangular ducts with relative high noise can stand any pressure level from micro- to high pressure, and oblate ducts are designed only for high pressure.

The specifications of circular ducts can be seen in Table 4.17. Circular ducts, whose specifications apply to steel (aluminium) ducts, dust removal ducts, airtight ducts, hard polyvinyl chloride ducts and inorganic FRP ducts, have basic and auxiliary series. The former should be adopted as the priority.

The specifications of rectangular ducts, which apply to steel (aluminium) ducts, hard polyvinyl chloride ducts, inorganic Fiber Reinforced Plastic (FRP) ducts, phenolic aluminium foil composite board ducts, polyurethane aluminium foil composite board ducts and fibreglass composite board ducts, are shown in Table 4.18.

In some projects of air-conditioning systems for public buildings, rectangular ducts with a uniform specification fail to be adopted due to the limitation of story and ceiling height, and the rectangular cross-sectional height therefore has to be reduced. To this end, the list of specifications of non-standard rectangular ducts (the length-to-width ratio of 4) of steel sheets is given out (see Table 4.19), whilst for designers, it is included into the list of the unit length's frictional pressure-loss calculation for steel sheet rectangular ducts.

The specifications of spiral circular ducts, which are provided by manufacturers due to the ducts and their fittings made by special processing machines and production lines, are shown in Table 4.20.

For metallic circular flexible ducts, when the diameter is $D\leq 250$ mm, the wall thickness is $\delta\geq 0.09$ mm; when the diameter is $D=250$–500 mm, the wall thickness is $\delta\geq 0.12$ mm; when the diameter is $D>500$ mm, the wall thickness is $\delta\geq 0.2$ mm. For aluminium foil polyester film composite flexible ducts, the wall thickness is $\delta\geq 0.021$ mm, with anti-corrosion coating on the steel surface (JGJ/T 141-2017).

The specifications of spiral oblate ducts and their fittings made by special processing machines are shown in Table 4.21.

Table 4.16 Allowable deviation of specification and size of integral common ducts (mm) (JC/T 646-2006)

Nozzle width (b)	Body			Flange						Nozzle width tolerance
	Wall thickness ≥	Number of glass cloth layers		Height		Thickness		Number of glass cloth layers		
		C1	C2	Value	deviation	Value	Deviation	C1	C2	
$b \leq 300$	3.0	4	5	40	−1.0+2.0	10	−0.5+2.0	7	9	±2
$300 < b \leq 500$	4.0	5	7	45		12		8	11	
$500 < b \leq 1,000$	5.0	6	8	45		14		9	12	±3
$1,000 < b \leq 1,500$	6.0	7	9	50		16		10	14	
$1,500 < b \leq 2,000$	7.0	8	11	50		18		14	18	±4
$b > 2,000$	8.0	9	12	55		20		16	21	

Table 4.17 Specification of circular duct (GB 50738-2011)

Diameter of circular duct (mm)			
Basic series	**Auxiliary series**	**Basic series**	**Auxiliary series**
100	80	500	480
	90	560	530
120	110	630	600
140	130	700	670
160	150	800	750
180	170	900	850
200	190	1,000	950
220	210	1,120	1,060
250	240	1,250	1,180
280	260	1,400	1,320
320	300	1,600	1,500
360	340	1,800	1,700
400	380	2,000	1,900
450	420	–	–

Table 4.18 Specification of rectangular air duct (JGJ/T 141-2017)

Side length of duct (mm)		**Side length of duct (mm)**	
120×120	630×500	320×320	1,250×400
160×120	630×630	400×200	1,250×500
160×160	800×320	400×250	1,250×630
200×120	800×400	400×320	1,250×800
200×160	800×500	400×400	1,250×1,000
200×200	800×630	500×200	1,600×500
250×120	800×800	500×250	1,600×630
250×160	1,000×320	500×320	1,600×800
250×200	1,000×400	500×400	1,600×1,000
250×250	1,000×500	500×500	1,600×1,250
320×160	1,000×630	630×250	2,000×800
320×200	1,000×800	630×320	2,000×1,000
320×250	1,000×1,000	630×400	2,000×1,250

Table 4.19 The size of non-standard rectangular steel plate duct (Lu, 2008)

Duct size (mm) length×width		**Duct size (mm) length×width**	
320×120	630×200	1,000×320	2,000×500
400×120	800×200	1,250×320	2,000×630
400×160	800×200	1,600×400	
500×160	1,000×250		

Table 4.20 The size of spiral round duct (Lu, 2008)

Duct diameter D (mm)	**Thickness t (mm)**	**Duct diameter D (mm)**	**Thickness t (mm)**	**Duct diameter D (mm)**	**Thickness t (mm)**	**Duct diameter D (mm)**	**Thickness t (mm)**
80	0.5	300	0.6	600	0.8	1,200	1.0
100	0.5	315	0.6	630	0.8	1,250	1.0
125	0.5	350	0.6	650	0.8	1,300	1.0
150	0.5	355	0.6	700	0.8	1,350	1.2
160	0.5	400	0.8	800	0.8	1,400	1.2
200	0.5	450	0.8	900	0.9		
250	0.6	500	0.8	1,000	0.9		
280	0.6	550	0.8	1,100	0.9		

Table 4.21 The size of spiral oblate air duct (Lu, 2008)

Thickness (mm)	Oblate air duct's nominal height, short axis A (mm)													
	75	100	125	150	175	200	225	250	300	350	400	450	500	600
0.6	200													
	275	265			220									
	315	300												
	350	340	325	310	300		260							
	390	375	360	350	325		305							
	440	415	400	390	375		340							
		425	410	400	390		355							
		500	490	470	450		425							
			525	510	490	475	460		415					
				550	540	520	500		460					
0.8				625	610	600	575	565	535					
				700	690	675	660	650	615					
				790	775	760	735	725	700	665	635			
				860	840	830	815	800	775	750	715			
				890	875	860	850	830	805	775	750	715		
				940	925	910	890	885	880	825	800	765	735	
				1,020	1,000	990	975	960	935	910	875	850	825	
				1,100	1,090	1,070	1,050	1,035	1,010	985	950	925	900	
					1,160	1,150	1,135	1,115	1,085	1,060	1,035	1,010	985	915
1.0					1,310	1,300	1,285	1,275	1,250	1,215	1,185	1,160	1,135	1,075
					1,475	1,455	1,440	1,435	1,400	1,375	1,350	1,315	1,285	1,225
					1,625	110	1,600	1,585	1,560	1,535	1,500	1,475	1,450	1,385
1.2					1,785			1,750	1,715	1,685	1,660	1,635	1,600	1,550
					1,940			1,900	1,875	1,850	1,815	1,785	1,760	1,700
														2,000

Table 4.22 Thickness of steel plate duct (unit: mm) (GB 50738-2011)

Diameter *D* or long side *b*	**Classification**			
	Circular duct	**Oblong air duct**		**The duct of dust removing system**
		Middle- and low-pressure system	**High-pressure system**	
$D(b) \leq 320$	0.5	0.5	0.75	1.5
$320<D(b) \leq 450$	0.6	0.6	0.75	1.5
$450<D(b) \leq 630$	0.75	0.6	0.75	2.0
$630<D(b) \leq 1{,}000$	0.75	0.75	1.0	2.0
$1{,}000<D(b) \leq 1{,}250$	1.0	1.0	1.0	2.0
$1{,}250<D(b) \leq 2{,}000$	1.2	1.0	1.2	By design
$2{,}000<D(b) \leq 4{,}000$	By design	1.2	By design	By design

Note: (1) The thickness of the steel plate of the spiral duct can be appropriately reduced by 10%–15%. (2) The thickness of the steel plate air duct of the smoke exhaust system can be according to the high-pressure system. (3) The thickness of the steel plate air duct of the special dust-removing system should meet the design requirements. (4) Not suitable for pre-embedded pipes of underground air defence and fire partition walls.

4.8.3 Duct sheet thickness

The duct sheet thickness (including the thickness of steel sheet duct, stainless-steel duct, aluminium duct, hard polyvinyl chloride duct and inorganic FRP duct) is determined by duct materials and sealing requirements (see Table 4.22). The duct thicknesses of different materials are shown in the following.

The thickness of steel sheet ducts applies to high-, medium- and low-pressure systems. See Table 4.22.

The thickness of stainless-steel ducts applies to high-, medium- and low-pressure systems. See Table 4.23.

The thickness of aluminium sheet ducts (see Table 4.26) applies to medium- and low-pressure systems.

The thickness of hard polyvinyl chloride (PVC) ducts (see Table 4.27) applies to medium- and low-pressure systems.

The thickness of inorganic FRP ducts (see Table 4.30) applies to medium- and low-pressure systems.

4.9 Reinforcement and sealing of ducts

4.9.1 Duct sealing

It is suggested that all transverse joints, longitudinal seams and penetrating holes, including ventilating door axis be closed. The longitudinal seams are those along the airflow, while the transverse seams are those vertical to the airflow direction. Transverse joints are connecting two ducts by the way of being vertical to airflow.

Table 4.23 Plate thickness of stainless-steel duct and fittings (mm) (GB 50243-2016)

Duct size *b* or diameter *D*	**Thickness of stainless-steel plate**
$100<D(b)\le 500$	0.5
$500<D(b)\le 1{,}120$	0.75
$1{,}120<D(b)\le 2{,}000$	1.0
$2{,}000<D(b)\le 4{,}000$	1.2

Note: (1) When the thickness of stainless-steel sheets is no more than 1 mm, their joint should adopt the way of locking or riveting; when it is larger than 1 mm, argon arc welding or arc welding rather than gas welding should be adopted. When welding, the welding material should match with the parent material, and splashes by welding should be prevented from staining the surface, while the splashes and welding slag should be cleaned up after welding. (2) When flanges are used to connect stainless-steel ducts, their materials and specifications for rectangular and circular ducts should conform to the regulations in Tables 4.24 and 4.25, respectively. When carbon steel is selected for flange materials, its surface should be treated by chrome plate or galvanization, while stainless-steel rivets should be adopted for the riveting of ducts. (3) When flanges made of thin steel sheet are selected to connect rectangular stainless-steel ducts, and fasteners are made of carbon steel, its surface should be treated by chrome plate or galvanization.

Table 4.24 The size of rectangular duct angle steel flange materials and the distance between screw bolt and rivet (mm) (GB 50738-2011)

Size of angle iron	**Size of screw bolt**	**Size of rivet**	**The distance between screw bolt and rivet**	
∠25×3	M6	Φ4	Middle- and low-pressure system	High-pressure system
∠30×3	M8		≤150	≤100

Table 4.25 Round duct flange and its specifications (mm) (GB 50243-2016)

Duct diameter	**Materials and specifications of flange**		**Bolt specifications**
	Flat steel	**Angle steel**	
$D\le 140$	20×4	–	M6
$140<D\le 280$	25×4	–	
$280<D\le 630$	–	25×3	
$630<D\le 1{,}250$	–	30×3	M8

The duct sealing should prioritize the sealing after the plate connection, followed by sealant caulking or other sealing methods. The performance of sealant should conform to the environmental requirements in usage, with the sealing surface set on the positive pressure side of ducts.

Table 4.26 Sheet thickness of aluminium duct (mm) (GB 50243-2016)

Duct size *b* or diameter *D*	**Sheet thickness of aluminium plate**
$100<D(b)\leq 320$	1.0
$320<D(b)\leq 630$	1.5
$630<D(b)\leq 2{,}000$	2.0
$2{,}000<D(b)\leq 4{,}000$	By design

Note: (1) When the thickness of aluminium sheets is no more than 1.5 mm, their joint should adopt the way of locking or riveting, rather than snap-fastener locking; when it is larger than 1.5 mm, argon arc welding or gas welding should be adopted. (2) When the riveting is used for connecting aluminium sheet ducts and flanges, rivets should be adopted. When flanges are made of carbon steel, its surface should carry out anti-corrosion treatment. (3) For rectangular aluminium sheet ducts, the connection of C- and S-shaped flat inserts are not allowed.

Table 4.27 Sheet thickness of hard PVC duct (mm) (GB 50243-2016)

Circle duct		**Rectangular duct**	
Duct diameter	**Sheet thickness**	**Long side of rectangular duct**	**Sheet thickness**
$D\leq 320$	3.0	$b\leq 320$	3.0
$320<D\leq 630$	4.0	$630<b\leq 2{,}000$	4.0
$630<D\leq 2{,}000$	5.0	$630<b\leq 2{,}000$	5.0
$630<D\leq 2{,}000$	6.0	$630<b\leq 2{,}000$	6.0
		$630<b\leq 2{,}000$	8.0

Note: (1) The allowable deviation of the duct sheet's thickness and inner diameter (or outer side length) should conform to Tables 4.28 and 4.29. (2) The four angles of rectangular ducts can adopt the connection of simmer angle or welding. When using the former, the longitudinal welds should be more than 80 mm from the simmer angles. (3) When using welding to connect ducts and flanges, the flange's end face should be vertical to the duct axis. At the connection between flanges and ducts with a diameter or a side length larger than 500 mm, triangular reinforcing boards should be placed, with the distance with each other being no more than 450 mm.

Table 4.28 Circle duct wall thickness of hard PVC and tolerance of diameter (mm) (GB 50243-2016)

Duct diameter	**Sheet thickness**	**Tolerance of the inside diameter**
$D\leq 320$	3	−1
$320<D\leq 630$	4	−1
$630<D\leq 1{,}000$	5	−2
$1{,}000<D\leq 2{,}000$	6	−2

Table 4.29 Rectangular duct wall thickness of hard PVC and tolerance of side length (mm) (GB 50243-2016)

Side length of duct	Sheet thickness	Tolerance of outer side length
$b \leq 320$	3	−1
$320 < b \leq 500$	4	−1
$500 < b \leq 800$	5	−2
$800 < b \leq 1{,}250$	6	−2
$1{,}250 < b \leq 2{,}000$	8	−2

Table 4.30 Thickness of Inorganic FRP duct (mm) (GB 50243-2016)

Diameter of circular duct or long side of rectangular duct	Duct wall thickness
$D(b) \leq 300$	2.5–3.5
$300 < D(b) \leq 500$	3.5–4.5
$500 < D(b) \leq 1{,}000$	4.5–5.5
$1{,}000 < D(b) \leq 1{,}500$	5.5–6.5
$1{,}500 < D(b) \leq 2{,}000$	6.5–7.5
$D(b) > 2{,}000$	7.5–8.5

The system should select an appropriate material for flange gaskets according to the air temperature and the mediums delivered.

For ordinary air supply, exhaust and air-conditioning systems (i.e., air ducts with the delivery temperature of less than 70 °C), rubber plates and closed sponge rubber plates should be selected.

For air supply and exhaust systems in drying rooms, smoke and dust removal systems for boiler rooms, and air exhaust systems of heating furnaces (i.e., air ducts with the delivery temperature of larger than 70 °C or for delivering smoke), asbestos rope or asbestos rubber plates should be used.

For plastic or stainless-steel systems for exhausting acid or alkaline gas, rubber plates resisting acid and alkali or soft polyvinyl chloride plates should be selected.

Ducts for dust removal systems should adopt rubber plates.

Flanges can ensure a great connection of metallic ducts.

The weld for duct flanges should be complete and have a great fusion, without tack welds and holes. The allowable deviation of the flange's outer diameter or outer side length and flatness should be no more than 2 mm. The screw holes for the flanges with the same specification processed in the same bitch should be consistently arrayed.

When the ducts and flanges are welded together, the welds should be lower than the end face of flanges. The ducts for dust removal systems should adopt a full weld for the inside, and a tack weld for the outside. When spot welding is adopted to connect ducts and flanges, it should have a great fusion, with the distance of no more than 100 mm; the flanges and ducts should be closely connected, without penetrated apertures and holes.

When carbon steel is selected for the flanges of stainless-steel or aluminium sheet ducts, it should carry out anti-corrosion treatment. The material for rivets should be the same as that for ducts, and the electrochemical corrosion should be avoided.

The non-flange connecting methods for rectangular ducts currently include S- and C-shaped inserts, vertical locking, overlapping vertical locking, thin steel sheet flange inserts, thin steel sheet flange spring clip and 90° flat inserts.

When the larger side length of rectangular ducts is 120–630 mm, the upper and lower facets (where the larger side lengths are located) adopt S-shaped connection, and the left and right facets (where the smaller side lengths are located) adopt U-shaped connection. When it is 630–800 m, the left and right facets still use the U-shaped connection and the upper and lower facets adopt vertical-rib, S-shaped connection for reinforcement.

When connecting two ducts, an insert is tightly inserted at the S-shaped or vertical-rib S-shaped flange insert on the two duct planes, a U-shaped flange insert is then inserted at the two vertical faces of the duct, and finally, the joint with a tongue is bent and fastened.

When using U-shaped flange inserts, a 10-mm laying-off folding margin at the duct end should be reserved for folding the edge into an angle 180°. The angle after that should be flat and tidy to ensure the shape and size of duct edge can firmly connect with the S-shaped insert.

When the C- and S-shaped insert connection is selected, the duct's longer side length should be no more than 630 mm. The insert should match with the width of the duct's folding edge, with an allowable deviation of no more than 2 mm. The connection should be tidy and tight, with a fixed folding length at the ends of four angles of no less than 20 mm.

The connections and accessories of rectangular sheet steel flange ducts should have a correct size, an inerratic shape and an excellent tightness, with a flat and straight folding edge of flanges having a curvature of no more than 5%. Flexible inserts or spring clips (whose thickness is no less than 1 mm and the duct thickness) should match with the folded width of flanges. The close fixation between corner fittings (whose thickness should be no less than 1 mm and the duct thickness) and the previous flanges should be done, with a flat and tidy end face and the continuous seaming at the connection being no more than 2 mm. When the flange spring clips are used to connect ducts, the side length of ducts should be no more than 1,500 mm. When relevant reinforcing measures are adopted for flanges, the side length of ducts should be no more than 2,000 mm *(GB 50243-2016).*

To ensure the tightness of connection by insert flanges, sealant, glass tape and aluminium duct tapes are generally used.

4.9.2 Reinforcement of ducts

Metallic ducts: For rectangular ducts and insulated ducts with a respective side length of no less than 630 mm and no less than 800 mm, when the duct length is larger than 1,250 mm, reinforcing measures should be taken. For ducts with a side length of no more than 800 mm, the rib and cord should be used for reinforcement. When circular ducts (excluding spiral ducts) have a diameter of no less than 800 mm, and a length of larger than 1,250 mm or total surface area of larger than 4 m^2, the reinforcement should be adopted. Metallic ducts can be reinforced by using angle irons, vertical folding locking, ribs, as well as flat steels, spiral rods and steel pipes as the inside supports (Figure 4.22) *(GB 50243-2016)*.

The alignment of ribs should be regular and have an even distance (the largest distance should be 300 mm) with each other, whilst the plate should be flat and tidy, with the deformation (unevenness) of no more than 10 mm.

The angle irons or reinforcing ribs by steel sheet, which should be tidily arranged, should have a height of no more than that for duct flanges. The riveting connection with ducts should be tight, with the largest gap of no more than 220 mm. The intersection of reinforcing ribs or between the reinforcing ribs and flanges should be fixed.

The inside supports and ducts should be firmly fixed, with sealing measures adopted at the positions of penetrating duct walls. The even distance between supporting points, or between duct edges and supporting points or between flanges should be no more than 950 mm.

When the single-facet areas of low-pressure ducts and medium- and high-pressure ducts are larger than 1.2 and 1.0 m^2, reinforcing measures should be adopted. When the duct length for high- and medium-pressure duct systems is larger than 1,250 mm, reinforcing frames should be adopted. The seaming of single

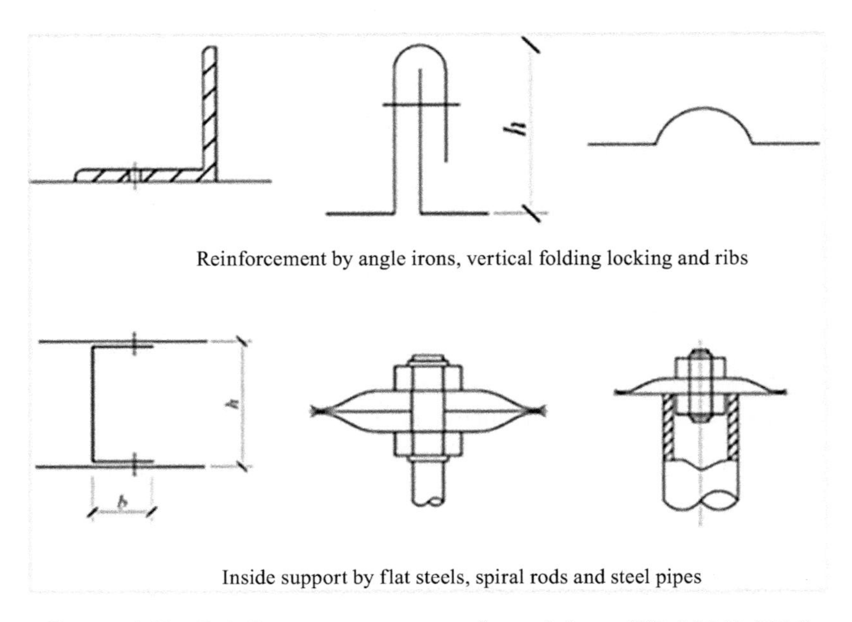

Figure 4.22 Reinforcement manners of metal ducts (GB 50243-2016)

folding locking for high-pressure system ducts should have reinforcement to prevent cracking by swelling. When the plate thickness of ducts is no less than 2 mm, the reinforcing range should be expanded.

Non-metallic ducts (such as hard polyvinyl chloride ducts): For rectangular ducts formed by four-angle welding with a side length of no less than 630 mm, rectangular ducts formed by heating folded angles with a side length of no less than 800 mm, and ducts with a length of larger than 1.2 m, reinforcing frames and ducts should be welded together for reinforcement.

Foil–insulant composite ducts: When the longer side length of positive pressure ducts is no less than 1,000 mm, external reinforcing frames need to be added. When composite fibreglass sheet ducts adopt metallic groove-shaped frames for external reinforcement, the specified inner support should be set and then tightened together with the frames. The reinforcement of negative pressure ducts should be set on the inner side of duct walls. At the positions of penetrating duct walls, sealing should be conducted, and the galvanized screw with a diameter of no less than 6 mm for the inside reinforcement should have no deformation under the working pressure.

Inorganic FRP ducts: The reinforcement of inorganic FRP ducts includes the reinforcement of integral and combination rectangular ducts. The reinforcing pieces, required to be an integral part of the duct after placement, should adopt the material the same as the duct perse or the material with the same anti-corrosive performances. Metal or other materials can be adopted for the reinforcement after the making of ducts, and anti-corrosion treatment should be carried out. When angular metallic fittings (such as steel or aluminium angles) are adopted to fix the four angles of combination ducts, the distance between these fasteners should be no more than 200 mm.

4.9.3 Installation of duct support and hangers

Supports and hangers that meet relevant weight-bearing requirements should be selected first before fixing them on the reliable building structures, whose safety would not be affected. Moreover, they are not allowed to be welded on the weight-bearing structure or on the steel bars for the room truss. If ducts have a function of thermal insulation, they should adopt thermal isolation gaskets, which should meet design requirements and relevant regulations.

The material of supports and hangers should be selected by the specification and weight of ducts, components and equipment, and the national standard drawings and regulations should be referenced to select the forms and specifications applicable to strength and intensity. For over-wide and over-weight supports and hangers for special ducts with a diameter or a side length of larger than 2,500 mm, they should be designed according to relevant regulations.

The reserved position for supports and hangers should be correct and reliable. The embedded part should be free of grease and paint.

During the horizontal installation of ducts with a diameter or a side length of no more than 400 mm or larger than 400 mm, the distance between supports and hangers should be no more than 4 or 3 m, respectively. Such distances can be extended to 3.75 and 5 m; for thin steel sheet flange ducts, the distance for their supports and hangers should be no more than 3 m. During the vertical installation of ducts, such a distance should be no more than 4 m, whilst the single straight pipe should have at least two fixation points *(GB 50738-2011)*.

Supports and hangers, which should not be set at tuyeres, dampers, inspection doors and auto-mechanisms, should be no less than 200 mm away from tuyeres or plugged ducts.

Supports and hangers should be no more than 1 m away from the end of ducts, no more than 0.5 m away from the bending part of horizontal elbows, and less than 1.2 m away from the main duct.

When the horizontally hanging main and trunk ducts have a length of over 20 m, at least one fixation point preventing swings should be set for each system.

The largest gap between supports and hangers for flexible ducts should be smaller than 1.5 m. Brackets and hoops with an even arc (which is consistent with the outer diameter of ducts) should be set on the circular ducts installed on the supports. Hoops, whose supports should have flat and straight folded angles, should be firmly fastened on the ducts.

Isolation or anti-corrosive isolation measures should be adopted at the contacting points of stainless-steel sheet or aluminium sheet ducts and carbon steel supports.

Separate supports and hangers should be set at elbows and tees with a side length (diameter) of more than 1,250 mm.

The screw holes for hangers should be machined, with flat and straight suspenders and smooth and intact threads. After installation, the force imposed on supports and hangers should be even, without obvious deformation. The extension or compression of adjustable, vibration–isolation supports and hangers for ducts or air conditioners should be adjusted by design requirements.

References

A. Li, X. Zhang, S. He, *et al.* Experiments on the pressure loss coefficient of pipe sections around elbow and tee junction. *HV&AC*, 2009, 39(10): 23–28.

A. Li, J. Li, R. Wang, *et al.* On the axial velocity method to determine the mean velocity of circular pipe. *Journal of Xi'an University of Architecture and Technology*, 1995(01): 53–56.

British Standards Institute, Ventilation for buildings. Ductwork. Strength and leakage of circular sheet metal ducts, BS EN 12237:2003.

China Institute of Building Standard Design and Research, Selection and Installation of Air Dampers: GJBT 1026-2007, China Planning Press, 2007.

J. Zhang, A. Li, D. Li. Modeling deposition of particles in typical horizontal ventilation duct flows. *Energy Conversion and Management*, 2008,49(12): 3672–3683.

Ministry of Housing and Urban-Rural Development of the People's Republic of China, Design code for heating ventilation and air conditioning of civil buildings: GB50736-2012, China Architecture and Building Press, 2015.

Ministry of Housing and Urban-Rural Development of the People's Republic of China, Code of acceptance for construction quality of ventilation and air conditioning works: GB 50243-2016, China Planning Press, 2016.

Ministry of Housing and Urban-Rural Development of the People's Republic of China, Code for construction of ventilation and air conditioning: GB 50738-2011, China Architecture & Building Press, 2011.

Ministry of Housing and Urban-Rural Development of the People's Republic of China, Design code for heating ventilation and air conditioning of industrial buildings: GB 50019-2015, China Planning Press, 2015.

Ministry of Housing and Urban-Rural Development of the People's Republic of China, Design standard for heating ventilation and air conditioning of coal industry: GB/T 50466-2018, China Planning Press, 2018.

Ministry of Housing and Urban-Rural Development of the People's Republic of China, Technical specification for air duct: JGJ/T 141-2017, China Architecture & Building Press, 2017.

L. Moody Friction factors for pipe flow. *Transactions of the American Society of Mechanical Engineers.* 1944, 66: 671–681.

L. F. Moody An approximate formula for pipe friction factors. *Trans. ASME.* 1947, 69(12): 1005–1011.

National Development and Reform Commission of the PRC, Glass fibre reinforced magnesium oxychloride cement ventiduct: JC/T 646-2006, China Building Materials Industry Press, 2006.

R. Gao, K. Liu, A. Li, *et al.* Study of the shape optimization of a tee guide vane in a ventilation and air-conditioning duct. *Building and Environment*, 2018, 132: 345–356.

R. Gao, S. Wen, A. Li, *et al.* A novel low-resistance damper for use within a ventilation and air conditioning system based on the control of energy dissipation. *Building and Environment*, 2019, 157: 205–214.

R. Jing, R. Gao, Z. Zhang, *et al.* An anti-channeling flue tee with cycloidal guide vanes based on variational calculus. *Building and Environment*, 2021, 205: 108271.

T. Wang, *Air Conditioning Equipment*, China. Beijing: Science Press, 2003.

A. Vedavarz, S. Kumar, M. I. Hussain. *HVAC: Handbook of Heating, Ventilation and Air Conditioning for Design and Implementation*, Industrial Press Inc, New York, US, 2007.

Y. Lu. *Practical Manual for Design of Heating Air Conditioning (Second Version)*, China Architecture & Building Press, 2008.

Y. Sun, *Industrial Ventilation*, China Architecture & Building Press, 2010.

Y. Liu, R. Gao, Z. Zhang, *et al.* Study on resistance reduction in a jugular profiled bend based on entropy increase analysis and the field synergy principle. *Building and Environment*, 2021, 203: 108102.

Chapter 5

Ventilation system: air cleaner technologies

Zhuangbo Feng[1]

5.1 Introduction

The indoor and outdoor particulate matters have significantly negative effect on human health. Therefore, different types of air cleaners are adopted to remove particles. The fibrous filter, electret filter, ion generator (IG), electrostatic precipitator (ESP) and hybrid electrostatic filtration system (HEFS) are widely utilized in the indoor environment.

The commonly used porous air filter (fibrous filter) is made up of dielectric fiber, such as glass fiber, PP (polypropylene) and PET (polyethylene terephthalate) (Thomas *et al.*, 2001; Hosseini and Tafreshi, 2010; Hasolli *et al.*, 2010). In the fibrous medium, the particles are captured by the fiber because of the Brownian diffusion, impaction, interception and electrostatic mechanisms (Lee and Liu, 1982; Kanaoka *et al.*, 1980; Wang., 2001). Some types of fibrous filters are highly efficient, such as HEPA (high efficiency particulate air filter). Different from the electrostatic precipitation, the filtration efficiency of the fibrous filter increases in the dust loading process (Thomas *et al.*, 2001; Joubert *et al.*, 2011). But the pressure drop across the filter medium also increases. The dirty filter medium should be replaced due to the added pressure loss. The operation cost is much higher than the ESP (Kim *et al.*, 2013). Due to its higher pressure loss and operation fee, electrostatic-enhanced air filtration systems (including ESP, HEFS, IG and electret filter) are drawing more and more attentions to overcome the limitations of fibrous filters.

The ESP could generate high-intensive electric field and charge the particles (Wen *et al.*, 2015; Kim *et al.*, 2013; Adamaik, 2013). Due to the electrostatic force, air drag force and Brownian diffusion force, the charged particles moving to the grounded collection plate were captured by ESP. Different from the fibrous filter (such as HEPA), the ESP should not be replaced regularly. But the dust deposited on the collection should be cleaned because of the efficiency decrease. Although the pressure loss of ESP could be ignored, it is difficult to obtain ultrahigh efficiency for the fine particles. And its efficiency decreases with the dust loading

[1]School of Architecture, Southeast University, Nanjing, China

process (Kim *et al.*, 2010; Li *et al.*, 2015). The corona discharge in ESP could generate some by-products (such as ozone, secondary aerosol or VOC) that have negative effect on the human health (Waring *et al.*, 2008; Bo *et al.*, 2010; Chen and Davidson, 2013; Huang and Chen, 2001). Overall, the advantages of ESP include low pressure loss and cheap operation fee; while the

In order to improve the ESP performance, the HEFS was developed (Feng *et al.*, 2020). The HEFS, combing the external electric field and fibrous filter, was utilized in the heating, ventilation and air-conditioning (HVAC) system. There are still two types of HEFS. One is the serial system consisting of the ESP and filter medium. The upstream ESP could charge and remove particles. The filter downstream of ESP is exposed to charged particles (Tu *et al.*, 2013). The second one is that the filter medium is placed in the corona discharge field. The electric force could drag the charged particle onto fibers (Lee *et al.*, 2001; Park *et al.*, 2011). The HEFS could utilize the pros and cons of the fibrous filter and ESP. In HEFS, the ions or ionized electric field could significantly improve filtration efficiency of fibrous filter without adding pressure drop. The corona discharge of HEFS was weaker than that of ESP, resulting in lower ozone generation. Overall, the HEFS has the potential to high-efficiently remove particles without adding pressure drop and generating harmful air pollutant (e.g., ozone).

The IG and electret filter are two other types of air cleaners. The IG would generate a large amount of ions and charge the particles. In the presence of the IG, more charged particles could deposit onto the indoor walls or duct surfaces. So the indoor particle concentration was reduced obviously. The disadvantage of IG was that its particle removal rate is much lower than those of fibrous filter/ESP. The electret filter is a special type of fibrous filter, which consists of fibers carrying electric charges. The charges carried by fibers could improve the filtration efficiency of fibrous filter without adding pressure loss. However, the amount of electric charge and filtration efficiency of an electric filter obviously decreases during the particle loading process. Although the IG and electret filter, respectively, have their limitations, they can be used as auxiliary air cleaners in indoor pollutant control.

The aim of this chapter is to provide an overview about the different patterns of an electrostatic air cleaner (including electret fibrous filter, IG, ESP and HEFS), as shown in Figure 5.1. The following issues are illustrated in detail: the system filtration performance (filtration efficiency, pressure loss), the electric characteristics of porous dielectric filter medium, the by-product generation, the bacteria inactivation effect by electric field and the effect of electrostatic air cleaner on indoor environment and human health. Finally, the analytical or numerical modeling of air filtration process was also summarized.

5.2 Electrostatic precipitator

ESP originates from industrial application. Currently, many studies proposed new types of ESP for indoor environment control. Different from industrial applications,

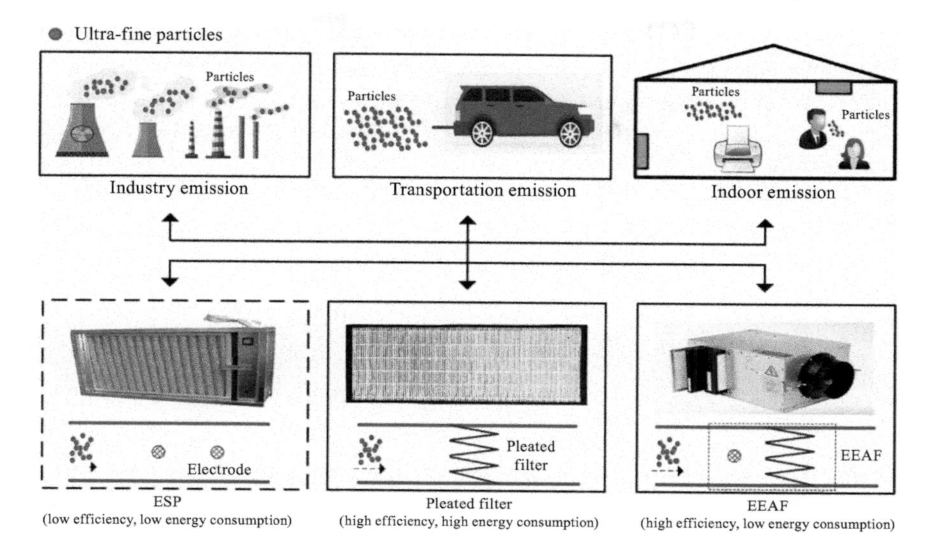

Figure 5.1 Description of different air purification systems, including ESP (electrostatic precipitator), pleat fibrous filter and EEAF (electrostatics enhanced air filter) (Feng et al., 2020)

the ozone generation of ESP should be lower than that of industrial ESP and satisfy some health standards (Feng *et al.*, 2020).

5.2.1 ESP design toward indoor environment applications

ESP consisting of discharge wires and collection plates was widely used in the dust removal process (Adamaik, 2013). If the applied high voltage was turned on, the discharge wire could generate a large amount of ions. The intensive electric field also existed between the discharge wire and collection plates (Hinds, 1992). The particle in ESP could be charged due to the diffusion and electric field mechanisms. The electrostatic force acting on the charged particle dragged it onto the collection plates and removed the particulate pollutant. The pressure loss of ESP could be ignored compared to the fibrous fiber with similar efficiency. Otherwise, the dust layer on the collection did not add the pressure drop. Different from the fibrous filter, the ESP should not be replaced if the operation time was long enough. But the ESP should be cleaned in order to maintain the high filtration efficiency because the dust cake layer deposited on the wires and collection plates reduced the filtration efficiency because of the back corona effect (Horenstein, 1995). The corona discharge generated ozone and other by-products (such as VOC, nitrogen oxides) which had negative effect on the human health (Rai *et al.*, 2015). Overall, the most important advantage of ESP was the energy-saving potential and economy characteristics.

In order to utilize the ESP in indoor environment, many studies developed new types of ESP with high particle removal efficiency and low ozone generation.

There were two categories of new ESP. One was the two-stage ESP combing the particle charging zone and collection zone (Kim *et al.*, 2013). In the particle collection zone with intensive electric field, there was no corona discharge and ozone generation. Another one attached the discharge wire directly on a dielectric plate, which is the so-called WOP-ESP (WOP-ESP, wire-on-plates ESP) (Li *et al.*, 2015). The two types of ESP could generate much lower ozone compared to the traditional ESP with the similar filtration efficiency.

5.2.2 WOP-ESP toward indoor application

A novel ESP (WOP-ESP, wire-on-plates ESP) was developed by previous literature (Li *et al.*, 2015). In WOP-ESP, the discharge wires were attached directly on a dielectric plate to ease the wire installation and minimize the particle deposition on wires. Compared to the traditional WIP-ESP (wire-in-plates ESP), the new WIP-ESP could have higher filtration efficiency and lower ozone generation. However, the ozone generation measured in the outlet of WOP-ESP was still high compared to the ozone concentration in commonly built environment (Yao and Zhao, 2021). If the efficiency for a 0.3-μm particle was higher than 80%, the net ozone concentration was 129 ppb. More filtration efficiency could lead to more ozone generation. This study also investigated the influence of dust loading on filtration efficiency. The dust deposited in WOP-ESP reduced the particle capture efficiency. If only cleaning the collection plates by water, the collection efficiency for the WOP-ESP could recover to the initial value. But the WIP-ESP had an obvious degradation in the collection efficiency, falling by more than 20% from the initial cleaning condition after the particle-loaded plates were cleaned. Otherwise, the numerical simulation was also conducted. The modeled and measured results agreed reasonably. Although the performance of novel WOP-ESP was much better than the traditional WIP-ESP, the most serious disadvantage was the high ozone generation (Figure 5.2). Using the carbon wires as the discharge wire in WOP-ESP could reduce the ozone generation without decreasing the efficiency.

5.2.3 Two-stage ESP toward indoor application

Another optimal ESP was the widely used two-stage type. The first stage was the particle charging zone with corona discharge, and the second stage was the particle collection zone with electric field. Kim *et al.* (2010) had developed a novel ESP that used an anticorrosive carbon brush precharger to charge particle and plastic collection plates onto which metallic films were inserted to collect particle. The system could remove 95% ultrafine particles with ozone generation lower than 5 ppb. The experimental results indicated that increasing the carbon brush numbers and applied voltage and decreasing the gap between the collection plates would improve the efficiency. The efficiency reduced to zero if the high-voltage generator did not work. The dust-loading process decreased the efficiency of the two-stage ESP obviously. Compared to the fibrous filter and the traditional one-stage ESP, the system was very complex and the initial cost was high. These three factors were the main disadvantages of the two-stage ESP. Yun *et al.* (2013) established a

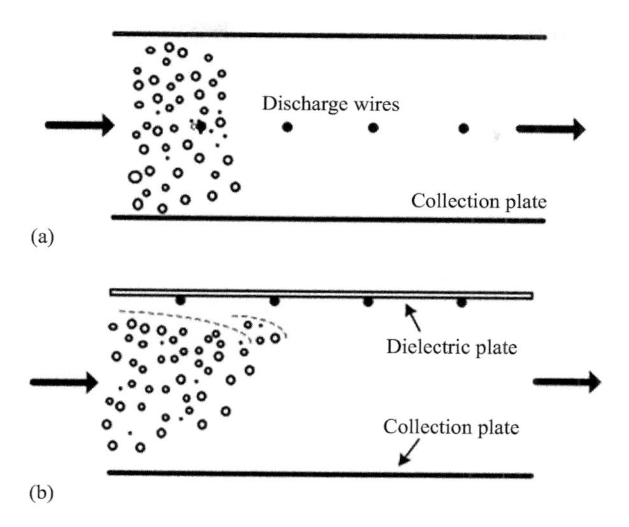

Figure 5.2 Schematic of the (a) traditional ESP, (b) proposed WOP-ESP (Li et al., 2015)

similar ozone-free ESP. The carbon fiber crush was used as the precharger. The PET film cover with the conductive ink was used as the collection plates. For a 0.3-μm particle, the efficiency could be 85% while the ozone generation could not be detected. Kim *et al.* (2013) tested the performance of the newly developed two-stage ESP, combing the ionization zone and collection zone in a real room. The efficiency for particle sizing from 0.2 to 0.35 μm could be higher than 85%. The test results indicated that emissions from the ESP cleaner resulted in ozone levels lower than 10 ppb. The ESP also achieved a high clean air delivery rate (CADR>12 m^3/min). This study also conducted a comparison between the HEPA and the ESP by utilizing a new index PC. PC was defined in (5.1). The PC value of ESP was much lower than the HEPA, indicating that the ESP could generate high CADR value with low pressure drop. In factor, this index was not very suitable because the ESP could consume the electric energy. So the total energy consumption rather than the pressure loss should be used in (5.1) to assess the filtration system performance. The ionization stage could capture some particles, but both of the studies did not provide the efficiency results. The influence of dust-loading process on particle charging and filtration efficiency of the ionization zone should be investigated in future study:

$$PC = \frac{\Delta P}{CADR} \tag{5.1}$$

where *CADR* is the clean air delivery rate, ΔP is the pressure loss.

Lee *et al.* (2001) established a two-stage ESP to remove indoor particle pollutants (Figure 5.3). The first stage was the traditional ESP consisting of pin-to-plate pattern, and the main function was to charge the particle. The second stage

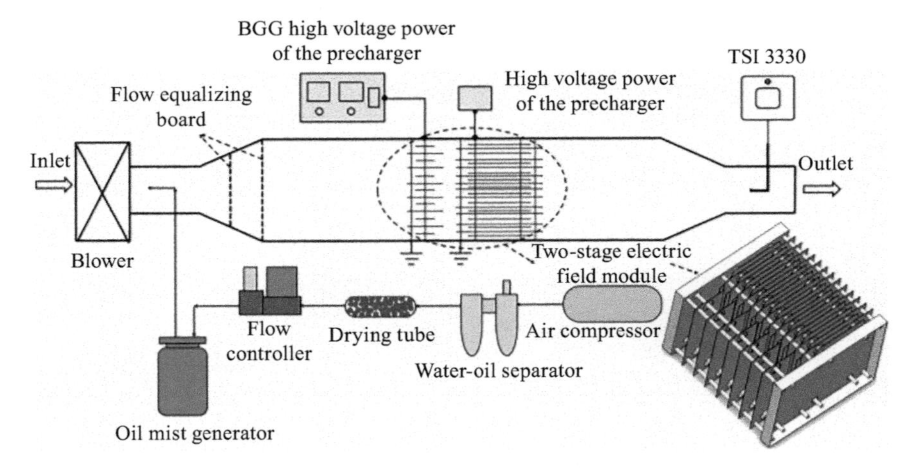

Figure 5.3 Schematic of a two-stage ESP (Li et al., 2019)

used the electrified fibrous filter to collect particles. The external electric field was applied across the filter medium thickness (about 1.4 kV/cm). In the experimental study, the dust ranging from 0.1 to 5.0 μm was utilized to test the filtration system performance. The total filtration efficiency rather than the particle fractional efficiency was adopted in the testing. The first ESP stage or polarized electric field across the filter medium could improve the filter efficiency from 70% to 84.5% or 82%, respectively. If both of the electric fields were turned on, the total efficiency could increase to 92.9%. The measured results illustrated that the particle-charging process and polarized electric field applied to a fibrous filter could improve the filtration efficiency obviously. This study also conducted field testing in an occupied space to prove the hybrid electrostatic filtration efficiency. However, the particle fractional efficiency was not provided, and the system performance for submicron particle was not known. Otherwise, the by-product generated by the HEFS was not measured.

Although the newly developed ESP could have high efficiency, low pressure loss and low ozone generation, there were still some problems. First, the by-products (such as VOC, nitrogen oxides) rather than ozone were not investigated sufficiently. The little amount of some by-products could have negative effect on human health. Second, the filtration efficiency of ESP decreased obviously in the dust-loading process. If the high-voltage equipment did not work, the efficiency may decrease to zero. The HEFS combing the ESP and fibrous filter may solve this problem (Lee *et al.*, 2001; Long and Yao, 2012). Third, there was no suitable index (considering the efficiency, pressure loss, ozone generation, initial cost and operation fee) to assess the filtration performance and economy of fibrous filter and ESP. The traditional index quality factor did not include the electric energy consumed by the ESP (Li *et al.*, 2015). Without the index, the designer could not make the suitable choice between the ESP and fibrous filter.

5.3 Hybrid electrostatic filtration system

To overcome the disadvantages of commonly utilized ESP, HEFSs are proposed. The HEFS consists of electric ozone and fibrous filter zone in series. For electric zone, the IG and ESP with corona discharge phenomenon are widely adopted. Different from corona discharge, IGs only emit a small amount of ions, and the electric field strength/ozone generations are lower than those of corona discharge. For fibrous filter zone in HEFS, it can be placed in (or out of) ionized electric field. If a fibrous filter is in electric field, filter efficiency can be improved due to electrostatic force effect. If the fibrous filter is placed out of ionized electric field, the interactions between charged particles and fibers can effectively improve air filter efficiency. In this section, HEFS types with IG or corona discharge will be illustrated as follows.

5.3.1 HEFS (with ion generator)

If IG is adopted as electric zone in HEFS, the filtration efficiency of fibrous filter will be improved due to electrostatic interactions (Figure 5.4). Although the filtration enhancement of an IG is lower than that of corona discharge, it is also widely used in HEFS due to its economy, especially for the indoor environments with relatively low pollutant concentrations.

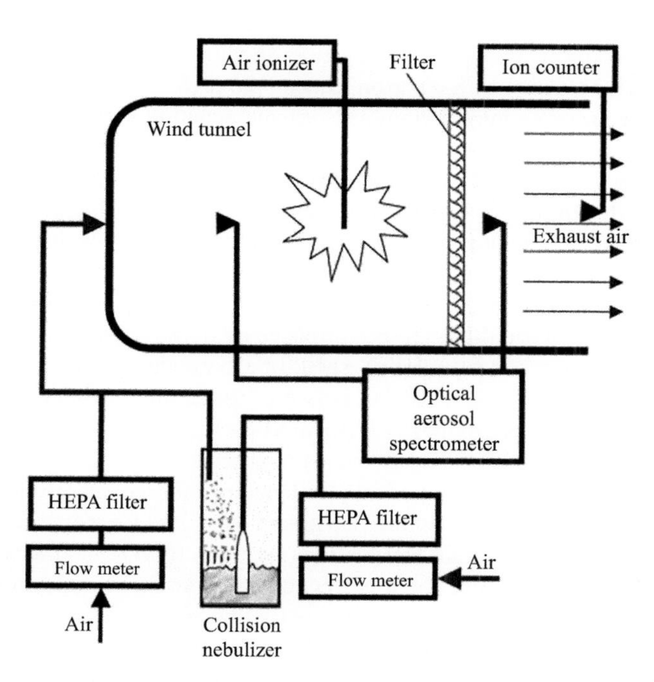

Figure 5.4 Schematic of HEFS consisting of an ion generator and a fibrous filter (Agranovski et al., 2006)

Lee *et al.* (2004a, 2004b) first used an ionizer installed in front of conventionally used filtering-facepiece respirators to improve the filtration efficiency without adding the pressure drop. The electrostatic efficiency enhancement could be 20% for 40-nm particles. Lee *et al.* (2005) conducted measurements to investigate the influence of system parameters (such as ion concentration, airflow rate, filter type and discharge polarity) on the filtration efficiency of the ionizer-assisted facepiece respirator. Based on their results, more studies focused on the ionizer-assisted air filter.

The previous literature investigated the effect of particle charging on the unsteady filtration performance (such as efficiency or pressure loss) of the fibrous filter by experiments (Huang, 2006). Different from the HEFS (Park *et al.*, 2011; Long and Yao, 2012), the filter medium was not placed in the external electric field and only exposed to the charged particles. The pressure loss of the particle cake layer was determined by the filtration velocity, particle charge status and filter medium type. For some types of filter medium (such as PP), the pressure loss of the cake layer consisting of charged particles was higher than the case with neutral particle cake. But for other filter types (such as polyester), the results may be contrary. Similar to the pressure drop, the filtration efficiency of the fibrous filter was also influenced by the three factors described earlier. The particle penetration ratio decreased obviously in the dust-loading process due to the blocking in the fibrous medium and cake layer formed on the filter surface (Endo *et al.*, 1998). For the PP medium, the penetration ratio for the neutral particle decreased faster than the charged particle. In the particle charging cases, the blocking process in filter medium was slower with the pressure drop increase. This may be the reason causing the high penetration ratio for charged particles. For the polyester medium, the particle penetration ratio for charged particles was much lower than the neutral particle if the filtration velocity was low. Increasing the filtration velocity would lead to a higher penetration ratio for charged particles. If the filtration velocity was high enough, the penetration ratio for the charged particle was higher than the neutral particle. So in this status, the particle-charging process was meaningless for electrostatic-enhanced filtration. The results by Huang (2006) indicated that the particle charge and external electric field had significant influence on the filtration performance of fibrous medium in HEFS. So the long-term measurement of the HEFS should be conducted to monitor the transient efficiency, by-product generation and pressure loss in the dust-loading process. But most of the available researches only focused on the initial performance of HEFS. The investigation of total service time was in absence.

The previous study developed a hybrid electrostatic air filtration system with a pair of carbon fiber ionizers installed upstream of a glass fiber air filter (Park *et al.*, 2011). The potassium chloride particle ranging from 10 to 1,000 nm was utilized to test the system performance. The ionizers could charge the particle and improve the fibrous filter efficiency. The ozone concentration of the system outlet was below 10 ppb. But the results indicated that the increase in efficiency (the difference between filter efficiency with and without ionizers) with electrostatic ionization gradually decreased as the particle size increased. The electrostatic efficiency increase was 20% for a 10-nm particle and approximately 0% for a 1,000-nm particle. So the influence of electrostatic force on large particles is very limited in the ionizer-assisted fibrous filter system. The filtration velocity in this study was from 0.4 to 0.8 m/s, which is much higher than the previous study focusing on the glass fiber air filter. We think that the experiments with lower filtration velocity were more reasonable. This study also conducted field tests in an office room with an ionizer-assisted glass fiber filter. The indoor particle concentration could be decreased by 17%–25% due to the ionizer effect. The study developed a mass balance model to simulate the indoor particle concentration. The modeled and measured results agreed well, although some error existed. The model did not consider the electrostatic force when calculating particle deposition velocity onto wall surface (Lai *et al.*, 2006). We think that this factor may be the main reason causing the model error.

The carbon fiber ionizer-assisted electret filter was proposed and tested in an automobile cabin (Park *et al.*, 2010). This study used total PM 2.5 efficiency rather than particle fractional efficiency to evaluate the filtration system performance. The increase in PM 2.5 removal efficiency with the use of carbon fiber ionizers was 11.1% at a filtration velocity of 0.6 m/s but decreased to 6.0% and 3.4% at 1.2 and 1.8 m/s, respectively. This study also conducted field tests in the automobile cabin. The particle concentration could be reduced by 10% in the presence of the ionizer installed in the HVAC system. The mass balance model was utilized to simulate the steady cabin particle concentration. Similar to the research (Park *et al.*, 2011), the electrostatic force was not included in the particle-deposition velocity. So, we think that this assumption could be the main reason causing the simulation error. Agranovski *et al.* (2006) also tested the performance of the ionizer-assisted air filter. The study investigated the influence of the ionizer to filter medium distance on the efficiency. The results illustrated that the shorter distance could lead to relative high electrostatic efficiency increase.

Noh *et al.* (2011) installed the ionizer upstream of the electret filter to increase the filtration efficiency. The DEHS particle ranging from 10 to 1,000 nm was utilized to test the system performance. The electret filter efficiency could be improved by about 10%–25% due to the ionizer effect. The influences of discharge polarity and filtration velocity were also investigated experimentally. Similar to other studies, the electrostatic efficiency decreased with the increase of air velocity. The collection efficiencies of the negative ionizers were 10%–15% lower than the values of the positive ionizers for the same velocities and applied voltages. The measured ozone concentration of the system outlet was less than 1 ppb. The particle size–filtration efficiency curve of the electret filter medium is quite different from the previous literature (Ji *et al.*, 2003). The efficiency did not vary with the particle size, and the saddle filtration efficiency curve or MPPS (most penetration particulate size) did not exist. Otherwise, the electrostatic filtration increase did not decrease with the particle size increase, which was in contrast of other ionizer-assisted air filter system. We think that the main reason may be the special fiber structure, and it should be investigated detailed in the future study.

The ionizer installed in front of the fibrous filter could improve the efficiency without adding the pressure loss and generating too much ozone (Lee *et al.*, 2004a, 2004b; Agranovski *et al.*, 2006; Park *et al.*, 2011; Shi *et al.*, 2012).

But the electrostatic efficiency enhancement due to the ionizer decreases with the particle size. If the particle size is larger than 2 μm, the electrostatic could be ignored. Compared to the corona discharge process (Feng *et al.*, 2014; Kim *et al.*, 2013), the discharged current value by an ionizer is much lower than the so-called pin. We think that it is the main disadvantage of the available ionizer-assisted air filter.

There are three reasons why the ionizer installed in front of the fibrous filter could improve the filtration efficiency. First, air ions with high mobility were deposited on the filter medium surface forming a macroscopic electric field, which shielded out some incoming unipolar charged particles or decreased the charged particle velocity due to repelling forces (Lee *et al.*, 2004a, 2004b; Agranovski *et al.*, 2006; Park *et al.*, 2011). Second, the ions would be collected by the filter, and electric field is generated in the fibrous filter medium. The electrostatic field is beneficial for filtration efficiency improvement (Wu *et al.*, 1999). Third, the image force between the uncharged fiber and charged particle could improve the particle collection efficiency. Unfortunately, nearly all of the studies focused on the experimental measurement. The theoretical investigation identifying which factor is dominant should be conducted in the future study.

The previous study investigated the influence of system parameters (such as filter medium type, ion concentration, filtration velocity and filter class) on the ionizer-assisted air filtration system experimentally (Shi *et al.*, 2012). The DEHS aerosol ranging from 10 to 1,000 nm was utilized to test efficiency. The ionizer in front of the filter medium could improve the filter efficiency. The electrostatic efficiency enhancement due to the ionization process decreased with the particle size increase, which is very similar to the previous studies (Park *et al.*, 2011). If the particle diameter was larger than 100 nm, the efficiency enhancement due to the electrostatic force could be ignored. The measured results show that the low filtration velocity and high ion concentration could lead to high filtration efficiency. However, at the same ion concentration and face velocity, the synthetic filters had an electrostatic-enhanced efficiency that was about 5%–10% units higher for submicron and ultrafine particles than the values for the glass fiber filters. The experimental efficiency enhancement due to the ionizer increases with the increasing filter class. The main reason could be that the electric field-induced force is higher in high-class filters than that in low-class filters. The ozone generated by the ionizer was lower than 5 ppb.

An HEFS combing an ionization part and an electret fibrous filter was developed and tested by the previous study (Shi and Zhao, 2014). In the ionization stage, the carbon fiber crush and the grounded metal plate were used as discharge polarity and collection plate. The particle could be charged in the ionization stage, and the filtration efficiency for electret filter could be improved due to the particle-charging process. This study conducted long-term measurement in a real building (about 240 days). With ionization, the filtration efficiency of the electret filter was increased by more than 20% units for PM 0.3–0.5 without adding the pressure drop. The measured efficiency enhancement due to the ionization phenomenon decreased with the particle size, which is similar to the previous studies (Park *et al.*, 2011). This study investigated the influence of some system parameters (relative humidity (RH), particle concentration, filtration velocity and air temperature) on the filtration performance.

High particle concentration and filtration velocity could lead to the lower efficiency. The influence of temperature on filtration efficiency could be ignored, and the filtration efficiency with or without the ionization is sensitive to the RH. The efficiency of the hybrid system decayed obviously in their lifetime because the electrostatic force decreases as the filter medium becomes more and more covered by collected dust. This is the main disadvantage of this hybrid electrostatic system. The hybrid air filtration system combing electric field and fibrous filter was utilized in the replacement of a traditional ESP because the filtration efficiency of ESP decreases with the dust loading. But for the fibrous filter, the filtration efficiency and pressure loss increases with dust-loading process (Thomas *et al.*, 2001; Xu *et al.*, 2013). So, new HEFS should be established to overcome this disadvantage. Otherwise, the particle zone consisting of carbon crush and collection plate was an ESP. It could charge and remove the particles. In the long-term measurement, the particle charging and removal effect of the ESP was not monitored and analyzed. The dust loaded on the carbon crush and collection plate may influence the ESP performance.

Overall, the HEFS consisting of an IG could obviously increase the filtration efficiency of the fibrous filter without adding pressure drop. The disadvantage of such system is that the filter efficiency enhanced by electrostatic force is quite limited (<30%). For the indoor environment without high cleanliness requirements, HEFS with an IG is a suitable choice. For the indoor environment with relatively high requirement, HEFS with corona discharge could be utilized to high-efficiently remove particles.

5.3.2 HEFS (with corona discharge) for nonindustrial indoor environment application

To improve the overall performance (including filtration efficiency, pressure loss, energy consumption and ozone generation) of HEFS, many available studies optimize the design of HEFS by using experimental method and numerical simulations (Figure 5.5). The key design variables include filter parameters (fiber size, solidity, etc.), pleat structure, discharge wire radius and wire-plate configurations.

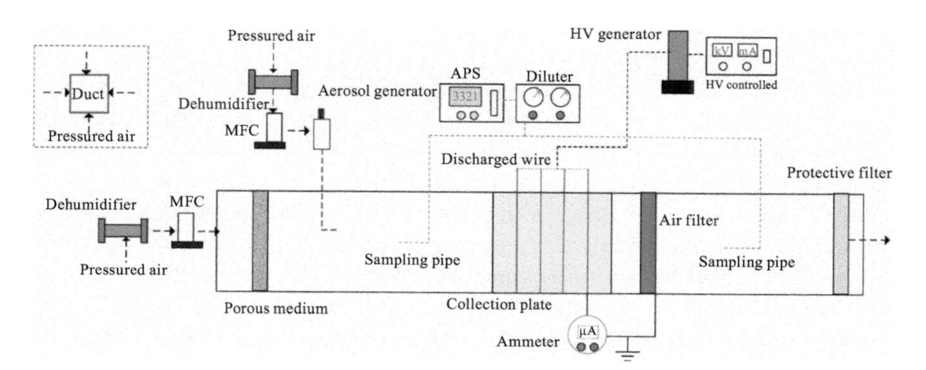

Figure 5.5 The experimental configuration of the hybrid electrostatic filtration system (HEFS) (Feng et al., 2016a, 2016b)

Feng *et al.* (2014) developed a needle-to-filter medium to conductive plate system to enhance the filtration efficiency of filter medium without adding the pressure drop. The needle was connected to the high-voltage generator, and the conductive plate was grounded. The filter medium was attached on the conductive plate. The needle to filter medium to conductive plate system was widely used in the filter-charging process (Horenstein, 1995). In the corona discharge field, the particles would be charged by the diffusion and electric field mechanisms. The intensive electric field was created in the porous medium due to the back corona effect. The charged particle could be captured easily by fiber because of the electric force. The experimental data indicated that the efficiency of filter medium could be improved obviously if the applied voltage was turned on. This study also investigated the influence of filter medium type, applied voltage and conductive plate type on the filtration efficiency. Otherwise, a mathematical model was established to predict the filtration enhancement due to the electrostatic force. The measured and predicted results agreed well.

A typical HEFS was developed, which combines an ESP and a fibrous filter installed downstream of the ESP (Feng *et al.*, 2016a). The particles escaping from the ESP carry a large amount of charge and this can increase the filtration efficiency of the fibrous filter. The filtration characteristics, including the efficiency, pressure drop and ozone generation, were investigated experimentally. The influences of system parameters, including the filter type, applied voltage and distance between the ESP and fibrous filter on the overall efficiency, were also studied. The measured results show that utilizing the non-high-efficient fibrous filter to remove the charged particle could provide much higher efficiency without adding the pressure drop due to the electrostatic force. In addition, the mathematical model was utilized to model the air filter efficiency in HEFS. The modeled and measured results agreed reasonably. After optimization by numerical simulations and experimental measurements, the filtration efficiency, ozone generation of HFES were 100% and lower than 10 ppb. The quality factor of HEFS was much higher than HEPA due to its lower energy consumption. Overall conclusion is that the HEFS could operate at high efficiency with the lower applied voltage, ozone generation and pressure drop.

A novel HEFS system was proposed, which could improve the filtration efficiency of the fibrous filter for fine particles without increasing the pressure drop (Feng *et al.*, 2016b). The system had a pin–filter medium-grounded conductive plate structure. The corona discharge field would be generated in the spaces among pin, filter medium and conductive plate when high voltage was applied. An experimental system was constructed to study the influences of many factors on the filtration efficiency, including filter medium type, applied voltage, pin–filter distance and dust loading. The ozone emission rate was also measured. Besides, two theoretical models for the voltage–current characteristics and the filtration efficiency of the HEFS were developed and validated by the experimental data. The results proved that the HEFS system could increase the filtration efficiency ($>$98%) of filter medium obviously without increasing the pressure drop ($<$40 Pa) and with low ozone generation ($<$20 ppb).

The previous literature developed a numerical model to simulate the performance of an HEFS with a pleated fibrous filter (Feng *et al.*, 2018). The model considered all the physical phenomena in a filtration system: the corona discharge, airflow, particle charging, particle motion and filtration. Measured data from previous studies were used for model validation. The validated numerical method was then used to investigate the influence of structure parameters of a pleated filter on the performance of HEFS. The filtration efficiency and energy consumption were evaluated. The effects of pleat distance, pleat height, applied voltage and discharge wire position were examined. The results show that HEFS with shorter pleat distance and greater pleat height has higher efficiency (close to 100%). An increase in applied voltage raises both the efficiency and the electrical energy consumption. Filter performance is also sensitive to discharge wire placement.

In order to high-efficiently remove ultrafine particles (<100 nm), the previous study adopted numerical method to optimize HEFS by adjusting filtration velocity and electric field in filter medium (Feng *et al.*, 2020). Due to that, the traditional Lagrangian-based model failed to predict ultrafine particles, so an Eulerian-based numerical method was used and validated by experiments. For 50-nm particles, the efficiency of optimized HEFS was enhanced to 96%, which was much higher than that of ESP (10%). The optimized EEAF could save a one-third of energy consumed by HEPA without sacrificing filtration efficiency. Overall, the numerical model developed here is a powerful tool for the design of electrostatic-enhanced pleated filtration system.

Besides, new types of ESP and HEFS have been proposed, which draw more and more attentions (Tian *et al.*, 2018a). Many advanced types of electrostatic-assisted air filtration systems were developed, including two-stage ESP (Li *et al.*, 2019), electrostatic-assisted metal foam coarse filters (Tian *et al.*, 2018b), ESP with dielectric coatings (Mo *et al.*, 2020) and others. The experimental results indicated that these newly developed systems can reduce energy consumption without sacrificing filtration efficiency. These electrostatic-assisted air filtration systems may be capable of high-efficiently disinfecting airborne species due to ion generation. It is necessary to evaluate these systems from the perspectives of disinfection efficiency and energy consumption. However, the numerical models developed in the current study could not completely predict the complex multiple physical fields (e.g., foam filter) in these newly developed electrostatic filtration systems. In future, complete experiments will be conducted to evaluate the disinfection performance of various electrostatic-assisted air filtration systems. Besides, numerical models will be improved to be capable of reasonably/accurately predicting the ion disinfection efficiency of different electrostatic filtration systems.

5.3.3 HEFS (with corona discharge) for industrial indoor environment application

The HEFS with corona discharge can also be efficiently used in industrial buildings. Compared to HEFS in nonindustrial buildings, the HEFS patterns have two different characteristics. First, the initial efficiency of fibrous filter is much lower due to a large size of particles in industrial environments. Second, applied voltage in the electric zone is adjusted to be higher to overcome the negative effects of high particle concentrations on ionized electric fields.

The previous literature investigated the filtration performance of HEFS consisting of ESP and filter medium experimentally (Tu *et al.*, 2013). In ESP, the perforated plate was utilized as the collection plate. Both of the front and reverse sides of the perforated plate would collect particles because the perforated plate was vertical to the airflow direction. Results indicated that higher filtration velocity decreased the ESP efficiency. The opening ratio of the perforated plate did not influence the voltage–current characteristics but determined the filtration efficiency of the ESP zone. Increasing the opening ratio reduced the effective particle collection area and ESP efficiency. But a high opening ratio made the air and particle velocity slower when passing through the perforated plate. A particle with lower velocity was more easily captured by the electric force. So the optimal opening ratio of a perforated plate existed. More particles deposited on the reverse side of the perforated plate upon increasing the opening ratio due to the weak inertia. This work only focused on the ESP zone and did not investigate the filter medium efficiency for charged particles. The ESP pattern would influence the particle and air filter charging status. The efficiency of filter-installed downstream of the ESP was determined by the charge amounts carried by the fiber and particles. This issue would be studied in future.

Numerical simulations were used to design the three-dimensional distributions of the electric field and the space charge in HEFS consisting of ESP and filter medium by finite-volume method (Long *et al.*, 2009a, 2009b; Long and Yao, 2010). The conductive perforated plate was utilized as the collection plate in ESP. The discharged current passed through the conductive plate and the filter medium. The numerical predictions of the current–voltage relations of model AHPC agreed well with the measurements. Numerical results show that the electric field and space charge density distributions on the perforated plate have the same number of peaks corresponding to the holes. The electric field on the bag plate surface was lower than that of the perforated plate. The surface of bag filter medium still has high space-charge density, although its current value was low. The presence of a conductive perforated plate would decrease the electric field and space charge obviously. The modeled results indicated that the filter medium downstream of the ESP was influenced by the corona discharge and was charged by external electric field. The filtration efficiency of filter medium for charged particles would be quite different from the neutral case. But this study did not provide the results of the charged filter efficiency.

The complex multiple physics fields were the modeled physical processes (including the corona discharge, the fluid flow, the particle charging and the filter medium filtration) of HEFS by CFD. Based on the results by Long and Yao (2010, 2012), the airflow and particle motion models were added. The collection process of the HPC was unsteady because the pressure drop across the bag filter increases with the deposition of the particles. The numerical results indicated that gas flow was strongly affected by the electric field in the hybrid particulate collector.

The effect of the hole diameter of the perforated plate on the collection efficiency of the electrostatic zone becomes weaker with the increasing applied voltage. But the collection efficiency of the electrostatic zone had no certain relation with the hole diameter of the perforated plate. However, this study only focused on the electrostatic zone efficiency and assumed that the efficiency of filter medium could be 100%. In fact, the charges carried by fiber and particle could have significant influence on the filter efficiency (Hinds, 1992). The filter medium efficiency enhancement due to the electrostatic force should be studied in the future work.

5.4 Ion generator

An IG installed in indoor environment could remove the particles (Lee *et al.*, 2004a, 2004b; Grabarczyk, 2001). The IG could emit large amount of ions and charge the particles, resulting in increasing the particle deposition velocity onto the walls. The energy consumption of IG was lower than HEPA or ESP because it did not cause the air pressure drop. The electric energy of IG could be ignored due to the weak discharge current value. There were a lot of literature focusing on the performance of the IG.

5.4.1 Air purification performance of an ion generator directly installed in indoor environment

The previous study tested the performance of ionic air purifiers that produce unipolar air ions and reduce aerosol exposure in confined indoor spaces (Grinshpun *et al.*, 2005). A relatively small (2.6 m^3) walk-in chamber where a human manikin was placed was utilized during the operation of an ionic air purifier. The concentration decay of respirable particles of different properties was monitored in real time inside the breathing zone of a human manikin. All five types of ionic air purifiers tested in this study were found capable of reducing the aerosol concentration effectively. The most powerful unit demonstrated an extremely high particle removal efficiency that increased sharply to almost 90% within 5–6 min, reaching about 100% within 10–12 min for all particle sizes (0.3–3 μm) tested in the chamber. The experimental results indicated that the influence of particle size, discharge polarity (positive or negative), aerosol type (NaCl, PSL, *Pseudomonas fluorescens* bacteria), manikin temperature and the breathing rate on the filtration performance were insignificant. But the ion emission rate determined the particle removal ability of the IG. The external ventilation and air mixing enhanced the air cleaning effect. There are two limitations in their study. First, the ozone and other by-product were not measured. The intensive ionization process and small enclosed space would lead to high ozone concentration. Second, the tested chamber volume was small (2.6 m^3). The IG would not be so effective in large space. Increasing the IG number could improve the particle removal efficiency, resulting in high ozone generation and initial cost.

Shiue and Hu (2011) investigated the influence of system parameters (such as particle size, ion emission rate, distance and air change rate) on the particle removal efficiency of IG. The particle removal efficiency increased with the increasing time of continuous emission of negative air ions (NAIs). The particle removal efficiency was significantly affected by the particle size (0.1–0.5 μm) and increased with the particle size. These observed results were quite different from the other studies (Lee *et al.*, 2004a, 2004b; Grinshpun *et al.*, 2005). Their measured data indicated that the particle removal efficiency was not influenced by particle size in this diameter range (0.1–0.5 μm). Similar to the results by Grinshpun *et al.* (2005), air mixing makes particle charging by air ions more efficient and consequently increases particle removal efficiency. The particle removal efficiency decreased with increasing distance between negative ionic air cleaner and sampling point. Experiments indicated that particles removal efficiency near the negative ionic air cleaner was better than at a long distance.

The effects of distance and height from the ion source on NAI concentration were also considered by Grinshpun *et al.* (2005). The NAI concentration dramatically declined with an increase in the distance and inclined with an increase in the height. Wu *et al.* (2006a, 2006b) investigated the concentration gradient of NAI at various RHs and distances from the source in indoor air. Increasing the distance caused the decrease of ion concentration. Experimental results indicated that the influence of RH on the concentration gradient of NAI was so complicated. The changes of NAI concentration with an increase in RH at different distances were quite steady (10–30 cm), strongly declining (70–360 cm), approaching stability (420–450 cm) and moderately increasing (560–900 cm). The authors also conducted fitting effort and found that a logarithmic linear relationship existed between the NAI concentrations and distances from the discharge electrode. The empirical equations were useful for the application of an air ionizer in indoor environments.

Wu *et al.* (2006a, 2006b) studied how wall surface materials influence the removal efficiency of airborne particles with IG in indoor environments. Five wall surface materials—stainless steel, wood, PVC (polyvinyl chloride), wallpaper and cement paint were adopted. Two monodispersed solid NaCl particle sizes (300 and 30 nm) were tested. The concentration of NAI was from 3,000 to 5,000/cm^3. The experimental data revealed that energized IG enhanced the removal efficiency of both 300- and 30-nm particles for each wall surface material. For the particle natural deposition case, the order for 300- and 30-nm particles under different wall surfaces was cement paint>PVC=wallpaper=wood>stainless steel, as perhaps determined by the roughness of the wall materials. If the IG was turned on, the order for 300-nm particles was wood>PVC>wallpaper>stainless steel>cement paint. The order for 30-nm particles was still quite different: wood>PVC>cement paint>wallpaper>stainless steel. The electrical characteristics and roughness of the wall materials may determine the order in the presence of IG. Overall, the performance of IG was related to the wall material in indoor environment.

5.4.2 Comparisons between ion generators and other air cleaning types

An experimental study was conducted to assess the performance of different air cleaners (including ESP, electret filter, HEPA and IG) by the previous literature

(Zuraimi *et al.*, 2011). The CADR of size-resolved particles (ranging from 10 to 10,000 nm) for the four air cleaning technologies were compared. ESP, HEPA and electret filter had the high CADR for various particle sizes ranging from 160 to 800 m^3. The IG generally had CADR values that were an order of magnitude lower than the other three cleaners ranging from 5 to 45 m^3/h. This study also established a mathematical model to predict the unsteady particle concentration in a typical bedroom if the four technologies were used to remove the cough or sneeze particles. The modeled results also indicated that the performance of IG was the worst. However, this study did not measure the ion concentration generated by the IG. The unsatisfied performance of IG may result from the low ion concentration. Increasing the applied voltage and ion concentration may improve the IG performance.

Waring *et al.* (2008) also tested different air cleaner types (including HEPA, ESP and IG). Similar to the results by Zuraimi *et al.* (2011), the CADR value of IG was much lower than the HEPA or ESP. This study also used air cleaner effectiveness (H) to evaluate the performance of an air cleaner in different spaces. H was defined as one minus the ratio of the indoor particle concentration with an operating air cleaner to the indoor concentration with no air cleaner operating. Equation (5.1) describes the detailed mathematical expression of H. The results described that the IG could reduce 40%–60% particles ranging from 50 to 500 nm in a room with 50 m^3. If the room volume increased to 350 m^3, the H value decreased to 10%–20%. This study reported that the ozone emission by ESP (3.8 mg/h) or IG (3.3–4.3 mg/h) was comparable. The negative effect of an IG on the indoor environment was investigated experimentally.

The basic limits existed for an IG. First, high electric field intensity resulting in undesired charging of objects in the room, relatively high ozone concentration and the settling of precipitated dust on surfaces bounding the room, furniture and on the human body. Second, portable corona ionizers are not effective because the volume of the space covered with ion and electric field streams was too small compared to typical indoor space, especially in the case of air movement. Increasing the ionizer number would solve this problem. Third, the particle-removing ability was not satisfied as other air cleaners (such as HEPA, ESP and electret filter). But the electric energy consumed by an IG was ignored compared to ESP. Finally, the IG generated by-product (such as ultrafine particles, ozone, gaseous pollutants). The influence of by-product on human health was not studied sufficiently. Therefore, the IG could be used as assisted particle removal equipment in indoor environment.

5.5 Electret filter

In an electret filter, the electric charges carried by fibers could improve the filtration efficiency of an fibrous filter without adding pressure loss (Kanaoka *et al.*, 2001; Tanthapanichakoon *et al.*, 2003; Hiragi, 1995; Baumgartner *et al.*, 1993; Wu *et al.*, 1993). However, two disadvantages of an electret filter limit are its practical applications. First, the amount of electric charges decreases during the operation process, leading to lower filter efficiency (Lathrachi *et al.*, 1986; Pich *et al.*, 1987).

Second, filter efficiency enhancement due to electric charges is lower than those of ESP and HEFS. Actually, it is necessary to develop new electret materials with better material properties (such as charge stability).

The previous study investigated the effect of particle charge on the glass beads granular filtration of nano-aerosol NaCl particles in the range of 10–100 nm experimentally (Givehchi *et al.*, 2015). Compared to the fibrous filter, a granular filter is a promising option for air filtration at high temperature and high pressure. The nanoparticles were generated and charged in an atomizer, and the neutralizer was utilized to neutralize the charged aerosol to Boltzmann equilibrium charge distribution. The charge polarities of particles and granular were different. The influence of particle charge on filtration efficiency could be obtained by measuring the efficiency for neutralized and charged NaCl nanoparticles. The filtration efficiency due to the particle charge increases with the particle size to the level of 30%, which is quite different from the ionizer-assisted filter. The reason may be the different particle-charging methods. In the study of granular filtration, the particle was charged in the bipolar ion environment. But the ionizer-assisted filtration system charged the particle by corona discharge. Otherwise, results also indicated positive correlation between the filtration efficiency due to electrostatic forces and the particle residence time in the granular filter. The study by Givehchi *et al.* (2015) is theoretical, which is far from the engineering application. First, the particulate pollutant from the atmospheric or indoor environment is not always charged ideally. Second, the charge carried by the granular and filtration efficiency would decay with dust-loading process. Therefore, the applied external electric field is more reliable for electrostatic enhanced filtration system.

The dependence of fibrous filter efficiency with the charge distribution of the aerosol (0.1–2 μm) was also investigated (Chazeleta *et al.*, 2011). No matter the electret filter was charged or neutralized, the particle-charging process decreased the particle penetration obviously. The effect of filter neutralization damages dramatically the performance of the filter. The experimental results also indicated that the penetration of the neutralized electret filter stagnates when the charge density of the particles increases, which is quite different from the traditional electrostatic filtration theory. This phenomenon should be investigated in the future study.

5.6 By-product

5.6.1 The by-product generated by an ion generator (directly installed in indoor environment)

The effects of an IG on indoor air quality (IAQ) were investigated by previous literature (in a residential room). Four experimental cases were established (with or without the carpet overlaying the floor and air freshener) in a 27-m^3 room (Waring and Siegel, 2011). A plug-in air freshener was used as a terpene source. Measurements included the airborne sampling of particulate matter (0.015–20 μm), terpenes and C1–C4 and C6–C10 aldehydes, ozone concentrations. The results show that particle number concentrations decreased with the operation of the IG in room without the air freshener but increased with the operation of the IG in room with the air freshener. The IG could remove particles due to charging effect. The majority of the particle number increase was in the ultrafine range because of ozone reactions with D-limonene. The ozone concentration increased for all the four cases in the presence of energized IG. The results implied that the addition of the air freshener may had increased the ozone reactivity of the room more than the addition of the carpet. The case with carpet and air freshener had the lowest ozone concentration. Though the study measured C1–C4 and C6–C10 aldehydes, only formaldehyde and nonanal were affected by the use of the IG. Concentrations of formaldehyde and nonanal increased due to the operation of the IG in all cases because these aldehydes were products of reactions initiated by ozone. Overall, the IG used in indoor environment generated ultrafine particles, ozone, aldehydes. Although the IG could remove the particle concentration due to particle charging, the negative effect on IAQ should not be ignored.

The negative effect of IG on the indoor environment was investigated experimentally (Waring *et al.*, 2008). The IG could generate gaseous or particulate pollutants. Five different IGs were operated separately in the presence of a liquid or solid air freshener, representing a strong terpene source. The five energized IGs elevated the ozone concentrations significantly. The reactions between the ozone and the terpene resulted in a reduction of ozone concentration from the corresponding case with only IG turned on. The screening experiments demonstrated a net increase in steady-state particle concentrations due to the use of an IG in the presence of a terpene source. The size ranges of particles generated by different IG types were different. The ranges of two types were from 4 to 157 nm, and 9 to 55.2 nm for other two types. This study also investigated the transient particle generation process due to the chemical reaction. After the air freshener was introduced to the chamber with energized IG, the particle concentrations initially increased sharply and then declined to their steady-state values. The use of IG could lower the concentration of D-limonene because of the ozone reaction. With or without the terpene source, the use of the IG increased formaldehyde concentrations slightly. No clear trend was observed for acetaldehyde, as described by others conducting testing on the products of air fresheners (Singer *et al.*, 2006). One limitation of the secondary by-product experiments was that the ozone and terpene concentrations in this study may quite different from the real indoor environment. Conducting the field testing in a real building is necessary.

5.6.2 The by-product generated by ESP and HEFS

The ESP could generate ozone and other by-products, which had negative effect on human health (Li *et al.*, 2015; Niu *et al.*, 2001). The ozone would react with the human skin and clothing surface generating a lot of by-product (such as secondary aerosol and VOC) (Rai *et al.*, 2013, 2014). In the HEFS, the electrostatic zone still generated ozone. But the chemical reaction between the ozone and filter medium occurred. The filter medium (especially a dirty filter) could remove ozone and generated a new type of by-product. Many studies investigated the reaction between the ozone and air filter medium.

The previous literature conducted experiments to study the interaction between the ozone and used air filter (Beko *et al.*, 2007). The ozone concentration was 75 ppb. The dirty glass fiber filter was cut from an HVAC system. Ozone initiated oxidation processes on the surface of loaded filters. The oxidation products mainly included semi-volatile compounds. The used filter could remove the ozone with initial efficiency between 40% and 60%. But the ozone removal efficiency decreased quickly to 10%–20% in 60 min. This result indicates that the zone exposure history of a filter subsequently influences the quantity of oxidation products generated when ozone-containing air flows through it. The increase of secondary organic aerosols downstream of the filter was observed when ozone was present in the air stream. The secondary organic aerosol particles had effect on the human health (Shi and Zhao, 2014). For new and slightly used filters, ozone removal appears to be dominated by organics remaining on the filter following manufacturing. For older filters, ozone removal appears to be dominated by organics associated with captured particles.

The ozone removal efficiency by different air filters with or without the deposited particles was investigated by experimental measurement (Lin and Chen, 2014). Two types of dirty used air filters were cut from the commercial or residual buildings. For most filters, the ozone removal efficiency declined rapidly but converged to a nonzero (steady-state) value. This steady-state ozone removal efficiency varied from 0% to 9% for clean filters. The mean steady-state ozone removal efficiencies for loaded residential and commercial filters were 10% and 41%, respectively. The continuous exposure of filters to ozone following a 24-h period of pause led to a regeneration of ozone removal efficiency. The results also indicated that HVAC filters were estimated to contribute 22% and 95% of total ozone removal in HVAC system for typical residential and commercial buildings. The high ozone removal efficiency may not be beneficial to IAQ due to the by-products generation. But this study did not measure the by-product (such as secondary aerosol and VOC).

The influence of filter medium type, loaded dust type, RH and exposure time on the removal of ozone by an air filter was tested experimentally (Hyttinen *et al.*, 2006). All types of the filter could remove ozone with different efficiencies, except for an unused prefilter made of polyester. Dust loading could enhance the reduction of ozone. The ozone removal efficiency of the filter with diesel soot was higher than the atmospheric dust. Increasing the RH resulted in a larger O^3 removal. The removal of O^3 was highest in the beginning of the test, but it declined within 2-h reaching almost a steady state as the exposure continued. The sooty filters continued to remove as much as 25%–30% of ozone in the steady stage. In this study, the formation of aldehydes, especially formaldehyde, was detected during exposure to ozone. The production of formaldehyde was highest in the beginning of ozone exposure and decreased to the level observed before ozone exposure. The nitrogen content of the filter dust is typically 1%–2% (Hyttinen *et al.*, 2002). In some tests, additional NO^2 formation was detected because ozone reacted with nitrogen-containing components of the dust.

Destaillats *et al.* (2011) measured the gaseous pollutant generated by the chemical reaction between the air filter and ozone. The ozone concentration was

150 ppb. The RH was 50%. Formaldehyde and acetaldehyde were the measured downstream of each filter sample. By-product emissions were consistently higher under humidified air than under dry conditions and were higher if the filters were loaded with particles, as compared with unused filters. But the effect filter media type on ozone reaction was not significant. Fiberglass filters heavily coated with impaction oil showed higher formaldehyde emissions than other samples. The results show that emission rates of formaldehyde and acetaldehyde were not found to be large enough to substantially increase indoor concentrations in typical building scenarios. But ozone reactions on HVAC filters cannot be ignored as a source of low levels of indoor irritants.

The experimental study by Lin and Chen (2014) indicated that filter materials, filter used time, filter thickness and the quality of air passing through the filters affected ozone removal efficiency. The results also reported that there was a positive correlation between ozone removal efficiency and carbonyls generation on filters. Positive correlation also existed among the three factors: the amount of organic carbon on filters, ozone removal efficiency and carbonyls generations. The influence of surface areas of dusts deposited on filters on ozone removal efficiency was not obvious. They concluded that chemical compositions of dusts on filters had significant influence on ozone removal efficiency and carbonyls generation.

Based on the experimental results described in the previous studies, the air filter could remove some ozone escaping from the ESP. In ESP and HEFS, some by-products were generated (such as VOC and secondary aerosol), which was harmful to human health. However, the existing researches focusing on the HEFS only measured the ozone generation (Lee *et al.*, 2001; Park *et al.*, 2011). Other types of by-products were not analyzed. One of concerns of the electrostatic filtration was the by-product. So complete product measurements should be the key points in the future work.

5.7 Bacteria inactivation

One of the advantages of the HEFS was that the intensive direct electric field in ESP or across the filter medium could inactivate microorganisms effectively. Yao *et al.* (2005) investigated whether electrostatic fields can be used to inactivate airborne bacteria. The results indicated that more than 90% of the bacteria (*P. fluorescens*) cells deposited on the surface of nonconductive filters were inactivated when fields of 15 kV/cm were applied for 30 min or longer. Similar results were observed when *P. fluorescens* were exposed to electric fields of 5 and 10 kV/cm for 2 h. If the exposure time was reduced to 30 s, the inactivation phenomenon did not occur. This research had shown that specific combinations of electrostatic field strength and exposure time can be used to effectively inactivate certain bacterial cells deposited on nonconductive surfaces. Although Yao *et al.* (2005) reported that the culturability of *P. fluorescens* was reduced upon exposure to an electric field, the experimental results by Hwang *et al.* (2014) proved that the electric field itself did not affect the inactivation of *Staphylococcus epidermidis* bacteria. The

S. epidermidis bacteria had more resistance to environmental stresses than *P. fluorescens*. Therefore, the inactivation effect of the electric field was influenced by the bacteria type. Bu *et al.* (2005) utilized a plane–plane electrode system to investigate the influence of electrostatic field on viability of *Bacillus subtilis* bacteria. It was found that the survival ratio of *B. subtilis* can be considerably affected by the electric field intensity and exposure duration. The viability of bacillus decreased with the increase of the duration time and electric field intensity.

Luo *et al.* (2006) treated the *Bacillus* coil by high-voltage electrostatic field. The positive correlation existed between the inactivation ratio and electric field intensity (or exposure time). If the distance between two plate electrodes was 30 mm, the applied electric field intensity was 20 kV and the exposure time was 45 min, the inactivation ratio of the *Bacillus* coil was 97%. After the treatment by intensive electric field, the temperature of bacteria liquid did not vary. The PH value decreased. The conductivity of bacteria liquid increased.

The literature described earlier indicated that the high-voltage electrostatic field could inactivate bacteria. But the inactivation effect was determined by the electric field intensity, exposure time and bacteria type. So the HEFS has the potential to inactivate bacteria. The electric field strength in HEFS was high enough compared to the bacteria inactivation study (Yao *et al.*, 2005; Luo *et al.*, 2006; Bu *et al.*, 2005). The exposure time could be ensured because of the long operation time of HEFS. But the HEFS may not inactivate some types of bacteria due to the resistance to environmental stresses.

5.8 The effects of electrostatic air cleaner on particle deposition in building

The ESP or the HEFS could charge and remove the particles. Although the filtration efficiency of some ESP was not high, the particles escaping from the ESP were charged. For the indoor environment with IG, ESP or HEFS, the charge carried by particles would influence the particle deposition rate and pattern.

Without the electrostatic force, the particle may deposit onto the indoor surface due to Brownian diffusion, turbulent diffusion, gravitational settling and turbophoresis effect (Zhao *et al.*, 2007; Lai and Nazaroff, 2000). The electrostatic force and image force between the wall surface and charged particles affect the particle deposition velocity. The electrostatic force existed between the charged particle and charged wall surface. The image force was caused by the interaction between the neutral wall surface and charged particle. Based on the traditional three-layer model, Chen and Lai (2004) developed a mathematical model to calculate the particle deposition velocity in the presence of the two types of electric force. The simulated results indicated that the electrostatic particle deposition velocity was higher than the mechanic deposition velocity by several orders of magnitude. The Coulombic force influenced the particle deposition significantly while the image force was important for high charge level particles in the presence of weak electric field. Lai (2006) conducted experiments to investigate the effect of different wall surface types and charge status on particle deposition velocity. The measured data reported that charge on different wall surface increased the particle deposition velocity obviously. This study also simulated the charged particle deposition velocity using model developed by Chen and Lai (2004). The modeled and measured data agreed well.

The experiment and simulation studies indicated that the charge could enhance the particle deposition velocity obviously. If the portable electrostatic cleaning equipment (such as IG, ESP and HEFS) was used, more charged particles deposited onto the indoor surface. This phenomenon was beneficial for reducing indoor particle concentration but polluted the surfaces of wall and furniture. It increased the exposure risk if someone touched the dirty surfaces. The effect of an electrostatic cleaner on the particle deposition rate should be studied in the future work.

The rate at which airborne particulate matter deposits onto HVAC components was important from both IAQ, system maintenance cost and energy perspectives. Waring *et al.* (2008) conducted modeling work and predicted size-resolved particle mass loading rates for residential and commercial filters, heat exchangers and supply and return ducts. A parametric analysis evaluated the impact of different outdoor particle distributions, indoor emission sources, HVAC airflow, filter efficiency, coils and duct system complexities. If IG, ESP or HEFS were utilized in the HVAC system or as an indoor portable cleaner, the charge carried by particle could increase the particle deposition velocity. The effect of electrostatic filtration system on the particle deposition rate for different positions in HVAC system should be investigated by modeling or experimental study. The model developed by Zhao *et al.* (2007), Lai and Nazaroff (2000) could be used directly.

5.9 Subjective or objective study

In the available researches, objective measurements (such as filtration efficiency, indoor particle concentration and by-product) were widely conducted (Park *et al.*, 2011; Shi and Ekberg, 2015). But the subject survey and human response recording were so limited. Different from the objective measurements, the subjective survey could reveal the effect on human health directly. Skulberg *et al.* (2005) conducted questionnaire study to investigate the effect of the HEFS consisting of ESP and carbon fiber filter on the human health. Eighty persons with airways symptoms were recruited and randomly assigned to an intervention or control group. In the control group, the non-electrostatic filter was used. Subjective symptoms were recorded using a questionnaire, and indexes calculated for general, irritation and skin symptoms. Objective respiratory health indicators were recorded, with acoustic rhinometry and peak expiratory flow meters. For all sizes of particles, the indoor particle concentration decrease (from 65 to 35 $\mu g/m^3$) of intervention group was more obvious than the control group (from 57 to 47 $\mu g/m^3$). The irritation and general symptom indices decreased in both groups, but there was no improvement in the intervention group, compared with the control group. In this study, no participants reported any unpleasant smell from the unit because the odor threshold of ozone was very low. However, the ozone concentration was not measured. It was difficult to relate the subjective phenomenon and objective parameters due to the absence of ozone measurement. For the electrostatic cleaner (including ESP, IG and HEFS), the objective measurement (such as filtration efficiency, indoor particle concentration and by-product) and subjective survey should be combined to investigate the human response to electrostatic cleaner quantitatively.

5.10 Advanced air cleaning toward COVID-19 control

The global outbreaks of corona-virus disease 2019 (COVID-19), induced by severe acute respiratory syndrome corona-virus 2 (SARS-CoV-2), have yet caused serious threats to human health and economic development. Airborne transmission of pathogens/aerosol is proven to be an important pathway. Toward the effective control of airborne SARS-CoV-2 aerosols in the ventilation system, some advanced air cleaning technologies have been developed, including UV+Filter system and electrostatic disinfector (ESD) **(Feng *et al.*, 2021a, 2021b)**, as shown in Figure 5.6.

For the widely used fibrous filter, its disadvantages include high pressure drop, energy consumption and replacement cost. Besides, bacterium can propagate in fibrous filter with loaded cake layer, therefore inducing health risk/secondary contamination during maintenance and replacement due to re-aerosolization. To avoid these shortcomings, the previous study **(Feng *et al.*, 2021a)** proposed an UV+Filter (ultraviolet and fibrous pleated filter) system to efficiently capture

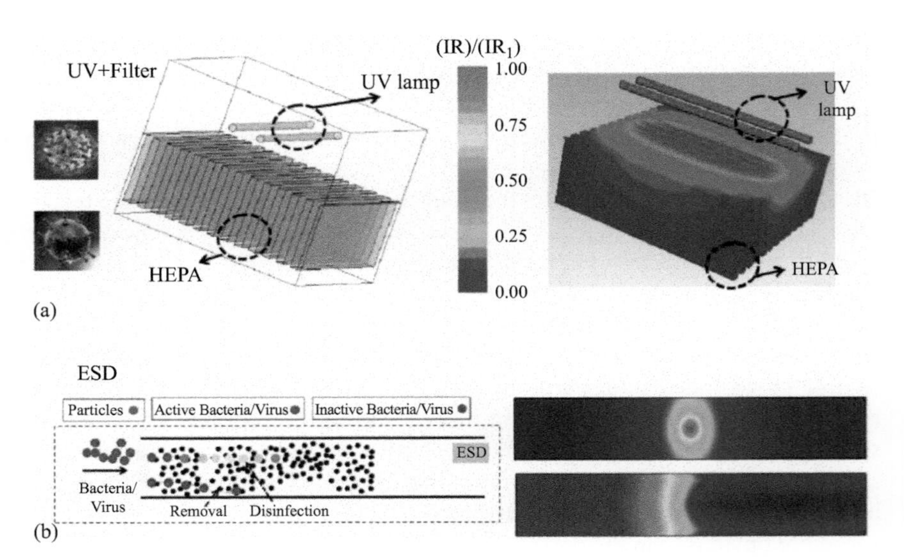

Figure 5.6 UV+Filter system (a) and electrostatic disinfector (ESD) (b) (Feng et al., 2021a, 2021b)

airborne SARS-CoV-2 aerosols and deactivate them in filter medium. Multiple physics model was established to simulate the complex characteristics in UV+Filter system (UV irradiance, turbulent airflow, porous medium and UV disinfection, etc.). After validation, this model was used for optimal design. The filtration efficiency and disinfection time of the optimized UV+Filter system were respectively >99% and ~50 min, respectively. Among various air cleaning types, UV+Filter is the most reliable and safest system.

As described above, electrostatic precipitator (ESP) is widely adopted in particle removal. Corona discharge in an electrostatic precipitator is capable of removing particulate matter and disinfecting biological aerosols to act as electrostatic disinfector (ESD). The ions generated by ESD and ionized electric field can effectively inactivate bacteria/viruses. However, the available research rarely studied the disinfection effect of ESD. The previous literature **(Feng *et al.*, 2021b)** developed an integrated numerical model to simulate disinfection performance of ESD, considering the ionized electric field, electrohydrodynamic flow and biological disinfection. After validation, this model was used to study the influences of essential design parameters (e.g. voltage, inlet velocity) of ESD on disinfection efficiency. The disinfection efficiency of well-designed ESD (with a space charge density of 3.6×10^{-06} C/m^3) could be as high as 100%. Compared with HEPA, ESD could save 99% of energy consumed by HEPA without sacrificing disinfection efficiency.

Overall, this chapter describes various advanced air cleaning systems, including electrostatics assisted air filtration system, electret filter, ion generator and UV devices, as shown in Figure 5.7. Based on physical mechanism, these systems can be divided into three categories: UV-based, electrostatic-based and ion-based system. In future, the following innovative points are research directions: 1) advanced material; 2) advanced modular and 3) advanced simulation tools. It is necessary to develop advanced materials of discharge electrode and collection plate in ESP, new non-woven material of fibrous filter (fiber characteristics, catalyst, etc.) and emitter materials of ion generator. For advanced modular, the stability, economical cost and size of UV lamp and high voltage generator should be reduced. The accuracy and computing cost of available multiple-physics field models should be further improved, toward effective design and optimization of various air cleaning systems.

5.11 Subjective or objective study

1. Two-stage ESP and HEFS have the potential to high-efficiently remove particulate matter (0.01–2.5 μm). The size-dependent filtration efficiency could be as high as 99%. The optimized ESP and HEPA could reduce 90% of energy consumed by the widely utilized HEPA. The net ozone concentration of HEFS was lower than 20 ppb, which satisfies the requirements of standards.
2. An IG directly installed in indoor environment could charge particles and remove these by increasing particle deposition velocity. Compared to ESP and HEFS, its particle removal efficiency/rate is much lower. An IG generates a small amount of by-product, which is harmful for human health.

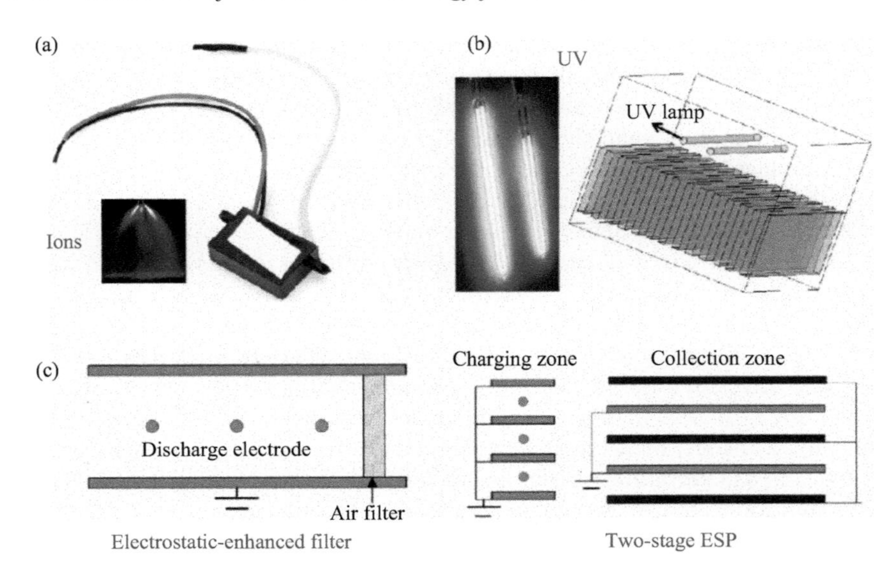

Figure 5.7 Summary of various air cleaning systems: (a) ion-based; (b) UV-based; (c) electrostatic-based technologies

3. Compared to a common fibrous filter, the electret filter has higher filtration efficiency due to electric charges carried by fibers. However, the amount of electric charges decreases during the operation process, leading to lower filter efficiency. Therefore, it is necessary to develop new electret materials with better material properties (such as electric charge stability).

References

Adamaik, K. Numerical models in simulating wire-plate electrostatic precipitators: A review. *J. Electrostat.* 2013, 71 (4): 673–680.

Agranovski, I. E.; Huang, R.; Pyankov, O. V.; Altman, I. S.; and Grinshpun, S. A. Enhancement of the performance of low-efficiency HVAC filters due to continuous unipolar ion emission. *Aerosol Sci. Technol.* 2006, 40 (11): 963–968.

Baumgartner, H.; Piesch, C.; and Umhauer, H. High-speed cinematographic recording and numerical simulation of particles depositing on electret fibres. *J. Aerosol Sci.* 1993, 24: 945–962.

Beko, G.; Clausen, G.; and Weschler, C. Further studies of oxidation processes on filter surfaces: Evidence for oxidation products and the influence of time in service. *Atmos. Environ.* 2007, 41: 5202–5212.

Bo, Z.; Yu, K.; Lu, G.; Mao, S.; Chen, J.; and Fan, F. Nanoscale discharge electrode for minimizing ozone emission from indoor corona devices. *Environ. Sci. Technol.* 2010, 44 (63): 37–42.

Bu, D.; Liu, Y.; Zhou, Y.; Xu, Z.; and Guo, L. Inactivation effects of electrostatic field on Bacillus subtilis. *J. Electrostat.* 2005, 63: 847–852.

Chazeleta, S.; Bemera, D.; and Grippari, F. Effect of the test aerosol charge on the penetration through electret filter. *Sep. Purif. Technol.* 2011, 79: 352–356.

Chen, F. and Lai, A. An Eulerian model for particle deposition under electrostatic and turbulent conditions. *J. Aerosol Sci.* 2004, 35: 47–62.

Chen, J. and Davidson, J. H. Ozone production in the negative DC corona: The dependence of discharge polarity. *Plasma Chem. Plasma Process.* 2013, 23: 501–518.

Destaillats, H.; Chen, W.; Apte, M.; Li, N.; and Spears, M. Secondary pollutants from ozone reactions with ventilation filters and degradation of filter media additives. *Atmos. Environ.* 2011, 45: 3561–3568.

Endo, Y.; Chen, D.; and Pui, D. Effects of particle polydispersity and shape factor during dust cake loading on air filters. *Powder Technol.* 1998, 98: 241–249.

Feng, Z.; Cao, S-J. and Haghighat, F. Removal of SARS-CoV-2 using UV+Filter in built environment. Sustain. *Cities Soc.* 2021a, 74: 103226.

Feng, Z.; Cao, S-J.; Wang, J.; Kumar, P. and Haghighat, F. Indoor airborne disinfection with electrostatic disinfector (ESD): Numerical simulations of ESD performance and reduction of computing time. *Build. Environ.* 2021b, 200: 107956.

Feng, Z.; Long, Z.; and Chen, Q. Voltage-current characteristics of needle-plate system with different media on the collection plate. *J. Electrostat.* 2014, 72 (2): 129–135.

Feng, Z.; Long, Z.; and Mo, J. Experimental and theoretical study of a novel electrostatic enhanced air filter (EEAF) for fine particles. *J. Aerosol Sci.* 2016a, 102: 41–54.

Feng, Z.; Long, Z.; and Yu, T. Filtration characteristics of fibrous filter following an electrostatic precipitator. *J. Electrostat.* 2016b, 83: 52–62.

Feng, Z.; Pan, W.; Zhang, H.; Cheng, X.; Long, Z.; and Mo, J. Evaluation of the performance of an electrostatic enhanced air filter (EEAF) by a numerical method. *Powder Technol.* 2018, 327: 201–214.

Feng, Z.; Yang, J.; and Zhang, J. Numerical optimization on newly developed electrostatic enhanced pleated air filters for efficient removal of airborne ultra-fine particles: Towards sustainable urban and built environment. *Sustain. Cities Soc.* 2020, 54: 102001.

Givehchi, R.; Li, Q.; and Tan, Z. The effect of electrostatic forces on filtration efficiency of granular filters. *Powder. Technol.* 2015, 277: 135–140.

Grabarczyk, Z. Effectiveness of indoor air cleaning with corona ionizers. *J. Electrostat.* 2001, 51–52: 278–283.

Grinshpun, S.; Mainelis, G.; Trunov, M.; Adhikari, A; Reponen, T.; and Willeke, K. Evaluation of ionic air purifiers for reducing aerosol exposure in confined indoor spaces. *Indoor Air* 2005, 15: 235–245.

Hasolli, N.; Park, Y. O.; and Rhee, Y. W. Filtration performance evaluation of depth filter media cartridges as function of layer structure and pleat count. *Powder Technol.* 2010, 237: 24–31.

Hinds, W. C. *Aerosol Technology: Properties, Behavior, and Measurement of Airborne Particles*; John Wiley & Sons, Inc., New York, NY, USA, 1992.

Hiragi, S. Dust collection performance of air filter mat with particles accumulation. Doctoral dissertation. Kanazawa University, Japan, (in Japanese), 1995.

Horenstein, M. Surface charging limit for a woven fabric on a ground plane. *J. Electrostat.* 1995, 35 (1): 31–40.

Hosseini, S. and Tafreshi, H. 3-D simulation of particle filtration in electrospun nanofibrous filters. *Powder Technol.* 2010, 201: 153–160.

Huang, B. Study on the effect of electrostatics in fibrous filtration of PM 10 (in Chinese). Tsinghua University, Beijing, 2006.

Huang, S. H. and Chen, C. C. Filtration characteristics of a miniature electrostatic precipitator. *Aerosol Sci. Technol.* 2001, 35: 963–968.

Hwang, G.; Park, H.; Bae, G.; and Jung, J. Effect of electric field strength on an antimicrobial filter. *Aerosol Air Qual. Res.* 2014, 14: 1028–1037.

Hyttinen, M.; Pasanen, P.; and Kalliokoski, P. Removal of ozone on clean, dusty and sooty supply air filters. *Atmos. Environ.* 2006, 40: 315–325.

Hyttinen, M.; Pasanen, P.; and Kalliokoski, P. VOC emissions from dusty air filters. *Indoor Air* 2002, 1: 368–373.

Ji, J.; Bae, G.; Kang, S.; and Hwang, J. Effect of particle loading on the collection performance of an electret cabin air filter for submicron aerosols. *J. Aerosol Sci.* 2003, 34: 1493–1504.

Joubert, A.; Laborde, J. C.; Bouilloux, L.; Chazelet, S.; and Thomas, D. Modelling the pressure drop across HEPA filters during cake filtration in the presence of humidity. *Chem. Eng. J.* 2011, 166: 616–623.

Kanaoka, C.; Emi, H.; and Myojo, T. Simulation of the growing process of a particle dendrite and evaluation of a single fiber collection efficiency with dust load. *J. Aerosol Sci.* 1980, 11: 377–389.

Kanaoka, C.; Hiragi, S.; and Tanthapanichakoon, W. Stochastic simulation of the agglomerative deposition process of aerosol particles on an electret fiber. *Powder Technol.* 2001, 118: 97–106.

Kim, H.; Han, B.; Kim, Y.; and Yo, S. Characteristics of an electrostatic precipitator for submicron particles using non-metallic electrodes and collection plates. *J. Aerosol. Sci.* 2010, 41 (11): 987–997.

Kim, H.; Han, B.; Kim, Y.; Oda, T.; and Won, H. Submicrometer particle removal indoors by a novel electrostatic precipitator with high clean air delivery rate, low ozone emissions and carbon fiber ionizer. *Indoor Air* 2013, 23 (5): 369–378.

Lai, A. and Nazaroff, W. Modeling indoor particle deposition from turbulent flow onto smooth surfaces. *J. Aerosol Sci.* 2000, 31: 463–476.

Lai, A. Investigation of electrostatic forces on particle deposition in a test chamber. *Indoor Built Environ.* 2006, 15: 179–186.

Lathrachi, R.; Fissan, H. J.; and Neumann, S. Deposition of submicron particles on electrically charged fibers. *J. Aerosol. Sci.* 1986, 17 (3): 446–449.

Lee, B.; Yermakov, M.; and Grinshpun, S. Removal of fine and ultrafine particles from indoor air environments by the unipolar ion emission. *Atmos. Environ.* 2004a, 38: 4815–4823.

Lee, J. K.; Kim, S. C.; Shin, J. H.; Lee, J. E.; Ku, J. H.; and Shin, H. S. Performance evaluation of electrostatically augmented air filters coupled with a corona precharger. *Aerosol Sci. Technol.* 2001, 35 (4): 785–791.

Lee, K. and Liu, B. Theoretical study of aerosol filtration by fibrous filters. *Aerosol Sci. Technol.* 1982, 1–2: 147–161.

Lee, U.; Yermakov, M.; and Grinshpun, S. Unipolar ion emission enhances respiratory protection against fine and ultrafine particles. *J. Aerosol Sci.* 2004b, 35: 1359–1368.

Lee, U.; Yermakov, M.; and Grinshpun, S. Filtering efficiency of N95- and R95-type facepiece respirators, dust-mist facepiece respirators, and surgical masks operating in unipolarly ionized indoor air environments. *Aerosol Air Qual. Res.* 2005, 5: 25–38.

Li, S.; Zhang, S.; Pan, W.; Long, Z.; and Yu, T. Experimental and theoretical study of the collection efficiency of the two-stage electrostatic precipitator. *Powder Technol.* 2019, 356: 1–10.

Li, Z.; Liu, Y.; Xing, Y.; Tran, T.; Le, T.; and Tsai, C. Novel wire-on-plate electrostatic precipitator (WOP-EP) for controlling fine particle and nanoparticle pollution. *Environ. Sci. Technol.* 2015, 49 (14): 8683–8690.

Lin, C. and Chen, H. Impact of HVAC filter on indoor air quality in terms of ozone removal and carbonyls generation. *Atmos. Environ.* 2014, 89: 29–34.

Long, Z. and Yao, Q. Evaluation of various particle charging models for simulating particle dynamics in electrostatic precipitators. *J. Aerosol Sci.* 2010, 41: 702–718.

Long, Z. and Yao, Q. Numerical simulation of the flow and the collection mechanism inside a scale hybrid particulate collector. *Powder Technol.* 2012, 215–216: 26–37.

Long, Z.; Yao, Q.; Song, Q.; and Li, S. A second-order accurate finite volume method for the computation of electrical conditions inside a wire-plate electrostatic precipitator on unstructured meshes. *J. Electrostat.* 2009b, 67: 597–604.

Long, Z.; Yao, Q.; Song, Q.; and Li, S. Three-dimensional simulation of electric field and space charge in the advanced hybrid particulate collector. *J. Electrostat.* 2009a, 67: 835–843.

Luo, Y.; Zhang, B.; and Wei, B. Sterilization effect of high voltage electric field on *E. coli*. *J. Shenyang Agric. Univ.* 2006, 37: 114–116.

Mo, J.; Tian, E.; and Pan, J. New electrostatic precipitator with dielectric coatings to efficiently and safely remove sub-micro particles in the building environment. *Sustain. Cities Soc.* 2020, 55: 102063.

Niu, J.; Tung, T.; and Burnett, J. Quantification of dust removal and ozone emission of ionizer air-cleaners by chamber testing. *J. Electrostat.* 2001, 51–52: 20–24.

Noh, K.; Lee, J.; Kim, C.; Yi, S.; Hwang, J.; and Yoon, Y. Filtration of submicron aerosol particles using a carbon fiber ionizer-assisted electret filter. *Aerosol Air Qual. Res.* 2011, 11: 811–821.

Park, J.; Yoon, K.; Noh, K.; Byeon, J.; and Hwang, J. Removal of PM 2.5 entering through the ventilation duct in an automobile using a carbon fiber ionizer-assisted cabin air filter. *J. Aerosol. Sci.* 2010, 41 (10): 935–943.

Park, K.; Yoon, K.; and Hwang, J. Removal of submicron particles using a carbon fiber ionizer-assisted medium air filter in a heating, ventilation, and air-conditioning (HVAC) system. *Build. Environ.* 2011, 46 (8): 1699–1708.

Pich, J.; Emi, H.; and Kanaoka, C. Coulombic deposition mechanism in electret filters. *J. Aerosol. Sci.* 1987, 18 (1): 29–35.

Rai, A.C.; Guo, B.; Lin, C. H.; Zhang, J.; Pei, J. and Chen, Q. Ozone reaction with clothing and its initiated particle generation in an environmental chamber. *Atmos. Environ.* 2013, 77: 885–892.

Rai, A.C.; Guo, B.; Lin, C. H.; Zhang, J.; Pei, J. and Chen, Q. Ozone reaction with clothing and its initiated VOC emissions in an environmental chamber. *Indoor Air* 2014, 24: 49–58.

Rai, A.C.; Lin, C. H. and Chen, Q. Numerical modeling of particle generation from ozone reactions with human-worn clothing in indoor environments. *Atmos. Environ.* 2015, 102: 145–155.

Shi, B. and Ekberg, L. Ionizer assisted air filtration for collection of submicron and ultrafine particles-evaluation of long-term performance and influencing factors. *Environ. Sci. Technol.* 2015, 49 (11): 6891−6898.

Shi, B.; Ekberg, L.; Truuschel, A.; and Gusteen, J. Influence of filter fiber material on removal of ultrafine and submicron particles using carbon fiber ionizer-assisted ventilation air filters. *ASHRAE Trans.* 2012, 118 (Part 1): 602−611.

Shi, S. and Zhao, B. Modeled exposure assessment via inhalation and dermal pathways to airborne semi volatile organic compounds (SVOCs) in residences. *Environ. Sci. Technol.* 2014, 48 (10): 5691–5699.

Shiue, A. and Hu, S. Contaminant particles removal by negative air ionic cleaner in industrial mini environment for IC manufacturing processes. *Build. Environ.* 2011, 46: 1537–1544.

Singer, C.; Coleman, B.; Destaillats, H.; *et al.* Indoor secondary pollutants from cleaning product and air freshener use in the presence of ozone. *Atmos. Environ.* 2006, 40: 6696–6710.

Skulberg, K.; Skyberg, K.; Kruse, K.; *et al.* The effects of intervention with local electrostatic air cleaners on airborne dust and the health of office employees. *Indoor Air* 2005, 15: 152–159.

Tanthapanichakoon, W.; Maneeintr, K.; Charinpanitkul, T.; and Kanaoka, C. Estimation of collection efficiency enhancement factor for an electret fiber with dust load. *J. Aerosol. Sci.* 2003, 34 (11): 1505–1522.

Thomas, D.; Penicot, P.; Contal, P.; Leclerc, D.; and Vendel, J. Clogging of fibrous filters by solid aerosol particles: Experimental and modelling study. *Chem. Eng. Sci.* 2001, 56 (11): 3549–3561.

Tian, E.; Mo, J.; and Li, X. Electrostatically assisted metal foam coarse filter with small pressure drop for efficient removal of fine particles: Effect of filter medium. *Build. Environ.* 2018a, 44: 419–426.

Tian, E.; Mo, J.; Long, Z.; Luo, H.; and Zhang, Y. Experimental study of a compact electrostatically assisted air coarse filter for efficient particle removal:

Synergistic particle charging and filter polarizing. *Build. Environ.* 2018b, 135: 153–161.

Tu, Y.; Song, Q.; Tu, G.; and Yao, Q. Experimental research on particulate collection performance by perforated plate in hybrid particulate collector. *Proc. CSEE* 2013, 33: 51–56.

Wang, C. Electrostatic forces in fibrous filters—A review. *Powder Technol.* 2001, 118: 166–170.

Waring, S.; Siegel, J.; and Corsi, R. Ultrafine particle removal and generation by portable air cleaners. *Atmos. Environ.* 2008, 42: 5003–5014.

Waring, S. and Siegel, J. The effect of an ion generator on indoor air quality in a residential room. *Indoor Air* 2011, 21: 267–276.

Wen, T.; Wang, H.; Krichtafovitch, I.; and Mamishev, A. L. Novel electrodes of an electrostatic precipitator for air filtration. *J. Electrostat.* 2015, 73: 118–124.

Wu, C.; Lee, G.; Cheng, P.; Yang, S.; and Yu, K. Effect of wall surface materials on deposition of particles with the aid of negative air ions. *J. Aerosol. Sci.* 2006a, 37: 616–630.

Wu, C.; Lee, G.; Yang, S.; Yu, K.; and Luo, C. Influence of air humidity and the distance from the source on negative air ion concentration in indoor air. *Sci. Total Environ.* 2006b, 370: 245–253.

Wu, Z.; Colbeck, I.; and Zhang, G. Deposition of particles on a single cylinder by a Coulombic force and direct interception. *Aerosol Sci. Technol.* 1993, 19 (1): 40–50.

Wu, Z.; Walters, J.; and Thomas, D. The deposition of particles from an air flow on a single cylindrical fiber in a uniform electrical field. *Aerosol Sci. Technol.* 1999, 30 (1): 62–70.

Xu, B.; Liu, J.; Ren, S.; Yi, S.; Yin, W.; and Chen, Q. Investigation on the performance of airliner cabin air filter throughout the lifetime usage. *Aerosol Air Qual. Res.* 2013, 13: 1544–1551.

Yao, M.; Mainelis, G.; and An, H. Inactivation of microorganisms using electrostatic fields. *Environ. Sci. Technol.* 2005, 39 (14): 3338−3344.

Yao, M. and Zhao, B. Ozone reactive compounds measured in skin wipes from Chinese volunteers. *Build. Environ.* 2021, 188: 107515.

Yun, S.; Min, B.; and Seo, Y. A novel polymer-arrayed electrostatic precipitator with Electrical resistance material for the removal of fine particles. *J. Aerosol. Sci.* 2013, 57: 88–95.

Zhao, P.; Siegel, J.; and Corsi, R. Ozone removal by HVAC filters. *Atmos. Environ.* 2007, 41: 3151–3160.

Zuraimi, M.; Nilsson, J.; and Magee, R. Removing indoor particles using portable air cleaners: Implications for residential infection transmission. *Build. Environ.* 2011, 46: 2512–2519.

Chapter 6

Air distribution in mechanical ventilation system

Risto Kosonen[1,2]

The air distribution in spaces is a major important factor for occupants' health, comfort and performance as well as for efficient energy use. In many buildings, mechanical systems are used to transport and supply clean and cooled/heated air to occupied spaces. Air distribution in such ventilated spaces in general depends on inertia and buoyancy forces, initial conditions of the supplied flow and boundary conditions in the space.

Different air distribution strategies based on the mechanical systems for air transportation are used to achieve the goal of thermal comfort and air quality. One of the strategies is to mix the supplied clean air and cool/warm air with the polluted air in the space. Dilution of the supplied air with room air is the aim of mixing ventilation. Another common strategy is displacement ventilation, where a stratified flow is created using the buoyance forces of heat sources. The air quality is then generally better than with mixing ventilation.

6.1 Introduction to air distribution

This chapter discusses methods of total volume ventilation by mixing ventilation and displacement ventilation. Other methods for achieving thermal comfort and good air quality in spaces which are now under development (including localized radiant and convective system, stratum ventilation) and advanced air distribution (such as personalized ventilation) are not considered in this chapter.

The aim of air conditioning and distribution into rooms is to maintain, in the most economical way (accounting for energy usage and cost efficiency), the desired thermal environment and air quality in the occupied zone so as to meet target levels during different operating conditions. Depending on the design criteria, the designer may choose different room air distribution methods in order to achieve the specified targets.

[1]Aalto University, Espoo, Finland
[2]Nanjing Tech University, Nanjing, China

The room air distribution method is critical for the air conditioning of spaces. Often air distribution in rooms is assisted by radiant or convective heating and cooling methods. However, it must be noted that in some cases, a strategy of room air distribution can also be fulfilled without any mechanical installations using only buoyancy forces. The classification of ideal room air distribution methods is summarized in Figure 6.1 (Hagström *et al.*, 2000).

The air distribution in a room with a fully developed turbulent flow can be described by the Archimedes number. It can be considered the ratio between the buoyancy force and the momentum force. In a given geometry, the Archimedes number (Ar) can be expressed as $Ar \sim \Delta T_o / u_o^2$ or as $\Delta T_o / q_o^2$ (Ar_{ratio}) because the supply area is constant, where ΔT_o = temperature difference between exhaust and supply air [K], u_o = supply air velocity [m/s] and q_o = air volume flow per supply area [m/s].

Apart from the Archimedes number, several boundary conditions, including geometry, supply and return air openings, different sources and sinks of heat load, including their strength and location, enclosure surface temperature, have influence on the air distribution in spaces. It can be very complicated to describe all details of the boundary conditions because they are individual for different rooms.

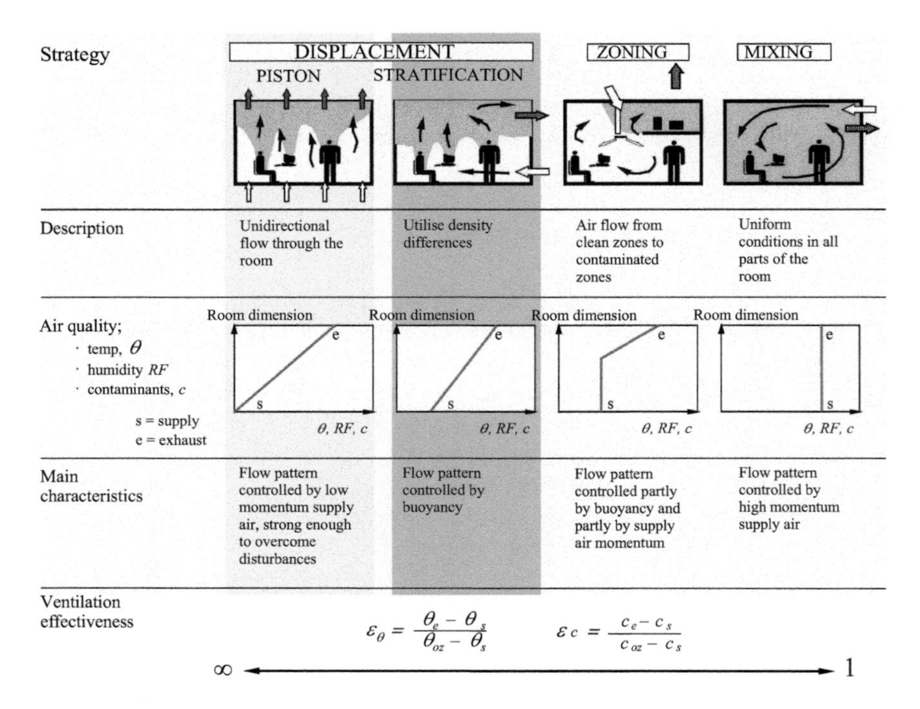

Figure 6.1 The ideal performance of the total volume room air distribution principles

But a few primary and common parameters, which are the important ones, shall be considered. These primary variables are as follows:

- Cooling or heating mode.
- Archimedes ratio $\Delta T_o/q_o^2$.
- The ratio between the total area of the supply openings and the wall/ceiling/floor area, a_o/A.
- Location of the air supply opening(s): close to the ceiling (high) or close to floor level (low).

The ratio between the total area a_o of the air supply openings and the surface area A of wall/ceiling/floor on/in which the supply openings are located, a_o/A, is an important parameter for the air distribution in the room. The ratio, a_o/A, is considered to be small for values smaller than 10^{-3}, medium for values between 10^{-3} and 0.3 and large for values larger than 0.3. The values smaller than 10^{-3} are typical for diffusers designed for mixing ventilation, and the value 6×10^{-3} is typical for displacement ventilation diffusers.

Figure 6.2 shows a principal scheme of the Archimedes ratio (as a function of q_s and $\Delta\theta_s$) for a constant value of a_o/A. The area on the right side of the curve defines momentum-driven flow while on the left side defines a flow driven by the buoyancy forces (Nielsen, 2011). The curve indicates the position of the critical Archimedes number where the air movement changes between the two different flow types. In practice, the convective flow is dominating with mixing ventilation, when the cooling power required is typically higher than 40–50 W/floor-m^2.

Buoyancy will influence the flow in a jet if air is supplied with a temperature T_o which is different from the return temperature T_e from the room. The flow is

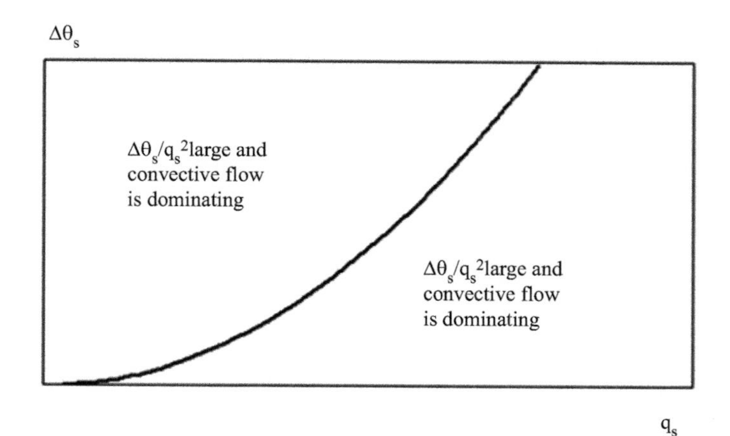

Figure 6.2 Principle determination of airflow in a room based on the critical Archimedes ratio. Convective flow is dominant on the left side of the graph while inlet momentum flow is dominant on the right

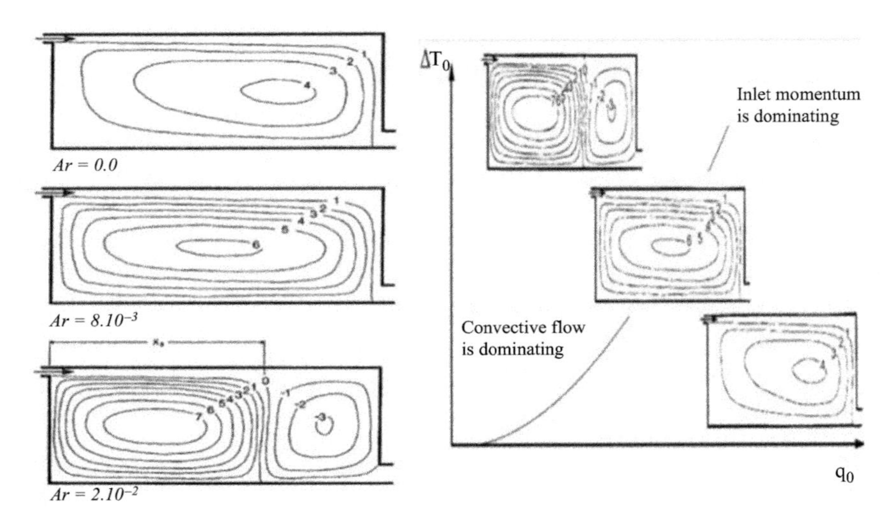

Figure 6.3 Left: streamline distribution in a room at different Archimedes numbers. Increasing load at floor results in an increasing air velocity and reduction of the penetration depth of the supply jet. Right: the effect of the critical Archimedes number

characterized by the Archimedes number *Ar*. Figure 6.3 (left) shows the two-dimensional flow in a room with a full-width slot. The room length is three times the room height *H* and the slot height $_o$ is 0.017*H*. The sketches show the change in streamline distribution in a room at increasing Archimedes number, and although the flow is two dimensional, the results may also give a qualitative expression of a general three-dimensional flow at increasing thermal load.

The streamlines in Figure 6.3 (left) contain an indirect statement of the velocity (Nielsen *et al.*, 1979). The flow between the two streamlines may be considered constant and the velocity is thus high where the streamlines are close and low where the distance between the streamlines is greater. The streamlines are normalized by the supply flow. A maximum value of, for example, four in the room means therefore a recirculating flow in the whole room which is four times the supply flow.

The flow in the upper sketch in Figure 6.3 (left) is isothermal and governed by the supply momentum flux. A heated floor will add a buoyancy force to the flow and increase the amount of recirculating flow in the room. The amount of recirculation is defined by the level of the Archimedes number *Ar*.

The heat load has reached such a level in the lowest sketch in Figure 6.3 (left) that the supply jet will separate from the ceiling at a distance from the diffuser and flow down into the occupied zone.

A further increase in temperature difference or in Archimedes number may result in a change in the direction of recirculation with the cold jet running down the wall below the supply opening. This effect of the critical *Ar* number is presented in Figure 6.3 (right). A very short penetration depth is undesirable because the jet may have a high velocity and a low temperature when it flows into the occupied zone.

6.2 Mixing ventilation

Mixing room air distribution aims for diluting of polluted and warm/cool room air with cleaner and cooler/warmer supply air. The air is supplied to the room with high initial mean velocity and the established velocity gradients generate high turbulence intensity aiming to promote good mixing and uniform temperature and pollution distribution in the occupied zone.

Mixing ventilation is an expression for an air distribution pattern, and not for a ventilation system. It can also be called an air distribution pattern with mixing effect or mixing air distribution. Mixing ventilation is traditionally considered to be the air distribution which is obtained by the use of diffusers with a high momentum supply flow.

6.2.1 Diffuser for mixing air distribution

The main objective of air supply diffusers in mixing ventilation is to dilute room air effectively with a supply of clean and conditioned air and at the same time to establish low velocity conditions in the occupied zone. Air diffusion may be achieved with high or low momentum air flux diffusers. The jet of high momentum flux diffusers dominates air distribution and they are strong enough to mix the air in the whole volume, whereas with low momentum flux diffusers, the driving force of the mixing is mainly buoyant force by heat sources.

Air terminal devices can be divided into four broad classes:

Class I devices from which the jet is essentially three dimensional (e.g. conical):

- nozzles,
- grilles and registers,
- ceiling diffusers with vertical discharge.

Class II devices from which the jet flows radially along a surface or as a free jet:

- ceiling diffusers.

Class III devices from which the jet is essentially two dimensional:

- linear grilles,
- slots and linear diffusers.

Class IV low velocity air terminal devices.

The other approach to classify air terminal devices is based on their structure and momentum flux. Diffusers with high momentum flux:

- perforated diffusers,
- conical diffusers,

- supply valves,
- multi-nozzle diffuser,
- linear slot diffusers,
- swirl diffusers,
- jet nozzles,
- grilles.

Diffusers with low momentum flux:

- textile terminals,
- floor diffusers.

6.2.1.1 Air diffusers with high momentum

Perforated air supply diffusers are usually circular or rectangle. Perforated diffusers can be installed flush to a suspended ceiling or can be exposed. Supply airflow rate is discharged partly or totally through the perforated front panel into room space. The diffuser is designed to deflect the flow over the ceiling in one to four directions depending on the application. A perforated diffuser is shown in Figure 6.4 (left).

Conical diffusers have a conical front panel with one or several slots. Conical diffusers can be installed either recessed into the ceiling or exposed. In some models, it is possible to control the throw automatically or to adjust it manually to be horizontal (typically in cooling mode) or vertical (typically in heating mode). A circular conical diffuser is shown in Figure 6.4 (right).

Supply valves are one-slot conical diffusers. The throw patterns of supply valves are radial (360°). In some models, it is possible to manually adjust the throw pattern in order to provide a horizontal or vertical supply air jet. The induction ratio of the supply valves is low. Thus, the throw pattern is relatively long. Figure 6.5 (left) shows an example of a supply valve.

With multi-nozzle diffusers, the throw pattern can be adjusted without affecting the pressure drop and the noise level. The adjustment is carried out by turning nozzles manually to the desired directions. The throw is possible to be directed to 1–4 directions or even swirl jet can be created. Multi-nozzle diffusers are either circular or rectangular. In Figure 6.5 (right) a circular multi-nozzle diffuser is shown.

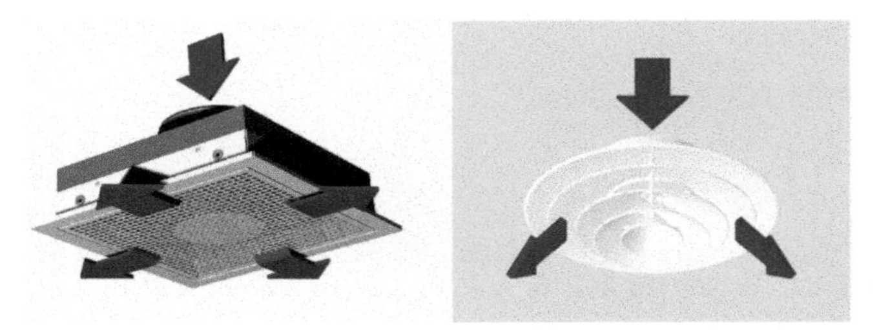

Figure 6.4 Left: rectangle perforated diffuser in ceiling installation. Supply air is discharged through the perforated front panel. Right: conical circular diffuser in exposed installation

Figure 6.5 Left: manually adjusted supply valve in ceiling installation. Right: multi-nozzle diffuser in ceiling installation

Figure 6.6 Swirl diffusers in ceiling installation. Left: adjustable compact throw pattern, right: radial throw pattern

In swirl diffusers, special blades create a rotating effect of the supply air inducing the room air effectively. High-induction swirl jet ensures efficient mixing and fast reduction of air velocity. In some models, it is possible to manually adjust or automatically to control the throw pattern in order to create a horizontal or vertical jet. The horizontal radial swirl jet is used mainly in cooling applications, whereas the vertical compact swirl jet is used in heating applications. In Figure 6.6, swirl diffuser with adjustable compact (for heating) and radial (for cooling) throw patterns are shown.

In linear slot diffusers, the supply air is discharged through one or several slots into the room space. Airflow is supplied either horizontally along the ceiling surface or vertically into the occupied zone. In some models, the throw pattern can be adjusted in one or two directions. Linear diffuser is also possible to use in wall installation where air is typically supplied horizontally along the ceiling surface. Slot diffusers can be installed in continuous rows in the ceiling or in the wall. A linear slot diffuser is shown in Figure 6.7 (left).

Supply grilles can be classified based on their structure to four categories:

- fixed vanes without throw pattern adjustment,
- adjustable horizontal vanes (adjustment of the jet inclination angle possible),
- fixed front vanes and adjustable rear vanes (with lateral adjustment possibility of the jet),
- adjustable vertical front vanes and adjustable horizontal rear vanes.

The induction ratio of grilles is low and thus throw length is long. The induction ratio describes the effectiveness of the discharged flow to draw

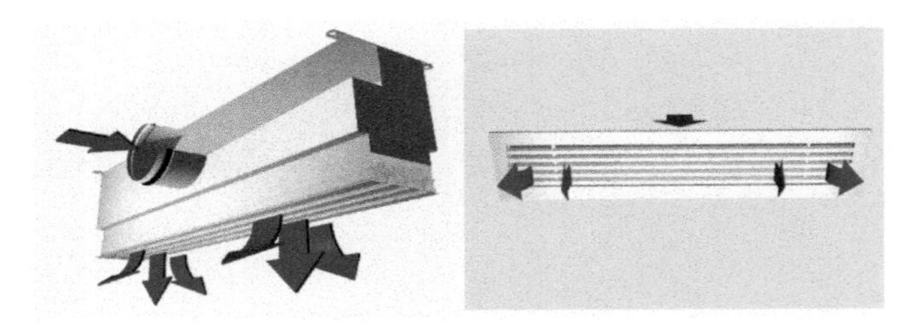

Figure 6.7 Left: linear slot diffuser in ceiling installation. Right: grille in wall installation

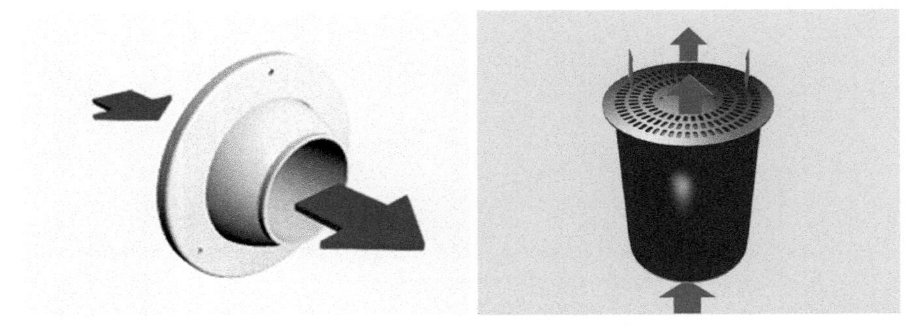

Figure 6.8 Left: nozzle diffuser in wall installation. Right: round floor supply unit in floor plenum installation

surrounding air into the jet. It is defined as the volume flow of induced room air divided by the volume flow of primary supplied flow. The supply grilles are installed in wall, ceiling or floor. A wall grille with adjustable rear vanes is shown in Figure 6.7 (right).

The initial air velocity of jet nozzles is high (>5 m/s). The compact jet of a nozzle is narrow and long. Typical applications of nozzles are industrial buildings, arenas, convention centres, etc., where long thrown patter is required. In some models, the throw length and pattern can be adjusted. In some models, it is possible to adjust manually or automatically the inclination angle of the supplied jet. A nozzle jet diffuser is shown in Figure 6.8 (left).

Floor supply units are used in displacement ventilation applications. However, in many cases in commercial buildings with ceiling height of about 3 m, the requested airflow rate is high and the thrown of diffuser is close to upper zone (1.8 m from floor level) or even exceeds it. Thus, the performance of the system is like a mixing ventilation system. Typical supply units of floor supply system are floor grilles and round supply units. In Figure 6.8 (right), a round floor supply unit is shown.

6.2.1.2 Air diffusers with low momentum

Mixing ventilation can be also implemented with low momentum supply units which perform together with buoyant forces of heat gains. In textile duct terminals (Figure 6.9),

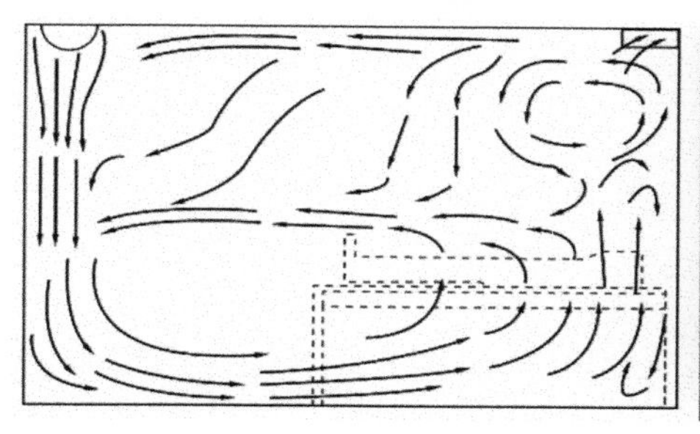

Figure 6.9 Left: thrown pattern of low impulse textile terminal. Right: ceiling mounted textile terminal for room air distribution

the room air is effectively mixed and supply air velocity is rapidly decreased. The thermal buoyant flow from heat loads rise up, mix with the supplied low impulse jet and spread the air within the whole room volume. In some terminals, it is possible to select the supply sectors (upwards, sideways and downwards) depending on the requested cooling capacity and the height of the space.

Air supplied from textile duct terminals is either directly downwards or follows full-circular air distribution. There is a risk of high velocity when cold air is supplied directly downwards. The location of the installed air supply terminal devices should be carefully analysed in order to minimize draught risk.

6.2.2 Turbulent free and wall jets

Several models are available for studying room air distribution at different stages of the design process. Simplified models such as flow elements are easy to use at the early stage. Flow elements are volumes or regions of the room flow where the air movement can be fully described by upstream and local conditions. An isothermal free jet is a typical example that can be described by upstream conditions such as the supply opening geometry and supply velocity. The flow in a room ventilated by mixing ventilation is recirculating.

The flow elements may be influenced by this recirculating flow and to some extent dependent on downstream conditions. The non-deflecting isothermal jet flow can be divided into a number of different basic types according to the boundary conditions and aerodynamics of the flow. Figure 6.10 is a representation of three-dimensional, two-dimensional, radial and swirling flows.

A jet supplied from a circular opening flowing into an, in principle, infinite space is called a circular free jet or a three-dimensional free jet. If the flow is supplied to a space with a wall parallel to the flow (half infinite space) a three-dimensional wall jet will be formed as indicated in Figure 6.10(b). The wall jet is called three-dimensional because it has a growth rate parallel to the wall, which is larger than the growth rate perpendicular to the wall. The expression 'wall' is any surface, e.g. a ceiling, floor or room wall.

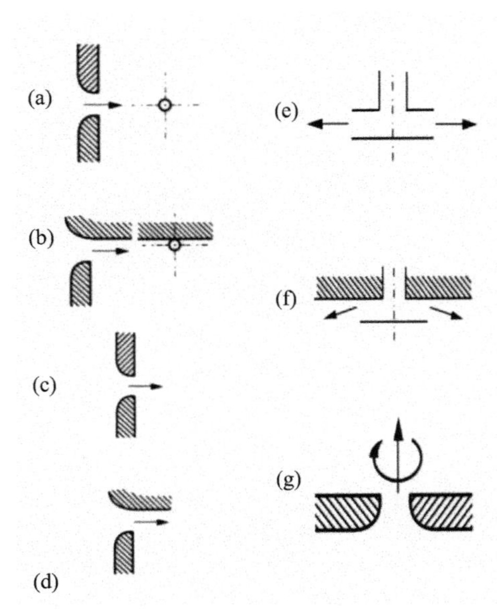

Figure 6.10 Basic types of jets for room air distribution: (a) three-dimensional free jet, (b) three-dimensional wall jet, (c) two-dimensional free jet, (d) two-dimensional wall jet, (e) radial free jet, (f) radial wall jet, (g) free jet with swirl

Two-dimensional jets or plane jets are jets which are, in principle, infinite in width. A two-dimensional free jet is a jet flowing into an infinite space (Figure 6.10(c)), and a two-dimensional wall jet (Figure 6.10(d)) is a jet flowing into a half infinite space. A two-dimensional jet flow can also be preserved if the width of the flow is limited to side walls in the whole flow regime.

Radial jets are jets which flow in a radial pattern away from the diffuser. They can also be divided into the free jet type and the wall jet type (Figure 6.10(e) and (f)). The last flow element in Figure 6.10(g) is a jet with swirl. The swirl is generated in the opening and is preserved in the forward flow of the jet. The jet is considered to be a free jet.

Air movement in a room may often be divided into more types of flow elements which can be treated independently of the surrounding flow and dimensions. Sketches (a)–(c) in Figure 6.11 show three kinds of supply jets (free jet, wall jet and a free jet changing to a wall jet), while the sketches (d) and (e) show disturbance of the jets as a deflection at an end wall or floor, and a flow around an obstacle. The flow close to an exhaust opening is indicated in sketch (f). The non-isothermal flow elements for a free jet with deflection due to gravity is shown in sketch (g), and a wall jet with restricted penetration depth in sketch (h).

Figure 6.12 shows a circular opening with a diameter d_o and a supply velocity u_o. There is an area immediately outside the opening which is called the constant

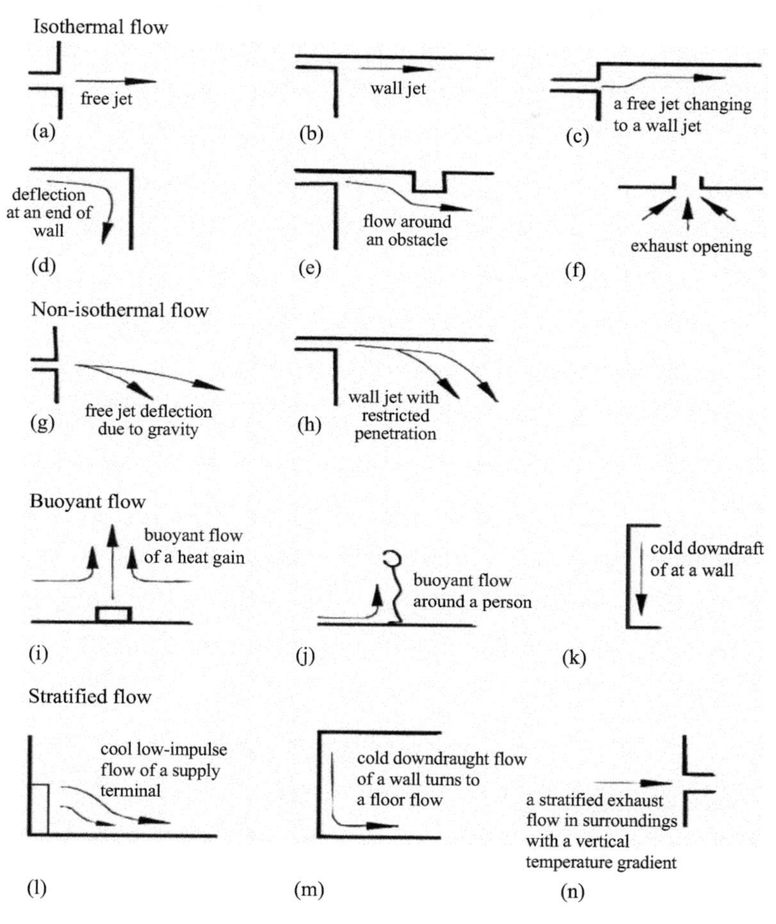

Figure 6.11 (a)–(h) Flow elements in room air movement

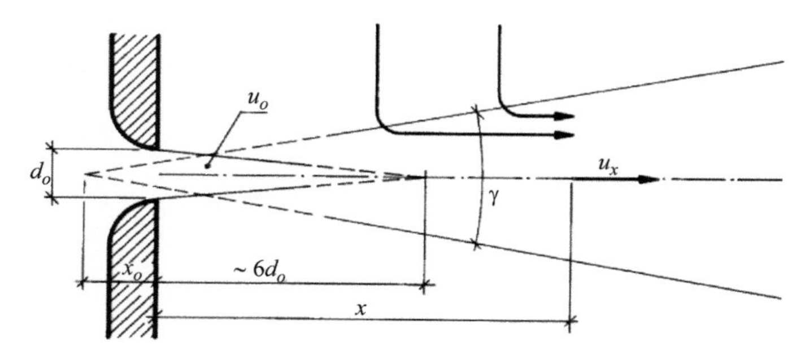

Figure 6.12 The airflow close to a circular opening

velocity core with velocity u_o. A turbulent mixing layer is formed between the core and the surrounding air. This mixing layer grows into the core at increasing distance from the opening, and the constant velocity core disappears at a distance of at ~$6d_o$. The jet will continue to entrain air from the surroundings in downstream direction, and the width of the jet increases proportional to the distance from a point called the virtual origin. This origin is in many situations located so close to the opening that the distance x_o is ignored in comparison with the distance x. The angle γ is ~24° for circular jets and ~33° for plane jets if the flow is fully turbulent. x_o is the virtual origin of the jet.

It is important to note that the development of the flow depends on the initial conditions at the supply diffuser. For example, an increase of the initial turbulence intensity of the supplied flow will reduce the length of the core.

The presence of a wall in the symmetry line – or symmetry plane – has only a small influence on the flow and the velocity profile. The boundary layer at the surface is thin compared to the turbulent mixing layer towards the free room. The flow described in Figure 6.12 does not change much when a surface is located in the symmetry line. A wall jet has theoretically a length of the constant velocity core which is up to ~$12h_o$, where h_o is the height of the opening, and the angle of the growth of the wall jet is approximately ~$\gamma/2$. x_o is the location of the virtual origin.

The width of a circular jet increases linearly to the distance $x+x_o$, which means that the area of the jet increases with the square root of the distance. The momentum flow is therefore preserved when the maximum velocity on the jet axis, u_x, is inversely proportional to the distance $x+x_o$:

$$\frac{u_x}{u_o} = \frac{K_a}{\sqrt{2}} \frac{\sqrt{a_o}}{x + x_o} \tag{6.1}$$

Equation (6.1) expresses that the velocity ratio u_x/u_o in the symmetry line is inversely proportional to the distance $(x + x_o)/\sqrt{a_o}$, where a_o is the area of the supply opening. K_a is constant in the case of high turbulent flow and it is individual for the different types of diffusers. It has a maximum level of ~10 for a nozzle with a contraction as the one shown in Figure 6.12.

The flow in a three-dimensional wall jet with a supply area of a_o is in practice identical to a flow in a free jet with a supply area of $2a_o$ (the opening and its mirror image). The velocity distribution in a three-dimensional wall jet is therefore given by the following expression:

$$\frac{u_x}{u_o} = K_a \frac{\sqrt{a_o}}{x + x_o} \tag{6.2}$$

where u_x is the maximum velocity in the flow close to the surface.

The velocity distribution in a radial free jet is given by the following expression:

$$\frac{u_r}{u_o} = \frac{K_r}{\sqrt{2}} \frac{\sqrt{a_o}}{r + r_o} \tag{6.3}$$

where K_r is an individual constant for different types of diffusers; r is the radial distance from the rim of the diffuser and r_o is the distance to the virtual origin of the radial jet. The maximum value which can be obtained for K_r is approximately 1.2. The velocity distribution in a radial wall jet is given by (method of images)

$$\frac{u_r}{u_o} = K_r \frac{\sqrt{a_o}}{r + r_o} \tag{6.4}$$

The width or the cross-sectional area of a plane (two dimensional) flow increases linearly to the distance $x+x_o$. The momentum flow is therefore preserved when the velocity is inversely proportional to the square root of the distance $x+x_o$. The velocity distribution is described by the following expression:

$$\frac{u_x}{u_o} = \frac{K_p}{\sqrt{2}} \sqrt{\frac{h_o}{x + x_o}} \tag{6.5}$$

where u_x/u_o is the velocity ratio in the symmetry line and $(x+x_o)/h_o$ is the distance from the virtual origin; h_o is the equivalent height of the supply slot and K_p is a constant. The maximum value which can be obtained for a slot with a contraction is K_p~4.

The velocity distribution in a plane wall jet is given by (method of images)

$$\frac{u_x}{u_o} = K_p \sqrt{\frac{h_o}{x + x_o}} \tag{6.6}$$

The buoyancy will give an increasing effect on the velocity level in a vertical jet with increasing Archimedes number and with increasing distance $y/\sqrt{a_o}$ from the opening. REHVA Guidebook (Mueller *et al.*, 2013) has shown the following correlation for a nozzle:

$$\frac{v_y}{u_o} = \frac{K_a}{\sqrt{2}} \frac{\sqrt{a_o}}{y} \cdot \left(1 \pm 2.7 \frac{\text{Ar}}{K_a} \left(\frac{y}{\sqrt{a_o}} \right)^2 \right)^{0.33} \tag{6.7}$$

where v_y is the maximum velocity in the vertical centre line of the flow and y is the distance from the opening. The distance to the virtual origin y_o is assumed to be zero. A downward-directed chilled jet will obtain a higher velocity level than an isothermal jet and a heated jet will decrease to lower velocities. The air movement will stop at a distance y_m in the case of a heated jet which corresponds to $v_y = 0$ in (6.7).

Ar number is calculated with the following equation:

$$Ar = \frac{\beta g h_o (T_e - T_o)}{u_o^2} \tag{6.8}$$

A non-isothermal free jet with a horizontal supply velocity will have a trajectory which is turned downward by gravity force in the case of a cool supply

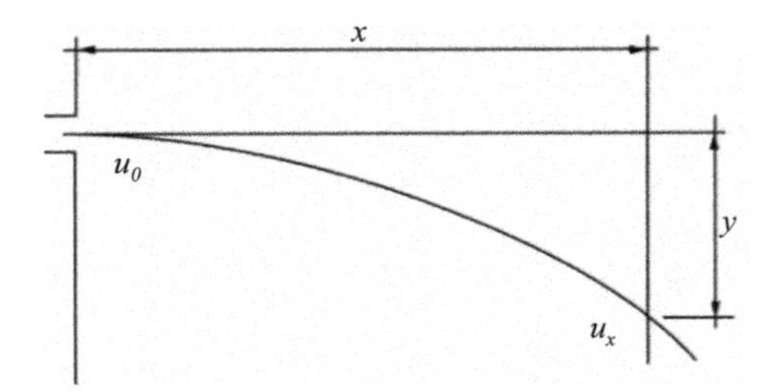

Figure 6.13 The airflow close to a circular opening

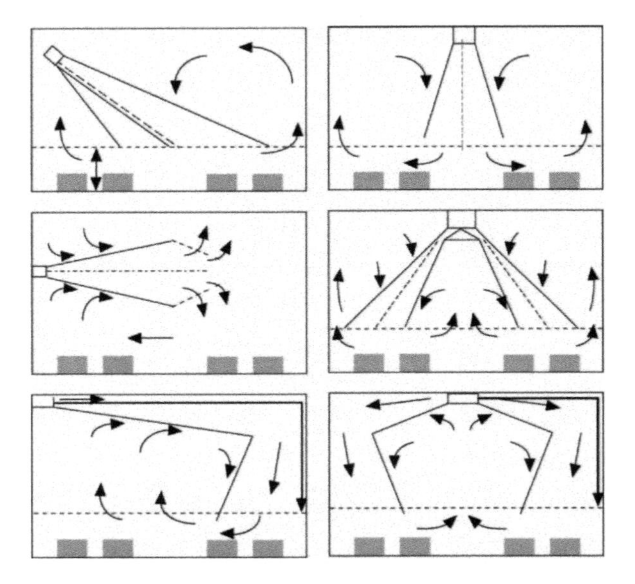

Figure 6.14 Typical air distribution schemes in large enclosures

temperature and turned upward by buoyancy in the case of a warm supply temperature, see Figure 6.13.

REHVA Guidebook (Mueller *et al.*, 2013) gives the following equation:

$$\frac{y}{\sqrt{a_o}} = 0.6 \cdot \frac{\mathrm{Ar}}{K_a} \cdot \left(\frac{x}{\sqrt{a_o}}\right)^3 \tag{6.9}$$

6.2.3 Design of mixing ventilation

6.2.3.1 Selection of air distribution method

Figure 6.14 presents some typical air distribution schemes for applications in large enclosures with high ceilings. By using high initial velocity and momentum flux, it

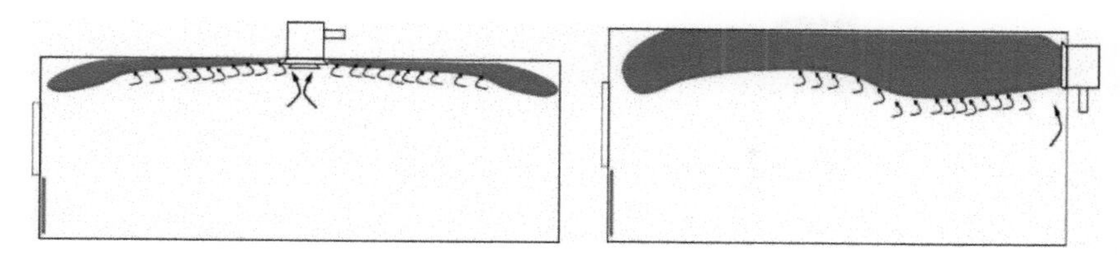

Figure 6.15 Left: ceiling supply air distribution method. Right: wall supply air distribution method

is possible to guarantee the required mixing in the occupied zone of large enclosures. Often air supply devices, generating three-dimensional airflow supplied directly to the occupied area, are used.

In applications where the ceiling height is about 3 m or less, it is practical to utilize surfaces (Coandǎ effect of ceiling and walls) for the installation of air supply diffusers in order to guarantee good mixing and low air velocities in the occupied zone. Typical air distribution design methods in commercial buildings with low ceiling height are ceiling supply (Figure 6.15 (left)) and wall supply (Figure 6.15 (right)).

Characteristics of the ceiling supply scheme are as follows:

- Suspended ceiling (exposed installation is also possible).
- High induction rate with short throw length in order to obtain high cooling capacity.
- Air distribution may be influenced by high heat gains like warm windows.
- Throw pattern control is needed to ensure good performance in heating mode and to prevent temperature gradient.

Characteristics of the wall supply scheme are as follows:

- During warm periods, thermal plumes may affect the performance causing early jet detachment and draught.
- Not suitable in spaces with high cooling loads.
- During cold periods, high velocities close to the floor can exist.
- In heating mode, continuous heating below window is required in order to avoid draught risk.

6.2.3.2 Selection of an air diffuser

When the diffuser type, number and location are selected, the following aspects should be considered:

- thrown length of the supplied flow,
- operation flow range of a diffuser,
- dimension of the room,
- possible location for a diffuser,
- need of thrown pattern control,
- internal obstacles of the flow,

- restrictions for installation,
- architectural aspects.

When the type and the location of diffuser are selected, the following aspects should be taken into account:

- The total airflow rate is distributed equally in the room space without draught. A sufficient number of the supply units should be used to ensure uniform air distribution.
- The throw length is analysed in cooling mode condition by using manufactures product data.
- The possibility of opposite directed jets and the interaction of jets and thermal plumes should be studied.
- Short circuit of supply air directly towards exhaust should be prevented, especially in heating mode.
- The supplied airflow should not be directed towards obstacles, e.g. structural beams and light fittings.
- Changes in the use of the spaces require adjustments of the airflow rates and throw pattern of the diffusers.
- Noise level should be according to the standards used during the design.

The throw should be adjusted to the room geometry. If the throw is too long, the jet can be deflected into the occupied zone. On the other hand in non-isothermal conditions, the throw can be too short. In this case, it is possible that the jet detaches from a ceiling and enters into occupied zone. In Figure 6.16, the situations with too short and too long throw length that increases draught risk (e.g. at the feet) are shown.

In practice, air distribution in a room will often be designed by product-related software or product-related design graphs. Figure 6.17 shows a typical design graph. The total pressure difference over the diffuser Δs_t is given as a function of the diffuser size a_o and volume flow q_o. The sound pressure level L_p and the throw length l_{02} curves shown in the graph together with the supply flow rate q_o and the pressure difference Δa_t are used for the selection of the diffuser.

The selection of the location and the number of diffusers, together with airflow rate and supply air temperature, has a significant effect on the air distribution. In Figure 6.18, there are presented examples of appropriate and impropriate air distribution design.

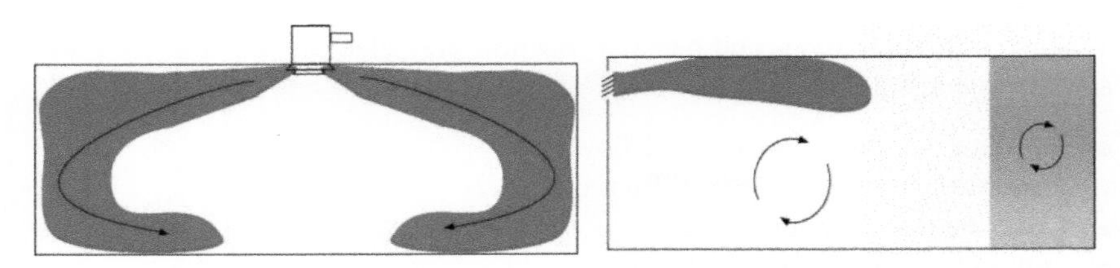

Figure 6.16 Design examples of air supply jets with too long (left) or too short (right) throw

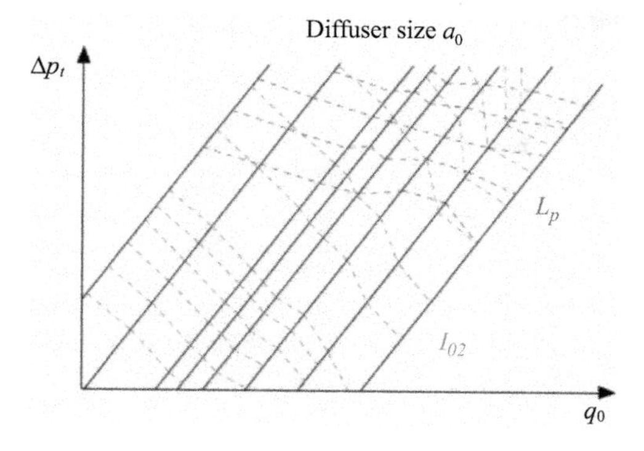

Figure 6.17 Graph for selecting a diffuser in a room with mixing air distribution

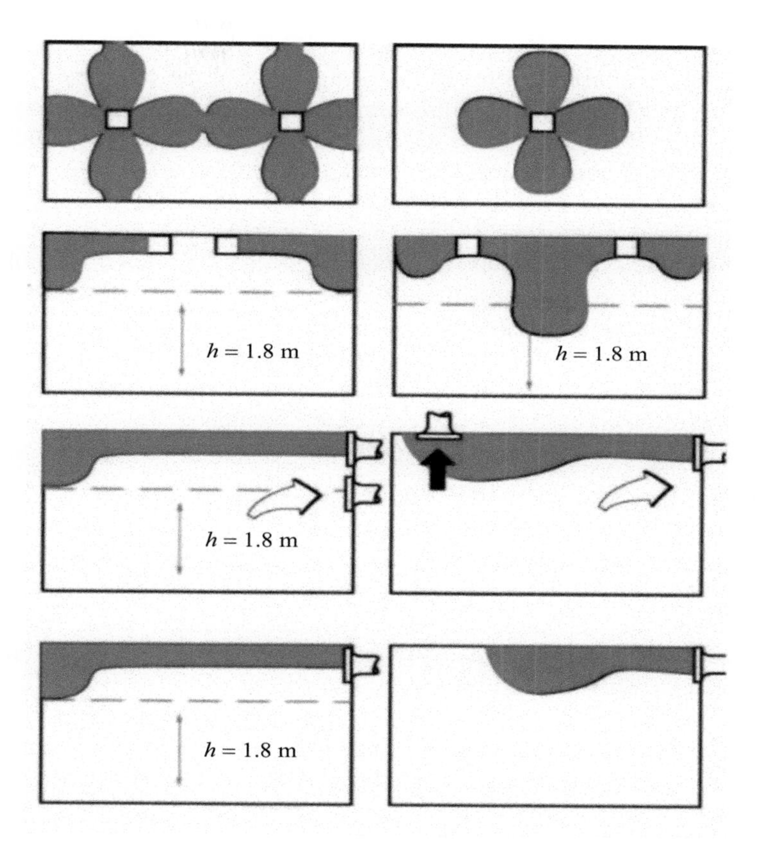

Figure 6.18 Examples of appropriate (left) and impropriate (right) air distribution schemes in isothermal conditions. Throw boundaries for a maximum velocity of 0.25 m/s are indicated (in blue)

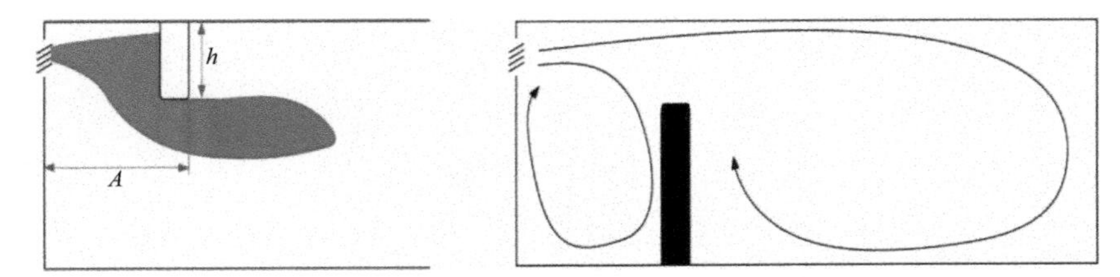

Figure 6.19 Left: the deflection of the supplied jet depends of the distance of the obstacle from the diffuser, the height of the obstacle and the temperature difference. Right: effect of partition on air distribution in room

Special attention should be paid on the effect of obstacles located on the ceiling that can deflect the jet into the occupied zone (Figure 6.19 (left)). A detached jet significantly increases the draught risk in the occupied zone. Also, flow obstacles (like partition walls) can affect in an unpredictable way the air distribution in the occupied zone and may decrease the ventilation efficiency in certain areas of the occupied zone (Figure 6.19 (right)).

In demand-based controlled ventilation, airflow rates vary according to the need. The variable volume flow systems are operated by one or more volume control units and should be combined with an intelligent fan speed control system preventing high pressure drops. A wide airflow control range can lead to draught problems at the minimum airflow rates (Figure 6.20). In order to avoid draught problems, it is important to maintain an appropriate throw pattern both at the minimum and the maximum supply airflow rates. The performance of a diffuser should always be checked over the required airflow range.

6.3 Displacement ventilation

In displacement ventilation, a stratified flow is created using the buoyancy forces in the room. The air quality in the occupied zone is then generally better than with mixing ventilation. Displacement ventilation has for many years been used in industrial premises with high thermal loads. Since the mid-1980s, it has also been used more extensively in non-industrial premises.

Displacement ventilation presents the opportunity to improve both the temperature effectiveness and the ventilation effectiveness. Displacement ventilation is considered to be the technique of supplying clean, cool air at floor level, letting warm air and contaminants rise to the ceiling and extracting the contaminated air at ceiling level.

Displacement ventilation is primarily a means of obtaining good air quality in occupied spaces that have a cooling demand. It has proved to be a good solution for the following:

- Gyms.
- Meeting rooms.
- Classrooms.
- Tall rooms: Convention centres, lobbies, sport arenas, auditoriums, theatres, museums, airports, shopping centres, etc.

6.3.1 Performance of displacement ventilation

The principle of displacement ventilation is based on air density differences where the room air separates into two layers: an upper polluted zone and a lower clean zone (Figure 6.21). This is achieved by supplying cool air with a low velocity in the lower zone and extracting the air in the upper zone. Free convection from heat sources creates vertical air movement in the room. When the heat sources in the

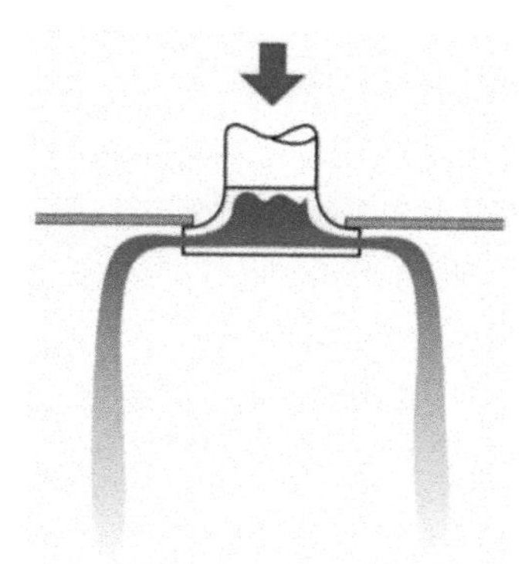

Figure 6.20 Performance of a diffuser at the minimum supply flow rate when the initial air velocity is too low to prevent detachment and drop of the supplied cold air

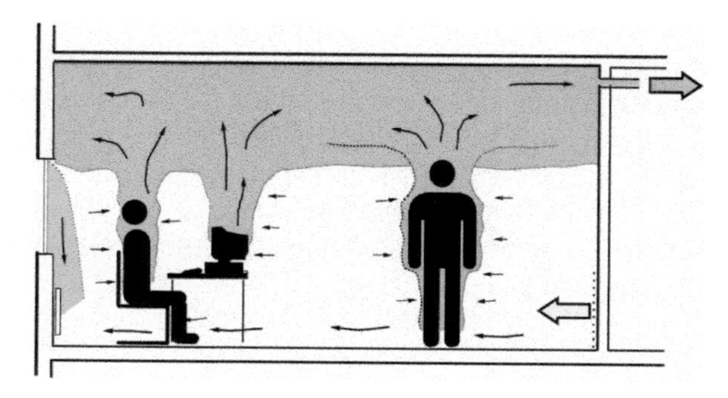

Figure 6.21 Schematic illustration of the airflow in a room ventilated by displacement ventilation

room are also the contamination sources, the convection flows transport the warm polluted air up to the upper zone.

The convection flow rates relative to the ventilation flow rate determine the height of the boundary between the two zones. The sum of the warm convection flow rates to the upper zone minus the downward directed flows from cold surfaces to the lower zone is equal to the ventilation airflow rate supplied to the room. An increased ventilation flow rate at fixed convection flow rates thus moves the boundary upwards and a decreased flow rate moves the boundary downwards.

The vertical air movement is caused by convection flows from warm sources or cold sinks. Warm objects such as people, computers, lamps and warm surfaces create rising convection flows. Depending on the power and geometry of the heat source, the convection flows will rise all the way to the ceiling or settle at a lower height.

Two principal methods can be used when the supply airflow rate of displacement ventilation system is calculated: (1) air-quality-based design, where the design criterion is the air quality in the occupied zone and (2) temperature-based design, where the design criterion is the air temperature in the occupied zone of the room. In Section 6.3.1.1, there is presented the main principle of air-quality-based design and in Section 6.3.1.2, there is shown the principle of temperature-based design, respectively.

6.3.1.1 Contamination distribution

Natural convection flows are the engines of displacement ventilation. A natural convection flow is the air current that rises above warm objects like people or computers, rises along a warm wall or descends from cold objects like windows or outer walls, due to buoyancy.

The convection flow rising above a hot object, including the human body, is called a thermal plume, or simply a plume. Empirical, analytical and computational fluid dynamics are commonly used methods to evaluate air temperatures, velocities and airflow rates in thermal plumes above different heat sources and convection flows at vertical surfaces.

The amount of air in the convection flow increases with height due to entrainment of the surrounding air. The amount of air transported in a natural convection flow depends on the temperature and the geometry of the source and the temperature of the surrounding air. As the driving force in convection flows is the buoyancy force caused by the density difference (i.e. the temperature difference), a temperature gradient in the room influences the plume rise height.

Thermal plumes are normally modelled as a point source. Analytical equations to calculate velocities, temperatures and airflow rates in thermal plumes over point source with given heat loads were derived based on the momentum and energy conservation equations and assuming Gaussian velocity and excessive temperature distribution in thermal plume cross-sections (Kosonen *et al.*, 2017).

However, in reality, heat sources are seldom a point. The most common approach to account for the real-source dimensions is to use a virtual source from which the airflow rates are calculated (Kosonen *et al.*, 2017) (Figure 6.22). The

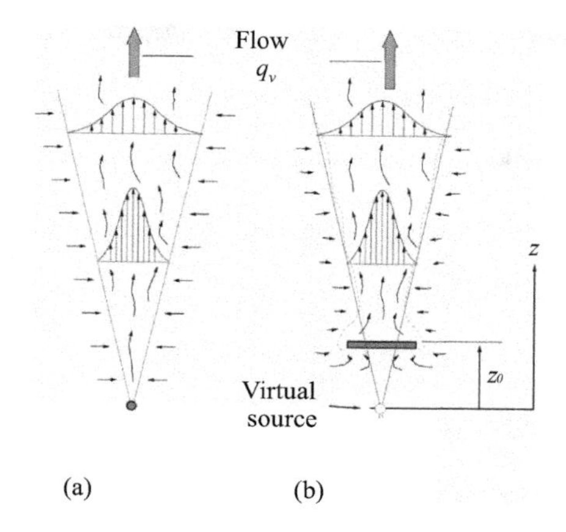

Figure 6.22 Thermal plume of the modelled point source by using the virtual origin: (a) point source and (b) extended source

virtual origin is located along the plume axis at a distance z_0 on the other side of the real-source surface.

The required airflow rate q_v [L/s] to rise up the convection flow (pollution) at the certain height is calculated with the following equation:

$$q_v = 5{\Phi_{cf}}^{1/3}(z + z_0)^{5/3} \tag{6.10}$$

where Φ_{cf} is the convective heat flux [W] from the heat source, z is the height above the level [m] of the heat source and z_0 is the distance of the virtual origin. For a flat heat source, REHVA Design Guide suggests the position of the virtual source to be located at $z_0 = 1.7-2.1D$ below the real source, where D is the hydraulic diameter of the source. D is calculated for the rectangular source (sizes $a \times b$) with $D = (2ab)/(a+b)$.

6.3.1.2 Vertical temperature distribution

In commercial buildings, the removal of the excess heat is likely to be the main concern. The cases where cooling is the main issue, the temperature-based design is the most commonly applied method. In the design process, the challenging task is to estimate vertical contaminant or temperature gradients in the room space.

While the contaminant stratification level is mainly affected by the relation of supply airflow rate and convective airflow rate, thermal stratification is also affected by thermal radiation exchange between different room surfaces. The thermal radiation from upper level surfaces warms lower level surfaces and thus affects the air temperature at floor level and in the occupied zone.

The vertical temperature distribution depends on the vertical location of the heat sources. When the heat sources are in the lower part of the room, the temperature gradient is larger in the lower part whereas the temperature gradient is more constant in the upper part. On the other hand, when the heat sources are located mostly in the upper zone, the temperature gradient is smaller in the lower part and increases in the upper part.

The type and location of the source have a significant effect on the relative temperature difference. Point sources and horizontal sources (warm floor) create a clear mixing layer.

In displacement ventilation, the air temperature increases from floor to ceiling. This means that the occupied zone is the coolest part of the room. In many cases, the vertical temperature gradient is assumed to be linear between the floor and ceiling level. However, it is valid only with some type of heat sources. Vertical temperature profiles measured with different individual types of heat load (occupants, warm floor, warm window and warm ceiling) are shown in Figure 6.23 (Kosonen *et al.*, 2016). With the load dominated by occupants or by a warm floor, the data indicates an obvious mixing layer. A two-layer structure is generated with heat and pollutants accumulated in the upper part of the room. Across the mixing layer, the room air temperature could be assumed to be constant or exhibiting just a slight increase. The warm window produces a near linear temperature profile with no clear two-layer structure. With the heated ceiling, the convection heat remains mainly in the upper portion of the room.

With displacement, the supply air temperature is typically about 3K–5K cooler than the room air temperature at a height of 1.1 m. In areas where people are

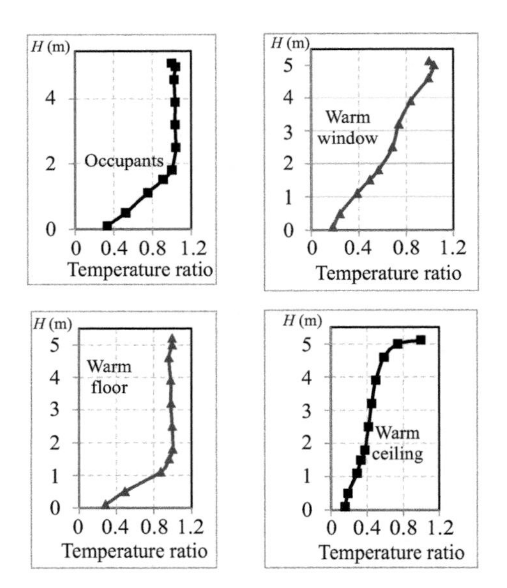

Figure 6.23 Vertical temperature profiles in room with displacement ventilation with different heat loads (temperature ratio = $(T-T_{supply})/(T_{exhaust}-T_{supply})$)

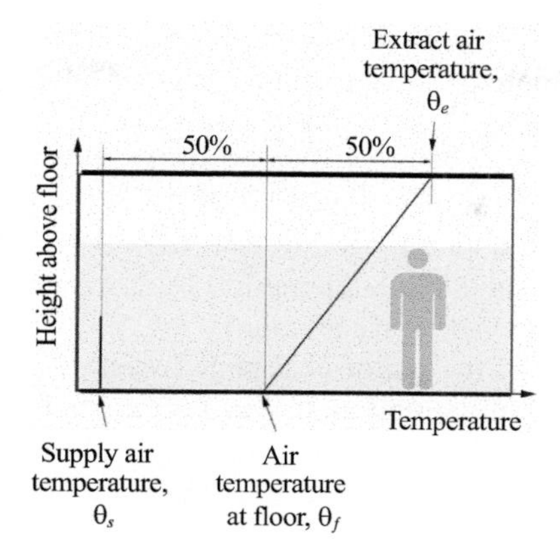

Figure 6.24 The '50%-rule' for vertical temperature distribution

moving, for example in shopping centres, the supply air could be 6–8K lower than the room air temperature. Depending on the particular design, the temperature difference between the supply and exhaust air is typically between 4 and 10K.

For most practical purposes, temperature profiles are assumed as shown in Figure 6.24. The '50%-rule' for the vertical temperature distribution indicates that the air temperature at the floor is half-way between the supply air temperature and the extract air temperature. This is a general experience that may be used as a first approximation for most normal rooms and normal air diffusers.

In rooms with higher ceilings than normal, it is often found that the temperature increase from supply air temperature to that of the air at the floor is less than 50% of the total temperature increase. In these cases, a '33%-rule' may be appropriate.

A calculation method should be used that is suitably accurate for the particular heat loads. Dynamic energy simulation should be used in order to accurately estimate the cooling load. Simplified steady-state calculations typically overestimate the actual demand and cannot take into account the effect of thermal mass on room air temperature. To get more accurate estimation for the vertical temperature gradient, multi-nodal models (Lastovets *et al.*, 2020). In demanding design cases, a full-scale mock-up is recommended together with CFD.

6.3.2 Diffusers for displacement ventilation

There are several types of diffusers used for displacement ventilation. The most commonly used types are integrated in the walls. Other types are placed at the walls or in a corner, free-standing on the floor, or integrated in the floor. The layout of the room should be considered in connection with the selection of the type of air diffusers.

The air is supplied in the lower part of the room and it is, therefore, convenient to locate the diffusers, near to corridors and other unoccupied areas to obtain an area in front of the diffusers where a higher velocity can be tolerated. Figure 6.25 shows circular and semi-circular diffuser.

Figure 6.26 (left) shows a wall-mounted diffuser with integral nozzles for the adjustment of the supplied flow pattern. The flow close to the diffuser can be directed parallel to the wall with only a small amount of forward directed flow. Figure 6.26 (right) shows a floor-mounted diffuser with swirl effect.

The air from a single wall-mounted diffuser flows over the floor and generates radial flow. Close to the air supply diffuser, there is a zone where the flow has relatively high velocity and low temperature. In this zone, the risk of draught may increase. This zone is called 'adjacent zone'.

Avoiding draught is the major challenge in selecting low-velocity air diffusers and the location of the supply unit. Most draught problems reported in rooms with displacement ventilation are due to high velocity in the zone adjacent to the diffuser (Figure 6.27). It is important to choose a diffuser that is suited for the application and only utilize diffusers from manufacturers that supply robust documentation together with the products.

To reduce the size of the adjacent zone, the number of diffusers in the room or diffuser face area should be increased. This also generates a more homogenous indoor environment in the occupied zone. Different diffuser types may be used. Typically those that supply air in only one direction will generate a longer adjacent

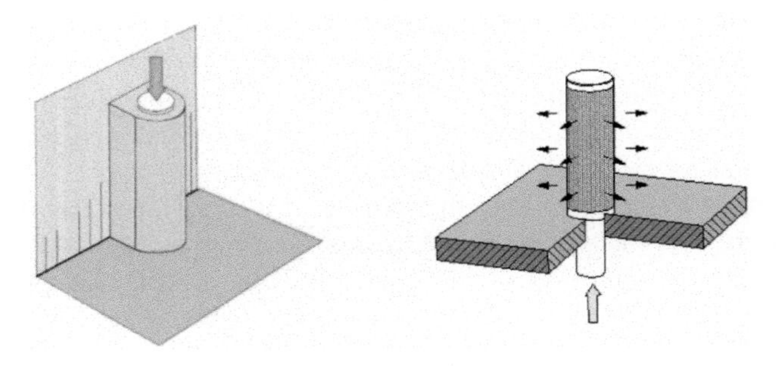

Figure 6.25 Left: semi-circular diffuser. Right: circular diffuser

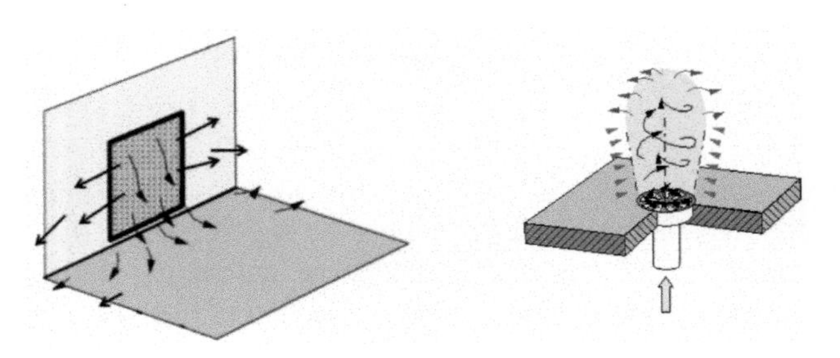

Figure 6.26 Left: wall-mounted low-velocity diffuser with integrated nozzles for the adjustment of the flow patterns. Right: floor-mounted diffuser with high-velocity swirl supply

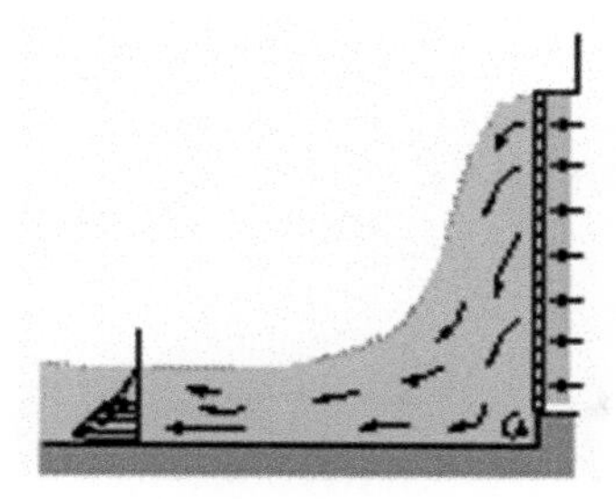

Figure 6.27 Diffuser has adjacent zone with high velocities where the risk of draught should be analysed

zone than the distributing supply air from semi-circular or circular diffusers. One way to reduce the draught in the occupied zone is to direct the supply air sideways to the wall outside the occupied zone.

6.3.3 Design of displacement ventilation

When the total required supply airflow rate is calculated (by using methods introduced in the previous sections), the diffuser type, number and locations could be preselected taking into account the following considerations:

- dimension of the room,
- the location of heat gains and/or pollutant sources,
- possible locations of units,
- internal obstacles of the flow,
- restriction of installation,
- architectural aspects.

Typical standard diffuser types are wall mounted, corner mounted, free standing and floor mounted. The diffusers require a certain wall area, or space in the floor or in the floor. With regard to the type and location of diffusers, the following aspects should be taken into account:

- The total airflow rate is distributed equally in the room space. A sufficient number of units should be used to ensure uniform air distribution.
- Special attention needs to be paid to the adjacent zone around diffuser.

Knowing the total airflow rate, it is possible to estimate the number of supply units required with Table 6.1 (Kosonen *et al.*, 2017). The table presents typical airflow rates and floor areas covered as a function of the nominal size of the supply unit.

Air distribution through large floor area is possible using low-velocity units. As a rule of thumb in open space layout, the maximum distance between supply units is 30 m (Figure 6.28). If the distance between the supply units is more than 30 m, an additional row of supply units between these units is needed.

If there are special requirements for interior design, supply units can be recessed in the structure. If is needed, the supply units could be covered with a

Table 6.1 Preselection of the number and size of supply unit

Nominal size (mm)	Airflow rate per unit (m^3/s)	Floor area per supply unit (m^2)
100	Up to 0.030	10–15
125	0.020–0.030	10–20
160	0.030–0.080	10–30
200	0.070–0.150	10–50
250	0.100–0.200	15–60
315	0.200–0.400	20–70
400	0.250–0.500	30–100
500	0.400–0.800	40–150
630	0.600–1.300	50–170

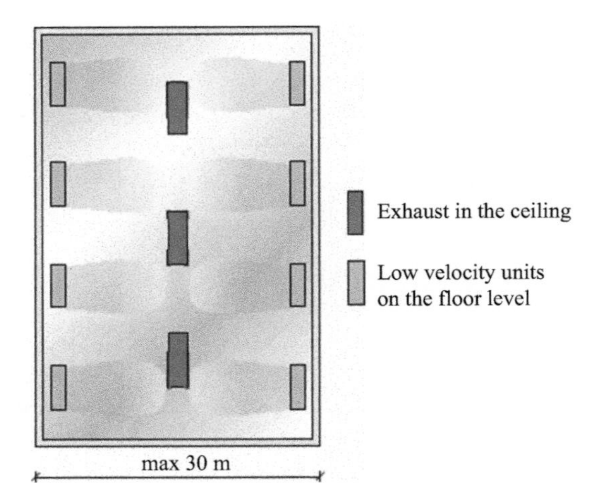

Figure 6.28 A rule of thumb of maximum distance between the supply units

special decorative panel. When the units are behind a decorative panel, the free supply area should be designed (over 50% of open area) to guarantee the normal performance of the supply units. Also, the units can be painted for special decoration demand if needed.

The selected supply units could serve as a visible element in the interior design. Those visual architectural elements could be, e.g. integrated with pillar structure (Figure 6.29 (left)) or free-standing units installed on the floor (Figure 6.29 (right)) somewhere in the space area. Using free-standing units makes design of equally air distribution for large spaces quite easy.

Figure 6.29 *Left: ceiling-ducted displacement units that are integrated with pillar structure. Right: floor-ducted displacement units that are architectural visual elements*

References

Hagström K., Sandberg E., Koskela H., and Hautalampi T. 'Classification for the room air conditioning strategies'. *Building and Environment* 2000; 35: 699–707.

Kosonen R., Lastovets N., Mustakallio P., da Graça G.C., Mateus N.M., and Rosenqvist M. 'The effect of typical buoyant flow elements and heat load combinations on room air temperature profile with displacement ventilation'. *Building and Environment* 2016; 108: 207–219.

Kosonen R., Melikov A., Mundt E., Mustakallio P., and Nielsen P.V. *Displacement Ventilation, REHVA Guidebook No 23*. Forssa: Federation of European Heating and Air-Conditioning Associations; 2017. p. 96.

Lastovets N., Kosonen R., Mustakallio P., Jokisalo J., and Li A. 'Modelling of room air temperature profile with displacement ventilation'. *International Journal of Ventilation* 2020; 19 (2): 112–126.

Mueller D., Kandzia C., Kosonen R., Melikov A., and Nielsen P.V. *Mixing Ventilation REHVA Guidebook No 19*. Forssa: Federation of European Heating, Ventilation and Air Conditioning; 2013. p. 114.

Nielsen P.V. 'The Family Tree of Air Distribution Systems'. Proceedings of ROOMVENT, 12th International Conference on Air Distribution in Rooms, June 2011, Trondheim, Norway. ISBN 978-82-519-2812-0.

Nielsen P.V., Restivo A., and Whitelaw J.H. 'Buoyance-affected flows in ventilated rooms'. *Numerical Heat Transfer* 1979; 2 (1): 115–127.

Chapter 7

Air ventilation system: air distribution in a natural ventilation system

Runming Yao[1] *and Jie Xiong*[2]

7.1 Introduction

The role of natural ventilation (NV) is to supply fresh outdoor air to an indoor environment whilst diluting and removing potentially harmful air contaminants generated from the space. NV systems aim to maintain indoor air quality (IAQ) whilst serving as an effective measure for passive cooling by removing excess heat and moisture. They aim at providing a thermally comfortable indoor environment without additional energy consumption.

Wind and thermal pressure are the two basic driving forces of NV. The outdoor wind fields, mostly relying on the urban morphology, influence the ventilation rate of a room. There are different wind pressures on different parts of the building facade, which create wind-driven NV. The thermal pressure caused by the differences between indoor and outdoor air temperatures drives the airflow through building openings. A good design of building openings will improve the effectiveness of NV performance.

NV is favoured by low-energy building design. However, its implementation is not only challenged by global warming but also by increased air pollution problems. Within the context of urban air pollution, the cleanliness of outdoor air cannot always be guaranteed. Polluted air entering into the indoor environment could negatively affect human health.

Good NV design will provide occupants with adaptive opportunities to take advantage of NV to meet their thermal comfort and air quality demands. The air distribution arising from NV is very complicated and is affected by a number of factors, including the shape of the building, the position and layout of openings, the wind field around buildings, the indoor and outdoor temperatures and the internal heat sources. This chapter will discuss how some of these factors affect the air distribution in NV.

[1]School of the Built Environment, University of Reading, Reading, United Kingdom
[2]Joint International Research Laboratory of Green Buildings and Built Environments, Chongqing University, Chongqing, China

7.2 Driving force of natural ventilation

The magnitude of the airflow through a building depends on the strength and direction of the natural driving forces and the resistance of the flow path.

7.2.1 Wind-driven natural ventilation

Wind is formed by pressure differences in the atmosphere. The wind flow close to the ground surface is restricted by the characteristics of the atmospheric boundary layer. The specific profile is subject to the terrain conditions (see Figure 7.1); the more congested the terrain, the slower will be the local wind speed relative to the meteorological wind speed.

Wind-driven NV uses the pressure differences formed by the airflow acting on the building facade as the driving force. When the wind encounters a building, the phenomenon of turbulence occurs, and the airflow will resume a parallel flow after a certain distance. Due to the obstruction of the building, the pressure distribution of the airflow around the building will change drastically. The airflow on the windward side will be blocked, the dynamic pressure will decrease and the static pressure will increase. The lateral side and leeward sides will form a local vortex and reduce the static pressure (see Figure 7.2). The flow field and wind pressure distributions around the building are related to the geometry of the building itself and the arrangement of its surrounding buildings.

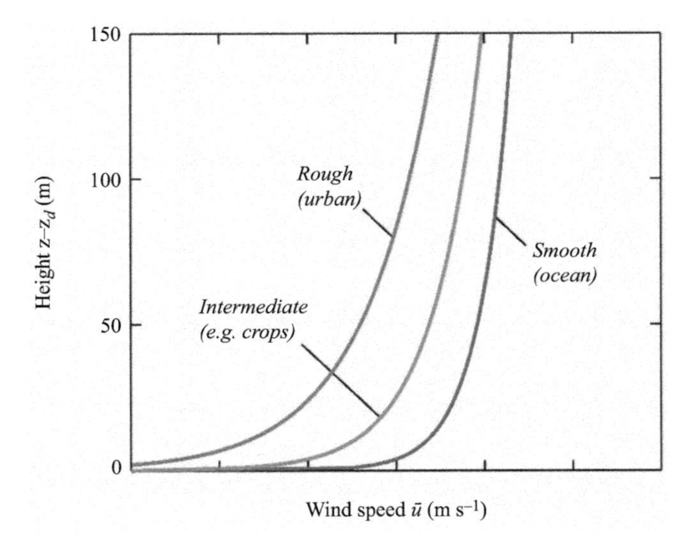

Figure 7.1 General form of the mean horizontal wind speed profile in the lower atmosphere over rough, intermediate roughness and smooth terrain in neutral stability (Oke et al., 2017)

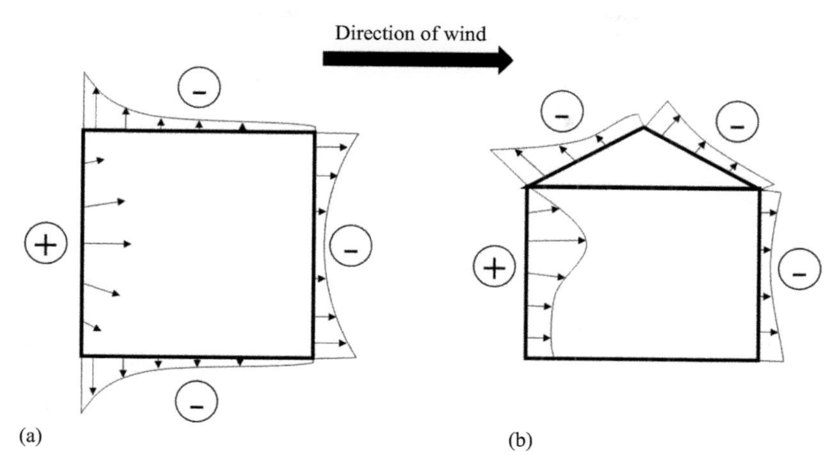

Figure 7.2 Wind pressure distribution on building envelopes: (a) horizontal plane and (b) vertical plane (Chartered Institution of Building Services & Engineers (CIBSE), 2019)

When the shape and arrangement of buildings and the wind direction are determined, the wind pressure value P_f at a certain point on the surface of a building envelope can be approximated by the following equation:

$$P_f = C_p \frac{\rho_\infty v_\infty^2}{2} \tag{7.1}$$

where ρ_∞ is the air density of incoming wind (kg m^{-3}); v_∞ is the velocity of incoming wind at a reference height z (m s^{-1}); C_p is the wind pressure coefficient (dimensionless), a positive value indicates that the wind pressure at this point is positive with a negative value indicating a negative wind pressure (see Figure 7.3). Buildings of different shapes have different distributions of wind pressure coefficients under different wind directions.

If there are multiple openings on the different building facades, the opening with the larger wind pressure coefficient will form the air inlet, and the window with the lower wind pressure coefficient will form the air outlet. As shown in Figure 7.3, under a wind speed of v_∞, the wind pressure of the opening on the windward side is P_{fa}, and the wind pressure of the opening on the leeward side is P_{fb}, where $P_{fa}>P_{fb}$. The NV rate driven by wind pressure q_w is

$$q_w = (C_d A)\sqrt{\frac{2(P_{fa} - P_{fb})}{\rho_i}} \tag{7.2}$$

where C_d is the discharge coefficient (dimensionless) and A is the total ventilation area (m^2).

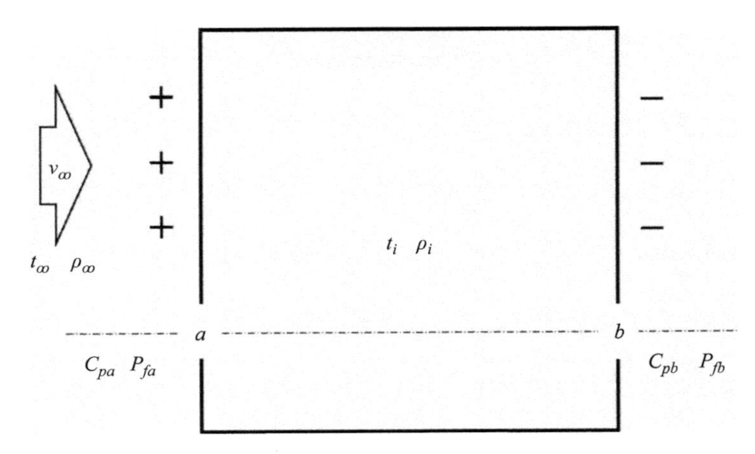

Figure 7.3 Wind-driven natural ventilation

It can be seen that the indoor flow field generated by the wind pressure is directly related to the differences between the wind pressures of the two openings on the building facades.

The NV can also be induced by wind turbulence, especially when the buoyancy effect is small, and the mean wind speed is not high. The wind pressure fluctuates with time, which constantly modifies the inflow and outflow pressure differences. However, this cannot be regarded as the main mechanism of NV since the air exchange rate is negligible when compared with other ventilation mechanisms.

7.2.2 Buoyancy-driven natural ventilation

Buoyancy-driven NV mainly uses the buoyancy or stack effect. Due to the temperature differences between indoor and outdoor air, a certain air density difference is produced at different positions, thus forming the driving force for airflow. Warmer air rises and leaves the building through openings near the top, causing an internal suction which draws fresh air into the building through openings near the ground.

As shown in Figure 7.4, there are openings *a* and *b* at different heights of building facades, the height difference (m) is *h*, the static pressure outside the openings (Pa) are P_a and P_b, respectively, and the static pressure inside the openings (Pa) area P_a' and P_b', respectively. The outdoor air temperature (°C) is t_o, and the outdoor air density (kg m^{-3}) is ρ_o; the indoor air temperature (°C) is t_i, and the indoor air density (kg m^{-3}) is ρ_i. Due to the presence of indoor heat sources, in general, $t_i > t_o$, so $\rho_i < \rho_o$. With the difference in air density, the air will flow in from the lower opening and flow out through the upper opening.

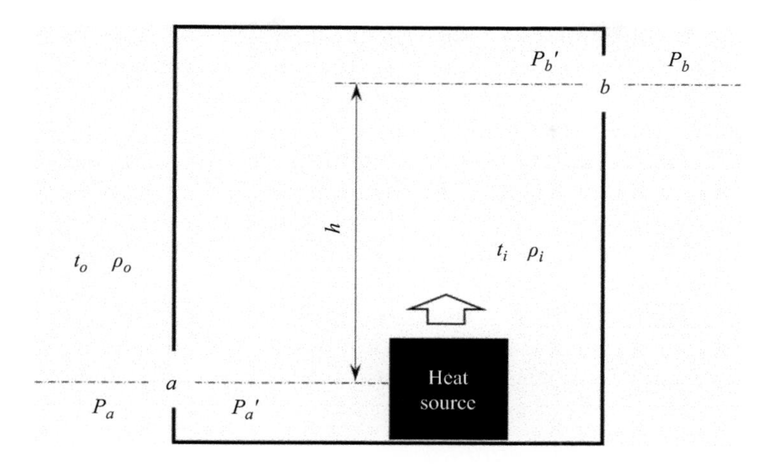

Figure 7.4 Buoyancy-driven natural ventilation

According to the principles of fluid mechanics, the absolute value of the pressure difference between the inlet opening and the outlet opening is related to the height difference between the two openings and the density difference between indoor and outdoor air, which is often referred to as the thermal pressure $gh(\rho_o - \rho_i)$. According to the law of mass conservation, the buoyancy-driven NV rate q_s is

$$q_s = (C_d A)\sqrt{\frac{2gh(\rho_o - \rho_i)}{\rho_i}} \tag{7.3}$$

where C_d is the discharge coefficient of the opening (dimensionless) and A is the area of the opening (m^2). The flow rate generated by the buoyancy effect is mainly affected by the indoor–outdoor temperature difference (and consequently the density difference) and the effective height difference between the openings. If there is no temperature difference between indoor and outdoor air or there is no height difference between openings, buoyancy-driven NV will not occur. However, a single opening can still create NV since the wind entering from the lower part of the opening and exiting through the upper part of the opening is equivalent to two openings with a small vertical separation which creates some buoyancy-driven NV.

7.3 Natural ventilation systems

There are many components that make up an NV system, including the ventilation inlet, the supply point, the flow path, the exhaust point and the ventilation outlet. The ventilation inlet is where the fresh air is drawn into the ventilation system, and the supply point is where the fresh air is delivered into the occupied space. For a window opening, the ventilation inlet and the supply point are the same component, but they are different for systems using duct distribution. The flow path is the pathway of the airflow from the supply point through the occupied space to the exhaust point. The relationship between the exhaust point and the ventilation outlet is equivalent to that between the supply point and the ventilation inlet with the air being exhausted outside the building through the ventilation outlet.

The arrangements of the ventilation inlet, supply point, exhaust point and ventilation outlet decide the air distribution of NV in the occupied space. There are three main types of NV: single-sided, cross and stack.

7.3.1 Single-sided ventilation

Single-sided ventilation relies on openings situated on only one side of a room. It often occurs in cellular buildings where some rooms only have one external wall with windows on. Single-sided ventilation is only effective for the room in which the openings are located. Specifically, there may be two openings, one inlet and one outlet, on one external wall (see Figure 7.5(a)), or a single opening serving as inlet and outlet simultaneously (see Figure 7.5(b)).

7.3.1.1 Multiple openings

If there are more than two openings on one facade in a room, they should be provided at different heights on that facade for enhanced stack buoyancy-driven ventilation. Due to the indoor heat gains, the upper air is warmer and will flow out through the upper opening. This forms a vacuum and draws fresh, colder air into the room through the lower opening (see Figure 7.6). The ventilation rate increases with the vertical separation of the openings, and the indoor/outdoor temperature difference. This form of ventilation can penetrate further into a room than single-opening ventilation with an effective depth considered to be 2.5 times the floor-to-ceiling height, around 6 m.

7.3.1.2 Single opening

For a room with only a single opening, a steady pressure difference cannot be developed across space, because there are no separate inlet and outlet openings with different wind pressures. Thus, the main driving force for NV is wind turbulence. The pressure will never be steady, so the pressure across the opening fluctuates rapidly, which makes a pressure difference across the opening, and it is continuously reversing its direction. Sometimes the air is forced into the space, and sometimes the air is extracted from it. A buoyancy-driven wind flow can also be found in single-opening ventilation, but only if the vertical dimension is reasonably large to create sufficient thermal pressure. The air flows from the lower part of the opening and leaves through the upper part of the opening. It is suggested that the

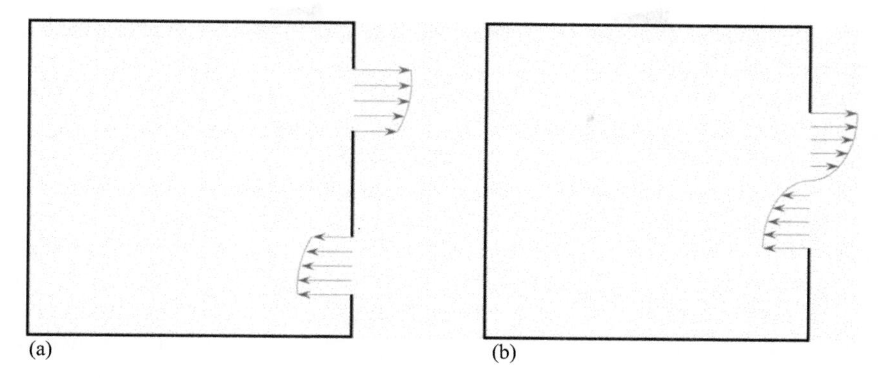

Figure 7.5 Single-sided natural ventilation through (a) the upper and lower openings and (b) a single opening

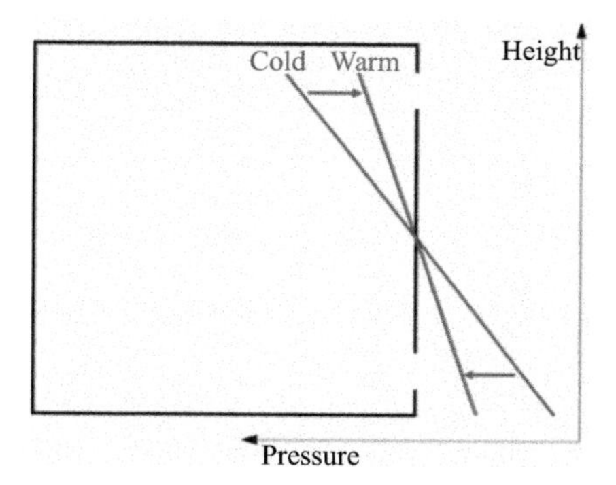

Figure 7.6 The flow path of double-opening, single-sided, natural ventilation

openable window area should be at least 1/20 of the floor area for rapid ventilation. Compared with other strategies, single-opening ventilation provides lower ventilation rates, and its depth of penetration is short, considered to be about two times the floor-to-ceiling height. Thus, locations far from the window may experience poor air quality.

Stack-induced airflow increases with a greater vertical window separation. To maximize the vertical distance, individual openings above and below windows could be used although controllable inlets at a lower position that could create low-level draughts in winter need to be carefully arranged.

7.3.2 Cross ventilation

When the ventilation openings are positioned on two or more different sides of a room, cross ventilation occurs due to the pressure difference between the different sides. Air flows in through the opening on one side of the room, flows over the whole space and leaves the room through the opening on the other side of the room. As heat sources and pollutant sources appear in the room, the air will be heated up and the air quality will progressively decrease. Hence it is necessary to properly arrange the floor plan according to the different functions of each room.

Cross ventilation is usually wind-driven, based on the wind pressure differences between the two facades. To enhance the airflow, care must be taken to design the building to ensure a significant difference in the wind pressure coefficient between the inlet and outlet openings. Because the wind develops a pressure difference across the building in the oncoming direction, it is better to design the flow path horizontally, with air entering and leaving the building at the same vertical level. The orientation of the building relative to the prevailing wind direction can maximize the potential of wind-driven NV. The maximum distance between two facades is five times the floor-to-ceiling height, so it usually occurs in buildings with a narrow plan.

The internal partitions will influence ventilation performance. There may be insufficient airflow, especially in summer when gentle breezes prevail, if the large partitions (particularly full height ones) block the airflow or if the window on one side of the building is closed. In this situation, single-sided NV occurs.

7.3.2.1 Windcatchers

There are also some variants of cross ventilation. One such is the wind scoop (see Figure 7.7), the so-called windcatcher. It captures the wind at a higher level where the dynamic pressure of the wind is higher. The supply points for all the rooms on each floor are connected to the high-level air inlet which improves the wind-driven ventilation on the lower levels. But the effect of stack pressure in the opposite direction to the airflow must be considered since it will influence the overall ventilation rate. The air inlet should be located according to the dominant prevailing wind direction. If the wind direction changes very frequently, then multiple inlets need to be prepared, combined with automatic controls to open the windward vents and close those to leeward.

Following the wind scoop concept, the roof-mounted ventilator uses the pressure difference across a device with partitions to drive airflow through the segment on the windward side into the space and sucks the air within the space using the negative pressure on the leeward side of the segment. The ventilation rate can be controlled using a damper to connect the duct to distribute the air to areas according to demand, or to those spaces which cannot be ventilated using window openings.

7.3.3 Stack ventilation

Stack ventilation is driven by the buoyancy arising from air density differences. The heat gains are a critical factor for the success of this NV strategy since it draws air from the inlet openings, across the ventilated space, and exhausts the air through a vertical flow path, like a chimney or an atrium.

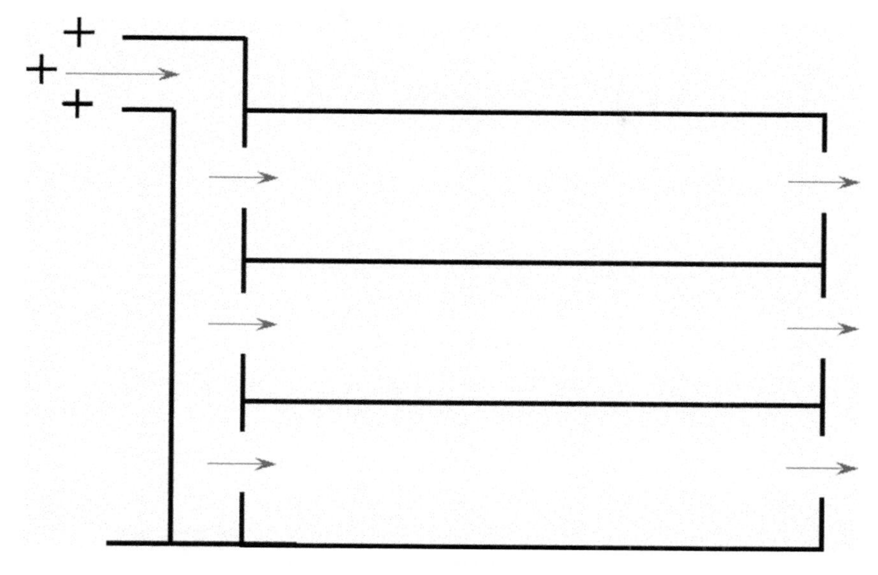

Figure 7.7 Wind scoop

The distance limit between the supply and exhaust points of a room is also around five times the floor-to-ceiling height. To avoid large-size outlet ventilators, the stack outlet needs to be at a height of at least one storey above the roof. Furthermore, it should be located in a wind-induced negative pressure area; otherwise, the wind-driven ventilation will offset the buoyancy effect, thereby affecting the NV performance.

7.3.3.1 Chimney ventilation

Nowadays, chimneys serve no functional purpose other than as a means of stack ventilation. The air in the chimney is required to be warmer than the ambient air. Warm air is less dense than colder air. Due to the existence of heat sources, the heated air will rise and leave the space, forming a partial vacuum area, which naturally sucks in cooler air from the outside atmosphere (see Figure 7.8). If the chimney has a large surface area exposed to the outside air, it should be well insulated to minimize heat loss and prevent flow reversal during cooler weather. In order to reduce the resistance, vertical air shafts are usually used in modern buildings, and inlet openings for air supply are retained.

The solar chimney, which is mainly composed of glazed elements as its main structure, can enhance stack pressure. The glazing elements, as transparent materials, absorb solar radiation and release the heat to the air in the chimney so that its temperature increases, thereby promoting the buoyancy effect. Glazing elements can be set in both upper and lower levels of a chimney. If they are set at the upper level, a pulling effect occurs and if they are at the lower level, a

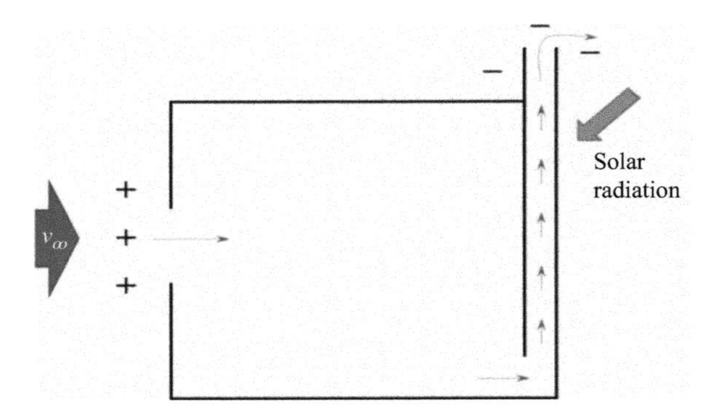

Figure 7.8 Stack ventilation driven by buoyancy and wind

pushing effect. However, the lower level glazing elements should avoid being shaded by other structures. Moreover, the glazing may cause heat loss in winter requiring a balance to be achieved between heat gains from solar radiation and the heat loss from convections. Otherwise, it may generate down draughts in the chimney that reverse the direction of the stack ventilation and lead to insufficient NV.

In order to enhance the combined effects of wind pressure driving and the thermal gradient within the chimney, the outlet should be located in a region of wind-induced negative pressure. The negative pressure zone can be created by careful building design or community morphology considering the prevailing wind direction. If the chimney outlet is not properly located, any positive wind pressure can disturb the upward airflow.

7.3.3.2 Atrium ventilation

The atrium has become a very common component in modern public buildings. Following the concept of chimney ventilation, the atrium also provides a vertical path for airflow. Usually, outlet openings are set at the top of the atrium, using indoor heat sources to heat the air, so that the air rises and discharges out of the space whilst, at the same time, the outdoor cold air is introduced into the room through the side windows (see Figure 7.9). The air distribution of atrium ventilation is generally more complicated.

The atrium can draw the air from both sides of the building to the central area, doubling the effective width of a building that can be naturally ventilated. The roof profile should be carefully designed, ensuring that the openings are in the negative pressure zone for all wind directions or using multiple vents with automatic controls to open the vents on the leeward side and close the vents on the windward side, which is opposite to the wind scoop. The main difference is that the atrium has more functions than the chimney, like vertical transportation and provision of space for social communications. To locate the atrium in a building, the ventilation is not

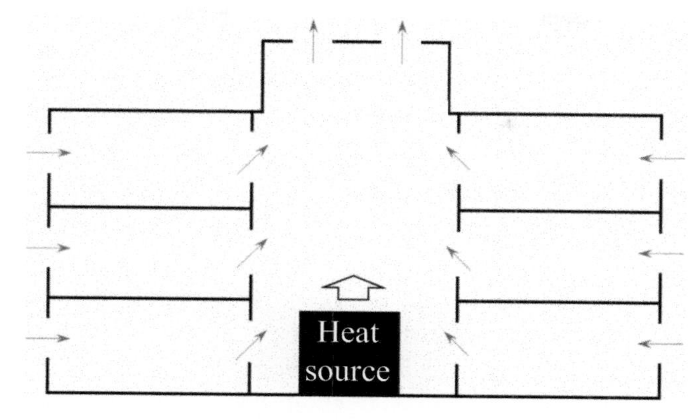

Figure 7.9 Atrium ventilation driven by buoyancy

the only factor to consider, the developer will take its commercial values into account.

The atrium also provides daylight into the centre of a deep plan building, so it provides both energy-efficient and environmental benefits. In the meantime, it can turn into a large solar-assisted chimney. But, in a very hot summer, there should be shading for the top glazing to avoid increasing the cooling load of the air-conditioning system.

7.3.3.3 Double-skin facade ventilation

The double-skin facade is a special variant of the solar chimney where the whole facade acts as an air duct. In most cases, the interspace between the two facade layers is driven by the buoyancy effect. It uses a transparent envelope to trap the solar gain in the space between the inner and outer skins. The heat is absorbed by the air in that space which rises and removes the heat to the external environment. A connection exists between the double-skin facade and the room space, so the updraft inside the double-skin facade will drive the airflow in the room. There are designs allowing ventilation air to be taken from, or exhausted to, the interspace, thereby providing a practical means of NV for tall buildings.

7.4 Design calculation methods

There is a range of methods to calculate the NV airflow rate, namely semi-empirical methods like envelope flow models, multi-zone models and computational fluid dynamics (CFD) and physical scale models. Different models can help the design in different stages. The simplest calculation tools, like envelope flow models and zonal models, are recommended for the initial design stage with other models being used for later stages when more detailed information is available.

7.4.1 Envelope flow models

Envelope flow models calculate the flow rates through openings in the envelope for given wind conditions. The single-cell model solves the flow rate in the room itself whilst the multi-cell model can be used for the flows between rooms (internal partitions are included).

There are three basic assumptions for the use of these models. First, the external wind pressure distribution will not be influenced by the presence of openings, which allows the pressure distribution to be obtained in the absence of the openings. Second, the air velocity inside the space is small enough for the internal pressure to be given by the hydrostatic equation. Third, the air density distribution inside the envelope is independent of the flow rate.

It is common to express the relationship between the flow rate through an opening and the pressure difference across it using the discharge coefficient and the geometric area:

$$q = C_d A \sqrt{\frac{2|\Delta p|}{\rho}} \tag{7.4}$$

where q is the flow rate through the opening (m^3 s^{-1}), C_d is the discharge coefficient (dimensionless) that is defined and measured under still-air conditions with uniform density with the flow being generated by a fan, A is the area of the opening (m^2), Δp is the pressure difference (Pa) and ρ is the air density (kg m^{-3}).

Some cases of different NV systems using the empirical envelope flow model are shown next.

7.4.1.1 Single-sided ventilation, two openings, buoyancy driven

The ventilation of a room with two openings in one envelope with different heights can be driven by the buoyancy effect. The ventilation flow rate q (m^3 s^{-1}) can be calculated by the following equation:

$$q = C_d A \sqrt{\frac{2gh(t_i - t_o)}{t_i + 273}} \tag{7.5}$$

where A is the area of the opening (m^2), t_i is the indoor temperature (°C), t_o is the outdoor temperature (°C), h is the height difference of two openings (m), and C_d is the discharge coefficient, 0.6 is a typical value for this case.

7.4.1.2 Single-sided ventilation, single opening, buoyancy driven

The ventilation of a room with only one opening in the envelope can also be driven by the buoyancy effect, following (Equation (7.5)), where h is the height of the opening. The discharge coefficient C_d is significantly reduced, with a typical value for this case of 0.25, due to only around half of the opening area being available for air inflow, and the middle part of the opening contributing little to the ventilation flow rate.

7.4.1.3 Single-sided ventilation, single opening, wind driven

The ventilation of a room with only one opening in the envelope can be wind driven. The ventilation flow rate q ($m^3\ s^{-1}$) can be calculated by the following equation:

$$q = CAU \tag{7.6}$$

where U is the wind speed ($m\ s^{-1}$), and C is a coefficient depending on the geometry of the opening, the position where the reference wind speed is measured and the flow around the building. The air inflow is caused by turbulent diffusion, so C is typically low. When U is measured at the same height as the building, the value of C is from 0.01 to 0.05 (Etheridge and Sandberg, 1996).

7.4.1.4 Cross ventilation, wind driven

The ventilation of a room with two openings, one in each facade, is usually driven by the pressure difference between the two openings. The ventilation flow rate q ($m^3\ s^{-1}$) can be calculated by the following equation:

$$q = C_d AU\sqrt{\frac{\Delta C_p}{2}} \tag{7.7}$$

where ΔC_p is the difference in the wind pressure coefficient for the two openings with pressure coefficients of C_{p1} and C_{p2}, respectively.

7.4.1.5 Stack ventilation, atrium, buoyancy driven

For single-cell buildings, the room spaces are all connected to an atrium. The fresh air is introduced into the occupied spaces through the inlet openings on the side of the space and is removed through the upper outlet at the top of the atrium (see Figure 7.10).

For this case, the fresh air enters the openings in the sides of each room, and all exhausts are through the upper outlet. The atrium is a large stack with the internal temperature assumed to be uniform. This means the neutral plan, where Δp_n=0, is located between the inlet opening on the top floor and the outlet opening at the top of the atrium with a height of z_n. Given that the only driving force is buoyancy and the oncoming wind speed U=0, the pressure difference for each opening Δp_i can be calculated by the following equation:

$$\Delta p_i = \Delta\rho_0 g z_n - \Delta\rho_0 g z_i \tag{7.8}$$

where $\Delta\rho_0$ is the density difference between indoor and outdoor air, and z_i is the height of the inlet opening i. The airflow rate for each opening can be calculated by the following equation:

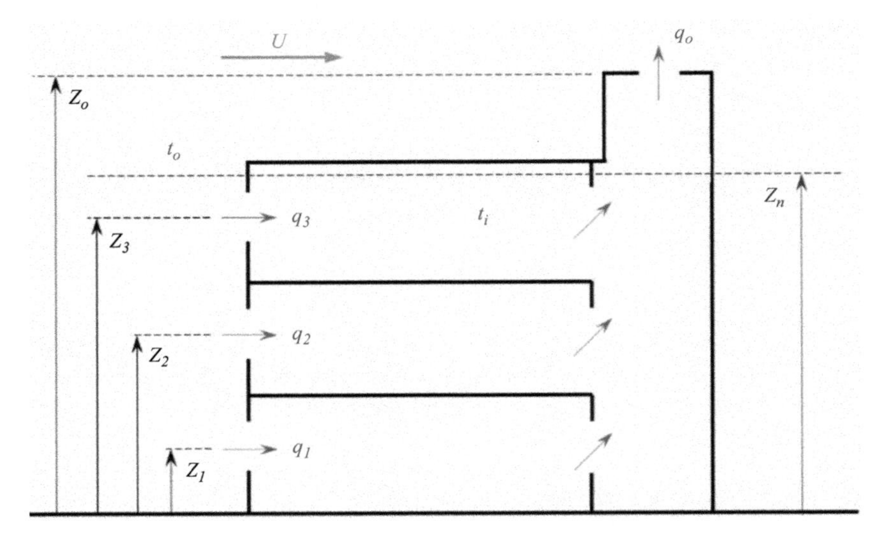

Figure 7.10 Properties of the single-cell building with an atrium

$$q_i = C_{di}A_i\sqrt{\frac{2|\Delta p_i|}{\rho_o}} \tag{7.9}$$

where C_{di} is the discharge coefficient of the inlet opening i, and A_i is the area of the inlet opening i. Simultaneously the mass conservation equation is

$$\sum q_i = q_o \tag{7.10}$$

which allows the airflow rate for each opening to be solved.

By using an envelope flow model, many assumptions for the simplification of calculations in the early design stages can be made.

7.4.2 Multi-zone models

Multi-zone models divide the space into a number of zones and use semi-empirical conservation equations for mass, momentum and energy. They can calculate the velocity and temperature fields inside a bounded space with given thermal boundary conditions and envelope flow rates.

Zonal models are divided into two types: primary flow and secondary flow. The primary flow models do not calculate the detailed structure of the secondary flow, by making simplifying assumptions about the secondary flow pattern and temperature distribution, and the number of zones is very small (usually only two). However, a large number of zones are applied in the secondary flow models, which use approximate forms of conservation equations to obtain velocity and temperature fields.

7.4.2.1 Primary flow models

One basic configuration for a primary flow model is a naturally ventilated room with a heat emitter and cooler wall (Howarth, 1985). There are two equal and opposite plumes on the opposite sides of the room, the model makes use of the interface between two stratified layers. It is assumed that no mass flow across the interface occurs, so that

$$\dot{m}_h = \dot{m}_c \tag{7.11}$$

where $\dot{m}_h$ is the mass flow rate of the positively buoyant plume heated by the heat emitter, and $\dot{m}_c$ is the mass flow rate of the negatively buoyant plume cooled by the cooler wall. The fundamental limitation of such a model is the assumption of stable stratification. Especially when envelope flows are stronger, the primary flow models are less justifiable.

7.4.2.2 Secondary flow models

The secondary flow models divide the room into a number of zones. The values of temperature and pressure of zone i are t_i and p_i, respectively. The mass flow rate across the interface between zone i and zone j is $\dot{m}_{ij}$. The equation given by mass conservation is

$$\sum \dot{m}_{ij} = 0 \tag{7.12}$$

and the equation for thermal energy is

$$\sum \dot{m}_{ij} C_p T = 0 \tag{7.13}$$

where C_p is the specific heat of air.

The mass flow across the interface in terms of pressure is given by the power law model:

$$\dot{m}_{ij} = C \Delta p_{ij}^{\beta} \tag{7.14}$$

where C is a constant flow coefficient and $\Delta p_{ij}=p_i-p_j$.

The secondary flow models can be regarded as a solution of the continuity equation.

7.4.3 *Computational fluid dynamics*

CFD is now widely used in building engineering with the rapid increase in computing power and the developments in numerical analysis. CFD provides practical solutions for problems that would otherwise be impossible or very time-consuming.

CFD solves some forms of the Navier–Stokes equations and thermal energy equations for a large quantity of small elements divided within the whole space.

Continuity equation:

$$\frac{\partial \rho u}{\partial x} + \frac{\partial \rho v}{\partial y} + \frac{\partial \rho w}{\partial z} = 0 \tag{7.15}$$

Momentum equations:

$$\rho\left(\frac{\partial u}{\partial t} + u\frac{\partial u}{\partial x} + v\frac{\partial u}{\partial y} + w\frac{\partial u}{\partial z}\right) = -\frac{\partial p}{\partial x} + \mu_0\left(\frac{\partial^2 u}{\partial x^2} + \frac{\partial^2 u}{\partial y^2} + \frac{\partial^2 u}{\partial z^2}\right) \tag{7.16}$$

$$\left(\frac{\partial v}{\partial t} + u\frac{\partial v}{\partial x} + v\frac{\partial v}{\partial y} + w\frac{\partial v}{\partial z}\right) = -\frac{\partial p}{\partial x} + \mu_0\left(\frac{\partial^2 v}{\partial x^2} + \frac{\partial^2 v}{\partial y^2} + \frac{\partial^2 v}{\partial z^2}\right) \tag{7.17}$$

$$\rho\left(\frac{\partial w}{\partial t} + u\frac{\partial w}{\partial x} + v\frac{\partial w}{\partial y} + w\frac{\partial w}{\partial z}\right) = -pg - \frac{\partial p}{\partial x} + \mu_0\left(\frac{\partial^2 w}{\partial x^2} + \frac{\partial^2 w}{\partial y^2} + \frac{\partial^2 w}{\partial z^2}\right) \tag{7.18}$$

Energy equation:

$$\frac{\partial T}{\partial t} + u\frac{\partial T}{\partial x} + v\frac{\partial T}{\partial y} + w\frac{\partial T}{\partial z} = \frac{k}{\rho C_p}\left(\frac{\partial^2 T}{\partial x^2} + \frac{\partial^2 T}{\partial y^2} + \frac{\partial^2 T}{\partial z^2}\right) \tag{7.19}$$

The methods based on the N–S equation include direct numerical solution, large eddy simulation (LES) and Reynolds-averaged N–S.

The choice of calculation domain and the grid is essential for the simulations. The computational domain for NV has two parts: external and internal. The external domain needs to be large enough, ensuring that the simulation results are independent of its size. It is suggested that the blockage ratio should be no more than 3%, the inlet, top and lateral boundaries should be set at least five times the tallest building height, and the outlet boundary should be set at least ten times the tallest building height (Tominaga *et al.*, 2008). There are several types of grid: structured, unstructured and adaptive, and the choice of the grid depends on the geometry and arrangement of the buildings and the problem-solving (Nielsen, 2004). The size of the cells is as fine as possible considering the computational power and time available.

Boundary conditions are fundamental to the CFD simulations. They should be specified at the domain surfaces and at the solid surfaces within the domain. The nature of the domain will determine its specific form. For example, the inlet boundary normally uses the inlet velocity, the outlet boundary uses the pressure outlet, the top and lateral boundaries use the symmetry, and the solid wall is considered to be non-slippery. However, there is some flexibility according to the actual situations.

CFD is mainly used to calculate the velocity and temperature fields in the room, the surface wind pressure coefficient, the envelope flows, etc. According to the purpose, the area of interest includes the external flow, the flow through the openings and the internal flow which helps to understand the detailed flow information for very complex NV systems.

7.4.3.1 Case of natural ventilation using a windcatcher

The windcatcher was designed for the retrofit of two multi-storey residential slab buildings in Hangzhou and Chongqing, China (Laetitia *et al.*, 2020). The CFD simulations are conducted to demonstrate its effect on enhancing the airflow rate through space.

The Qiushi Building in Hangzhou, constructed between 1995 and 1997, is a six-storey reinforced concrete slab building with a total height of 18.3 m, oriented 20° from north, located within an array of similar buildings. The retrofit scheme for the Qiushi Building used two multi-opening windcatchers (see Figure 7.11(a) and (d)). The outdoor air is introduced into the indoor space through ducts having openings in the internal rooms. The indoor opening to the duct is located at ceiling height in the central room of the flat, such that the cross-contamination between rooms is less likely.

The Shuishixiang building in Chongqing, constructed between 1995 and 1997, is a nine-storey reinforced concrete slab building with a total height of 25.4 m, oriented −5° from north, surrounded by buildings of similar type and height, particularly on the north, east and south sides. A cluster of tall buildings lies to the west of the Shuishixiang building at a distance of around 300 m. The retrofit scheme for the Shuishixiang building is based on multi-opening windcatchers and external chimneys (see Figure 7.11(b), (c) and (e)). The outdoor air is introduced into the living room by the windcatcher. The chimney, linked to the kitchen, exhausts polluted air generated when cooking. Each flat is linked to an individual chimney, thus preventing cross-contamination between flats. In both buildings, the windcatchers have multi-openings and internal blades guiding the air towards the indoor spaces.

The CFD simulations are conducted using Fluidity (AMCG, 2015), open-source finite-element CFD software. The geometry is set up as shown in Figure 7.11(d) and (e). The flow field is computed from the 3D N–S equations. The turbulence model is based on an LES approach. The mesh adaptivity capability with unstructured meshes is applied. The inlet profiles of the mean velocity, mean Reynolds stresses and turbulence length scales are given to set the velocity inlet. A zero-pressure condition is set at the outlet boundary, while the no-slip wall boundary condition is set at the building facades and the ground. These configurations have been validated using wind tunnel experiments.

The results for temperature and the velocity field in the seventh floor for the Shuishixiang building and the fourth floor for the Qiushi Building are obtained as shown in Figures 7.12 and 7.13, respectively. It can be seen that the spatial distribution of the temperature as well as the indoor air movement and exchanges between rooms are solved in the three-dimensional space. For the Shuishixiang building, as expected, the outdoor air enters the windcatcher and reaches the living room, and the indoor air escapes through the chimneys connected to the kitchen.

7.4.4 Physical scale models

Wind tunnels can be used for measuring the ventilation rate of buildings. The building should be large enough to ensure that the volume of the envelope at model scale allows a tracer gas measurement inside the model. If these requirements can be met, a wind

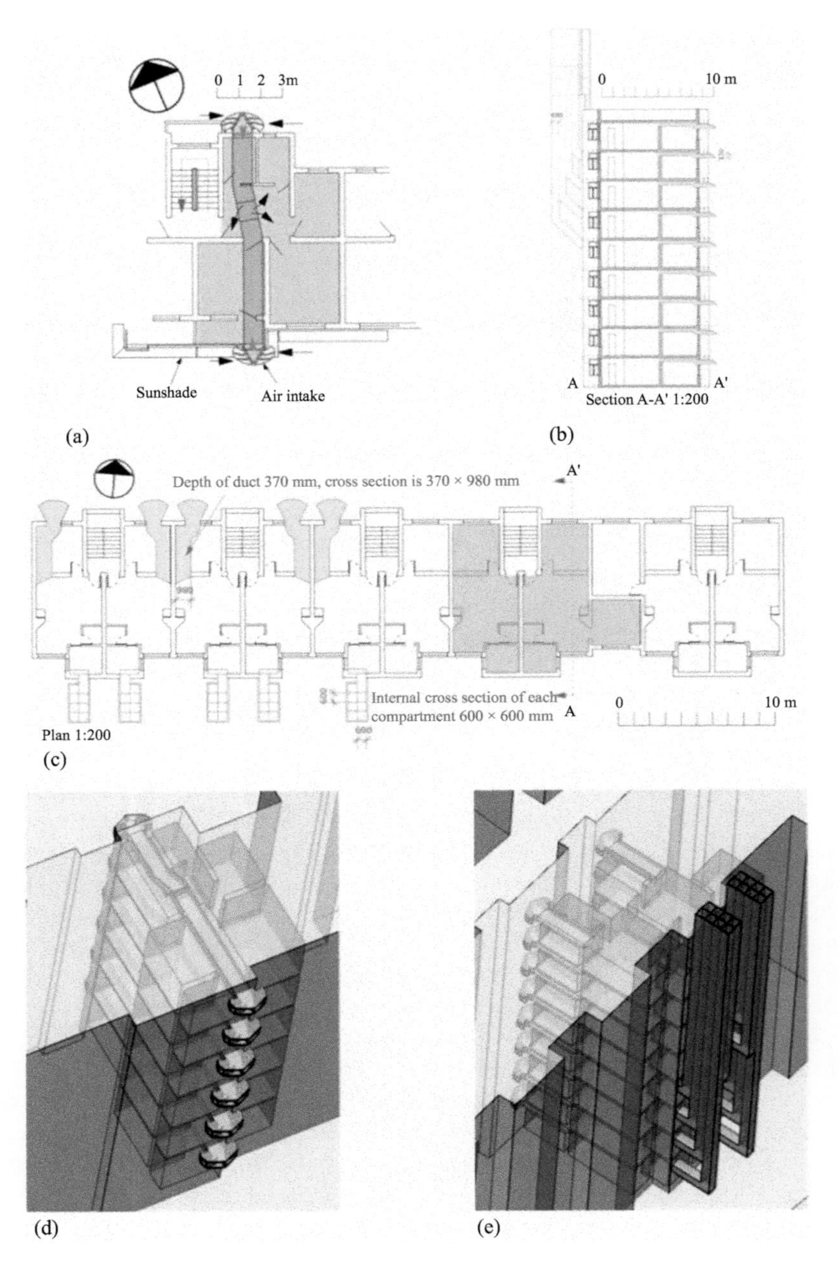

Figure 7.11 Adaptation scheme to enhance natural ventilation for (a) the Qiushi Building and (b) and (c) the Shuishixiang building. 3D view of the schemes as implemented in the CFD geometry for (d) the Qiushi Building and (e) the Shuishixiang building (Mottet et al., 2020)

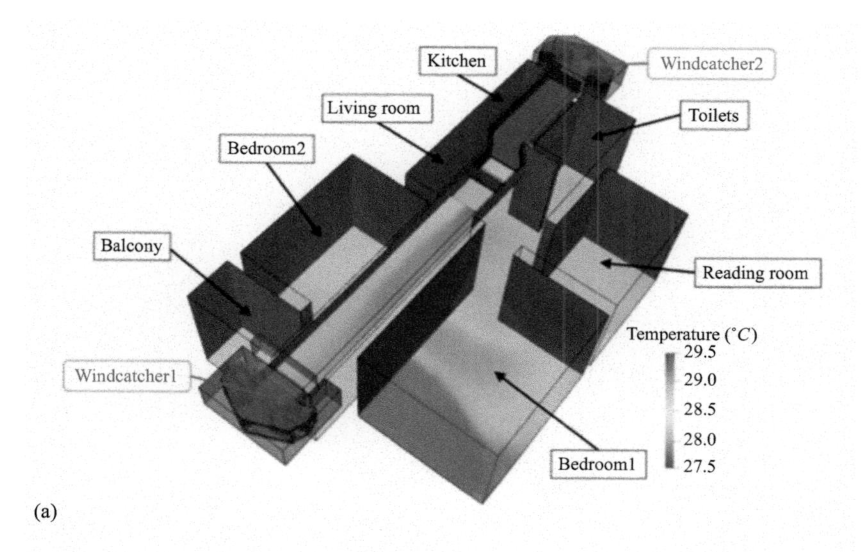

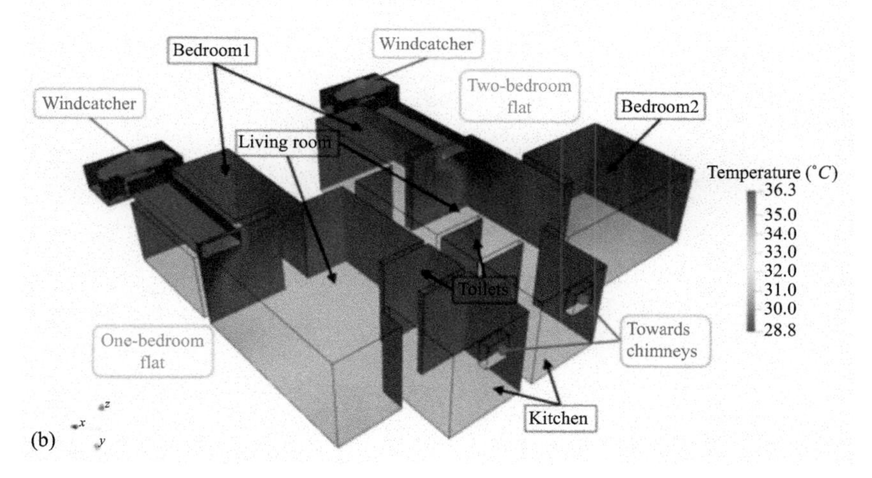

Figure 7.12 Spatial distribution of the temperature field at 02:00 on Day 2 of the hottest month TMY when under SE wind condition for (a) a two-bedroom flat at Level 4 in the Qiushi Building; and (b) a two-bedroom flat and a one-bedroom flat at Level 7 within the Shuishixiang building (Mottet et al., 2020)

tunnel test provides relatively high accuracy for the ventilation rate induced by wind. The scaling laws can be applied to calculate the full-scale ventilation rate.

Wind tunnels are also the main source of information on the wind pressure coefficient. The results of measurements are available as design data for NV in generic building forms.

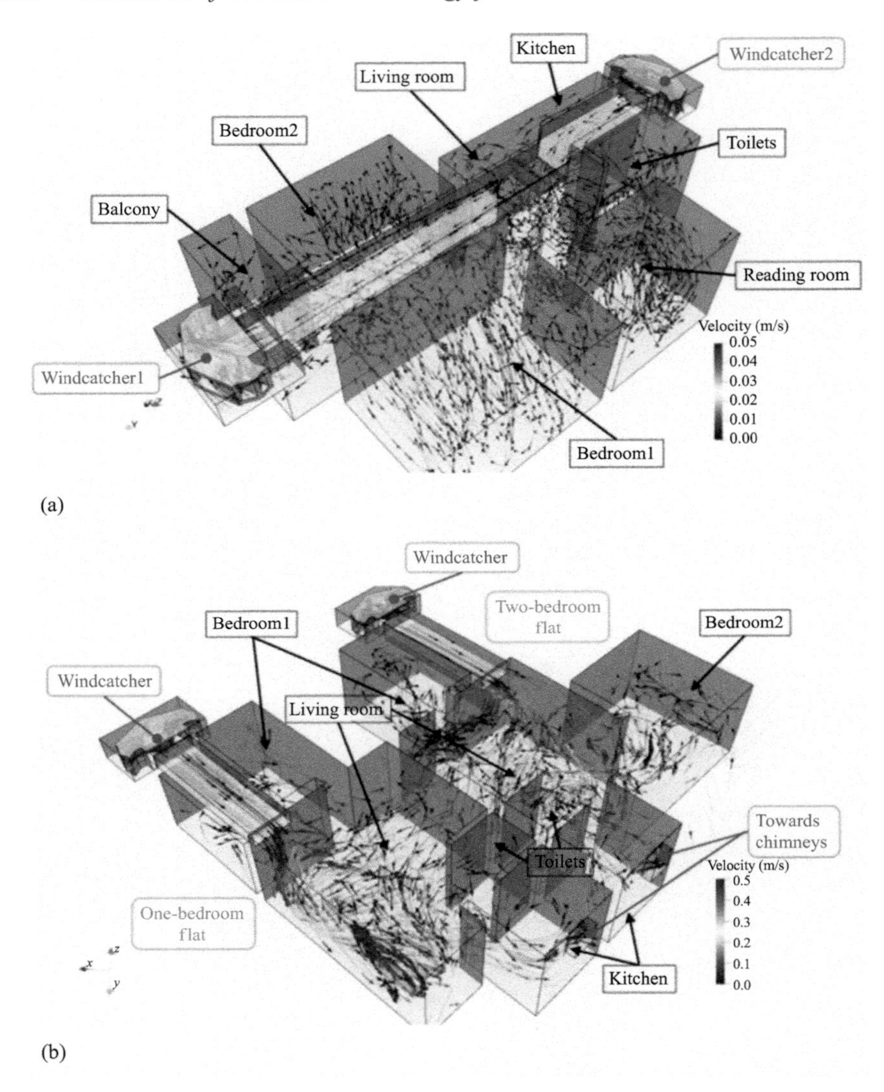

Figure 7.13 Spatial distribution of the velocity field at 02:00 on Day 2 of the hottest month TMY when under SE wind condition for (a) a two-bedroom flat at Level 4 in the Qiushi Building; and (b) a two-bedroom flat and a one-bedroom flat at Level 7 within the Shuishixiang building (Mottet et al., 2020)

7.5 Natural ventilation potential

The NV potential evaluates the possibility of ensuring acceptable IAQ and thermal comfort by using an NV system. Benefits are constrained by the conditions of the outdoor air: its thermal conditions reflecting the climate characteristics, its flow conditions reflecting the local urban context and its composition reflecting the air pollution levels. In order to appropriately design and operate NV, an evaluation of the NV potential is important.

7.5.1 Climate

Climate characteristics are considered one of the most important factors in passive building design, as they have considerable impact on energy consumption (Yao and Short, 2013). The principle of passive design is based on climate-responsive strategies, which take advantage of natural resources, like wind and sunlight, while taking some measures to avoid exposure to heat, cold and excessive radiation. Thus, effective passive design requires a detailed understanding of the local climatic conditions (Yao *et al.*, 2006).

The potential of NV varies in different climates. NV hours, defined as the number of hours in a typical year when the outdoor weather conditions are suitable for the utilization of NV, are often used as the evaluation criteria of NV potential. One possible way to estimate the NV hours is based on the outdoor dry-bulb temperature, the outdoor dew-point temperature and the outdoor air velocity (Tong *et al.*, 2017). The upper threshold of the outdoor dry-bulb temperature T_{up} (°C) is based on the adaptive thermal comfort model defined by de Dear and Brager (2002):

$$T_{up} = 0.31T_{out} + 17.8 + \frac{1}{2}\Delta T_{80\%} \tag{7.20}$$

where T_{out} is the monthly average outdoor air temperature (°C), and ΔT is the temperature band of the mean comfort zone (°C), $\Delta T_{80\%}$ is 3.5°C for 80% thermal acceptability. The lower threshold of the outdoor dry-bulb temperature T_{out} is determined by the lowest air supply temperature specified in ASHRAE 55 to avoid occupants experiencing unpleasant drafts, usually 12.8°C. The threshold of the dew-point temperature is chosen when T_{dew} is below 17°C for the sake of humidity control (Hiyama and Glicksman, 2015). The maximum allowable indoor air velocity $u_{in,\max}$ is 0.8 m s^{-1} according to ASHRAE 55 (ASHRAE, 2017). The corresponding upper threshold for the outdoor wind speed $u_{out,up}$ can be calculated by the following empirical equation (Phaff *et al.*, 1980):

$$u_{in,\max} = \sqrt{C_1 u_{out,up}^2 + C_2 h \Delta T_{\max} + C_3} \tag{7.21}$$

where h is the height of the opening (m), $\Delta T_{\max}$ is the hourly maximum temperature difference between outdoor and indoor temperatures (°C); C_1 is the wind speed coefficient, usually 0.001; C_2 is the buoyancy coefficient, usually 0.0035 m s^{-2} K^{-1}; and C_3 is the turbulence coefficient, usually 0.01 m^2 s^{-2}.

The NV hours in 1,854 locations around the world have been calculated by using the previous method (Chen *et al.*, 2017). The results show that the subtropical highland climates, including South-Central Mexico, the Ethiopian Highlands and Southwest China, are most favourable for NV; Mediterranean climates, which occur not only near the Mediterranean Sea, but also in California, Western Australia, Portugal and Central Chile, are also conducive to the use of NV; In some desert climates and areas with a large diurnal temperature range, including the Middle East and Central Australia, greater-than-expected NV hours are observed due to the significant potential for night-purge ventilation; Countries in Southeast Asia, including Singapore and Malaysia, have hardly shown any potential for NV, due to the hot and humid climate throughout the year (see Figure 7.14). It can be seen that due to different climate conditions in different regions, the potential for using NV varies considerably (see Figure 7.15).

Data from 259 meteorological stations in Europe are used to calculate degree-hours based on a variable building temperature within a standardized range of thermal comfort and the passive cooling effect of ventilated buildings at night analysed (Artmann *et al.*, 2007). The climatic cooling potential (CCP) is defined as a summation of the products between indoor temperature T_i (°C) and outdoor temperature T_o (°C) difference and time interval:

$$CCP = \frac{1}{n}\sum_{n=1}^{N}\sum_{h=h_i}^{h_f} m_{n,h}\left(T_{i,n,h} - T_{o,n,h}\right) \tag{7.22}$$

$$\begin{cases} m = 1h \text{ if } T_i - T_o \geq \Delta T_{crit} \\ m = 0 \text{ if } T_i - T_o < \Delta T_{crit} \end{cases}$$

where h is the time of day, where h=0, 1, ..., 22, and 23, h_i and h_f are the initial and final times of ventilation, and ΔT_{crit} is the threshold of the temperature difference for effective convection, usually 3K. The temperature of the building varies throughout the day when it stores or releases energy, the building temperature T_i is assumed to oscillate harmonically:

$$T_{i,h} = 24.5 + 2.5\cos\left(2\pi\frac{h - h_i}{24}\right) \tag{7.23}$$

The results show that the Nordic region has great potential for NV utilization, with significant potential found in Central, Eastern and even some regions of Southern Europe (see Figure 7.16). Oropeza-Perez and Østergaard (2014) simulate a household located in Vejle, Denmark with a thermal-airflow programme during June, July and August and calculate the indoor air temperatures during this period. Results show that a reduction of 90% in the mechanical ventilation hours is possible, showing the feasibility of achieving thermal comfort within the house by using NV.

The potential for using NV in 65 cities in North America is evaluated based on climate conditions and mass flow rates from wind-driven NV (Cheng *et al.*, 2018). For a given internal cooling load (kW) denoted by CL, the indoor air temperature T_i (°C) can be obtained based on energy conservation:

$$T_i = \frac{CL}{\dot{m}}C + T_o \tag{7.24}$$

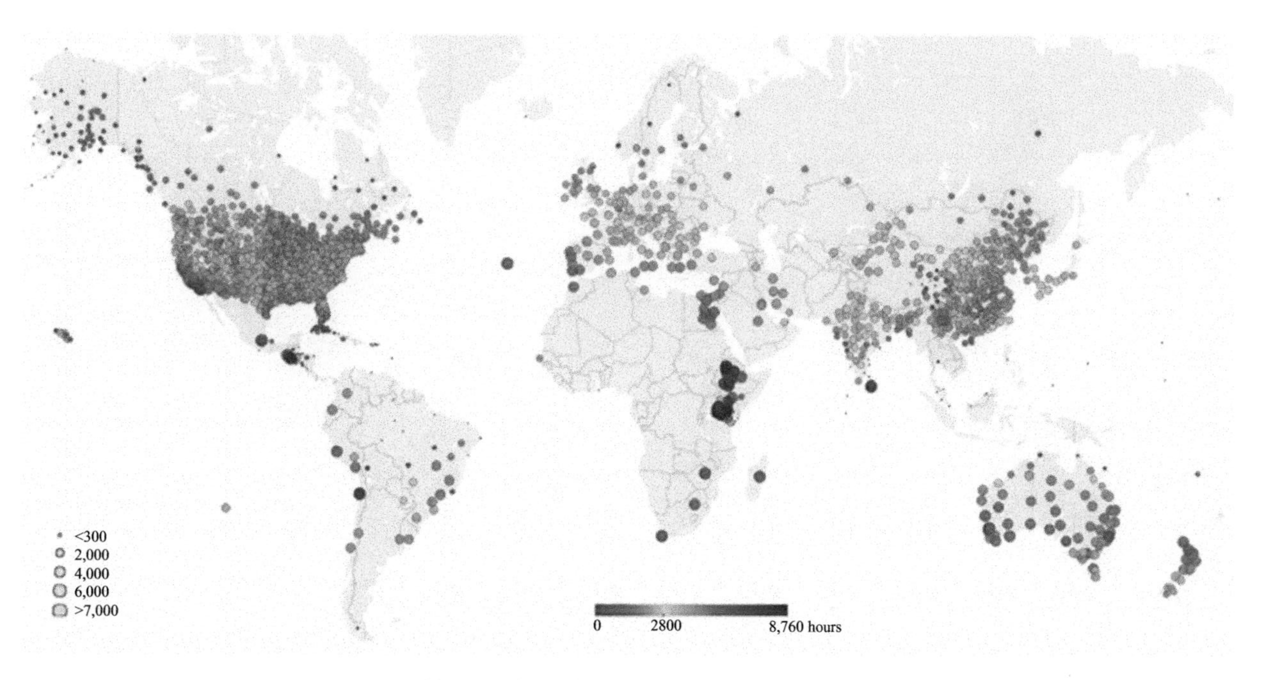

Figure 7.14 World map of NV hours in 1,854 locations (Chen et al., 2017)

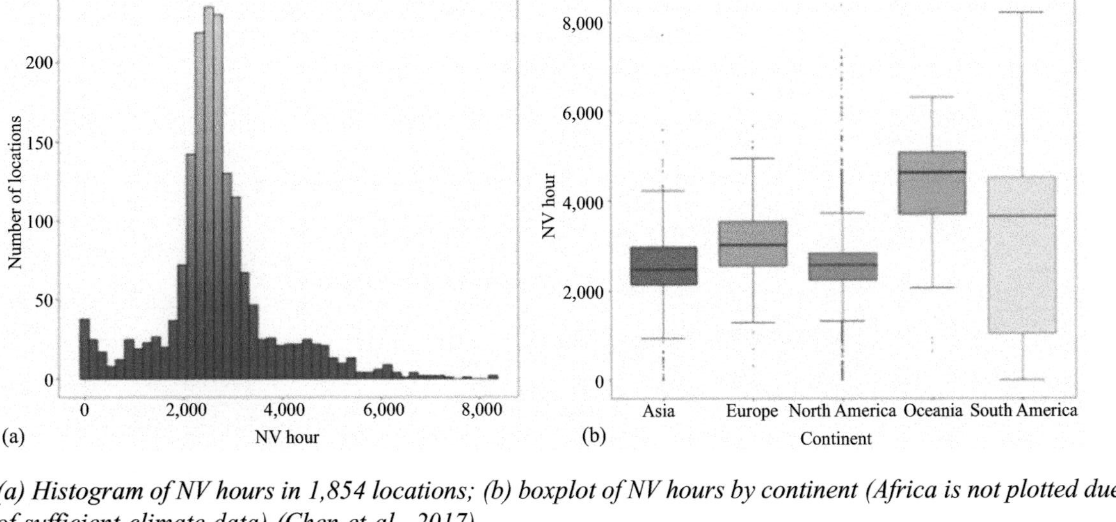

Figure 7.15 *(a) Histogram of NV hours in 1,854 locations; (b) boxplot of NV hours by continent (Africa is not plotted due to the lack of sufficient climate data) (Chen et al., 2017)*

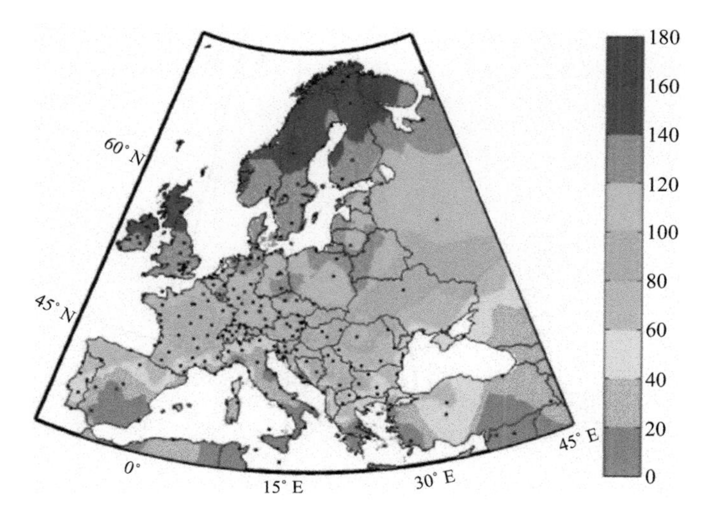

Figure 7.16 Map of mean climatic cooling potential (Kh/night) in July (Artmann et al., 2007)

where T_o is the outdoor air temperature (°C), $\dot{m}$ is the mass flow rate from the NV (kg s^{-1}) obtained by CFD simulations and C is the specific heat of air (kJ kg^{-1} °C^{-1}). The annual NV potential is quantified by the total number of hours in a year when the indoor air temperature T_i and outdoor air relative humidity (RH) fall within the acceptable ranges for occupants of the buildings:

$$\begin{cases} 21.5°\text{C} \le T_i \le 27.8°\text{C} \\ 20\% \le RH \le 70\% \end{cases} \tag{7.25}$$

Then, the total NV potential hours of all the selected cities in North America are counted for a given indoor heat gain (see Figure 7.17). The greatest NV potential can be achieved in the south-west of the United States, where the climate conditions are moderate. The maximum value is 1,563h, which occurs in California. However, NV can rarely be used in Miami with its hot and humid conditions in summer, and Montreal with its high mid-summer temperatures and severe cold conditions in winter.

China has been divided into five climate zones for the thermal design of buildings: Severe Cold, Cold, Hot Summer and Cold Winter, Hot Summer and Warm Winter and Temperate (or Mild) (The Ministry of Housing and Urban-Rural Development of the People's Republic of China, 2016). The NV cooling potential (NVCP) index is proposed to evaluate NV potential by considering internal heat gain levels (high, medium and low), ventilation types (cross ventilation or single-sided ventilation), a building's thermal mass (heavyweight, medium and light-weight) and thermal comfort levels (high with 90% satisfaction and low with 80% satisfaction) based on the ASHRAE Standard (ASHRAE, 2017) for a typical office building. This allows the calculation of the NV potential of the five climate zones

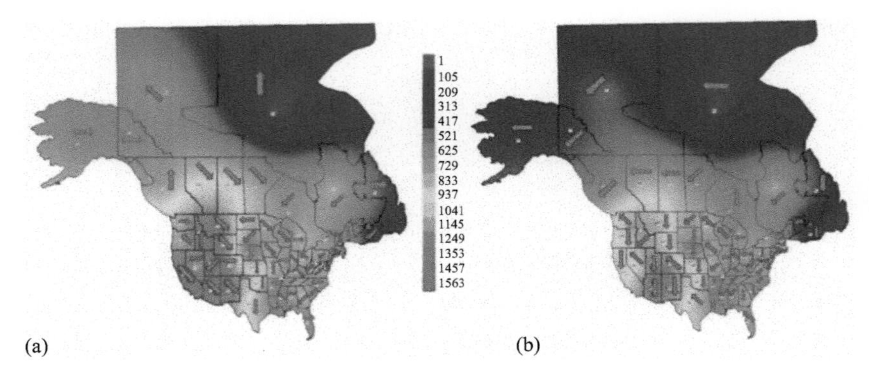

Figure 7.17 Maximum annual NV potentials (h year^{-1}) across North America: (a) single-sided and (b) cross ventilation for 70 W m^{-2} internal cooling load (Cheng et al., 2018)

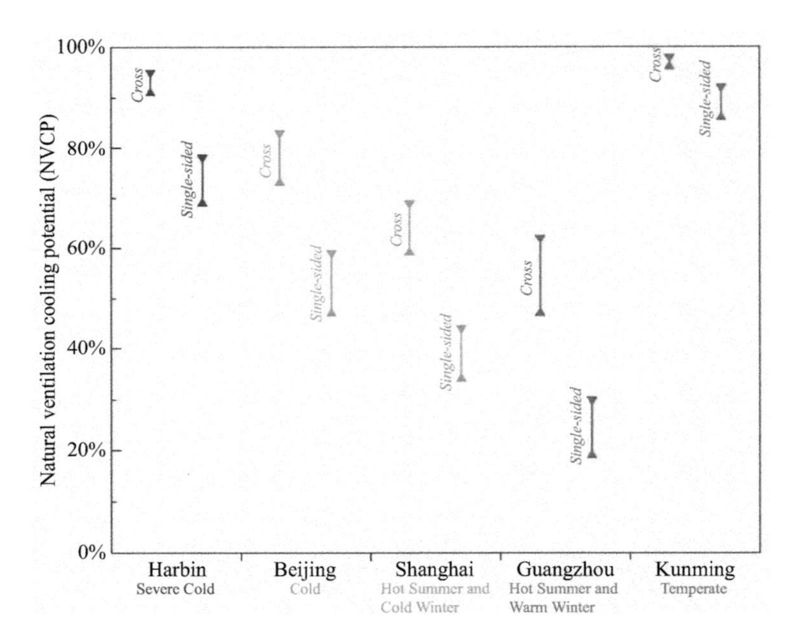

Figure 7.18 NVPC for buildings with medium heat gain in cities from five climate zones by Yao et al. (2009): the upper triangle represents 80% satisfaction and the lower triangle represents 90% satisfaction

in China (Yao *et al.*, 2009). Figure 7.18 shows the NVCP for buildings with medium heat gain in cities from five climate zones: Harbin from the Severe Cold zone, Beijing from the Cold zone, Shanghai from the Hot Summer and Cold Winter zone, Guangzhou from the Hot Summer and Warm Winter zone, and Kunming from the Temperate zone. The climate with lower external air temperature in summer has a higher NVCP. For example, Kunming from the Temperate zone and Harbin from the Severe Cold zone have lower external air temperatures than other cities, and therefore they have higher NVCPs. In the Cold zone and the Hot Summer and Cold Winter Zone, NV is not sufficient for space cooling in summer. In the Hot Summer and Warmer Winter zone, NV cannot provide space-cooling, especially for single-sided ventilation. So NV design in this area should be considered prudently, and mechanical cooling or air conditioning is desirable.

To provide more refined guidance for NV design, a new climate zoning method, with two-tier classification, considering temperature, humidity, solar radiation and wind, is proposed (Xiong *et al.*, 2019). Also, the Hot Summer and Cold Winter zone, where the climatic characteristics are still quite diverse, is divided into seven sub-zones (see Figure 7.19).

The building energy simulation with EnergyPlus is conducted to estimate the NV potential of each sub-zone under different ventilation rates from March to November. Take Nanjing city from zone A1 with the highest demand for both cooling and heating as an example (see Figure 7.20), there is a very slim opportunity to naturally ventilate a building in winter, the NV potential is 59% with a ventilation rate of 5ACH during the period from March to November, but only 19% in July and August. But for the city of Chengdu from zone C2 with lower demand for both cooling and heating (see Figure 7.21), some opportunity to naturally ventilate a building in early and late winter presents itself, the NV potential is 80% with a ventilation rate of 5ACH during the period from March to November and is 58% even in July and August.

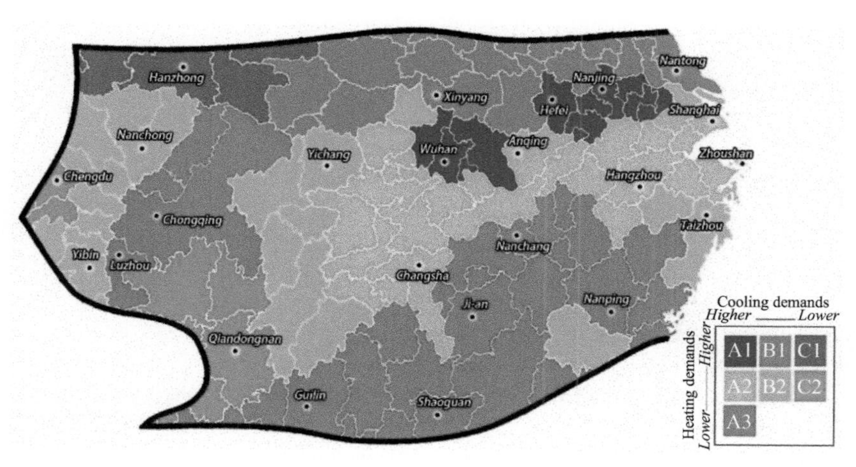

Figure 7.19 Sub-zones of the Hot Summer and Cold Winter zone identified using heating and cooling demands for passive strategies (Xiong et al., 2019)

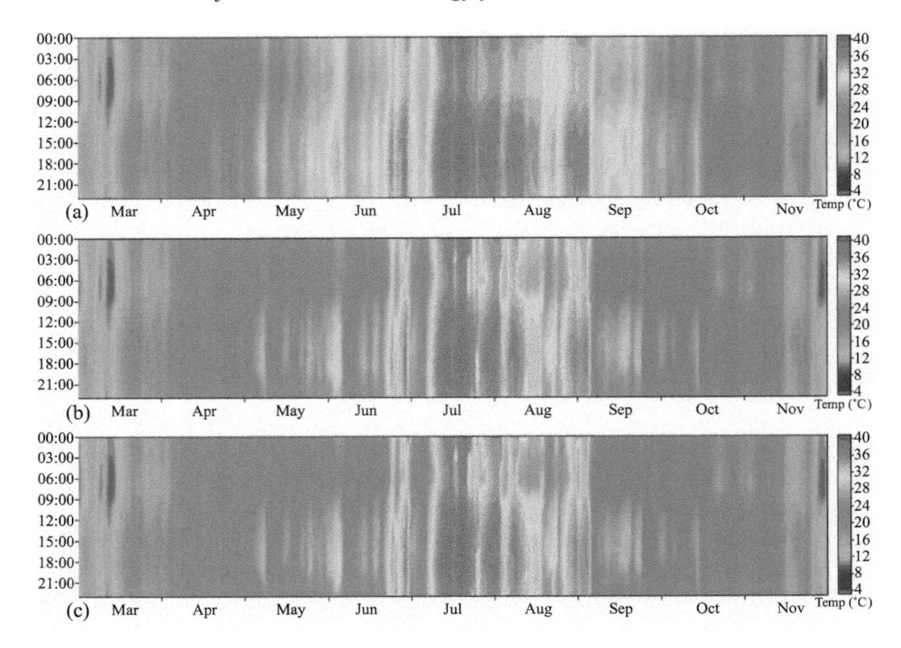

Figure 7.20 Annual temperature variation of a baseline building under natural ventilation in the representative city of Nanjing in climate zone A1: (a) infiltration ventilation of 1ACH; (b) natural ventilation of 5ACH; (c) natural ventilation of 10ACH

7.5.2 Building characteristics

NV capacity is very much climate dependent. Those designing the NV strategies should cooperate to consider the building characteristics in order to maximize the use of climate resources.

Internal heat gain has a significant impact on NV potential. Appropriate internal heat gain is necessary for the stack ventilation driven by the buoyancy effect arising from temperature differences. However, a higher internal gain will limit the capacity of NV. According to the CIBSE AM10, if the internal gains are greater than 30–40 W m^{-2}, such as in highly equipped office buildings, NV is not encouraged as a method of controlling summer overheating (CIBSE, 2005).

As mentioned earlier, Yao's NVCP research also compares the NV potential for buildings with different internal heat gain levels (low, medium and high) (Yao *et al.*, 2009). Take a typical building in Shanghai as an example, the NVCP decreases with higher internal heat gain (see Figure 7.22). When the internal heat gain rises from 15 to 35 W m^{-2}, the NVCP decreases from 72% to 62% for cross ventilation with 80% satisfaction, and it decreases from 51% to 35% for single-sided ventilation.

The effective thermal mass with night ventilation can increase the NV cooling capacity. Night ventilation makes full use of the diurnal temperature variations.

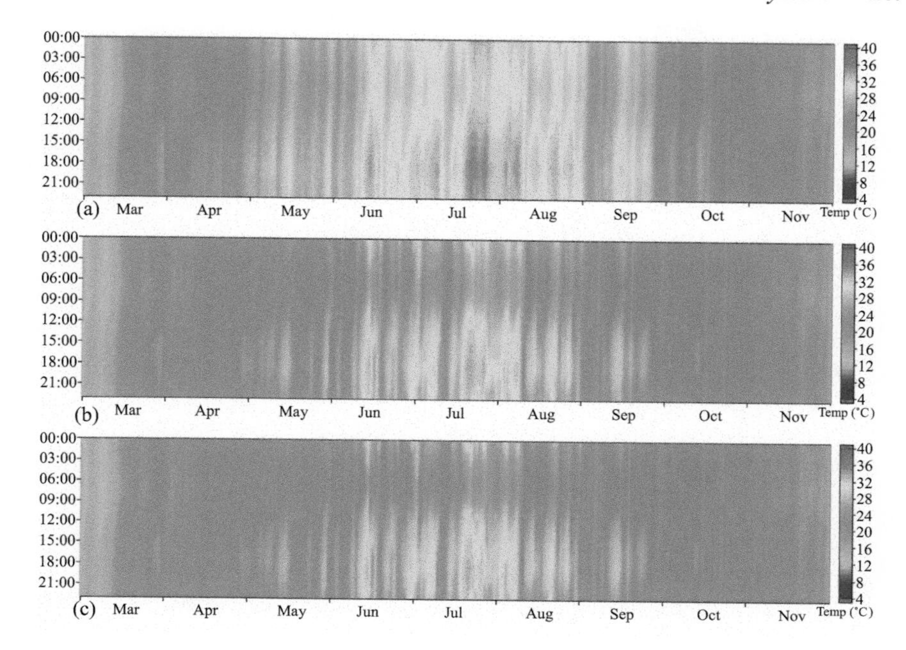

Figure 7.21 Annual temperature variation of a baseline building under natural ventilation in the representative city of Chengdu in climate zone C2: (a) infiltration ventilation of 1ACH; (b) natural ventilation of 5ACH; (c) natural ventilation of 10ACH

The outdoor temperature at night-time is lower than in daytime, making the differences between indoor and outdoor temperature larger, which enhances the flow rate of buoyancy-driven ventilation. The air with a lower temperature will cool the thermal mass of the building fabric, which reduces the mean radiant temperature of the internal space, so improving the comfort perceptions on the following day. However, there should be appropriate controls to avoid over-cooling which can lead to occupants feeling discomfort in the morning. As shown in Figure 7.23, a heavyweight structure significantly increases NV performance. The peak temperature dropped 4°C to around 31°C without night ventilation and dropped 2°C to around 30°C with night ventilation.

7.5.3 Air pollution

The implementation of NV is constrained by ambient air pollution. Occupants will close the windows when the outdoor air quality is unsatisfactory. Thus, it is essential to be aware of local outdoor pollution levels and the need to implement interventions to control them.

Tong *et al.* (2016) also calculated the NV hours for 35 major Chinese cities spread over 5 climate zones considering air pollution problems. There are clear

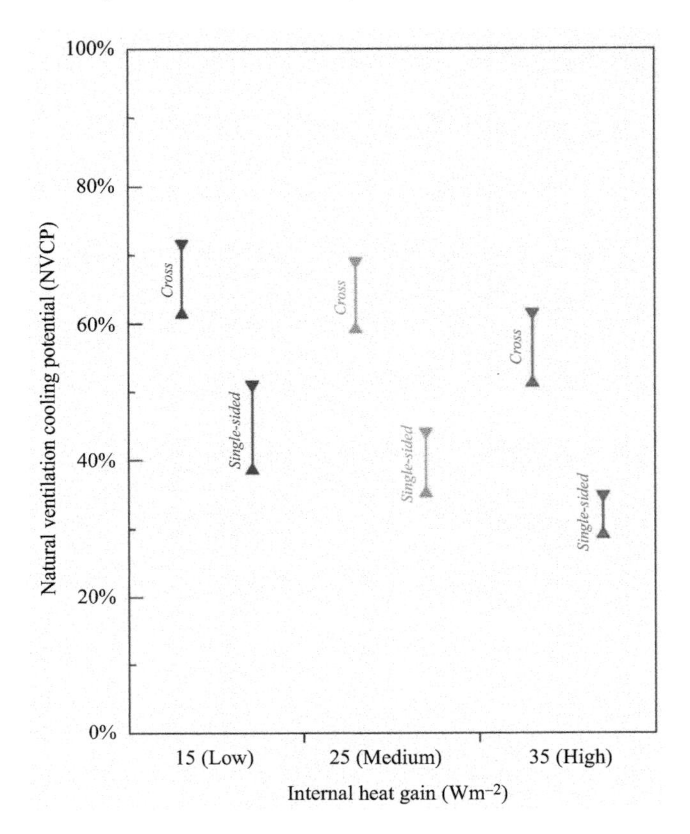

Figure 7.22 NVPC for a typical building in Shanghai from the Hot Summer and Cold Winter zone with various levels of internal heat gain (Yao et al., 2009): the upper triangle represents 80% satisfaction and the lower triangle represents 90% satisfaction

differences in NV hours across the five climate zones (see Figure 7.24). The mild zone includes the cities with most NV hours, which makes them most suitable for the utilization of almost year-round NV. On the contrary, the Hot Summer and Cold Winter zone is the least favourable for NV due to its extreme climate characteristics in both summer and winter, which is consistent with the conclusion by Yao *et al.* (2009). However, the NV potential is found to decrease considerably for all the climate zones when air pollution is considered.

There are two main approaches to acquire outdoor air pollution levels around buildings: on-site measurements and modelling predictions.

On-site measurement is highly accurate as it directly reflects the true value at the sampling point (ignoring any system errors). Some studies use the data from the nearest observation station to estimate the influence on NV. However, the pollution level is unevenly distributed due to the urban setting, including community

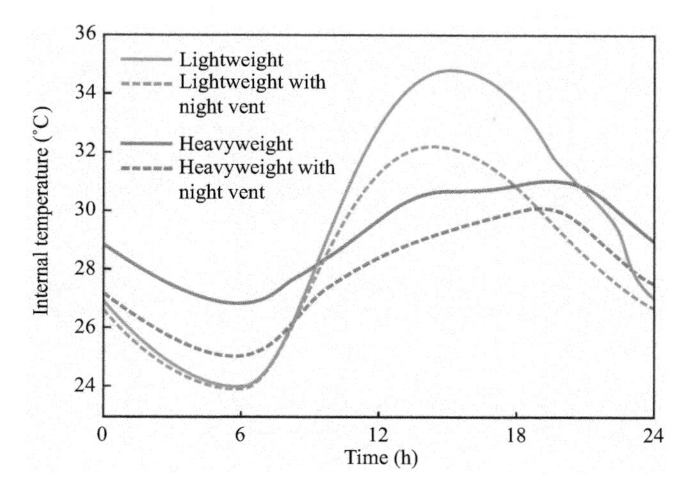

Figure 7.23 Effect of thermal mass and ventilation rate on peak indoor temperature (CIBSE, 2005)

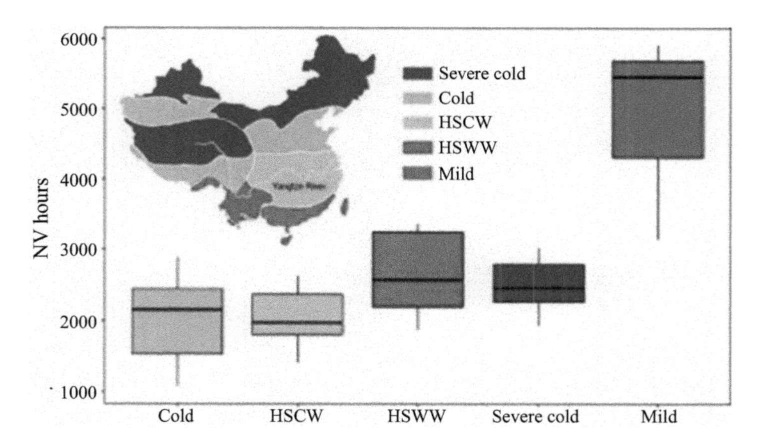

Figure 7.24 NV hours considering air pollution in five climate zones of China by Tong et al. (2016)

morphology and local pollutant sources. The cost of on-site measurement, including sensors, maintenance and labour, is very high, which makes it impractical to take measurements everywhere. Park *et al.* (2014) carried out a field measurement of particles in 15 single-family apartments located in the city of Seoul, South Korea. For all ventilation types, indoor particle concentrations were found to be lower than outdoors because of the infiltration process through the building envelope and deposition onto indoor surfaces. However, no assumption is made about the surrounding area, so that it is not possible to understand how the urban layout affects the results of the monitoring campaign.

There is a need to develop local-scale modelling to estimate pollutant concentrations, which can be further divided into two types: high-dimension, process-driven, physical models and low-dimension, data-driven, statistical models.

The physical models simulate the dispersion process based on basic CFD theory and the mass-transfer mechanism; they demand sufficient knowledge of microclimate conditions, particle emission sources and an explicit description of physical deposition and chemical transformation processes (Lateb *et al.*, 2016; Li *et al.*, 2006). Studies using CFD simulations to predict pollutant concentrations have focused on the street canyon (Blocken *et al.*, 2012; Tominaga *et al.*, 2011; Vicente *et al.*, 2018). This method is mostly used to analyse the pollutant dispersion around buildings from certain known sources (Ai and Mak, 2013; Short *et al.*, 2018).

A more detailed and fully coupled approach has been employed in a study about the impact of traffic-related pollution on IAQ for a naturally ventilated office building (Tong, *et al.*, 2016). Most people live in buildings that are in close proximity to roadways, indeed the modelled road-building geometry (see Figure 7.25(a)) is a representative and generalizable scenario in an urban setting. Assuming the roadway is a well-mixed emission zone, with a known on-road emission rate and vehicle-induced turbulence (Hagler *et al.*, 2012). The modelled road represents a major roadway with annual average daily traffic of 38,000 and a speed limit of 45 mph. A small office building with a distance *D* to that roadway is created. The building geometry employed in this study is chosen from the US Department of Energy commercial reference building database (Deru *et al.*, 2011). The relationship between the linear pollutant source (the road) and the study building is explored by varying the distance from the roadway, the size and location of the window and the oncoming wind speed (see Figure 7.25(b)–(e)). The baseline

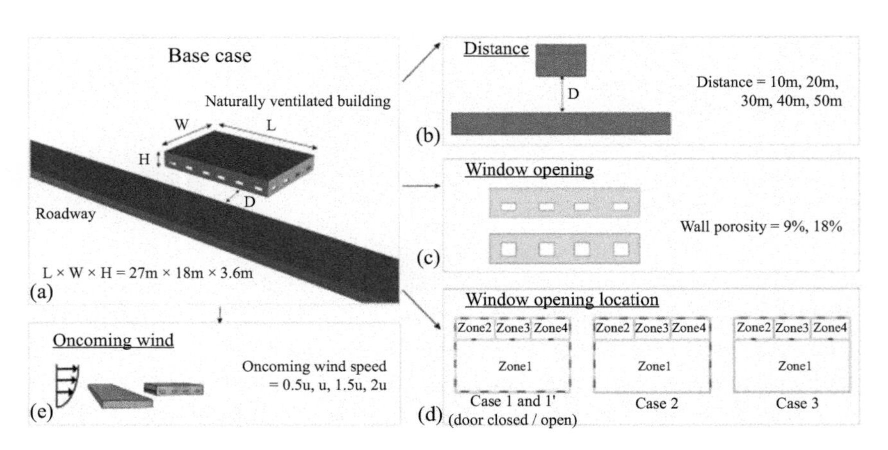

Figure 7.25 Schematics of (a) the baseline geometry and test building parameters, including (b) distance, (c) window size, (d) window opening locations and (e) oncoming wind speed (Tong et al., 2016)

scenario is where $D=10$ m, wall porosity=9%, window opening location=Case 1 and wind speed=u for five distances between the building and the road: 10, 20, 30, 40 and 50 m. The options for the window-to-wall ratio are 9% and 18%. There are three scenarios for window opening locations: Case 1 represents the scenario where the interior doors are shut; Case 1′ has the same configuration as Case 1 except the interior doors are open; Case 2 represents the scenario where only the windows facing the roadway are closed; Case 3 represents the scenario where both the side windows and windows facing the roadway are shut. The computational domain has a dimension of 200 m × 200 m × 40 m, and it is divided into 3.3–4.5 million unstructured elements. The boundary conditions in this case are similar to those introduced in Section 7.4.3.

The results show that the IAQ strongly depends on the distance between the roadway and the building, the wind conditions and the size and location of the windows. Figure 7.26 shows the exponential curve fitting based on concentrations at five simulated building-road distances. The indoor concentration at $D'=4$ declines to approximately 82% of the baseline scenario.

Low-dimensional, data-driven modelling is favoured due to its highly efficient simulation based on the established relationships between variables and responses, while ignoring the limited knowledge of the processes involved. Multiple linear regression (MLR) and artificial neural networks (ANNs) are mainstream approaches to handle the estimation of pollutant concentrations. Honarvar and Sami (2019) considered the road network structure data to predict the PM concentration based on a transfer learning perspective in which a neural network and MLR were leveraged to provide the core of the prediction. The urban morphology influences the dispersion of particles; de Gennaro *et al.* (2013) developed the ANN model to

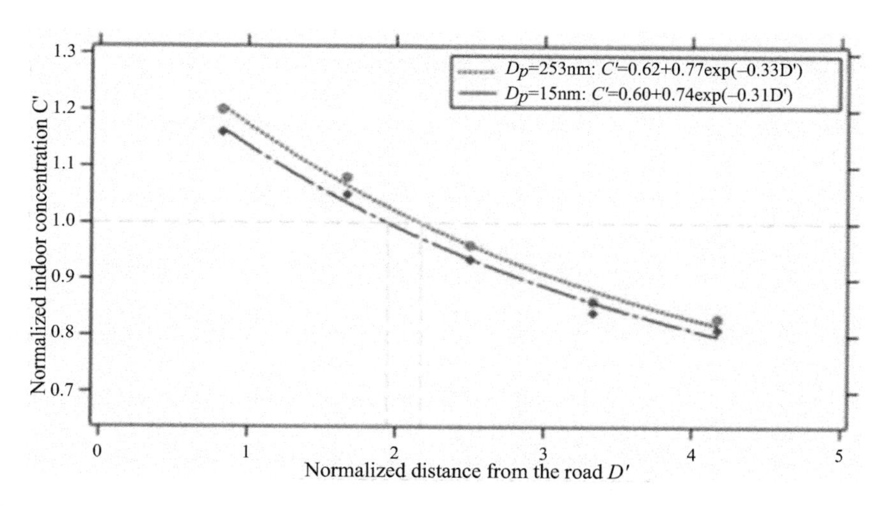

Figure 7.26 Normalized average indoor concentration C' as a function of distance D' away from the roadway (Tong et al., 2016)

forecast PM_{10} daily concentrations in two contrasting environments: a regional background site and an urban background site, with local meteorological data and information about the origin of air masses being used as inputs.

A data-driven ANN model that can rapidly estimate particulate concentrations based on the background pollution level, weather conditions, urban morphology and local pollution sources has been developed (Xiong *et al.*, 2020) (see Figure 7.27). The data for the predictors is collected from a publicly available database: (1) background particle pollution data is obtained from official observation sites in the studied areas, containing hourly PM_{10} and $PM_{2.5}$ concentration (*PM*) data from a number of scattered locations; (2) meteorological condition data is obtained from the state meteorological administration, including ground-level (2-m height) air temperature (*Temp*), relative humidity (*RH*), wind speed (*WS*) and precipitation (*RF*) from a local weather station; (3) urban morphology is obtained based on satellite images and the principle of solar shadow, the building coverage ratio (*BCR*) is calculated at different heights to express the horizontal compactness of the infrastructure, and the coverage-area-weighted building height (*BH*) is calculated to express the extension in the vertical direction; (4) the local pollution sources information, containing emissions from road traffic and construction sites, is obtained from the navigation software and satellite images. The distance to the nearest main road (D_t), the speed limit of the nearest main road (SL_t), the lane-count of the nearest main road (LC_t), the average congestion status in a land lot of 500 m × 500 m (CS_t), the single-lane road length per unit area in a land lot of 500 m × 500 m (*SLRL*), the area of construction sites within 500 m (A_{cs}) and the distance to the nearest construction site (D_{cs}) are all calculated. Then, the real-time field measurements of particle concentrations are conducted in selected locations and the measured value at the specific location represents the predicted variable. After the data for the predictors and the predicted variable have been prepared, they are randomly divided into two subsets: one for model training and the other for testing, in a ratio of 3:1. And the training dataset is fed into the ANN to develop the particle estimation model that is verified with the testing dataset for its accuracy evaluations. Finally, this model can be used to predict the PM_{10} and $PM_{2.5}$ concentrations for any location at any time with known data for the predictors.

7.5.4 Urban context

The shape and arrangement of buildings in an array, which influences the flow field around the target building, constitutes the urban context of NV. For most buildings in cities, the oncoming wind has been disturbed by the upstream buildings, which makes the flow characteristics more complicated. This problem needs to be studied on a larger scale, involving factors such as the spacing between multiple buildings and the orientation and direction of the street.

According to the different ratios of building height (*H*) to building spacing (*W*), three basic flow patterns are identified (Oke, 1988): (a) isolated roughness flow, when the building spacing is wide enough ($H/W<0.35$), will generate flow wakes similar to an isolated building, but the difference is that the flow field of

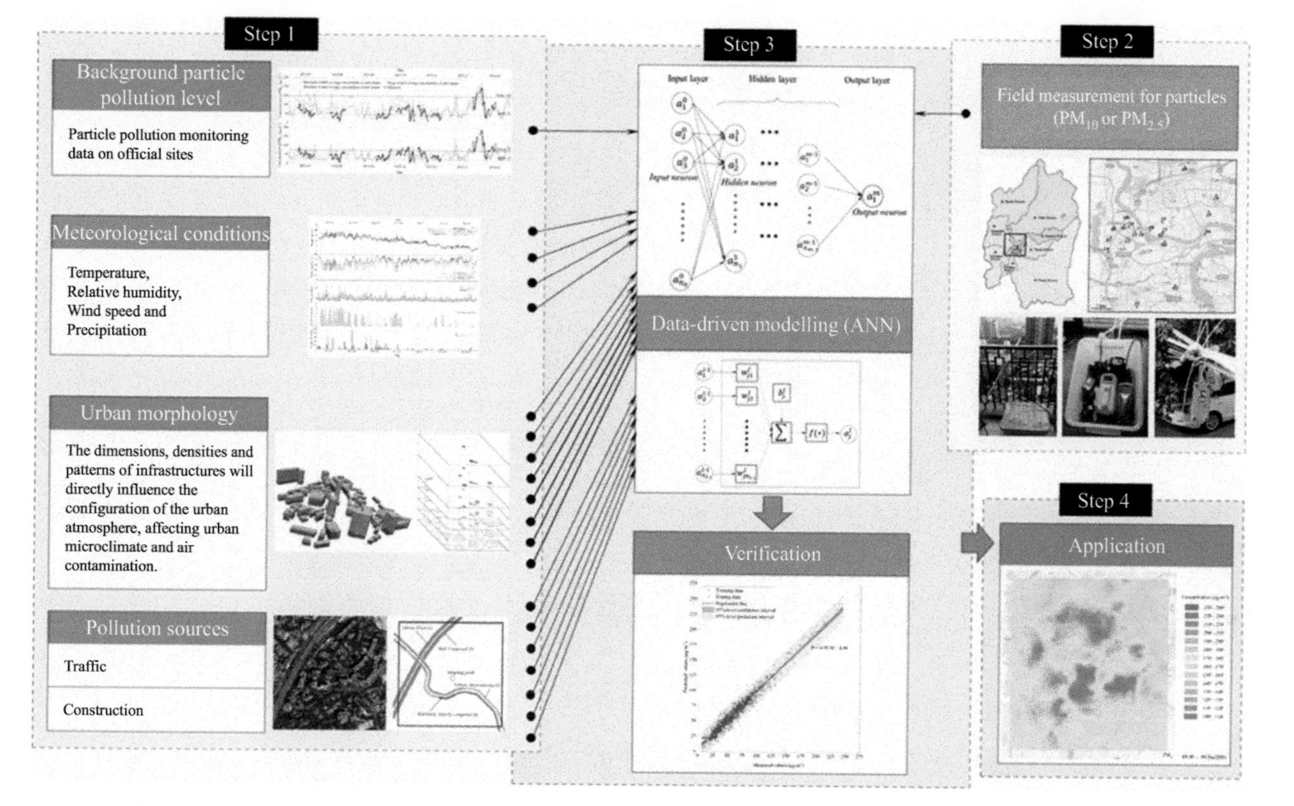

Figure 7.27 The framework of the ANN model for the estimation of particle pollution levels based on the background pollution level, weather conditions, urban morphology and local pollution sources (Xiong et al., 2020)

each building in the city will be disturbed by the residual effects of the upstream buildings; (b) wake interference flow, where the arrangement of buildings is denser ($0.35<H/W<0.65$) and the building spacing along the mainstream direction is equivalent to the size of the cavity area of a building, will generate a vortex in the cavity area behind the upstream building which is enhanced by the downward flow on the downstream building's windward side; (c) skimming flow, where the distance between the buildings is closer ($H/W>0.65$) so that the airflow at the top directly skips the roof of the building and is less likely to enter the street valley, and the airflow above the building is basically separated from the airflow in the street valley.

Although abstract, the summary of these flow patterns provides a basic framework for describing the general influence of the urban context on airflow. Later scholars have proposed six more complex basic block forms: pavilions, slabs, terraces, terrace-courts, pavilion-courts and courts (see Figure 7.28) for the daylight and NV analysis (Ratti *et al.*, 2003). Many of studies have also been carried out on the flow field distribution of different block shapes (Asfour, 2010; Hu and Wang, 2005; Toparlar *et al.*, 2017; Zhang *et al.*, 2005).

The NV potential of the case study of the Shuishixiang building in Chongqing, China (as mentioned in Section 7.4.3) is evaluated considering its urban context (Costanzo *et al.*, 2019). The study area for CFD simulation consists of 1,200 m × 2,100 m around the target building (see Figure 7.29).

A series of CFD simulations was conducted with a step of 45° for wind directions so that the pressure coefficient at different heights on each surface of the target building under different wind directions is obtained (see Figure 7.30). The highest value of around 0.552 occurs on the windward side for the predominant north-west (315°) wind direction. The case study building is surrounded by buildings of similar height. Under this circumstance, it can be considered the representative of these buildings in terms of the peak values achievable. A consistently lower distribution of the wind pressure coefficients for all these buildings is identified, being always in the range of −0.170–0.120 for every wind direction.

After obtaining the wind pressure coefficient C_p from steady-state CFD simulations, the external coupling approach passes these values to the thermal simulations as the exchange variables.

7.5.5 *Systematic evaluation*

To comprehensively evaluate the NV potential, a systematic method concerning various environmental factors, including climate characteristics, surrounding wind fields and outdoor air pollution, is desired. It is helpful in both the early design stages and for the control of building operations.

Based on the discussions of various factors in the previous sections, a possible systematic framework to estimate NV potentials can be proposed (see Figure 7.31). The climate characteristics for a given building thermal performance will result in different indoor thermal conditions under NV, the comfort hours are counted for each NV potential. The urban context will influence the surrounding wind field of

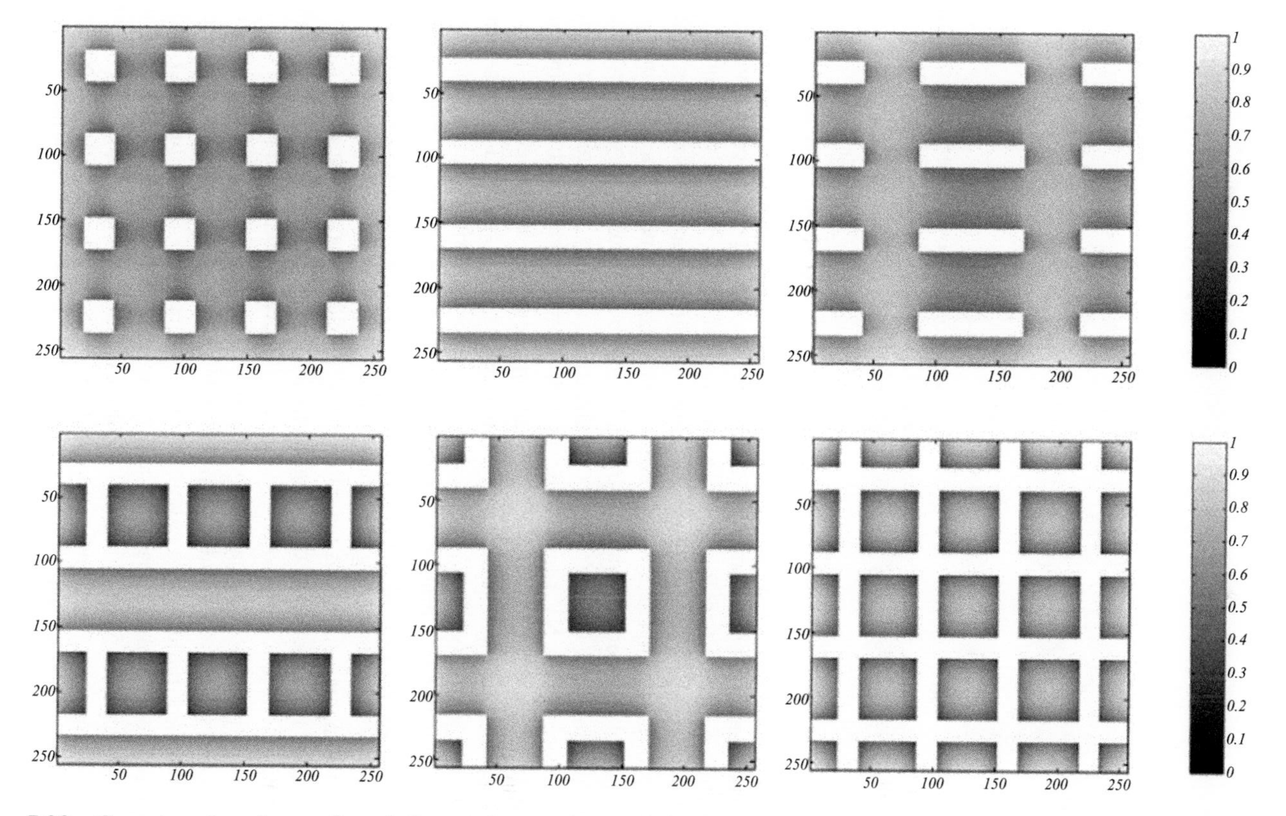

Figure 7.28 Generic urban forms, from left to right: pavilions, slabs, terraces, terrace-courts, pavilion-courts and courts (Ratti et al., 2003)

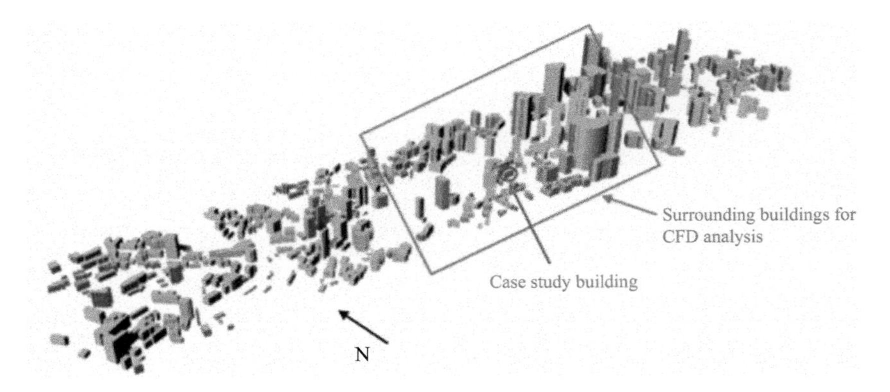

Figure 7.29 Axonometric view of the 3D model of the district with the case study building highlighted in red and the CFD computational domain highlighted in green (Costanzo et al., 2019)

the target building, which is a determinant for the flow rate of NV affecting the effective ventilation performance. The appropriate urban context can provide more ventilation opportunities, for example, a very dense urban context will limit the use of natural resources. Occupants in buildings within the urban area suffer from all kinds of air contaminants, even within an indoor environment. The local emissions make the pollutant concentrations unevenly distributed, and they will penetrate the indoor environment via NV. If the concentrations are large enough to threaten the health of occupants, NV should be restricted. After considering these contributory factors, the actual NV potentials can be comprehensively evaluated.

Take a benchmark building in Chongqing as an example. The non-heating-and-cooling period is around 40.9% of the time during spring, summer and autumn in this area with an infiltration of 1.0ACH. The NV potential during spring, summer and autumn soars to 78.5% when NV of 5.0ACH is applied (see Figure 7.32(a)). However, when considering the impact of urban forms on the surrounding wind fields, the actual air exchange rate changes with time and locality. The NV potential of the case-study dwelling drops to 69.3% (see Figure 7.32(b)), as it is located in the old urban area with dense multi-storey slab buildings.

Using the rapid estimation model of particle concentrations proposed by Xiong *et al.* (2020), the outdoor PM_{10} and $PM_{2.5}$ concentrations in the location of the benchmark building are calculated. Combined with the assumption that the coefficient of infiltration is 0.70, and the average particle concentrations from indoor sources are 5 $\mu g\ m^{-3}$ in Chongqing based on the statistics derived from field measurement, the time-series indoor particle concentration is obtained. Filtering the periods when indoor concentrations exceed the threshold in the standard – the threshold for PM_{10} is 150 $\mu g\ m^{-3}$ and for $PM_{2.5}$ is 75 $\mu g\ m^{-3}$ – the practical period for NV during spring, summer and fall is only 58.8% (see Figure 7.33).

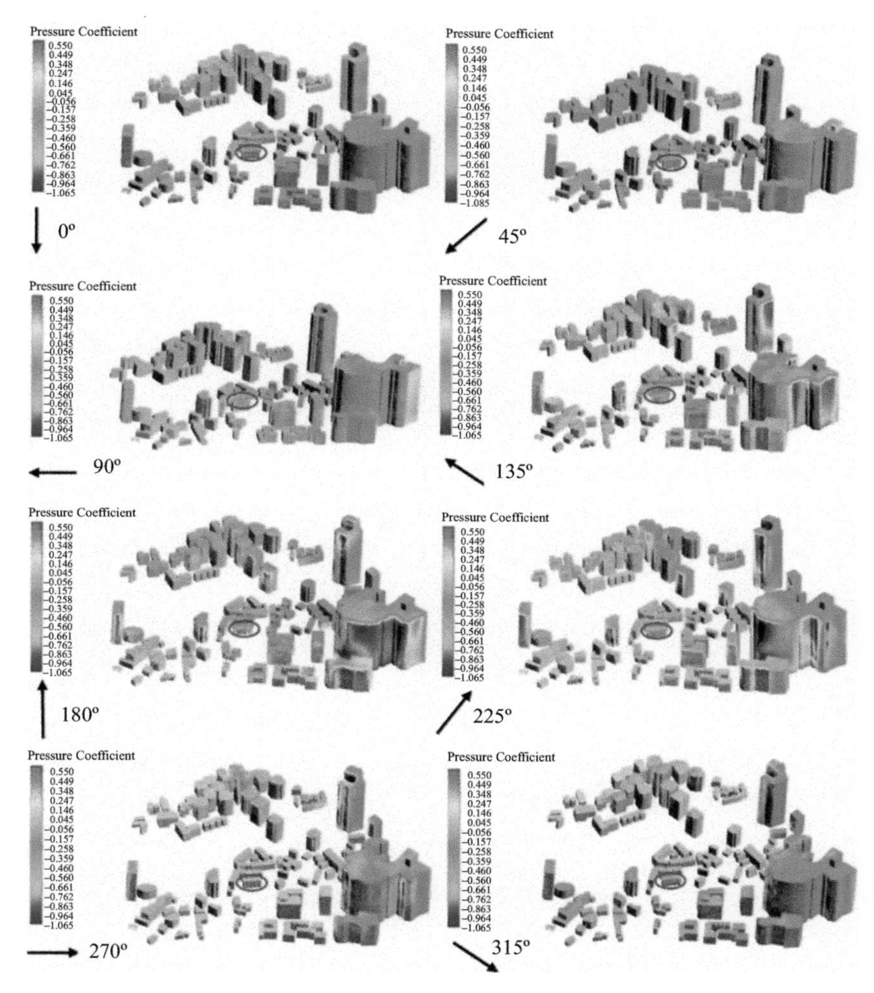

Figure 7.30 Coefficient of pressure distribution in the urban setting according to different wind directions (Costanzo et al., 2019)

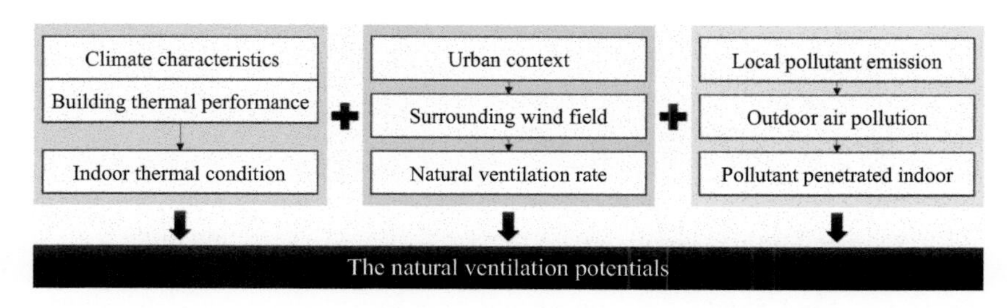

Figure 7.31 A systematic framework to estimate natural ventilation potentials

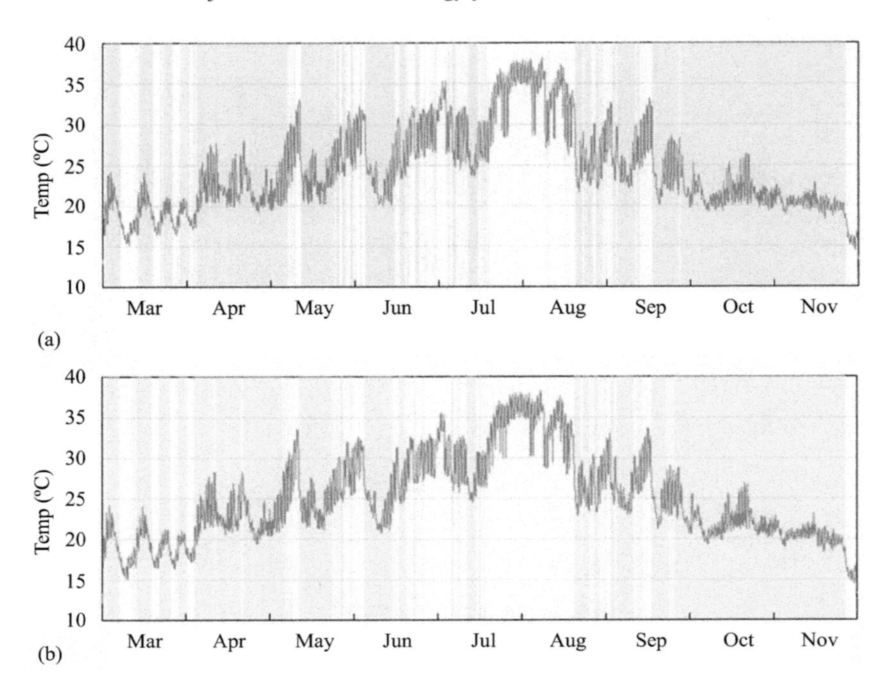

Figure 7.32 The simulated indoor temperature during spring, summer and autumn of the case-study building in free-running mode by EnergyPlus based on (a) the scenario infiltration of 1.0ACH and (b) the scenario where natural ventilation is 5.0ACH

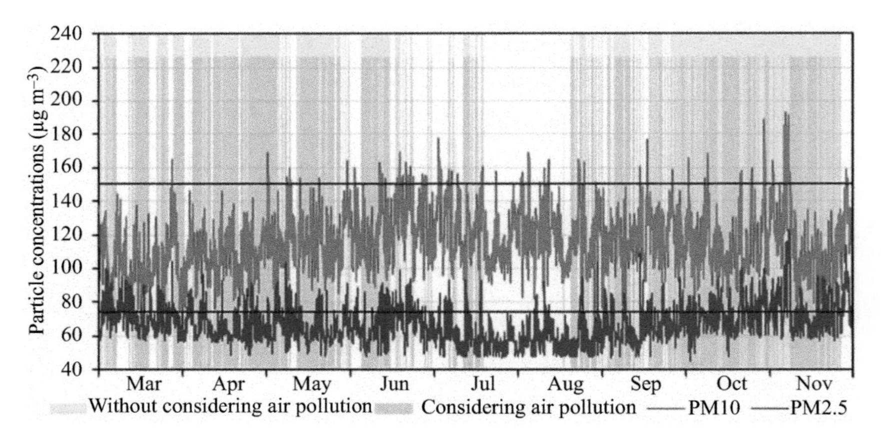

Figure 7.33 The simulated indoor particle concentration level and practical natural ventilation hours during spring, summer and autumn

7.6 Summary

A building should be properly ventilated to support the health of its occupants by supplying fresh air into the living space. NV can fulfil this purpose without energy consumption. The basic strategies for NV include single-sided ventilation, cross ventilation, stack ventilation and other types. In reality, buildings combine these basic schemes leading to a very complicated airflow so that CFD techniques must be applied to analyse the air distribution. When the outdoor climate is favourable, NV can provide a potential cooling effect for the space and remove the heat and moisture released by people and equipment. The actual capacity relies on the climatic conditions and building characteristics and is constrained by the quality of the outdoor air and the urban context. The occasions for NV should be comprehensively evaluated to enhance the design and operation of ventilation strategies.

References

Ai, Z. T., and Mak, C. M. (2013). CFD simulation of flow and dispersion around an isolated building: Effect of inhomogeneous ABL and near-wall treatment. *Atmospheric Environment*, 77, 568–578.

AMCG. (2015). *Fluidity Manual v4.1.12*.

Artmann, N., Manz, H., and Heiselberg, P. (2007). Climatic potential for passive cooling of buildings by night-time ventilation in Europe. *Applied Energy*, 84 (2), 187–201.

Asfour, O. S. (2010). Prediction of wind environment in different grouping patterns of housing blocks. *Energy and Buildings*, 42(11), 2061–2069.

ASHRAE. (2017). *Thermal Environmental Conditions for Human Occupancy*. Atlanta, GA, US: ASHRAE. Pub. L. No. ANSI/ASHRAE Standard 55-2017.

Blocken, B., Stathopoulos, T., Carmeliet, J., and Hensen, J.L.M. (2011). Application of computational fluid dynamics in building performance simulation for the outdoor environment: an overview. *J. Build. Perform. Simul.*, 4, 157–184.

Chartered Institution of Building Services & Engineers (CIBSE). (2019). *CIBSE Guide A: Environmental Design*. Suffolk: Lavenham Press Ltd.

Chen, Y., Tong, Z., and Malkawi, A. (2017). Investigating natural ventilation potentials across the globe: Regional and climatic variations. *Building and Environment*, 122, 386–396.

Cheng, J., Qi, D., Katal, A., Wang, L. (Leon), and Stathopoulos, T. (2018). Evaluating wind-driven natural ventilation potential for early building design. *Journal of Wind Engineering and Industrial Aerodynamics*, 182, 160–169.

CIBSE. (2005). *Natural Ventilation in Non-Domestic Buildings*. East Sussex, UK: Berforts Ltd. Pub. L. No. CIBSE Applications Manual AM10.

Costanzo, V., Yao, R., Xu, T., Xiong, J., Zhang, Q., and Li, B. (2019). Natural ventilation potential for residential buildings in a densely built-up and highly polluted environment. A case study. *Renewable Energy*, 138, 340–353.

de Dear, R. J., and Brager, G. S. (2002). Thermal comfort in naturally ventilated buildings: Revisions to ASHRAE Standard 55. *Energy and Buildings*, 34(6), 549–561.

de Gennaro, G., Trizio, L., Di Gilio, A., *et al.* (2013). Neural network model for the prediction of PM10 daily concentrations in two sites in the Western Mediterranean. *Science of the Total Environment*, 463–464, 875–883.

Deru, M., Field, K., Studer, D., Benne, K., Griffith, B., and Torcellini, P. (2011). *US Department of Energy Commercial Reference Building Models of the National Building Stock.*

Etheridge, D. W., and Sandberg, M. (1996). *Building Ventilation—Theory and Measurement*. Chichester: John Wiley & Sons.

Hagler, G. S. W., Lin, M.-Y., Khlystov, A., *et al.* (2012). Field investigation of roadside vegetative and structural barrier impact on near-road ultrafine particle concentrations under a variety of wind conditions. *Science of the Total Environment*, 419, 7–15.

Hiyama, K., and Glicksman, L. (2015). Preliminary design method for naturally ventilated buildings using target air change rate and natural ventilation potential maps in the United States. *Energy*, 89, 655–666.

Honarvar, A. R., and Sami, A. (2019). Towards sustainable smart city by particulate matter prediction using urban big data, excluding expensive air pollution infrastructures. *Big Data Research*, 17, 56–65.

Howarth, A. (1985). The prediction of air temperature variations in naturally ventilated rooms with convective heating. *Building Services Research & Technology*, 6(5), 169–175.

Hu, C.-H., and Wang, F. (2005). Using a CFD approach for the study of street-level winds in a built-up area. *Building and Environment*, 40(5), 617–631.

Laetitia, M., Jiyun, S., Alan, S. C., *et al.* (2020). The Hot Summer-Cold Winter region in China: Challenges in the low carbon adaptation of residential slab buildings to enhance comfort. *Energy and Buildings*, 223, 110181.

Lateb, M., Meroney, R. N., Yataghene, M., Fellouah, H., Saleh, F., and Boufadel, M. C. (2016). On the use of numerical modelling for near-field pollutant dispersion in urban environments – A review. *Environmental Pollution*, 208, 271–283.

Li, X.-X., Liu, C.-H., Leung, D. Y. C., and Lam, K. M. (2006). Recent progress in CFD modelling of wind field and pollutant transport in street canyons. *Atmospheric Environment*, 40(29), 5640–5658.

Nielsen, P. V. (2004). Computational fluid dynamics and room air movement. *Indoor Air*, 14(Suppl. 7), 134–143.

Oke, T. R. (1988). Street design and urban canopy layer climate. *Energy and Buildings*, 11(1–3), 103–113.

Oke, T. R., Mills, G., Christen, A., and Voogt, J. A. (2017). *Urban Climates*. Cambridge: Cambridge University Press.

Oropeza-Perez, I., and Østergaard, P. A. (2014). Potential of natural ventilation in temperate countries – A case study of Denmark. *Applied Energy*, 114, 520–530.

Park, J. S., Jee, N.-Y., and Jeong, J.-W. (2014). Effects of types of ventilation system on indoor particle concentrations in residential buildings. *Indoor Air*, 24(6), 629–638.

Phaff, J., de Gids, W., Ton, J., van der Ree, D., and Schijndel, L. (1980). *The Ventilation of Buildings: Investigation of the Consequences of Opening One Window on the Internal Climate of a Room*. Delft.

Ratti, C., Raydan, D., and Steemers, K. (2003). Building form and environmental performance: Archetypes, analysis and an arid climate. *Energy and Buildings*, 35(1), 49–59.

Short, C. A., Song, J., Mottet, L., Chen, S., Wu, J., and Ge, J. (2018). Challenges in the low-carbon adaptation of China's apartment towers. *Building Research & Information*, 46(8), 899–930.

The Ministry of Housing and Urban-Rural Development of the People's Republic of China. (2016). *GB 50176-2016 Code for Thermal Design of Civil Building*. Beijing: China Architecture & Building Press.

Tominaga, Y., Mochida, A., Yoshie, R., *et al.* (2008). AIJ guidelines for practical applications of CFD to pedestrian wind environment around buildings. *Journal of Wind Engineering and Industrial Aerodynamics*, 96(10–11), 1749–1761. https://doi.org/10.1016/j.jweia.2008.02.058.

Tominaga, Y., and Stathopoulos, T. (2011). CFD modeling of pollution dispersion in a street canyon: Comparison between LES and RANS. *J. Wind Eng. Ind. Aerodyn.*, 99, 340–348.

Tong, Z., Chen, Y., and Malkawi, A. (2017). Estimating natural ventilation potential for high-rise buildings considering boundary layer meteorology. *Applied Energy*, 193, 276–286.

Tong, Z., Chen, Y., Malkawi, A., Adamkiewicz, G., and Spengler, J. D. (2016). Quantifying the impact of traffic-related air pollution on the indoor air quality of a naturally ventilated building. *Environment International*, 89–90, 138–146.

Tong, Z., Chen, Y., Malkawi, A., Liu, Z., and Freeman, R. B. (2016). Energy saving potential of natural ventilation in China: The impact of ambient air pollution. *Applied Energy*, 179, 660–668.

Toparlar, Y., Blocken, B., Maiheu, B., and van Heijst, G. J. F. (2017). A review on the CFD analysis of urban microclimate. *Renewable and Sustainable Energy Reviews*, 80, 1613–1640.

Vicente, B., Rafael, S., Rodrigues, V., et al. (2018). Influence of different complexity levels of road traffic models on air quality modelling at street scale. *Air Qual. Atmos. Heal.* 11, 1217–1232.

Xiong, J., Yao, R., Grimmond, S., Zhang, Q., and Li, B. (2019). A hierarchical climatic zoning method for energy efficient building design applied in the region with diverse climate characteristics. *Energy and Buildings*, 186, 355–367.

Xiong, J., Yao, R., Wang, W., Yu, W., and Li, B. (2020). A spatial-and-temporal-based method for rapid particle concentration estimations in an urban environment. *Journal of Cleaner Production*, 256, 120331.

Yao, R., Li, B., Steemers, K., and Short, A. (2009). Assessing the natural ventilation cooling potential of office buildings in different climate zones in China. *Renewable Energy*, 34(12), 2697–2705.

Yao, R., and Short, A. (2013). Energy efficient building design. In *Design and Management of Sustainable Built Environments* (pp. 179–202), London: Springer-Verlag.

Yao, R., Steemers, K., and Li, B. (2006). Introduction to sustainable urban and architectural design. In *Introduction to Sustainable Urban and Architectural Design*. Beijing: China Architecture and Building Press.

Zhang, A., Gao, C., and Zhang, L. (2005). Numerical simulation of the wind field around different building arrangements. *Journal of Wind Engineering and Industrial Aerodynamics*, 93(12), 891–904.

Chapter 8
Ventilation system and heating and cooling

Moon Keun Kim[1]

8.1 Ventilation strategies

Main aims of ventilation are to provide fresh and clean outdoor air to dilute indoor air pollutants such as dust, volatile organic compounds, odor, and carbon dioxide (CO_2), and to promote indoor thermal comfort with heating and cooling systems (Nathanson, 1995; Awbi, 2003; Kim and Choi, 2019). Generally, air ventilation strategies in buildings are categorized into three main methods: natural, mechanical, and hybrid (Kim and Baldini, 2016; Heiselberg, 2002).

8.1.1 Natural ventilation

Natural ventilated systems allow building envelopes with several openings and utilize stack and wind effect and the temperature differences between indoor and outdoor air to supply fresh outdoor air without mechanical or electrical system devices (Howell, 2017). It is a very effective passive cooling method to save ventilation energy and improve indoor air quality and thermal comfort in a room. However, natural ventilation strategy shows significant disadvantages as it is affected by surrounding environmental conditions, especially in urban areas (Awbi, 2003; Heiselberg, 2002). Its applicability is not acceptable in the conditions such as ambient severe air pollution, exceedingly hot or cold weather, heavy rain, or exposed heavy traffic or environmental noise (Kim and Leibundgut, 2014a, 2014b, 2014c; Kim *et al.*, 2014). The daily temperature and wind speed variations limit stable usage rates for using natural ventilation. Therefore, to overcome these disadvantages of using natural ventilation system, hybrid ventilation (HV) or mechanical ventilation system has been suggested. Figure 8.1 shows a conventional single-sided natural ventilation strategy. Temperature and pressure differences between indoor and outdoor air conditions lead to fresh outdoor air supply into indoor space. Typically, most residential buildings have used natural ventilation system. The window size and opening positions primarily impact room ventilation rate and efficiency. Therefore, the room depth can be designed and determined based on the ventilation efficiency. Natural ventilation methods also categorize into

[1]Department of Civil Engineering and Energy Technology, Oslo Metropolitan University, Oslo, Norway

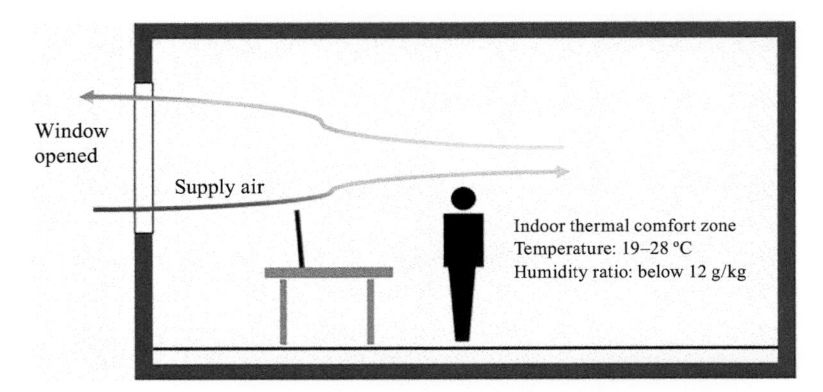

Figure 8.1 Single-sided natural ventilation strategy

several strategies: single-sided ventilation, cross-ventilation, and sub-slab distribution (Awbi, 2003; Lechner, 2014). Even though natural ventilation presents many advantages, buildings need to install additional mechanical ventilation systems just to adapt to extreme weather conditions. Natural ventilation strategy relies on surrounding local environments; therefore, occupants should also consider the decision of availability of natural ventilation and the mechanical cooling or heating demands. These steps lead to results in a building that is sustainable and economical.

8.1.2 Hybrid ventilation

The HV strategy combines natural ventilation and fan-assisted mechanical system in overcoming the disadvantages of the natural ventilation system depending on surrounding thermal conditions (Heiselberg, 2002). In extreme weather conditions such as too cold or hot season, the mechanical ventilation system has to be suggested. During mild weather conditions, fan-assisted HV system can be effectively used as natural ventilation modes for saving air ventilation energy in buildings. Figure 8.2 illustrates a hybrid natural ventilation strategy with fan-assisted system. There are three main HV system types: stack and wind-assisted ventilation, fan-assisted natural ventilation, and economizer mechanical ventilation system. Figure 8.3 illustrates these ventilation strategies. Natural ventilation system hardly adjusts airflow rate because surrounding environmental conditions such as temperature, humidity ratio, and wind speed have been changed every hour or second, and local weather condition mainly affects the availability of using natural ventilation system; however, fan-assisted system can simply adjust the amount of airflow rate. Therefore, it can well adapt to the variable indoor thermal conditions required. Figure 8.3 introduces different types of HV strategies. Depending on room height, room depth, or ventilator location, an adequate HV system can be selected and applied in a building. The HV strategy is well mixed with natural ventilation. However, it is also limited using an HV system and directly supplying outdoor air with fan-assisted system in extreme weather conditions. Temperature

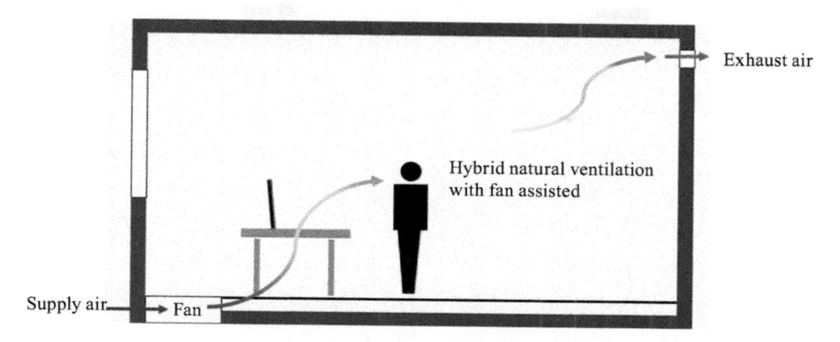

Figure 8.2 Hybrid natural ventilation strategy with fan-assisted system

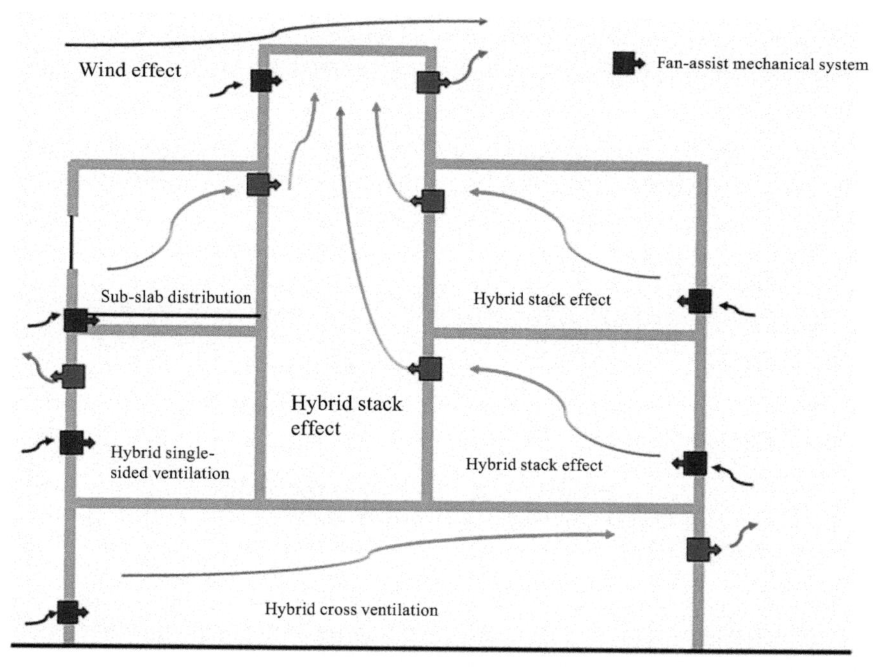

Figure 8.3 Different types of hybrid ventilation

and humidity ratios are main parameters to determine whether hybrid natural ventilation system can be used (Menassa *et al.*, 2013; Brager and de Dear, 2000). And occupancy behavior and internal heat gains can also affect to select the fan-assisted ventilation. Based on occupancy behavior and internal heat gains, acceptable supply parameters, air temperature, and humidity ratios are calculated. The design method to determine the adequate temperature and humidity ratio is suggested for unoccupied and occupied hours.

Based on Brager and Dear's literature and ASHARE Standard 55, a minimum indoor air temperature allowed (T_{ia}) for occupied hours calculates as follows:

$$T_{minia} = 0.25 \times T_{mean} + 16.4\ ^{\circ}\mathrm{C} \quad (8.1)$$

where T_{minia} is the minimum temperature T_{ia} for occupied hours, and T_{mean} is the mean outdoor air temperature during the previous 30 days, with a minimum value of 10 °C (Zhang and Niu, 2003).

The suggested maximum T_{ia} for both unoccupied and occupied hours is as follows:

$$T_{maxia} = 0.255 \times T_{mean} + 21.4\ ^{\circ}\mathrm{C} \quad (8.2)$$

where T_{maxia} is the maximum T_{ia} (ANSI/ASHRAE, 2013).

The supply air temperature should be lower than the indoor air temperature because internal heat gains must be considered (Menassa *et al.*, 2013).

Menassa *et al.* (2013) and Taylor and Menassa (2012) suggested allowed supply outdoor air temperatures for natural ventilation for occupied or for unoccupied time, followed by

$$T_{minia,\ unoccupied} \leq T_{oa} \leq T_{maxia} - T_{gene} \quad (8.3)$$

$$T_{minia,\ occupied} - T_{gen} \leq T_{oa} \leq T_{maxia} - T_{gene} \quad (8.4)$$

where T_{gene} is a reference and given temperature generation for large public spaces, and T_{oa} is the accepted supply outdoor air temperature (Menassa *et al.*, 2013).

Internal heat gains can impact the determination of supply outdoor air temperature. Based on the internal heat gain sources, e.g., occupant behavior, lighting, electronic devices, and work stations from ASHARE handbook fundamentals (ASHRAE, 2005), the energy balance equation can calculate the acceptable supply outdoor air temperature by

$$Q_{Tsa} = m_a C_{pa}(T_{oa} - T_{ea}) = Q_{ish} \quad (8.5)$$

where Q_{Tsa} is the airside-sensible load (kJ/h), C_{pa} is the airside-specific heat (kJ/kg), T_{oa} is the outdoor air temperature (°C), T_{ea} is the exhaust indoor air temperature (°C), and Q_{ish} is internal sensible heat gain (kJ/h).

ASHRAE Standard 55, 2013 has suggested a comfort indoor humidity ratio below 12 g/kg. The moisture balance equation determines the designed supply outdoor humidity ratio in a room (equation (8.6)) based on occupant's moisture generation and infiltration effect (Kim and Baldini, 2016; Zhang and Niu, 2003; Niu *et al.*, 2002; Meggers *et al.*, 2013):

$$\frac{V_r \rho_r dw_r}{d\tau} = -Q_s \rho_s (w_r - w_s) + Q_i \rho_o (w_o - w_r) + m_s \quad (8.6)$$

where V_r is the volume of room (m^3), ρ is the air density (kg/m^3), w is the humidity ratio (kg/kg), τ is time (h), Q_s is the volumetric flow rate of supply air (m^3/h), and m_s is the indoor humidity generation rate (kg/h) (Zhang and Niu, 2003; Niu *et al.*, 2002).

8.1.3 Mechanical ventilation

Mechanical ventilation strategies mainly categorize into three methods, which are typically centralized ventilation (CV), semi-decentralized, and decentralized ventilation (DV) (Kim and Baldini, 2016; Kim and Leibundgut, 2014a, 2014b, 2014c; Kim *et al.*, 2014; Meggers *et al.*, 2013). Typical CV system has been commonly used in buildings as a major HVAC strategy. Most modern commercial buildings have used CV system. Major CV systems are introduced: all-air system, displacement ventilation, underfloor ventilation system, stratum ventilation system, and personalized ventilation system.

Fresh outdoor air is filtrated and heated or chilled in air-handling unit (AHU), and then the air is distributed by air ducts and fans and is finally supplied to indoor space. Some exhaust air is returned to AHU in all-air system and mixed to fresh air to save energy, and others are exhausted (Nathanson, 1995; Awbi, 2003; Heiselberg, 2002). This ventilation strategy shows a steady-state airflow rate, and this system could not be affected by extreme weather conditions. However, this system has long supply, and exhaust air distribution passages with ducts, and the mechanical and air duct systems consume a lot of indoor space as well (Kim and Baldini, 2016; Baldini *et al.*, 2014; Kim and Leibundgut, 2015; Kim *et al.*, 2015, 2018; Kim and Cui, 2018; Saber *et al.*, 2014). Therefore, it has high fan energy consumption and shows ventilation complications to adjust supply airflow rate depending on occupancy ratios in a room (Meggers *et al.*, 2013; Baldini *et al.*, 2014; Kim and Cui, 2018; Kim *et al.*, 2018, 2019). Figure 8.4 illustrates mechanical ventilation strategies: centralized, semi-decentralized, and decentralized system

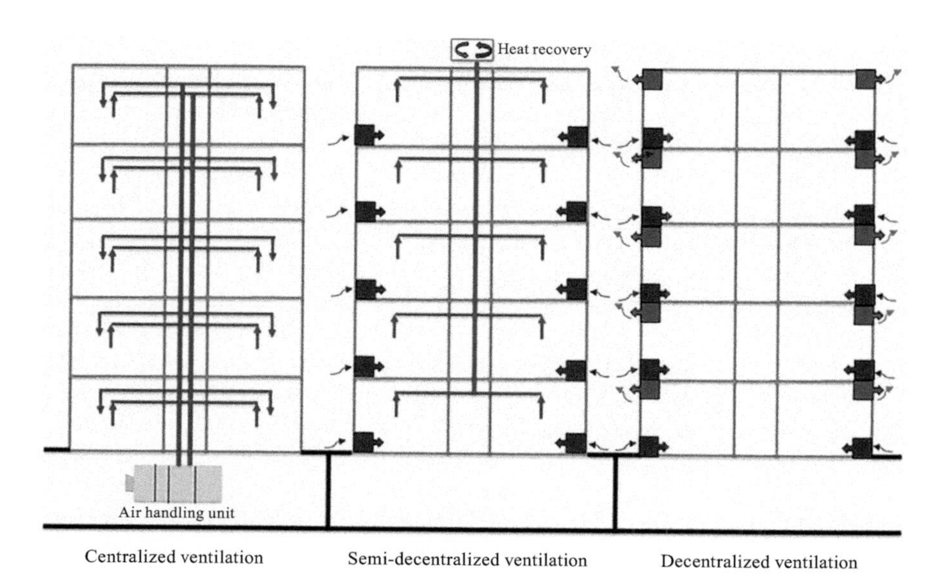

Figure 8.4 Schematic mechanical ventilation strategies: centralized, semi-decentralized, and decentralized ventilation systems

(Kim and Baldini, 2016; Baldini *et al.*, 2014). A DV system designed to reduce ventilation energy demand and to adapt to variable occupants' requirements. DV system has shortest air passages using compact AHU called Airbox convector (Kim and Baldini, 2016; Kim and Leibundgut, 2014a, 2014b, 2014c; Kim *et al.*, 2014; Meggers *et al.*, 2013; Baldini *et al.*, 2014). However, this system cannot use a heat recovery system. To use a heat recovery system for saving energy, a semi-DV system is designed. Semi-decentralized and DVs are only utilized for air ventilation (Kim and Baldini, 2016; Kim and Leibundgut, 2014a, 2014b, 2014c; Kim *et al.*, 2014; Baldini *et al.*, 2014). To adapt to extreme weather conditions, additional mechanical heating and cooling units should be installed, such as radiant heating or cooling panel system in a room.

DV strategy is started as one type of conventional mechanical ventilation used in a similar ventilation strategy to fan-assisted natural ventilation system. However, a compact decentralized AHU called Airbox convector supplies and distributes fresh outdoor air into the space utilized directly with short air passages (Kim and Baldini, 2016; Kim and Leibundgut, 2014a, 2014b, 2014c; Kim *et al.*, 2014; Baldini *et al.*, 2014). Many pieces of literature have illustrated and discussed this system's potential as a variable replacement for centralized mechanical ventilation systems. A compact air ventilator's DV performance is analyzed and compared to conventional CV systems in variable climates. As a case study, selected weather conditions can be analyzed to adapt to acceptable conditions for the operation of fan-assisted natural ventilation systems and evaluate the performance of the DV system's cooling, dehumidification, and heating loads with a compact AHU.

Compared with a typical CV system, the DV system has shorter air distribution passage since the compact AHU is installed near a building envelope (Kim and Baldini, 2016; Baldini *et al.*, 2014; Kim and Leibundgut, 2015; Kim *et al.*, 2015). Therefore, it entails lower air pressure losses and less fan energy consumption. In the system, the airflow rate and fan can be adjusted simply and effectively based on occupancy ratios and indoor thermal conditions. The DV system highlights ventilation performance integration aspect and space-wise functionality.

Conventional CV system is mostly used in commercial buildings. The main disadvantages of the system are that it consumes space volume to install duct system and has long air distribution passages with ducts compared with DV system. Therefore, it consumes more air fan energy (Kim and Baldini, 2016; Kim and Leibundgut, 2014a, 2014b, 2014c; Kim *et al.*, 2014; Baldini *et al.*, 2014; Eke and Senturk, 2013). In a semi-DV system, outdoor air filtered, chilled, or heated at a compact AHU (Airbox) is supplied at the façade or under the floor, and indoor air is exhausted to an exhausting centralized system for heat recovery. DV implements an Airbox unit with supply and exhaust air control for each room's air conditioning (Kim and Baldini, 2016; Kim and Leibundgut, 2014a, 2014b, 2014c; Kim *et al.*, 2014; Baldini *et al.*, 2014; Manz *et al.*, 2001). A DV system effectively combines with radiant cooling or heating systems providing thermal load to minimize ventilation energy demand in space. DV has four main strengths compared to the CV and natural ventilation systems (Kim and Baldini, 2016; Kim and Leibundgut, 2014a, 2014b, 2014c; Kim *et al.*, 2014; Baldini *et al.*, 2014; Manz *et al.*, 2001). Initially, DV can simplify individual thermal space zoning control in a room. Depending on occupancy ratios and preferences, the Airbox fan can effectively adjust supply air volume. For example, a CO_2 sensor detects occupancy ratios and activates or adjusts the air fan speed of the Airbox unit to supply enough fresh air in a space (Kim and Baldini, 2016; Kim and Leibundgut, 2014a, 2014b, 2014c; Kim *et al.*, 2014; Baldini *et al.*, 2014). The natural ventilation system is highly influenced by outdoor environmental conditions (thermal load, wind, and stack effect) (Nathanson, 1995; Awbi, 2003; Heiselberg, 2002; Brager and de Dear, 2000; Kulpmann, 1993). However, DV systems are less affected by surrounding ambient conditions. Second, the DV system can reduce mechanical equipment and air duct space; therefore, it saves construction volume and materials in buildings. A floor height is generally different from the ceiling height since space is needed to install mechanical equipment and air duct between the ceiling and floor space. DV system can minimize the space volume between ceiling and floor. Hence, the DV system can save construction materials, time, and energy in a building (Kim and Baldini, 2016; Kim and Leibundgut, 2015; Kim *et al.*, 2015, 2018, 2019; Kim and Cui, 2018). For instance, if a height of 20–30 cm per floor is reduced in a high-rise building or multistory building, one extra story volume could be reduced for every 10–11 story building. Additionally, the DV system includes an air damper to reduce outside environmental noise (traffic, wind, or construction noise) and air fan noise.

Figure 8.5 illustrates schematics of the DV system of a floor type (upper) and a façade type (bottom). The system has successfully implemented in many projects. Especially, Figure 8.6 shows that Switzerland company, BS2 group, and the research group of Building Systems at ETH Zurich have installed the system in Switzerland (Kim and Baldini, 2016; Meggers *et al.*, 2013; Kim and Leibundgut, 2014a, 2014b, 2014c; Kim *et al.*, 2014; Baldini *et al.*, 2014). Façade-type ventilation system implements a compact Airbox unit to install near a building envelop. It is a relatively simple installation type compared to floor-type DV system. However, it cannot supply fresh air deeply into the room. Floor-type DV system has somewhat higher construction time and cost and consumes more space in a room; however, it can distribute fresh air deeply into space. The uses of natural ventilation and hybrid natural ventilation system are limited under extreme weather conditions (Heiselberg, 2002; Abdullah and Wang, 2012; Liu *et al.*, 2017); however, the DV system is well adapted to the conditions since a compact Airbox unit can heat or chill air to supply into indoor space. This system can be utilized to the HV system in moderate weather conditions without heating and cooling demand by using an air fan and an air filter in the Airbox unit.

In order to improve HVAC energy efficiency, the DV system accompanies radiant panel heating and cooling system. Generally, the chilled radiant cooling panel system has a moisture condensation risk on the radiant panel surface, especially in hot and humid climates. Therefore, the DV system needs high cooling and dehumidification capacity with heat exchanges to supply outdoor air into a space in a hot and humid climate. In a hot and dry conditions such as Europe countries' summer season, relatively there is less moisture condensation risk on the surface of the cooling panel because the average dew point temperature of outside air is the cooling panel's surface temperature (Kim and Baldini, 2016; Baldini *et al.*, 2014;

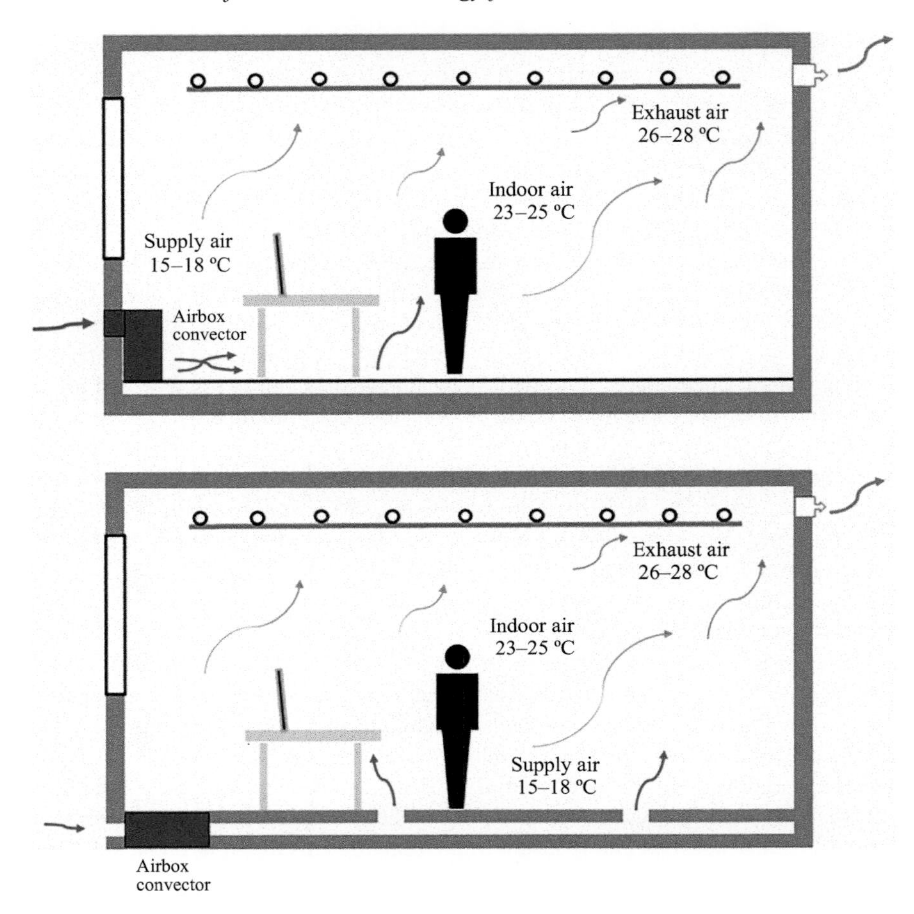

Figure 8.5 Schematic of decentralized ventilation systems: envelope type (top) and floor type (bottom)

Kim and Leibundgut, 2015; Kim *et al.*, 2015). Hence, the DV system in European countries does not need high cooling and dehumidification potential capacity. In order to adapt to hot and humid climate, Kim and Leibundgut (2014a, 2014b, 2014c), Kim *et al.* (2014), and Baldini *et al.* (2014) had designed a novel Airbox unit with three heat exchangers connected in parallel to improve energy efficiency in the air cooling and dehumidification process for adapting to hot and humid climates.

Kim and Baldini (2016) represented that DV systems can reduce around 76% of mechanical fan energy consumption than those of CV systems for a designed room.

8.1.4 Performance evaluation

AHU's heating and cooling loads can be defined and evaluated based on the simulation and measurement results. The total heating and cooling loads are

Figure 8.6 Pictures of a floor-type decentralized ventilation system (B35 in Zurich, left) (Kim and Baldini, 2016) and a façade-type decentralized ventilation system (HPZ building at ETH Zurich, right)

categorized into two parts: sensible and latent. Equations (8.7)–(8.10) define the thermal loads.

The total thermal load for airside ventilation system Q_{Tair} is the sum of sensible load (8.7), which is basically based on temperature differences between in-and-out of air supply in an AHU and latent load that is based on humidity ratio difference between in-and-out of air supply as given by

$$Q_{Tair} = Q_{sen} + Q_{lat} \tag{8.7}$$

where Q_{Tair} is the total load of airside (kJ/h), Q_{sen} is the sensible load of air (kJ/kg), Q_{lat} is the latent load of air (kJ/kg).

The total thermal energy load is being defined as the enthalpy differences of heating and cooling process. Therefore, the total heating and cooling load of the AHU is calculated based on the enthalpy value differences of the air across the heating or cooling coils as shown by

$$Q_{Ta} = m_a(h_{airin} - h_{airout}) \tag{8.8}$$

The sensible thermal load of air, Q_S, is

$$Q_S = m_a C_{Pa}(T_{oair} - T_{inair}) \tag{8.9}$$

and the latent thermal load of air, Q_L, is

$$Q_L = m_a(W_{oair} - W_{inair})h_{fg} \tag{8.10}$$

The reheating load of air in the cooling process, Q_{re}, is

$$Q_{re} = m_a C_{Pa}(T_{oair} - T_{inair}) \tag{8.11}$$

The thermal load of the hot or chilled water, Q_{Tcw}, which is the same as total thermal load of the hot or chilled air, is

$$Q_{Tcw} = m_{hcw} C_{pw}(T_{ow} - T_{iw}) \tag{8.12}$$

Kim and Leibundgut (2014a, 2014b, 2014c), Kim *et al.* (2014) defined the enthalpy effectiveness to present the performance of total thermal load increased or removal of heat exchangers. It is given as

$$E_h = \frac{(h_{inair} - h_{oair})}{(h_{inair} - h_{hcwi,sa})} \tag{8.13}$$

where C_{pa} is the airside-specific heat (kJ/kg), T_{oair} is the outdoor air temperature (°C), T_{inair} is the exhaust indoor air temperature (°C), w is the humidity ratio (kg/kg), m_a is mass flow rate of supply air (m^3/h), and m_{hcw} is the mass flow rate of water supply to radiant panel (kg/h), h is the enthalpy value (kJ/kg).

8.1.5 Low-exergy system

All building energy system analyses such as HVAC, lighting, building thermal material performance, and heat pump system are based on thermodynamics' first law as energy balance method (energy conservation law) (Moran and Shapiro, 2008; Bejan, 2016). However, the energy balance method has represented energy quantity but not evaluated the energy quality (Bejan, 2016; Shukuya, 2013). Exergy analysis additionally has considered both energy quantity and quality because it combines both the first and the second laws of thermodynamics, which assesses energy conservation, equilibrium, and entropy generation. Therefore, it explains energy irreversibility, and it involves energy potential and availability (Torio and Schmidt, 2010; Bejan *et al.*, 1996; Shukuya and Hammache, 2002; Torio *et al.*, 2009; Meggers and Leibundgut, 2011).

Exergy accounts for the available theoretical work from the interaction in system progress with its ambient environment until equilibrium is obtained (Shukuya and Hammache, 2002; Torio *et al.*, 2009; Shukuya, 2009). As a result, exergy represents the potential of an energy flow to be obtained into high-quality energy (Torio and Schmidt, 2010; Meggers *et al.*, 2012; Shukuya, 2019), which defines energy availability. Based on the first and second thermodynamics law, the maximal amount of work is extracted and linked within the temperature gradient. Literature (Shukuya, 2013, 2019; Meggers *et al.*, 2012; Shukuya and Hammache, 2002; International Energy Agency, 2009) presented how exergy could evaluate heat fluxes achievement across different temperature and pressure gradients. The exergy consumption of the heat flux can be maximized for high-temperature differences and minimized for small temperature difference with respect to the energetic value (Kim and Leibundgut, 2014a, 2014b, 2014c; Kim *et al.*, 2014). Today, fossil fuel consumption has been substantially reduced to CO_2 emissions. Many low-exergy system strategies have been proposed in the built environment. Shukuya and Schmidt's works presented the exergetic approach as case studies (Torio and Schmidt, 2010; Shukuya and Hammache, 2002; Torio *et al.*, 2009; Wei and Zmeureanu, 2009).

For this reason, as a low-exergy strategy, high-temperature cooling system and low-temperature heating system could maximize the coefficient of performance (COP) in the built environment with low-exergy values destroyed. The low-exergy system in the built environment has three main benefits. First, the system can minimize the environmental impact due to the reduction of potentially damaging. Second, it reduces the CO_2 emission and thus contributes to minimizing the amount of greenhouse gas consumption because the low-exergy system does not use fossil fuels that consume huge exergy amounts in the combustion process. And low-exergy approaches present a better understanding of the thermal load calculation with both energy quantity and quality analysis to design efficient HVAC system in built environment.

8.2 Heating and cooling system

Active heating and cooling strategy in buildings is categorized into several mechanical systems. A schematic of a typical all-air heating and cooling system is illustrated in Figure 8.7. The main thermal energy source of all-air heating and cooling system is from the supply air. In order to supply enough thermal energy in a room, a large amount of air volume is needed for heating and cooling in a building. However, this system has lower energy efficiency than a radiant heating and cooling system with displacement ventilation system because it deals with large air volume and air duct system. Therefore, it has relatively higher air pressure losses. It has a lower cooling impact ratio and a higher temperature gradient at which the higher temperature heating and lower temperature cooling system operates, and a lower COP can be achieved (Niu *et al.*, 2002; Kim and Cui, 2018; Kim *et al.*, 2018; Novoselac and Srebric, 2002). Around 70% of the supply airflow is recirculated and mixed with fresh outdoor air for heating and cooling demand. Supplied air is not only for heat transfer of heating and cooling sources but also for ventilation. Compared with radiant heating and cooling system with a displacement ventilation system, this system has some disadvantages: relatively lower energy efficiency, lower thermal comfort, high air pressure losses, and fan noise due to a large amount volume of air. However, this system still is widely used in commercial buildings since it simply adapts to local environments and no moisture condensation risk on the radiant cooling panel, and relatively short reaction time to active the system (Kim and Baldini, 2016; Kim and Cui, 2018; Kim *et al.*, 2018).

Figure 8.8 shows typical radiant heating and cooling system with a displacement ventilation system. Radiant ceiling panels release the main sensible heating and cooling, and displacement ventilation system supplies fresh air into a room. Compared with the all-air system, the radiant heating and cooling system shows

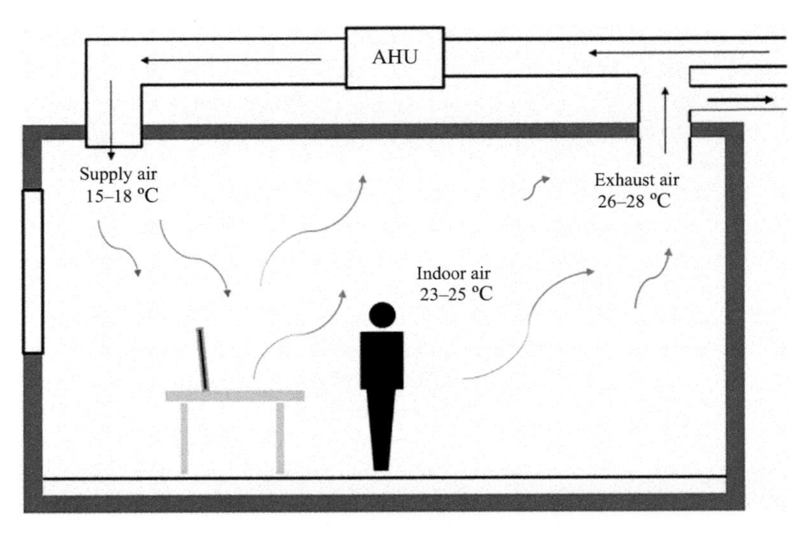

Figure 8.7 Schematic of a conventional centralized all-air heating and cooling system

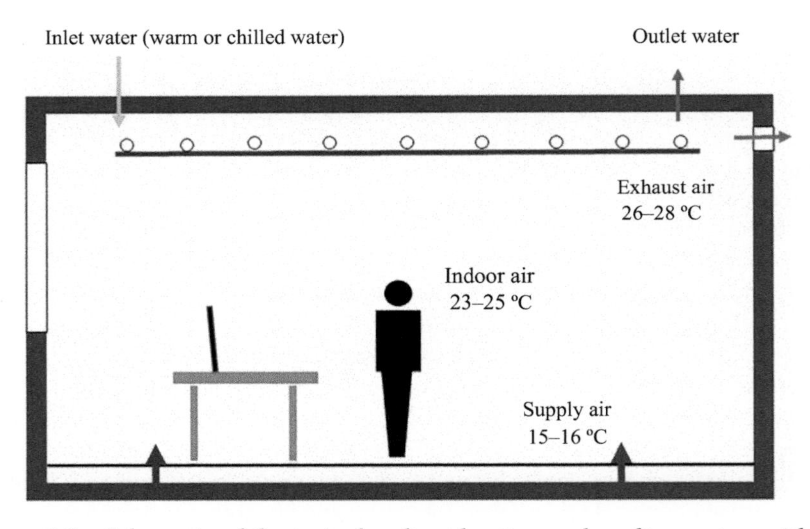

Figure 8.8 Schematic of the typical radiant heating and cooling system with a centralized displacement ventilation system in an office room

energy-saving strategies because the system can minimize supply air volume; therefore, it can reduce air pressure losses and fan energy consumption. And high-temperature radiant cooling and low-temperature heating system using a heat pump unit can maximize the COP as a low-exergy system (Kim and Leibundgut, 2014a, 2014b, 2014c; Kim *et al.*, 2014; Baldini *et al.*, 2014). The typical radiant heating and cooling system has a higher thermal comfort value in a room compared with all-air heating and cooling system. However, this system has also disadvantages. For cooling, it has moisture condensation risk on the radiant panel surface when a room is exposed to high infiltration rate in hot and humid surrounding weather condition or high internal water vapor gains. And in highly polluted indoor air condition, it has difficulty to dilute the air pollutant quickly because ventilation rate is designed by the displacement unit. And radiant heating and cooling system has relatively a long reaction time to release heating or cooling output. Generally, in order to remove the moisture condensation risk, this system should involve significant dehumidification system to minimize the humidity ratio of supply air and to increase the indoor supply airflow rate (Kim and Leibundgut, 2014a, 2014b, 2014c; Kim *et al.*, 2014, 2018; Kim and Cui, 2018; McQuiston and Parker, 1982; Jeong and Mumma, 2003, 2004).

The novel hybrid radiant heating and cooling modeling strategy designed by Kim and Leibundgut (Kim and Leibundgut, 2014a, 2014b, 2014c; Kim *et al.*, 2014, 2018; Kim and Cui, 2018) at the group of Building Systems in Swiss Federal Institute of Technology in Zurich is illustrated in Figure 8.9. This novel cooling strategy is designed as a low-exergy system to adapt to hot and humid climate conditions. This system can prevent moisture condensation risk on the radiant cooling panel and provide additional cooling output by the thermal heat exchanger in the Airbox convector and enhanced mixed natural and mechanical forced convection effect (Kim and Leibundgut, 2014a, 2014b, 2014c; Kim *et al.*, 2014, 2018; Baldini *et al.*, 2014; Kim and Cui, 2018). The hybrid radiant heating and cooling system consists of two units: one is a typical radiant panel system and the other is an Airbox convector. The Airbox unit comprises several divisions: compact heat

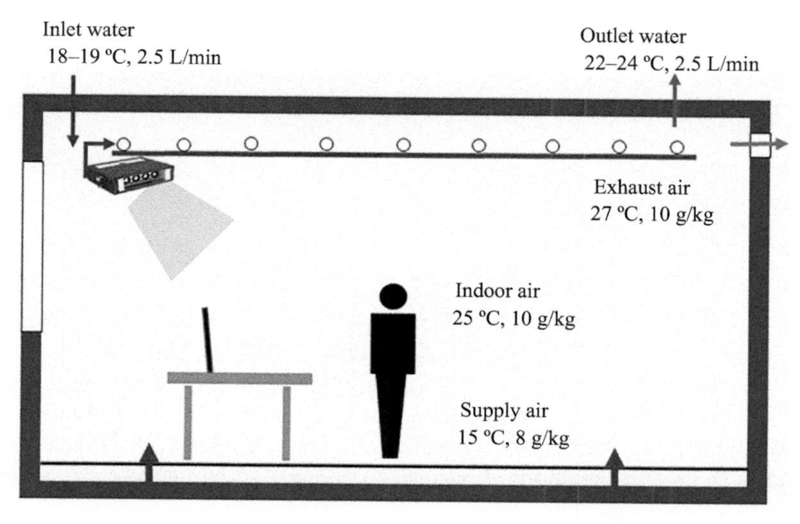

Figure 8.9 Schematic of the hybrid radiant cooling system connected hydronically in series with an Airbox convector

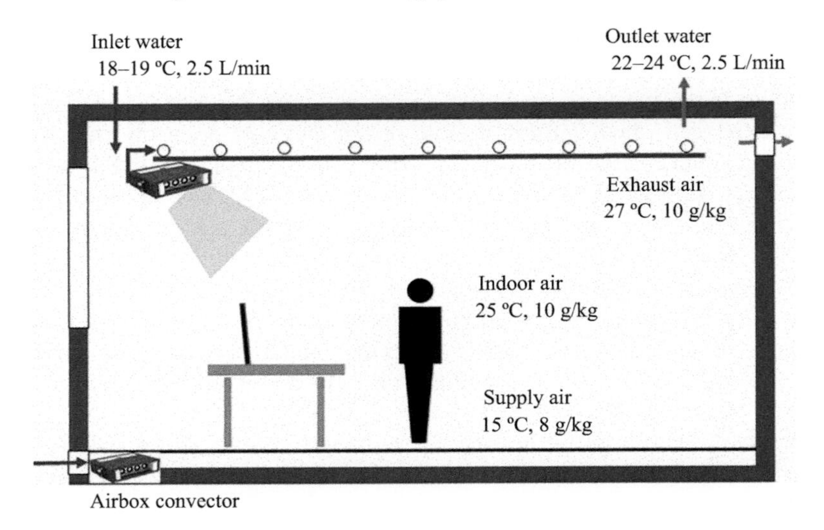

Figure 8.10 Schematic of the performance of the hybrid radiant cooling system with a decentralized ventilation unit

exchangers, hydronic pipe, compact fans, air filter, and drain tube. And the Airbox unit and radiant panel system are hydronically connected in series (Kim and Leibundgut, 2014a, 2014b, 2014c; Kim *et al.*, 2014, 2018; Kim and Cui, 2018). First, the supply water passes through the Airbox unit, and then the water flow passes into the radiant panel system. When the Airbox fan is off, the system heating and cooling outputs are the same as that of a typical radiant panel system. However, when the Airbox fan is on, the active thermal output is increased since the Airbox unit generates airflow. The heat exchangers can release additional thermal energy and dehumidify indoor air (Kim and Leibundgut, 2014a, 2014b, 2014c; Kim *et al.*, 2014). In cooling performance, the supplied water temperature rises in the Airbox unit, and thus, the actual surface temperature of the radiant ceiling panel is also increased. Therefore, this system can minimize moisture condensation risk. Schematics of the hybrid radiant cooling system with DV systems to adapt to hot and humid climate condition are illustrated in Figure 8.10. The main benefit of using the hybrid radiant heating and cooling system with a DV system is energy saving as a low-exergy system. It can be used as a hybrid natural ventilation system in shoulder season and has short distribution passage (Kim and Baldini, 2016; Kim and Leibundgut, 2014a, 2014b, 2014c; Kim *et al.*, 2014). Additionally, it merely adjusts a supply airflow rate depending on the occupancy ratio using a CO_2 sensor. And the Airbox unit connected to a radiant panel can also adjust heating and cooling output in a room based on the indoor thermal condition. However, this system has the disadvantage installing a heat recovery system adapting to extreme cold weather conditions.

The radiant heating and cooling system with CV system have about 20% higher total costs than all-air systems because radiant panel systems in each room affect capital cost (Kim and Baldini, 2016; Kim and Leibundgut, 2014a, 2014b, 2014c; Kim *et al.*, 2014); however, the total cost of all-air systems is dramatically increased with increasing heating and cooling load because it has lower energy efficiency than other radiant systems (Novoselac and Srebric, 2002; Jeong and Mumma, 2003). Literature (Kim and Baldini, 2016; Kim and Leibundgut, 2014a, 2014b, 2014c; Kim *et al.*, 2014; Baldini *et al.*, 2014) described that radiant panel system with a DV unit has the lowest capital costs because it minimizes air distribution space and can reduce construction time and materials. However, it has higher maintenance cost compared with radiant panel with a CV system due to the need to clean Airbox units and to replace air filters regularly. The DV system saves energy consumption as a fan-assisted natural ventilator using the Airbox units. In cooling or heating loads season, short supply air passages minimize fan energy consumption. The fan energy is significantly reduced, while the pump energy consumption is grown by around two–three times. The hybrid radiant heating and cooling systems have a higher COP due to using a higher temperature cooling source and a lower temperature heating source. Especially, in mild climates, a radiant panel with DV units has many benefits to adapt to the surrounding environments. Figure 8.11 shows a comparative analysis of total energy consumption for cooling and ventilation systems in buildings. Compared with the all-air system, radiant panel systems could save around 25%–33% of cooling and more than 93%

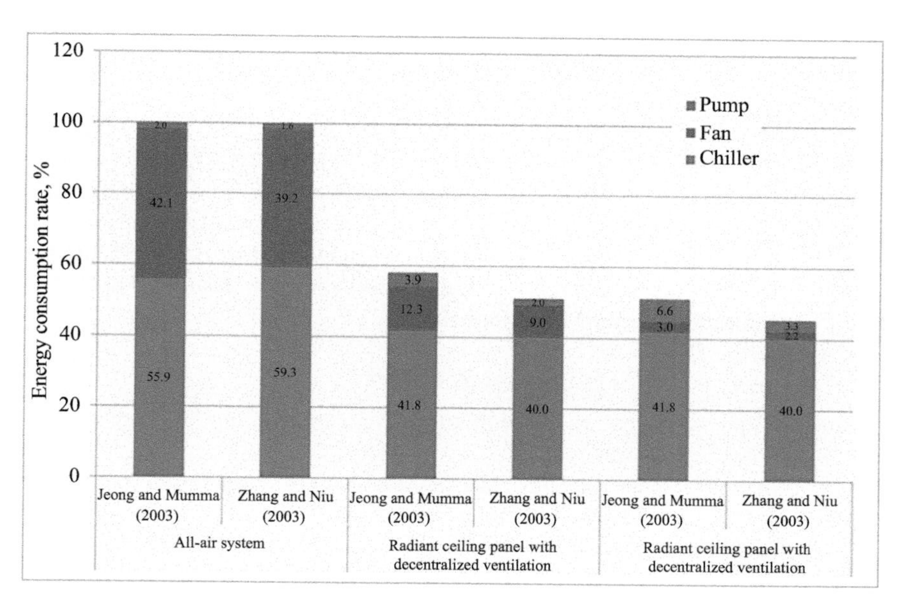

Figure 8.11 Comparative analysis of the yearly total energy consumption for all-air, radiant panel with centralized ventilation, and radiant panel with decentralized ventilation system based on the results of Jeong and Mumma (2003, 2007), Zhang and Niu (2003), Kim and Cui (2018), Kim et al. (2018), and Niu et al. (2002)

of fan energy, while it consumes around three times more hydronic pump energy. And DV system can save about 6%–7% additional total cooling energy (Zhang and Niu, 2003; Kim and Cui, 2018; Kim *et al.*, 2018; Novoselac and Srebric, 2002; Jeong and Mumma, 2003). Actual cooling output of the radiant panel system is affected by the natural convection effect between the radiant panel and indoor air. The hybrid radiant cooling system releases more cooling output because the Airbox convector enhances mixed convection effect that combines natural air convection performance with mechanical forced convection performance by the Airbox fan.

The cooling output of the radiant panel system is affected by supply water temperature. Higher supply inlet water temperature leads to a higher COP but less cooling outputs; however, the Airbox convector connected to the radiant panel system can compensate for cooling output losses due to the increase in mixed convection effect. And the Airbox convector simply adjusts the indoor humidity ratio. If the indoor humidity ratio is suddenly increased, moisture condensation occurs in the Airbox convector, and the moisture condensed is drained out. One of the main challenges of using the radiant panel system is that it is hard to adjust cooling output because lower supply inlet water temperature leads to increase moisture condensation risks and the operating reaction time is quite slow; however, the hybrid radiant cooling panel system easily and simply adjusts indoor cooling output and air movement using an Airbox convector and reduces moisture condensation risk on the surface of the chilled radiant panel. Therefore, compared with the typical radiant cooling system, the hybrid radiant cooling panel system can reduce around 9%–10% of the cooling energy consumption due to the highest cooling supply water temperature.

Additionally, the international standard describes that while indoor air movement increases, thermal comfort zone can be expanded to higher temperature because the increase of air velocity can offset around 2 °C of air temperature. Therefore, using the hybrid radiant cooling system, the operating temperature can be raised to save cooling energy in a room with equal thermal comfort. This higher operating temperature with equal thermal comfort can save around 9.3% of cooling energy consumption in built environment.

8.2.1 Exergy performance evaluation

In terms of the exergy load calculation, the heating and cooling loads are defined by two parts: physical and chemical.

Two main elements determine the physical exergy: temperature (T) and pressure (p) of the procedure, and chemical exergy for humidification or dehumidification in latent energy load is determined by the chemical potential μ_i (Kim and Leibundgut, 2014a, 2014b, 2014c; Kim *et al.*, 2014; Torio and Schmidt, 2010). The detailed formula derivations are described in literature (Shukuya, 2009).

In supply air heating or cooling systems, the total exergy is

$$Ex_{tot} = Ex_{phys} + Ex_{ch} \tag{8.14}$$

where Ex is exergy (kJ/h).

The physical exergy of humid air as a sensible energy load procedure is

$$Ex_{phys.a} = \left(C_{p.a} + \omega C_{p.v}\right)\left[(T - T_o) - T_o \ln\frac{T}{T_o}\right] + (1+\omega)R_o T_o \ln\frac{p}{p_o}$$

where Ex is exergy (kJ/h), while the chemical exergy of humid air for humidification or dehumidification as a latent load assessment is

$$Ex_{ch.a} = R_a T_o \left[(1+\omega)\ln\frac{1+\omega_o}{1+\omega} + \omega \ln\frac{\omega}{\omega_o}\right] \tag{8.16}$$

where R is specific ideal gas constant (J/kg K), and the physical exergy of liquid water for hydronic system such a radiant heating or cooling system is

$$Ex_{phys.w} = \left(C_{p.w}\right)\left[(T - T_o) - T_o \ln\frac{T}{T_o}\right] + v(p - p_o) \tag{8.17}$$

and the chemical exergy for hydronic system is

$$Ex_{ch.w} = (P - P_{sat})v - R_w T_o \ln\phi_o \tag{8.18}$$

where ϕ is relative humidity (%).

Literature (Torio and Schmidt, 2010; Torio *et al.*, 2009; Shukuya, 2009; International Energy Agency, 2009) suggested the reference formula applying building systems in the surrounding environment. The references are defined as the various environmental conditions as exergy applications. The energy and exergy analysis presents more complex procedures than steady-state assessment because the exergy value has been changed surrounding environmental conditions: temperature, humidity, and pressure.

Fan and pump energy is analyzed electric motor power performance and electricity consumption and electricity power is 100% exergy. And pump and fan exergy analysis is shown by Niu *et al.* (2002) calculated as

$$Fan\ (pump)\ power = V\Delta p/3,600\ \eta_f \tag{8.19}$$

where V is the air (water) flow rate (m^3/h), Δp is the total pressure difference (Pa), and η_f is the fan (pump) efficiency.

The COP value of a heat pump system for heating or cooling is determined by the temperature difference between the condenser and evaporator units. This is expressed in the following formula:

$$COP_{heating\ or\ cooling} = \frac{Q_{t\ \mathrm{or}\ c}}{W} = \eta\frac{T_{c\ \mathrm{or}\ e}}{T_c - T_e} = \eta\frac{T_{c\ \mathrm{or}\ e}}{\Delta T} \tag{8.20}$$

where Q_c is the heat removal, W is the work (electricity) consumption by compressor, η is the efficiency of the system, and T_e and T_c are the temperatures at the evaporator and condenser unit in the device, respectively, which are the cold and hot reservoir, respectively.

8.2.2 Exergy efficiency

Exergy efficiencies are categorized into two different types: simple and rational. Simple exergy is shown by Torio *et al.* (2009) and Torio and Schmidt (2010) that the estimation of how close the system processes were to the ideal performance. On the other hand, rational efficiency presents how much the potential of exergy was lost to provide a specific output in the processes. The overall system impact ratio is defined as the ratio between the actual heating or cooling energy output achieved for the supply air or hydronic energy source and the total exergy consumed in a heating and cooling system (Kim and Leibundgut, 2014a, 2014b, 2014c; Kim *et al.*, 2014; Torio and Schmidt, 2010; Torio *et al.*, 2009).

Each formula is illustrated as follows:

$$\psi_{simple} = \frac{Ex_{out}}{Ex_{in}} \tag{8.21}$$

$$\psi_{rational} = \frac{Ex_{des,out}}{Ex_{in}} \tag{8.22}$$

$$OSR_{system} = \frac{Q_{ac.c}}{Ex_{e.c.}} \tag{8.23}$$

Literature (Shukuya, 2013; Shukuya and Hammache, 2002; Meggers and Leibundgut, 2011; Shukuya, 2009, 2019) described that radiant ceiling panel systems have higher simple and rational exergy efficiency values than those of all-air systems. And these values are sensitively affected by ambient thermal conditions, i.e., temperature and humidity ratio. And especially in extreme weather conditions, both exergy efficiencies have the highest simple and rational values. The simple efficiency ratio represents how much close the system to ideal performance, and rational efficiency illustrates how the environment impacts the system performance (Shukuya *et al.*, 2013; Torio and Schmidt, 2010; Torio *et al.*, 2009). Based on the exergetic analysis, the hybrid radiant panel system has a higher cooling output and relatively lower exergy losses. And the potential can be maximized for mild conditions because lower temperature and humidity differences between indoor and outdoor environment conditions strongly impact the exergy minimization and system performance.

8.3 Conclusion

This chapter has introduced variable air ventilation strategies with heating and cooling system and evaluated the systems' energy and exergy performance in a building. The natural and HV systems had good energy-saving strategies; however, these are quite limited to use in extreme weather conditions and had to combine other mechanical systems to adapt to local conditions. Radiant heating and cooling systems with a displacement ventilation unit had lower energy consumption and better thermal comfort than the all-air system. Hybrid radiant heating and cooling system presented more energy efficiency and reduced moisture condensation risk on the surface of the radiant cooling panels. And it shows higher COP values due to higher temperature cooling and lower temperature heating source supply. And it has a relatively short reaction time to operate the system. DV system can reduce energy consumption for shoulder season using fan-assisted natural ventilation and for summer and winter season because a decentralized compact AHU can minimize air pressure loss. Even hybrid radiant heating and cooling system can combine with a DV unit. These combinations reduce thermal energy consumption, higher thermal comfort, less construction cost, time, and materials compared with the centralized HVAC system. Exergy analysis evaluated how each system or process is closed to ideal performance and how much energy availability is remained or is destroyed. Exergy efficiency is quite sensitive, and the value is highly affected by the ambient conditions.

References

Abdullah, A.H. and F. Wang. (2012). Design and low energy ventilation solutions for atria in the tropics. *Sustainable Cities and Society*, 2(1), 8–28.

ANSI/ASHRAE. (2013). Thermal environmental conditions for human occupancy. ASHRAE Standard *55*.

ASHRAE. (2005). *ASHRAE Handbook-Fundamental SI Edition*. Atlanta, GA: ASHRAE Inc.

Awbi, H.B. (2003). *Ventilation of Buildings*. Online access with purchase. London; New York, NY: Spon Press Taylor & Francis.

Baldini, L., M.K. Kim, and H. Leibundgut. (2014). Decentralized cooling and dehumidification with a 3 stage LowEx heat exchanger for free reheating. *Energy and Buildings*, 76, 270–277.

Bejan, A. (2016). *Advanced Engineering Thermodynamics*. Online access with JISC subscription agreement: Knovel. Newark, NJ: Wiley.

Bejan, A., G. Tsatsaronis, and M.J. Moran. (1996). *Thermal Design and Optimization*. New York, NY: John Wiley and Sons, Inc.

Brager, G.S. and R. de Dear. (2000). A standard for natural ventilation. *ASHRAE Journal*, 42(10), 21–23+25–28.

Eke, R. and A. Senturk. (2013). Monitoring the performance of single and triple junction amorphous silicon modules in two building integrated photovoltaic (BIPV) installations. *Applied Energy*, 109, 154–162.

Heiselberg, P. (2002). *Principles of Hybrid Ventilation*. Aalborg Univ. (Denmark), Hybrid Ventilation Centre. p. Medium: X; Size: 73 pages.

Howell, R.H. (2017). *Principles of Heating Ventilating and Air Conditioning: A Textbook With Design Data.* 8th edition, Based on the 2017 ASHRAE Handbook of Fundamentals. Atlanta, USA: ASHRAE.

International Energy Agency. (6 September 2009). *ECBCS Annex 49, Low Exergy Systems for High-Performance Building and Communities*. Newsletter.

Jeong, J.W. and S.A. Mumma. (2003). Impact of mixed convection on ceiling radiant cooling panel capacity. *HVAC&R Research*, 9(3), 251–257.

Jeong, J.W. and S.A. Mumma. (2004). Simplified cooling capacity estimation model for top insulated metal ceiling radiant cooling panels. *Applied Thermal Engineering*, 24(14–15), 2055–2072.

Jeong, J.W. and S.A. Mumma. (2007). Practical cooling capacity estimation model for a suspended metal ceiling radiant cooling panel. *Building and Environment*, 42(9), 3176–3185.

Kim, K.M., *et al.* (2019). Simplified neural network model design with sensitivity analysis and electricity consumption prediction in a commercial building. *Energies*, 12(7).

Kim, M.K. and H. Leibundgut. (2014a). A case study on feasible performance of a system combining an Airbox convector with a radiant panel for tropical climates. *Building and Environment*, 82, 687–692.

Kim, M.K. and H. Leibundgut. (2014b) Advanced Airbox cooling and dehumidification system connected with a chilled ceiling panel in series adapted to hot and humid climates. *Energy and Buildings*, 85, 72–78.

Kim, M.K. and H. Leibundgut. (2014c). Evaluation of the humidity performance of a novel radiant cooling system connected with an Airbox convector as a low exergy system adapted to hot and humid climates. *Energy and Buildings*, 84, 224–232.

Kim, M.K. and H. Leibundgut. (2015). Performance of novel ventilation strategy for capturing CO_2 with scheduled occupancy diversity and infiltration rate. *Building and Environment*, 89, 318–326.

Kim, M.K. and J.-H. Choi. (2019). Can increased outdoor CO_2 concentrations impact on the ventilation and energy in buildings? A case study in Shanghai, China. *Atmospheric Environment*, 210, 220–230.

Kim, M.K. and S. Cui. (2018). Evaluation of PMV and PPD Model of a Hybrid Radiant Cooling System. in 4th International Conference On Building Energy, Environment, *COBEE2018*, Melbourne, Australia.

Kim, M.K. and L. Baldini. (2016). Energy analysis of a decentralized ventilation system compared with centralized ventilation systems in European climates: Based on review of analyses. *Energy and Buildings*, 111, 424–433.

Kim, M.K., Baldini, L., Leibundgut, H., Wurzbacher, J.A., and Piatkowski, N. (2015). A novel ventilation strategy with CO_2 capture device and energy saving in buildings. *Energy and Buildings*, 87, 134–141.

Kim, M.K., H. Leibundgut, and J.H. Choi. (2014). Energy and exergy analyses of advanced decentralized ventilation system compared with centralized cooling and air ventilation systems in the hot and humid climate. *Energy and Buildings*, 79, 212–222.

Kim, M.K., J.Y. Liu, and S.J. Cao. (2018). Energy analysis of a hybrid radiant cooling system under hot and humid climates: A case study at Shanghai in China. *Building and Environment*, 137, 208–214.

Kulpmann, R.W. (1993). Thermal comfort and air quality in rooms with cooled ceilings-results of scientific investigations. *ASHRAE Transactions*, 99(2), 448–502.

Lechner, N. (2014). *Heating, cooling, lighting: Sustainable design methods for architects*. 4th Edition, Hoboken, NJ: John Wiley and Sons, Inc.

Liu, G., Xiao, M., Zhang, X., *et al.* (2017). A review of air filtration technologies for sustainable and healthy building ventilation. *Sustainable Cities and Society*, 32, 375–396.

Manz, H., H. Huber, and D. Helfenfinger. (2001). Impact of air leakages and short circuits in ventilation units with heat recovery on ventilation efficiency and energy requirements for heating. *Energy and Buildings*, 33(2), 133–139.

McQuiston, F.C., Parker, J.D., and Spitler, J.D. (2005). *Heating, Ventilating, and Air Conditioning: Analysis and Design*. Sixth Edition, Hoboken, NJ: John Wiley & Sons, Inc.

Meggers, F. and H. Leibundgut. (2011). The potential of wastewater heat and exergy: Decentralized high-temperature recovery with a heat pump. *Energy and Buildings*, 43(4), 879–886.

Meggers, F., Ritter, V., Goffin, P., Baetschmann, M., and Leibundgut, H. (2012). Low exergy building systems implementation. *Energy*, 41(1), 48–55.

Meggers, F., Pantelic, J., Baldini, L., Saber, E.M., Kim, M.K. (2013). Evaluating and adapting low exergy systems with decentralized ventilation for tropical climates. *Energy and Buildings*, 67, 559–567.

Menassa, C.C., N. Taylor, and J. Nelson. (2013). A framework for automated control and commissioning of hybrid ventilation systems in complex buildings. *Automation in Construction*, 30, 94–103.

Moran, M.J. and H.N. Shapiro. (2008). *Fundamentals of Engineering Thermodynamics*. Hoboken, NJ: Wiley.

Nathanson, T., Federal-Provincial Working Group on Indoor Air Quality in the Office Environment. (1995). *Indoor Air Quality in Office Buildings: A Technical Guide*. Minister of National Health and Welfare, Minister of Supply and Services Canada: Minister of National Health and Welfare.

Niu, J.L., L.Z. Zhang, and H.G. Zuo. (2002). Energy savings potential of chilled-ceiling combined with desiccant cooling in hot and humid climates. *Energy and Buildings*, 34(5), 487–495.

Novoselac, A. and J. Srebric. (2002). A critical review on the performance and design of combined cooled ceiling and displacement ventilation systems. *Energy and Buildings*, 34(5), 497–509.

Saber, E.M., Iyengar, R., Mast, M., Meggers, F., Tham, K.W., and Leibundgut, H. (2014). Thermal comfort and IAQ analysis of a decentralized DOAS system coupled with radiant cooling for the tropics. *Building and Environment*, 82, 361–370.

Shukuya, M. (2009). Exergy concept and its application to the built environment. *Building and Environment*, 44(7), 1545–1550.

Shukuya, M. (2013). Exergy: theory and applications in the built environment. *Green Energy and Technology*. London: Springer, c2013.

Shukuya, M. (2019). Exergetic approach to the understanding of built environment-state-of-the-art review. *Japan Architectural Review*, 2(2), 143–152.

Shukuya, M., A. Hammache. (2002). *Introduction to the Concept of Exergy – for a Better Understanding of Low-Temperature-Heating and High-Temperature-Cooling Systems*. Espoo, Finland: VTT Technical Research Centre of Finland.

Taylor, N. and C.C. Menassa. (2012). An Experimental Approach to Optimizing Natural Ventilation in Public Spaces of Complex Buildings. in Construction Research Congress 2012. p. 1651–1661.

Taylor, N., C.C. Menassa, and J.S. Nelson. Automated Hybrid Ventilation Control in Complex Buildings. *International Conference on Computing in Civil Engineering*, June 17–20, 2012, Clearwater Beach, Florida, United States.

Torio, H. and D. Schmidt. (2010). Development of system concepts for improving the performance of a waste heat district heating network with exergy analysis. *Energy and Buildings*, 42(10), 1601–1609.

Torio, H., A. Angelotti, and D. Schmidt. (2009). Exergy analysis of renewable energy-based climatisation systems for buildings: A critical view. *Energy and Buildings*, 41(3), 248–271.

Wei, Z. and R. Zmeureanu. (2009). Exergy analysis of variable air volume systems for an office building. *Energy Conversion and Management*, 50(2), 387–392.

Zhang, L.Z. and J.L. Niu. (2003). Indoor humidity behaviors associated with decoupled cooling in hot and humid climates. *Building and Environment*, 38 (1), 99–107.

Chapter 9
Natural ventilation system design: predictive methods

Zhengtao Ai[1] *and Guoqiang Zhang*[1]

9.1 Introduction

Natural ventilation is a well-established branch discipline of ventilation under the subject of heating, ventilating, and air-conditioning engineering. It can be simply classified into two categories: single-sided and cross natural. Single-sided natural ventilation is a more common ventilation strategy particularly in densely populated urban areas, while cross natural ventilation is more efficient to provide high ventilation rates. General descriptions of natural ventilation have been documented in several books (Allard and Santamouris, 1998; Etheridge, 2011; Ghiaus *et al*., 2006; Etheridge and Sandberg, 1996; Awbi, 2003). However, many aspects of natural ventilation system are still not fully understood so far, and many efforts are still devoted to explore its basic mechanisms and to achieve its reliable application. Although the advantages of natural ventilation are widely known and most people prefer operable windows with natural ventilation, its application with intentional designs is surprisingly very limited (Carrilho da Graca and Linden, 2016) and few international or national ventilation standards (ASHRAE Standard 62, 2013; WHO, 2006; CIBSE, 2005) involve natural ventilation design (Liddament, 2009).

One important barrier of the application of natural ventilation in buildings is that it strongly relies on the local outdoor microclimate, including particularly wind speed, air temperature, pollutant concentration, and noise level (Liddament, 2009; Ai *et al*., 2015). Wind speed in urban areas, especially in high-density cities, is seriously decreased when compared to that in the atmosphere above the buildings or in suburban areas (Georgakis and Santamouris, 2006; Assimakopoulos *et al*., 2006). Such a decrease is augmented with the increase of building density due to, for example, rapid urbanization. Hong Kong Observatory reported that the average wind speed in urban area of Hong Kong has decreased considerably from 2.5 to 1.5 m/s from 1994 to 2004 (HKPD, 2005). Lower wind speeds in urban areas not only lead to accumulated high concentrations of pollutants in street canyons (Yim *et al*., 2009) but also accelerate the urban heat island effect (Cheng *et al*., 2010).

[1]College of Civil Engineering, Hunan University, Changsha, China

This urban canopy layer where buildings immerse is totally different from the atmospheric boundary layer in an open space, which would definitely lower the natural ventilation performance (Ghiaus *et al.*, 2006; Etheridge and Sandberg, 1996; Awbi, 2003; Carrilho da Graca and Linden, 2016; ASHRAE Standard 62, 2013; WHO, 2006; CIBSE, 2005; Liddament, 2009; Ai *et al.*, 2015; Georgakis and Santamouris, 2006) and even hinder its use. However, past studies on natural ventilation were mostly limited to using simple, isolated, building models, without taking into account the influence of surrounding buildings in an urban context. There is no doubt about the importance of such studies in revealing the basic fluid mechanics and the influence of various parameters, which, however, provide a limited guidance on the natural ventilation design of urban buildings.

The basic theories of natural ventilation and some typical theoretical studies, such as Bernoulli theory, envelope theory, and fluid mechanics, can be found in the references (Allard and Santamouris, 1998; Etheridge and Sandberg, 1996; Awbi, 2003; Linden, 1999; Li, 2000; Chaplin *et al.*, 2000; Hunt and Linden, 2001; Lishman and Woods, 2009; Narasaki *et al.*, 1989; Aynsley, 1997; Li *et al.*, 2000, 2001; Li and Delsante, 2001; Etheridge, 2009), which are not included in this book. Analytical studies (Narasaki *et al.*, 1989; Aynsley, 1997; Li *et al.*, 2000, 2001; Li and Delsante, 2001; Etheridge, 2009) are useful for the understanding of ventilation mechanism (Chen, 2009), which, however, are also not included in this book, as the resulted models can only be applicable to very simple geometries. This book centers on reviewing experimental, computational fluid dynamics (CFD) and empirical studies of natural ventilation, with a special attention paid to some important issues, such as the limitations of major research methods (i.e., experimental measurements, CFD simulations, and empirical predictions) and the oversimplification of physical models (i.e., building envelope, building, and surroundings). Discussions on the needs toward a reduced design risk are made alongside the review.

9.2 Experimental measurements

9.2.1 On-site measurements

On-site measurements in real buildings (see, e.g., Figure 9.1(a)) for single-sided ventilation (e.g., Chen, 2009; Warren, 1977; Hughes *et al.*, 2011; Etheridge, 2015; BRECSU, 2000; Oropeza-Perez and Ostergaard, 2014a, 2014b; Aflaki *et al.*, 2015; Heiselberg *et al.*, 2001; Phaff and De Gids, 1982; Caciolo *et al.*, 2011; Ai *et al.*, 2015; Li *et al.*, 2014a, 2014b; Niachou *et al.*, 2005, 2008; Omrani *et al.*, 2017a, 2017b; O'Sullivan and Kolokotroni, 2017) and for cross ventilation (e.g., Li *et al.*, 2014a, 2014b; Niachou *et al.*, 2005, 2008; Omrani *et al.*, 2017a, 2017b; Mochida *et al.*, 2008; Qian *et al.*, 2010; van Hooff and Blocken, 2012; Wang *et al.*, 2014; Park, 2013; Lo and Novoselac, 2012) provide the first-hand data of the ventilation conditions in buildings. However, on-site measurements of natural ventilation are limited in several aspects: (a) the experimental data is of very low consistency and thus of low repeatability due to the constantly varying outdoor environments; (b) it is extremely difficult to perform parametric analysis of influential factors; (c) they

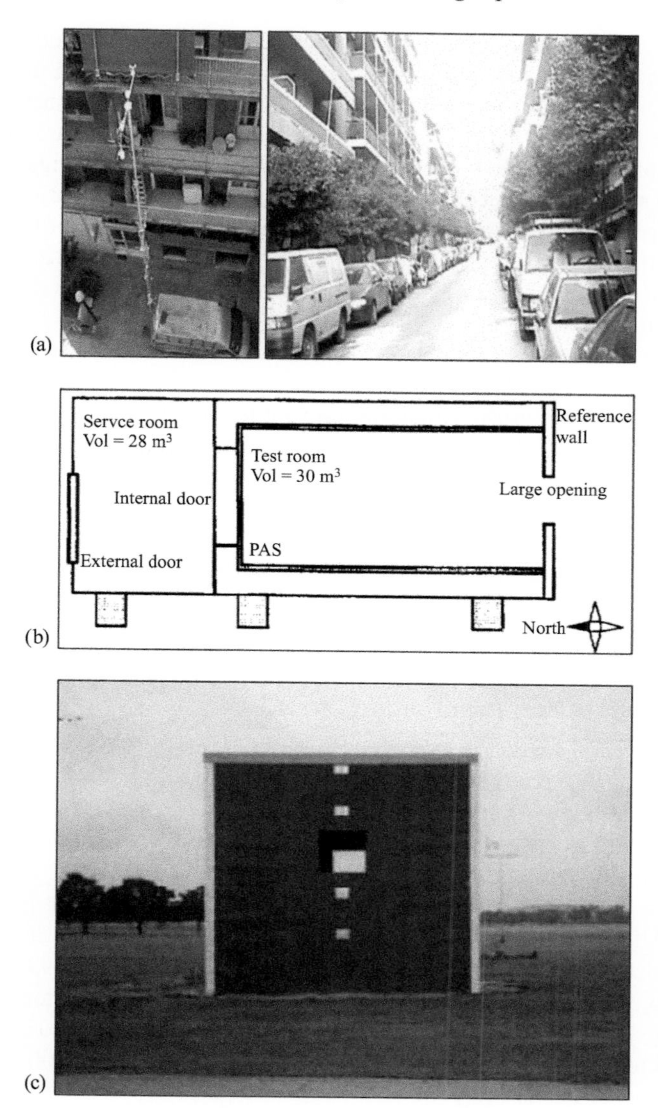

Figure 9.1 Some building and building models used for the measurements of natural ventilation: (a) naturally ventilated multistory residential buildings in an urban area (Georgakis and Santamouris, 2006; Niachou et al., 2005); (b) a full-scale test chamber of single-sided ventilation in an open space (Eftekhari, 1995); (c) a full-scale test chamber of cross ventilation in an open space (Straw et al., 2000)

are expensive in both instrumentations and labors; and (d) they are strongly dependent on building availability.

Full-scale measurements using experimental chamber constructed in an open space were conducted to investigate single-sided ventilation (Eftekhari, 1995; Dascalaki *et al.*, 1996) (see Figure 9.1(b)) and cross ventilation (Straw *et al.*, 2000) (see Figure 9.1(c)). Compared to on-site measurements in real buildings, such measurements exclude the influence of surrounding buildings and they are flexible to study the influence of opening configuration and location. Long-term measurements with careful statistical treatments are able to improve the repeatability of the experimental data, which can be therefore used to validate CFD models. However, they are still expensive and suffer from varying outdoor environments.

Another common limitation of the measurements in both real buildings and full-scale environmental chambers is the difficulty of evaluating the influence of room location in a multistory building on ventilation performance of a room. The reason is that the simultaneous measurements in several different rooms post a high demand on the quantity of the available building, instrumentations, and labors as well as on the time consistency of these several measurements. Recent studies (Ai *et al.*, 2011a, 2011b, 2011c, 2013, 2015, 2016; Ai and Mak, 2014a, 2014b, 2014c, 2014d) of single-sided natural ventilation performance in multistory buildings show that the ventilation performance of a room is highly dependent on its three-dimensional location in a building.

9.2.2 Wind tunnel experiments

As an alternative method, wind tunnel experiments using reduced-scale building models can take a full control over boundary conditions. They were widely used in fundamental studies to increase the understanding of flow structures, evaluate influential factors, and generate data for CFD validations. Wind tunnel experiments for single-sided ventilation were conducted mostly to reveal basic flow characteristics (Heiselberg *et al.*, 2001; Eftekhari *et al.*, 2003; Katayama *et al.*, 1992; Jiang *et al.*, 2003; Etheridge, 2004; Kato *et al.*, 2006; Chu *et al.*, 2011) and to examine the influence of opening configurations (Carey and Etheridge, 1999; Larsen and Heiselberg, 2008; Daish *et al.*, 2016) and external wind direction (Larsen and Heiselberg, 2008). Wind tunnel experiments for cross ventilation were mostly conducted to reveal basic flow characteristics (Heiselberg *et al.*, 2001; Ohba *et al.*, 2001; Karava *et al.*, 2011; Tominaga and Blocken, 2015; Kobayashi *et al.*, 2010; Larsen *et al.*, 2011; Lo *et al.*, 2013; Bangalee *et al.*, 2013) and to examine the influence of opening configurations (Kobayashi *et al.*, 2010; Tecle *et al.*, 2013; Stavridou and Prinos, 2013; Yang *et al.*, 2014), external wind condition (Karava *et al.*, 2007; Tecle *et al.*, 2013; Ji *et al.*, 2011), building depth (Chu and Chiang, 2014), and roof shape (van Hooff *et al.*, 2011).

As long as the reduced-scale models are used, a key issue is to ensure the achievement of similarity between small models and their full-scale prototype so that the results obtained from the small models can be used to represent the full-scale data. With regard to natural ventilation, similarity should be taken care over not only the model geometry but also the surrounding atmospheric boundary layer as well as the envelope flow and internal flow (Cermak *et al.*, 1984; Meroney, 2004; Snyder, 1981). Satisfying all similarity criteria is practically impossible, but key criteria must be strictly obeyed. For example, the important similarity criterion for wind-induced natural ventilation is the independence of Reynolds number. Snyder (1981) suggested that building Reynolds number based on building height and approaching reference wind speed should be larger than 15,000 to ensure a similar surrounding flow. Cermak *et al.* (1984) suggested that, for natural ventilation, both the building Reynolds number and the opening Reynolds number (based on the maximum velocity and minimum dimension of an opening) should be larger than 20,000. The building Reynolds number is easy to achieve, whereas the opening Reynolds number is not difficult to violate. The reason for the latter is that, in a highly reduced-scale model, the opening(s) is too small to ensure a fully developed turbulent flow through the opening(s) (Sherman, 1990).

In order to achieve Reynolds-number independence of the opening flows, the opening(s) should be as large as possible. However, in wind tunnel experiments, the dimensions of building models (thus the opening dimensions) are restricted by the cross-sectional area of a wind tunnel, as the resulted blockage ratio (i.e., the ratio of the largest area of the cutting plane of a building model that is perpendicular to the streamwise direction to the cross-sectional area of the test section of a wind tunnel) of the occupied wind tunnel should be sufficiently small, say 3%–5% (Franke *et al.*, 2007), to lower the influence of physical boundaries on the flow development. The building model used in wind tunnel experiments of natural ventilation must therefore meet the requirements of both the blockage ratio and the Reynolds-number independence of the opening flows. Owing to these considerations, past wind tunnel studies on natural ventilation were mostly limited to very simple building models, such as single-room building models (see, e.g., Figure 9.2).

9.2.3 Experimental determination of ventilation rate

Ventilation rate can be simply determined by integrating velocities over an opening area (Jiang *et al.*, 2003), which is called integration method. However, an accurate measurement of velocities at an opening is very difficult, because (a) the velocity at an opening is always varying over time, (b) the area-averaged velocity can only be estimated from the measurements at a limited number of locations, and (c) the presence of anemometric probes would disturb the flow development at the opening. A more accurate and well-established method of determining ventilation rates is tracer gas technique (Laussmann and Helm, 2011; McWilliams, 2002; Sherman, 1990), including concentration decay method, constant injection method, and constant concentration method. Each method has its advantages and disadvantages as well as applicable conditions. In general, the concentration decay method is the simplest and thus the most widely used method. The constant injection method is very suitable for long-term investigations of the relationship between ventilation rate and indoor pollutant level. The constant concentration method is relatively expensive, because of the large volume of tracer gas required. Commonly used tracer gases are nitrous oxide (N_2O), sulfur

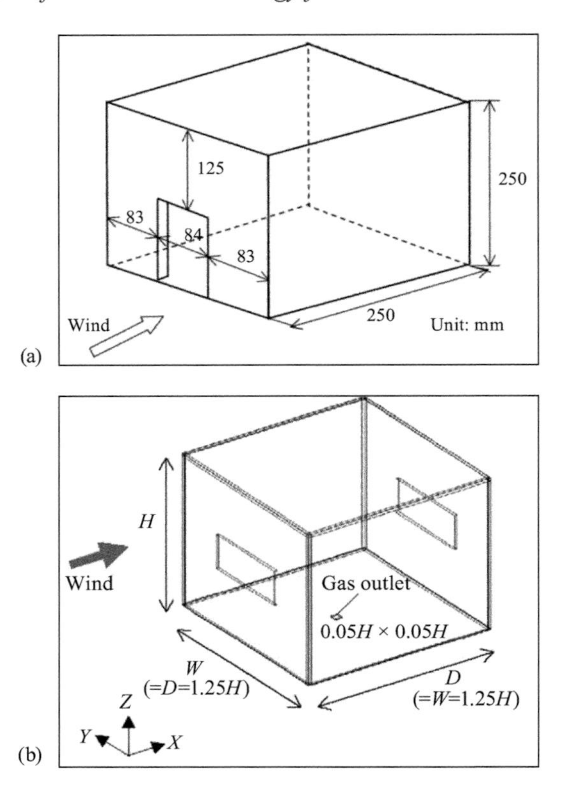

Figure 9.2 Some building models used in past wind tunnel experiments to study natural ventilation: (a) a building model (250×250×250 mm) with one opening (125×84 mm) for single-sided ventilation (Jiang et al., 2003); (b) a building model (160×160×160 mm) with two openings (92×36 mm) for cross ventilation (Karava et al., 2011; Tominaga and Blocken, 2015)

hexafluoride (SF_6), and halogenated hydrocarbons. Carbon dioxide (CO_2) is also widely used, as it is cheap and easy to obtain. When a stable CO_2 release can be ensured, such as a certain number of sleeping persons (Li *et al.*, 2014a, 2014b; Ai *et al.*, 2016), the constant injection method can be cheaply used. It was reported that the tracer gas technique is difficult to use in wind tunnel experiments of natural ventilation, because first there are additional similarity criteria required to be achieved and second it would involve uncertainties regarding the mixing between tracer gas and air (Chiu and Etheridge, 2007; Birdsall and Meroney, 1995).

9.3 CFD simulations

CFD simulations are widely used to investigate natural ventilation, as they have many advantages compared to experimental methods (van Hooff and Blocken, 2010a,

2010b, 2013). The major advantage of CFD simulations is that, without similarity constraints and flow disturbances, they provide information of the relevant flow variables in the whole computational domain under well-controlled boundary conditions. Particularly, for natural ventilation of complex building configurations, such as multistory buildings, CFD simulations can easily examine the ventilation performance of all rooms, while this is a challenge for full-scale measurements and for reduced-scale wind tunnel experiments.

9.3.1 RANS and LES models

Reynolds-averaged Navier–Stokes (RANS) models are the most widely used turbulence models to predict both single-sided and cross natural ventilation. For single-sided natural ventilation, RANS studies are centered mainly on numerical model validations and verifications, parametric studies (Gan, 2000; Chou *et al.*, 2008; Prianto and Depecker, 2003; Favarolo and Manz, 2005; Visagavel and Srinivasan, 2009;Omrani *et al.*, 2017a, 2017b; Sun *et al.*, 2017; Wang *et al.*, 2017), and revealing flow characteristics (Ai *et al.*, 2013, 2015; Jiang and Chen, 2001; Jiang *et al.*, 2003; Bu and Kato, 2011a, 2011b; Li *et al.*, 2014a, 2014b). For cross natural ventilation, similar studies on numerical model validations and verifications (Seifert *et al.*, 2006; Meroney, 2009; Ramponi and Blocken, 2012; Martins *et al.*, 2016; van Hooff *et al.*, 2017), parametric studies (Visagavel and Srinivasan, 2009; van Moeseke *et al.*, 2005; Horan and Finn, 2008; Nikas *et al.*, 2010; Cheung and Liu, 2011; Peren *et al.*, 2016; Wang *et al.*, 2007; Wang and Wong, 2007; Stavrakakis *et al.*, 2012; Gao and Lee, 2011a, 2011b; Shetabivash, 2015; Cui *et al.*, 2016), and flow characteristics (Larsen *et al.*, 2011; Lo *et al.*, 2013;Bangalee *et al.*, 2012, 2013; Li *et al.*, 2014a, 2014b; Nikas *et al.*, 2010; Stavrakakis *et al.*, 2008; Kao *et al.*, 2009; Nikolopoulos *et al.*, 2012; Faggianelli *et al.*, 2014) were conducted. In addition, some applied studies on relatively complex building configurations (Mochida *et al.*, 2008; van Hooff and Blocken, 2010a, 2010b, 2013) were also reported. Although RANS models provide effective time-averaged solutions, they are inherently based on the Reynolds-averaged treatment and thus significantly cancel out the contributions of temporal fluctuations.

Some studies (Caciolo *et al.*, 2012; Ai and Mak, 2014a, 2014b, 2014c, 2014d; Jiang and Chen, 2001; Jiang *et al.*, 2003; Bu and Kato, 2011a, 2011b; Caciolo *et al.*, 2013) have compared the performance of RANS models and LES models in the prediction of natural ventilation, which indicated that LES models are more accurate than RANS models at the expense of consuming more computational resources. LES models calculate directly the governing equations of large eddies and model-only small eddies. Compared to RANS models, LES models can capture the flow intermittencies and separations in and around naturally ventilated buildings in a more accurate manner. Using fluctuating inflow algorithms (Tabor and Baba-Ahmadi, 2010), LES models can provide not only the time-averaged flow patterns but also the transient flow patterns at various moments, where the latter cannot be produced by steady-state RANS models. Figure 9.3 shows the distribution of velocity vectors and pressure contours on

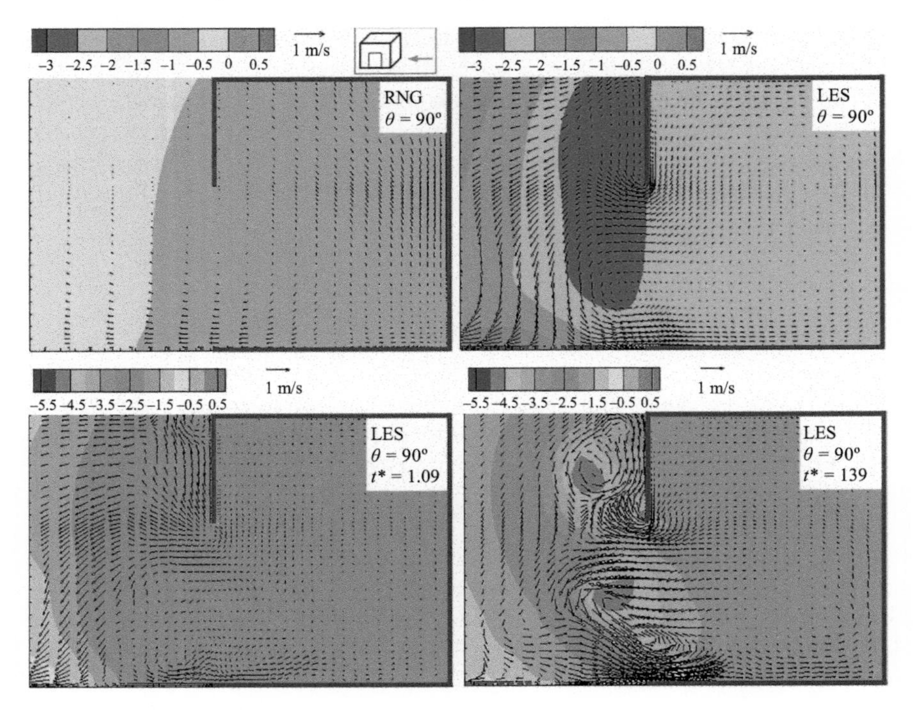

Figure 9.3 Distribution of velocity vectors and pressure contours on the vertical center plane of the opening of a single-room building model, where the wind direction is parallel to the plane of the opening ($\theta = 90°$) (Ai and Mak, 2014a, 2014b, 2014c, 2014d), where the θ denotes the angle between the wind direction and the normal of the plane of the opening and the t^* normalized time

the vertical center plane of the opening given by RANS and LES models. It can be seen that the LES results reveal clearly the moments of strong interactions between indoor and outdoor flows. More and more LES studies on natural ventilation were performed in the past decades. For single-sided ventilation, LES models were mainly used to reveal the more profound basic flow characteristics and ventilation mechanisms (Jiang *et al.*, 2003; Chu *et al.*, 2011; Caciolo *et al.*, 2012; Ai and Mak, 2014a, 2014b, 2014c, 2014d; Jiang and Chen, 2001; Jiang *et al.*, 2003; Bu and Kato, 2011a, 2011b; Jiang and Chen, 2002; Yamanaka *et al.*, 2006; Hasama *et al.*, 2008; Heiselberg and Li, 2009; Bu and Kato, 2011a, 2011b; Ai and Mak, 2014a, 2014b, 2014c, 2014d). For cross ventilation, in addition to revealing basic flow characteristics (Jiang *et al.*, 2003; Chu *et al.*, 2011; Jiang and Chen, 2001; Jiang and Chen, 2002; Wright and Hargreaves, 2006; Hu *et al.*, 2006, 2008; Chu and Chiang, 2013), LES models were also used for parametric studies (van Moeseke *et al.*, 2005; Kao *et al.*, 2009).

9.3.2 CFD prediction of ventilation rate

Although all the experimental determination methods of ventilation rate (see Section 9.2.3) can be reproduced in CFD simulations, the integration method and the tracer gas decay method are the most widely used two methods. The integration method can be used in both steady and unsteady simulations, but the tracer gas decay method can only be used in unsteady simulations. For single-sided natural ventilation, Ai and Mak (2014a, 2014b, 2014c, 2014d) have systematically compared four combinations of methods in the calculation of single-sided ventilation rates, namely, the integration method and the tracer gas decay method based on either the RANS or the LES models, respectively. It was found that the LES-based tracer gas decay method provides overall the most accurate results. This should be attributed to that, compared to other combinations, the LES models establish accurate flow and turbulence fields and the tracer gas decay method takes into account comprehensively the ventilation mechanisms. In contrast, the RANS models cannot provide an accurate turbulence field and the integration method cannot capture the part of ventilation rate contributed by turbulent fluctuations (explained later).

Using the best combination of method, the unsteady ventilation rate through a single opening was thoroughly investigated (Ai and Mak, 2014a, 2014b, 2014c, 2014d). It was reported that the ventilation rate through a single opening is always fluctuating, but the fluctuating intensities could be very different under different wind directions. The single-sided ventilation performs better in cases with a parallel (90°) and a leeward (180°) opening than that with a windward (0°) opening. Owing to the turbulent effect of wind and thus the fluctuating characteristics of ventilation rate, a total ventilation rate can be divided into the mean-flow-induced component and the turbulence-induced component. Figure 9.4 presents the transient evolution of ventilation rates under three wind directions, which suggests that the airflow exchange is dominated by the mean flow convection for windward ventilation and is dominated by the turbulent flow fluctuation for the parallel ventilation, while the turbulent flow still has an important impact on airflow exchange for leeward ventilation. Similar findings on the unsteady characteristics were obtained for LES simulations of cross ventilations (Hu *et al.*, 2008), which are also supported by previous full-scale and wind tunnel experiments by Straw *et al.* (2000) and Larsen *et al.* (2011), respectively.

These characteristics of single-sided ventilation rate may lead to three observations. First, the relationship between ventilation rate and wind direction contradicts the common sense of the ventilation rates under different wind directions, where the latter is that the highest one should be with windward ventilation. However, this common sense is supported only by steady-state RANS simulations. Second, the influence of some factors on single-sided natural ventilation, such as the wind direction, should be very different from that on cross natural ventilation, as their ventilation mechanisms are different. Third, in view of that a large percentage of ventilation rate is contributed by turbulent fluctuations, it is very important to accurately predict and take into account this turbulence-induced component.

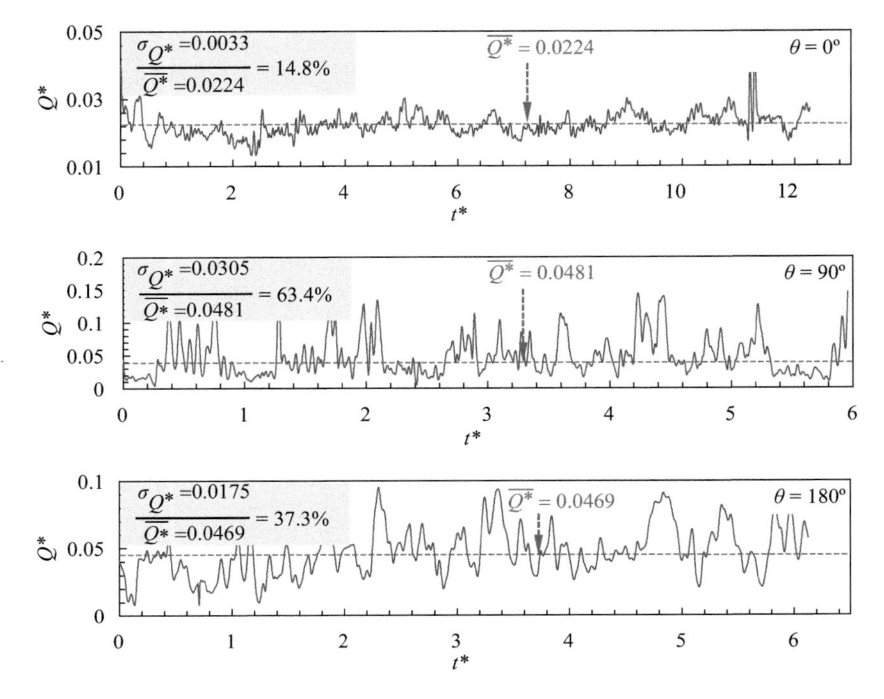

Figure 9.4 Transient evolution of ventilation rates under different wind directions (Kurabuchi et al., 2004), where Q^* denotes normalized ventilation rate, $\overline{Q^*}$ time-averaged Q^*, and σ_{Q^*} the standard deviation of Q^*; note that the σ_{Q^*} can be regarded as the turbulence-induced ventilation rate and the ratio $\sigma_{Q^*}/\overline{Q^*}$ denotes the role of the turbulence-induced component in the total ventilation rate

9.3.3 Lack of best practice guidelines

Given the complexity of natural ventilation that involves multiple length scales, i.e., from urban area, building, room, down to ventilation opening, any inappropriate settings of computational parameters would result in incorrect solutions. Detailed guidance for CFD simulations of urban aerodynamics has been provided by some best practice guidelines (BPGs) (Franke *et al.*, 2007; Chu *et al.*, 2009; Heiselberg and Li, 2009; Andersen, 2002). These BPGs provide recommendations of the whole process of conducting CFD simulations in order to ensure accurate results, which include mainly the acceptable model validations and the appropriate selections of turbulence model, geometrical simplifications, computational domain, boundary conditions, computational grid, numerical approximations, time step size, and iterative convergence criteria.

However, these guidelines are limited to outdoor flows, which do not provide specific guidelines for natural ventilation that contains outdoor flows, indoor flows, and their interactions at ventilation openings. There were a few studies attempting to examine the influence of computational parameters on CFD simulations of natural ventilation and then to recommend the appropriate selections. Ramponi and Blocken (2012) have conducted generic sensitivity analyses of six computational parameters for CFD simulations (RANS models) of cross ventilation. Ai and Mak (2014a, 2014b, 2014c, 2014d) tested the sensitivity of near-wall mesh density and compared the RANS and LES simulations of single-sided natural ventilation in terms of flow pattern and ventilation rate. Herein, more studies are required to support the establishment of BPGs for CFD simulations of natural ventilation. In addition, it is worth to point out that an acceptable validation of CFD simulations of natural ventilation should include the validations of the outdoor flow around a building, the indoor flow, and the interaction of indoor and outdoor flows at a ventilation opening.

9.4 Empirical predictions

9.4.1 Single-sided ventilation: empirical models

Different from experimental and CFD methods, empirical models were developed to provide a rapid estimation of natural ventilation rate, especially in the design stage. Warren (Chen, 2009) proposed two simple equations to calculate the single-sided natural ventilation rates due to wind and buoyancy effect, respectively; the larger of these two was used as the overall ventilation rate due to the combined effect of wind and buoyancy. This model has been widely used in later studies (e.g., Etheridge and Sandberg, 1996; Warren and Parkins, 1985; Etheridge, 2002) and has even been used for natural ventilation design (BS 5925, 1991). Phaff and De Gids (1982) considered simultaneously the contributions of wind and buoyancy effects in their empirical model. However, the window characteristics were not taken into account in this model. Based on network models and field measurements, Dascalaki *et al.* (1995) proposed an empirical model to predict single-sided ventilation rate. These three models, however, all ignored the influence of wind direction (Larsen *et al.*, 2003). Considering the influence of wind direction, Larsen and Heiselberg (2008) developed a new equation, although the relationship between ventilation rate and wind direction was found to be unclear. Caciolo *et al.* (2011) compared all these empirical models with their full-scale experiments. They found that the model by Warren (Chen, 2009) showed overall the best performance, although all of the models underperformed for leeward ventilation. Caciolo *et al.* (2013) then developed a new equation for leeward ventilation, but this equation did not include the conditions where the ventilation is dominated by wind. In order to predict the unsteady ventilation rates, Wang and Chen (2015) derived the contributions of pulsating flow (Haghighat *et al.*, 1991) and eddy penetration via spectrum analysis and LES simulations. However, the wind direction was not included in their model. In addition, a uniform velocity distribution was assumed to be on a window. Based on full-scale measurements and LES simulations, Wang *et al.* (2015) further developed semiempirical models to predict single-sided ventilation rates through hopper, awning, and casement windows. In general, these empirical models were established on the basis of either a single-room building or a specific room of a building.

Freire *et al.* (2013) verified three empirical models, including two models (Larsen and Heiselberg, 2008; Phaff and De Gids, 1982) for single-sided ventilation and one for cross ventilation using both wind tunnel and full-scale measurements. They found that different models predict very distinctive ventilation rates, and they then proposed some improvements for Larsen and Heiselberg's model. Recently, Ai and Mak (2014a, 2014b, 2014c, 2014d) evaluated the performance of these empirical models by comparing with their on-site measurements in high-rise buildings (Ai *et al.*, 2015). The predictions given by the previous empirical models showed large discrepancies with the on-site measurements. The main reason for this was attributed to that these models did not contain a parameter that accounts for the influence of room location in a multistory building. In fact, different rooms in a multistory building could experience very different envelope airflow patterns, which in turn result in distinctive ventilation rates among these rooms (Ai *et al.*, 2011a, 2011b, 2011c).

9.4.2 Cross ventilation: discharge coefficients

A simplified method to predict the flow rate through a cross-ventilation opening is the use of orifice equation, which is based on the Bernoulli assumption for incompressible flow. The orifice equation can be written in the following form:

$$Q = C_d \cdot A \cdot \sqrt{\frac{2|\Delta P|}{\rho}} \tag{9.1}$$

where Q is the ventilation rate, C_d the discharge coefficient of the opening, A the opening area, ΔP the mean pressure difference across the opening, and ρ the air density.

The discharge coefficient is empirically determined. The common discharge coefficients suggested in the literature are in the range of 0.6–0.65 for sharp-edged openings (Ohba *et al.*, 2001), 0.9–0.95 for round edge openings (Freire *et al.*, 2013), 0.25–0.65 for oblong louver window, 0.25–0.6 for pivoted window (Ohba *et al.*, 2001), 0.65–1 for side hung window (Heiselberg *et al.*, 2001), and 0.8–1 for bottom hung window (Heiselberg *et al.*, 2001). However, discharge coefficients are far more than constant values, which are affected by many factors (Heiselberg and Sandberg, 2006), such as wind direction (Carey and Etheridge, 1999; Sawachi *et al.*, 2004; Kurabuchi *et al.*, 2004; Ohba *et al.*, 2004; Jensen *et al.*, 2002a, 2002b; Sawachi, 2002; Nishizawa *et al.*, 2004; Chu *et al.*, 2009), pressure difference across the opening (Heiselberg *et al.*, 2001; Sawachi *et al.*, 2004; Sawachi, 2002; Nishizawa *et al.*, 2004; Heiselberg *et al.*, 2002), opening shape (Heiselberg *et al.*, 2002; Andersen, 1996; Axley and Chung, 2006; Kato *et al.*, 1992; Kato, 2004; Sandberg, 2002, 2004; Jensen *et al.*, 2002a, 2002b), opening area (Heiselberg *et al.*, 2002; Kato *et al.*, 1992; Kato, 2004; Murakami *et al.*, 1991), wall porosity (opening to wall ratio) (Sandberg, 2002, 2004; Jensen *et al.*, 2002a, 2002b), and Reynolds number (Etheridge, 2004; Carey and Etheridge, 1999; Chu *et al.*, 2009; Andersen, 1996). Particularly, it was reported that C_d varies largely with the Reynolds number, even though the Reynolds number was sufficiently large to make the flow fully turbulent. Apart from the influential factors mentioned earlier, many other reasons, such as different physical simplifications and even different research objectives, could lead to different C_d. It is then suggested that C_d values should be used after carefully considering their applicable conditions.

Kurabuchi *et al.* (2004) and Ohba *et al.* (2004) proposed a "local dynamic similarity model" to replace the orifice equation and discharge coefficient. However, its performance needs further investigations. In addition, there are some simplified models developed for a rapid prediction of cross-ventilation flow and ventilation rate (Carrilho da Graca and Linden, 2003; Carrilho da Graca *et al.*, 2015). Recently, Shirzadi *et al.* (2018) developed an adaptive discharge coefficient as a function of the geometry and location of a ventilation opening, which was shown to be able to improve the predictive accuracy of cross ventilation by up to 28% when compared to that using constant discharge coefficient. However, the uses of constant discharge coefficients were still the most widely employed treatment in recent studies of natural ventilation (e.g., Tecle *et al.*, 2013; Chiu and Etheridge, 2007; Chu *et al.*, 2009).

In fact, the orifice equation (namely, using discharge coefficients) was established based on the sealed-body assumption, which consists of the following: (a) the flow problem is fully developed turbulent flow; (b) the presence of openings does not influence the pressure distribution on building surfaces; and (c) the kinetic energy is dissipated at the windward opening. Many studies (Karava *et al.*, 2011; Kobayashi *et al.*, 2010; Kato *et al.*, 1992; Sandberg, 2004; Murakami *et al.*, 1991; Choiniere *et al.*, 1992; True *et al.*, 2003; Karava *et al.*, 2006; Kobayashi *et al.*, 2009) reported that, in the case of flow through large openings, the presence of openings has an obvious influence on pressure distribution on building surfaces and the turbulent kinetic energy is definitely not dissipated at the windward opening, where the orifice equation is therefore no longer correct. Several studies have indicated that the orifice equation can be acceptable if the wind direction is nearly perpendicular to the wall, less than 45°, and if the wall porosity is less than 30% (Sandberg, 2002) or 23% (Vickery and Karakatstanis, 1987) or 20% (True *et al.*, 2003). However, it is reasonable to infer that the orifice equation is less applicable for very small openings (Sandberg, 2002), as the assumption of fully developed turbulent flow tends to be violated.

9.5 General issues

9.5.1 Effect of envelope features

Building models used in past studies on natural ventilation mostly have flat facades. However, in practice, buildings usually have non-flat facades. There are always envelope features, such as balconies, wing walls, and shading devices, attached on the building facades. Normally, these envelope features are not constructed intentionally for natural ventilation. However, their presence increases the roughness of the building facades and modifies the envelope airflow pattern, which thus has a strong potential to modify the natural ventilation performance of buildings.

As an important envelope feature, balconies provide occupants an easy access to the outdoor environment, which have a series of influences on the building physics (Papamanolis, 2004). Both wind tunnel and CFD simulations (Omrani *et al.*, 2017a, 2017b; Chand *et al.*, 1998; Ai *et al.*, 2011a, 2011b, 2011c; Montazeri and Blocken, 2013) showed that the presence of balconies considerably changes the pressure distribution on building surfaces and the near-façade airflow pattern (see, e.g., Figure 9.5), which, in turn, alters the driving forces of natural ventilation. Further studies using unsealed building models (Ai *et al.*, 2011a, 2011b, 2011c) suggested that the presence of balconies introduces more turbulence to the near-wall flow through

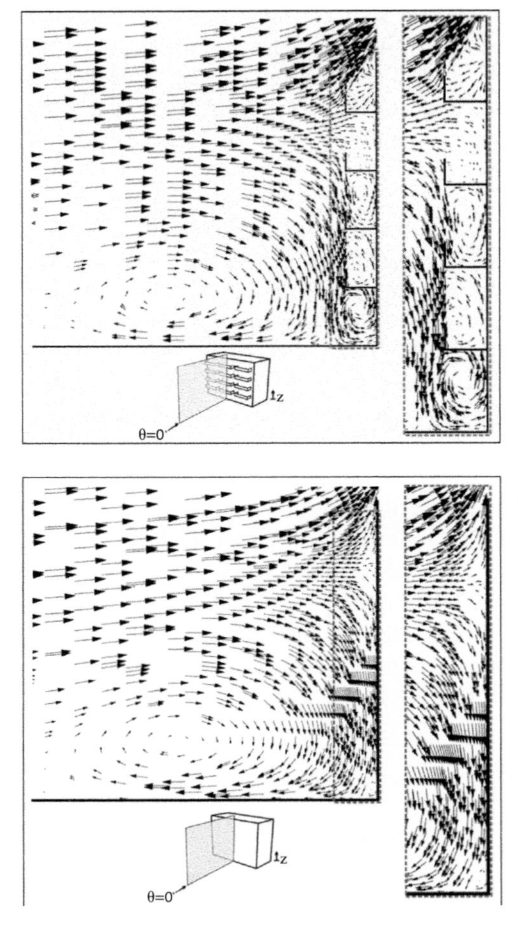

Figure 9.5 Comparison of envelope airflow pattern around buildings with balconies (left) and without balconies (right) (Montazeri and Blocken, 2013)

the enhanced interaction between approaching flow and building facades. This eventually results in a totally different distribution of natural ventilation rate among rooms in a multistory building (see Figure 9.6), indicating the importance of considering the balconies when evaluating natural ventilation of buildings. Similarly, wing walls (Khan *et al.*, 2008; Mak *et al.*, 2007), ventilation shafts (O'Sullivan and Kolokotroni, 2017; Prajongsan and Sharples, 2012), shading devices (Argiriou *et al.*, 2002; Hien and Istiadji, 2003; Al-Tamimi and Fadzil, 2011), and window configurations (Wang *et al.*, 2015; Ai and Mak, 2018) have complex influences on natural ventilation performance of buildings.

Figure 9.7 presents a quantitative comparison of ACH values given by the presence of different types of envelope features. Generally, ignoring the envelope

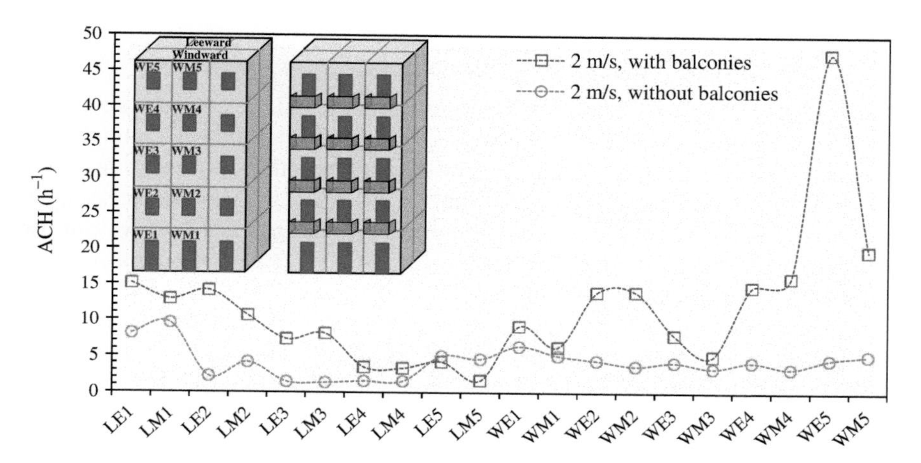

Figure 9.6 Comparison of ACH values of rooms in buildings without and with balconies, where the "2 m/s" denotes the reference wind speeds of the perpendicularly approaching wind (Ai et al., 2013)

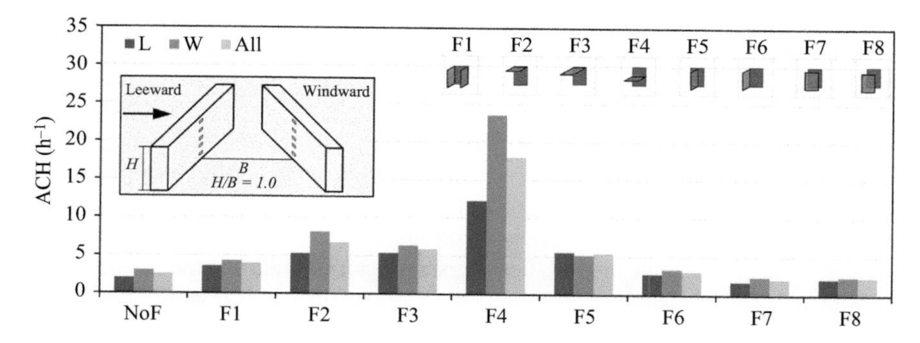

Figure 9.7 Average ACH values of rooms at leeward facade (L), windward facade (W) and both facades (All) for the street canyon of aspect ratio (H/B) = 1.0, where "NoF" represents "no envelope feature" (Ai and Mak, 2018)

features could result in large deviations between the predicted and the real-life natural ventilation performance in buildings. It is possible to use these envelope features to improve natural ventilation performance through careful designs to ingeniously utilize the building aerodynamics and to adapt to the local wind environment. However, the presence of envelope features does not necessarily improve natural ventilation performance for all rooms in a multistory building; with certain combinations of wind speed, wind direction, and opening dimensions, their presence could also lower largely natural ventilation performance. Understanding their influences is therefore the first step before successfully incorporating them into design. It should also be noted that the optimal design of the envelope features for natural ventilation may be compromised by their other important environmental and architectural functions.

9.5.2 Single-story versus multistory building model

Most past studies on natural ventilation were limited to using very simple building models, e.g., a single-room model. However, urban buildings are mostly multistory buildings, including high-rise buildings. Two important questions would therefore be as follows: to what extent the findings obtained from single-story (mostly single-room) building models are applicable to multistory buildings and how we can use the findings obtained from single-story building models to design natural ventilation of multistory buildings.

Some CFD simulations (Georgakis and Santamouris, 2006; Gao and Lee, 2011a, 2011b; Gao and Lee, 2012; Ai and Mak, 2018; Fung and Lee, 2014) and on-site measurements (Georgakis and Santamouris, 2006; Ai *et al.*, 2015; Omrani *et al.*, 2017a, 2017b; Prajongsan and Sharples, 2012; Gilkeson *et al.*, 2013; Santamouris *et al.*, 2008; Zhou *et al.*, 2014) of natural ventilation were based on multistory buildings, but they did not evaluate the influential factors and perform the parametric analysis. In addition, these studies did not analyze (a) the differences in the ventilation characteristics of multistory buildings from those of single-story (mostly single-room) buildings and (b) the differences in ventilation characteristics of the rooms in a multistory building. They can therefore provide limited implications to a systematic understanding of natural ventilation characteristics of multistory buildings and thus their natural ventilation design.

Two earlier studies (Ai and Mak, 2014a, 2014b, 2014c, 2014d; Ai *et al.*, 2015) analyzed comparatively the different envelope flow patterns around the openings of a single-story building and of the rooms in a multistory building. According to the bluff body aerodynamics revealed in wind engineering (Liu, 1991), the general external airflow patterns around both a single-story building and a multistory building are basically similar. However, for a single-story building, the two rooms enjoy the whole external airflow field, whereas, for a multistory building, many rooms share such an airflow field (see Figure 9.8). Obviously, different rooms in a multistory building would experience distinctive envelope flow characteristics. Such diversified interactions between the outdoor flow and the openings of

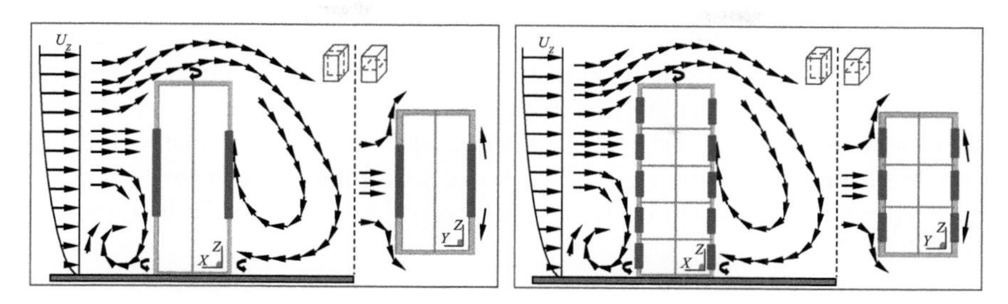

Figure 9.8 A similar outdoor flow field is experienced around a single-story and a multistory building, but the rooms in these two types of buildings experience different envelope flow characteristics; the geometry of the single-story building is stretched for the purpose of comparing schematically the flow patterns around the two types of buildings (Yang et al., 2014)

different rooms, which have different three-dimensional locations in a multistory building, cannot be investigated by using single-story building models.

Some studies (Ai *et al.*, 2013; Ai and Mak, 2014a, 2014b, 2014c, 2014d) employed generic building models to investigate the natural ventilation performance of the rooms in a multistory building under different wind directions. It was reported that, owing to the distinctive envelope flow patterns, different rooms in the same multistory building show very different ventilation performance (see Figure 9.6). This is also an important reason why past empirical models based on a single-room building or a specific room in a multistory building are not reliable to predict the ventilation performance of all rooms in a complex multistory building. In order to enhance the understanding of natural ventilation of multistory buildings and then to provide design guidelines, it is necessary to use multistory building models in future studies.

9.5.3 Isolated buildings versus urban buildings

As reviewed, past fundamental studies conducted using wind tunnel and CFD techniques mostly focused on revealing basic flow characteristics and examining the influential factors of natural ventilation performance. These studies have provided insights into the flow mechanics of natural ventilation in buildings and the optimization of various parameters. However, they were mostly based on very simplified building configurations without surrounding buildings, such as a standalone one-room building model. The findings of these studies cannot provide direct guidelines for natural ventilation of urban buildings, due to very different building configurations and complex surroundings of real urban buildings. Many studies have revealed the significant influence of urban canyon effect on natural ventilation performance in buildings, which mainly includes decreased and inconstant ventilation rate, increased pollutant penetration and ingress of traffic noise. In practice, the outdoor wind condition is always inconstant. The presence of street canyons creates unique local wind characteristics. Particularly, in a street canyon, normal-to-facade wind speed is very low, although vertically and/or horizontally parallel wind speed is still low compared to ambient wind speed above the canyon. Within an urban context, the flow behaviors of a naturally ventilated building are far from well understanding. Future fundamental studies should use more complex building configurations, such as a group of multistory buildings. In this respect, with the improvement of computational power, CFD simulations using advanced models like LES will show a great superiority.

A few applied studies on natural ventilation in urban buildings were conducted using CFD method (Georgakis and Santamouris, 2006; Gao and Lee, 2012; van Hooff and Blocken, 2010a, 2010b) and on-site measurements (Georgakis and Santamouris, 2006; Gilkeson *et al.*, 2013; Santamouris *et al.*, 2008) were considered directly the surrounding conditions. These studies increased quantitatively our understanding on the influence of surrounding buildings on natural ventilation performance of a specific building and on the real-life ventilation performance of urban buildings. However, because every building has its unique surroundings and again experiences a varying wind condition, different past studies reported inconsistent percentage decreases of ventilation rate caused by the presence of surrounding buildings (Georgakis and Santamouris, 2006; Gao and Lee, 2012; van Hooff and Blocken, 2010a, 2010b). In addition, these studies did not perform parametric analysis of influential factors. It is therefore difficult to use extensively their findings (based on a specific building community) to others.

King *et al.* (2017) and Tominaga and Blocken (2015) evaluated the natural ventilation performance of a cubical building that is located in an ideal building array. Figure 9.9 presents the experimental setup used in the study by King *et al.* (2017). Very distinctive ventilation performance was found in the cases with and without surrounding buildings. Compared to the isolation case, the relationship between ACH and wind direction was much weaker in the array case. Ai and Mak (2018) examined the natural ventilation performance in buildings near a street canyon, where the street configuration (i.e., aspect ratio) was an important variable. Figure 9.10 shows the large influence of street configuration on the flow field inside a street canyon as well as the resulted influence on natural ventilation performance of the nearby buildings. These studies not only took into account the urban context but also allowed a convenient parametric analysis of parameters. More such kind of

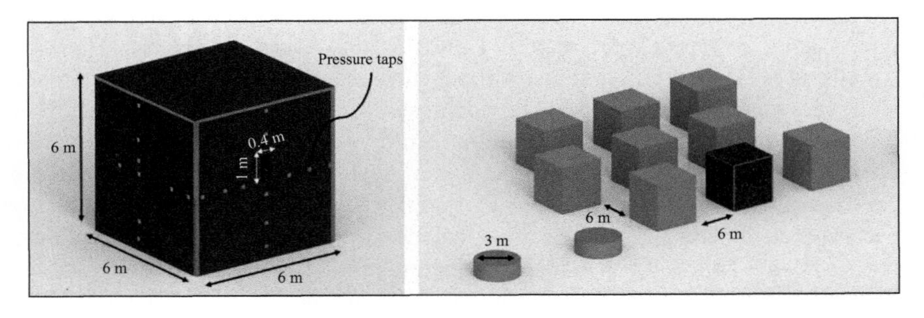

Figure 9.9 Experimental setup of the cubic building in isolation and array format

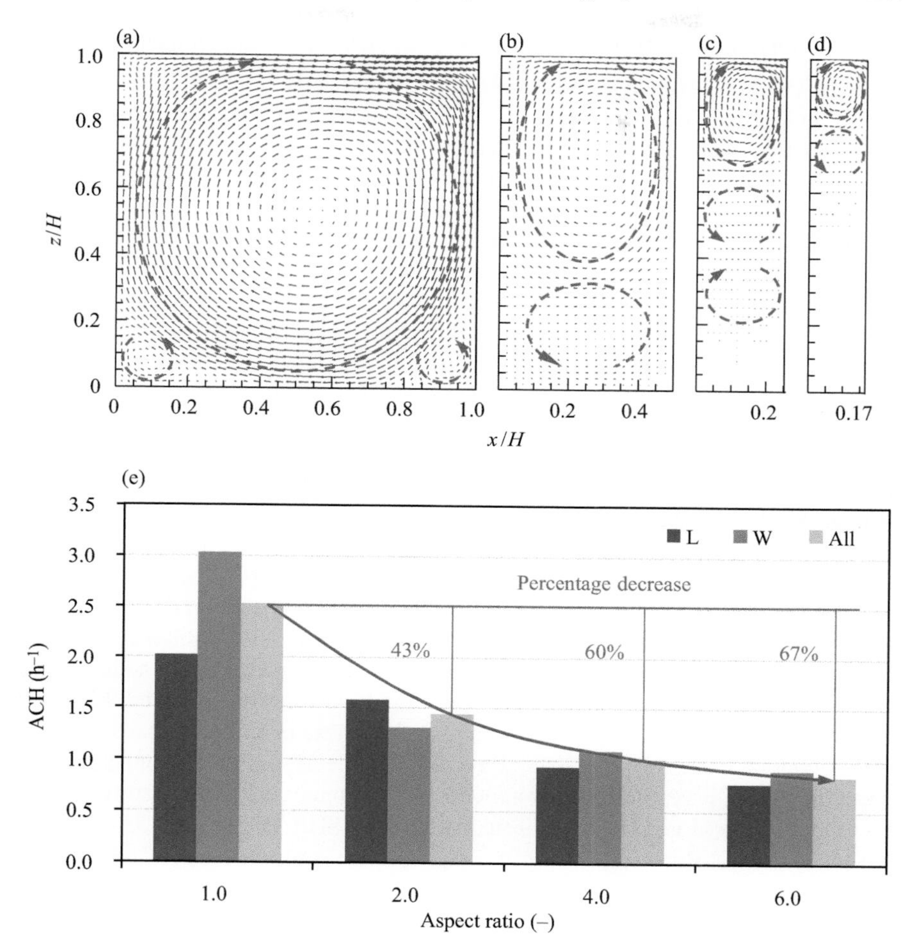

Figure 9.10 Flow fields on the vertical centerplane of a long street canyon under different aspect ratios when the approaching reference wind speed at the height of 2H was 4.2 m/s: (a) AR = 1.0, (b) AR = 2.0, (c) AR = 4.0, and (d) AR = 6.0, as well as (e) the ACH values of the rooms in the nearby buildings (see Figure 9.7 for the description of the building configurations)

studies using generic urban geometries should be performed in the future in order to increase the understanding of natural ventilation performance in urban buildings and its influential factors from an urban perspective. Combining these generic studies with those applied studies, the eventual goals would be first to find solutions for maximizing natural ventilation rate based on optimized urban and building designs and second to develop design guidelines for natural ventilation of urban buildings.

9.5.4 *Thermal comfort criteria for naturally ventilated spaces*

Despite various advantages of natural ventilation, it is impossible to provide a stable indoor environment, as it is strongly relied on the inconstant outdoor microclimate. However, on the one hand, a very good and stable IEQ as stipulated in air-conditioned design standards is not expected in naturally ventilated buildings, particularly in residential buildings where no serious office works are needed to handle. On the other hand, human beings have a strong adaptation to the natural environment and can tolerate a tougher indoor environment in naturally ventilated buildings than that in air-conditioned buildings. Such adaptation and tolerance are important supports of the successful application of natural ventilation.

Evaluation of thermal comfort in naturally ventilated buildings has been conducted in various climatic regions. A large number of questionnaire surveys concluded that people in naturally ventilated buildings accept a wider range of air temperature and air speed. In many climatic regions, the acceptable temperature limits are even over 30 °C (Heidari and Sharples, 2002; Wang *et al.*, 2010; Feriadi and Wong, 2004; Indraganti, 2010; Indraganti and Rao, 2010), which is definitely unacceptable in an air-conditioned environment. Wong *et al.* (2002) reported that a large proportion of people experiencing sensations of PMV that equals +2 and +3 still found the conditions to be comfortable. Fanger and Toftum (2002) explained that the main reason for the predicted bias of PMV in naturally ventilated buildings is the overestimation of the occupant's activity level and expectation. In addition, optimal air velocities for thermal comfort were found to be between 0.3 and 0.7 m/s for moderate and low activities, which are higher than those in air-conditioned buildings. These important findings that extend the comfortable scopes of air temperature and speed in naturally ventilated environments were summarized as an adaptive thermal comfort model and was included in standards like ASHRAE Standard 55 (2017) and EN15251 (2007). However, further investigations are required to extend this model to other conditions, such as with hybrid ventilation (Yuan *et al.*, 2018) and with occupant-controlled natural ventilation (Roetzel *et al.*, 2010).

9.6 Summary

A substantial number of past studies dealt with natural ventilation in buildings, but the application of natural ventilation with intentional designs was very limited. In order to explore the possible reasons behind the disconnection between research and practical design, this chapter performs a literature review of past studies on natural ventilation, with a focus on prediction methods. Some important issues are highlighted and the needs toward a reduced design risk are discussed.

Each research method of natural ventilation has its advantages and limitations. On-site measurements reveal the real-life ventilation conditions in buildings, but it is difficult to perform parametric analysis on influential factors due to the strong dependence on the varying outdoor environments. Wind tunnel experiments have a full control over the boundary conditions, but they are restricted by similarity criteria for both outdoor flow and flow at opening due to the use of reduced-scale models. Empirical models provide a rapid estimation, but they are limited by the building configurations used to develop the models. CFD simulations overcome the limitations of experiments and are very suitable for natural ventilation studies, but special attentions must be paid to their predictive quality. For CFD simulations of natural ventilation, there is a lack of BPGs, and further studies are therefore required. Overall, taking the advantages of each method and combining them together would be the best way to investigate natural ventilation.

From the viewpoint of turbulent flow, the natural ventilation rate is contributed by both mean flow convection and turbulent fluctuation. Particularly for single-sided natural ventilation with the bidirectional airflow interaction at the opening, the turbulence-induced component of ventilation rate can be even higher than the mean-flow-induced component in some wind directions. Ignoring the turbulence-induced component, as did in most past studies, would cause a large error and should thus be avoided. Tracer gas technique in combination with transient turbulence models, such as LES model, is recommended to predict not only the mean ventilation rate but also its fluctuating characteristics.

Past studies are mostly limited to using very simple building models, such as an isolated single-room building with flat facades, which are useful to reveal the fluid mechanics and the influence of parameters. However, they cannot provide direct guidelines for the natural ventilation design of urban buildings, as they are normally non-flat-façade, multistory, and surrounded by many other buildings. The presence of envelope features changes the envelope flow patterns and thus the natural ventilation performance of buildings. A single-room building enjoys the whole outdoor flow field, which is, however, shared by all rooms in a multistory building. A stand-alone building experiences a strong flow impingement and the follow-up flow separations around building facades. An urban building, however, is surrounded mostly by parallel-to-facade wind flows, while the normal-to-façade wind is usually very weak. All these differences suggest that more realistic building models should be used for natural ventilation studies in order to generate effective design guidelines.

References

Aflaki A, Mahyuddin N, Mahmoud ZAC, and Baharum MR. A review on natural ventilation applications through building facade components and ventilation openings in tropical climates. *Energy Build* 2015; 101: 153–62.

Ai ZT. *Numerical Investigation of Single-Sided Natural Ventilation and Interunit Dispersion in Multistory Buildings*. PhD thesis, The Hong Kong Polytechnic University, Hong Kong, 2015.

Ai ZT and Mak CM. A study of interunit dispersion around multistory buildings with single-sided ventilation under different wind directions. *Atmos Environ* 2014a; 88: 1–13.

Ai ZT and Mak CM. Analysis of fluctuating characteristics of wind-induced airflow through a single opening using LES modeling and the tracer gas technique. *Build Environ* 2014b; 80: 249–58.

Ai ZT and Mak CM. Modeling of coupled urban wind flow and indoor air flow on a high-density near-wall mesh: Sensitivity analyses and case study for single-sided ventilation. *Environ Model Softw* 2014c; 60: 57–68.

Ai ZT and Mak CM. Determination of single-sided ventilation rates in multistory buildings: Evaluation of methods. *Energy Build* 2014d; 69: 292–300.

Ai ZT and Mak CM. Comparison of single-sided ventilation characteristics between single-storey and multi-storey buildings due to wind effect. *Int J Vent* 2015a; 14(2): 181–9.

Ai ZT and Mak CM. From street canyon microclimate to indoor environmental quality in naturally ventilated urban buildings: Issues and possibilities for improvement. *Build Environ* 2015b; 94: 489–503.

Ai ZT and Mak CM. Large eddy simulation of wind-induced interunit dispersion around multistory buildings. *Indoor Air* 2016; 26: 259–73.

Ai ZT and Mak CM. Wind-induced single-sided natural ventilation in buildings near a long street canyon: CFD evaluation of street configuration and envelope design. *J Wind Eng Ind Aerodyn* 2018; 172: 96–106.

Ai ZT, Mak CM, and Cui DJ. On-site measurements of ventilation performance and indoor air quality in naturally ventilated high-rise residential buildings in Hong Kong. *Indoor Built Environ* 2015; 24(2): 214–24.

Ai ZT, Mak CM, and Niu JL. Numerical investigation of wind-induced airflow and interunit dispersion characteristics in multistory residential buildings. *Indoor Air* 2013; 23: 417–29.

Ai ZT, Mak CM, Cui DJ, and Xue P. Ventilation of air-conditioned residential buildings: A case study in Hong Kong. *Energy Build* 2016; 127: 116–27.

Ai ZT, Mak CM, Niu JL, and Li ZR. Effect of balconies on thermal comfort in wind-induced, naturally ventilated low-rise buildings. *Build Serv Eng Res Technol* 2011b; 32(3): 277–92.

Ai ZT, Mak CM, Niu JL, and Li ZR. The assessment of the performance of balconies using computational fluid dynamics. *Build Serv Eng Res Technol* 2011c; 32(3): 229–43.

Ai ZT, Mak CM, Niu JL, Li ZR, and Zhou Q. The effect of balconies on ventilation performance of low-rise buildings. *Indoor Built Environ* 2011a; 20(6): 649–60.

Allard F, and Santamouris M. *Natural Ventilation in Buildings: A Design Handbook*. London: James and James Ltd; 1998.

Allocca C, Chen Q, and Glicksman LR. Design analysis of single-sided natural ventilation. *Energy Build* 2003; 35: 785–95.

Al-Tamimi N and Fadzil SFS. The potential of shading devices for temperature reduction in high-rise residential buildings in the tropics. *Procedia Eng* 2011; 21: 273–82.

Andersen KT. Inlet and outlet coefficients: A theoretical analysis. *Proceedings of ROOMVENT 1996*, Yokohama, Japan, July 17–19, 1996.

Andersen KT. Friction and contraction by ventilation openings with movable flaps. *Proceedings of ROOMVENT*, Copenhagen, Denmark, September 8–11, 2002.

Argiriou AA, Balaras CA, and Lykoudis SP. Single-sided ventilation of buildings through shaded large openings. *Energy* 2002; 27: 93–115.

ASHRAE Standard 55. *Thermal Environmental Conditions for Human Occupancy*. Atlanta, GA: American Society of Heating, Refrigerating and Air-Conditioning Engineers, Inc.; 2017.

ASHRAE Standard 62. *Ventilation for Acceptable Indoor Air Quality*. Atlanta, GA: American Society of Heating, Refrigerating and Air-conditioning Engineers, Inc.; 2013.

Assimakopoulos VD, Georgakis C, and Santamouris M. Experimental validation of a computational fluid dynamics code to predict the wind speed in street canyons for passive cooling purposes. *Sol Energy* 2006; 80(4): 423–34.

Awbi HB. *Ventilation of Buildings*, 2nd Edition. London, UK: Routledge, Taylor & Francis Group; 2003, Chapter 7.

Axley JW and Chung DH. Well-posed models of porous buildings for macroscopic ventilation analysis. *Int J Vent* 2006; 5(1): 89–104.

Aynsley RM. A resistance approach to analysis of natural ventilation airflow networks. *J Wind Eng Ind Aerodyn* 1997; 67–68: 711–19.

Bangalee MZI, Lin SY, and Miau JJ. Wind driven natural ventilation through multiple windows of a building: A computational approach. *Energy Build* 2012; 45: 317–25.

Bangalee MZI, Miau JJ, Lin SY, and Yang JH. Flow visualization, PIV measurement and CFD calculation for fluid-driven natural cross-ventilation in a scale model. *Energy Build* 2013; 66: 306–14.

Birdsall JB and Meroney RN. Model scale and numerical evaluation of tracer gas distribution due to wind forced natural ventilation. *Proceedings of the 9th International Conference on Wind Engineering*, New Delhi, India, January, 1995.

Blocken B and Gualtieri C. Ten iterative steps for model development and evaluation applied to computational fluid dynamics for environmental fluid mechanics. *Environ Model Softw* 2012; 33: 1–22.

BRECSU. *Energy Consumption Guide 19: Energy Use in Offices*. Garston, Watford, UK: British Research Establishment Conservation Support Unit; 2000.

BS 5925. *Code of Practice for Ventilation Principles and Designing for Natural Ventilation*. London, UK: British Standards Institution; 1991.

Bu Z and Kato S. Investigation of ventilation effectiveness for wind-driven single-sided ventilated buildings located in an urban environment. *Int J Vent* 2011a; 10(1): 19–30.

Bu Z and Kato S. Wind-induced ventilation performances and airflow characteristics in an areaway-attached basement with a single-sided opening. *Build Environ* 2011b; 46: 911–21.

Caciolo M, Cui S, Stabat P, and Marchio D. Development of a new correlation for single-sided natural ventilation adapted to leeward conditions. *Energy Build* 2013; 60: 372–82.

Caciolo M, Stabat P, and Marchio D. Full scale experimental study of single-sided ventilation: Analysis of stack and wind effects. *Energy Build* 2011; 43: 1765–73.

Caciolo M, Stabat P, and Marchio D. Numerical simulation of single-sided ventilation using RANS and LES and comparison with full-scale experiments. *Build Environ* 2012; 50: 202–13.

Carey PS and Etheridge DW. Direct wind tunnel modelling of natural ventilation for design purposes. *Build Serv Eng Res Technol* 1999; 20(3): 131–42.

Carrilho da Graca G and Linden P. Ten questions about natural ventilation of non-domestic buildings. *Build Environ* 2016; 107: 263–73.

Carrilho da Graca G and Linden PF. Simplified modeling of cross-ventilation airflow. *ASHRAE Trans* 2003; 109(1): 4605–19.

Carrilho da Graca G, Chen Q, Glicksman LR, and Norford LK. Simulation of wind-driven ventilative cooling systems for an apartment building in Beijing and Shanghai. *Energy Build* 2002; 34(1): 1–11.

Carrilho da Graca G, Daish NC, and Linden PF. A two-zone model for natural cross-ventilation. *Build Environ* 2015; 89: 72–85.

CENEN15251. *Indoor Environmental Input Parameters for Design and Assessment of Energy Performance of Buildings Addressing Indoor Air Quality, Thermal Environment, Lighting and Acoustics*; Brussels, Belgium: European Committee for Standardization. 2007.

Cermak JE, Poreh M, Peterka JA, and Ayad SS. Wind tunnel investigations of natural ventilation. *J Transp Eng* 1984; 110: 67–79.

Chand I, Bhargava PK, and Krishak NLV. Effect of balconies on ventilation inducing aeromotive force on low-rise buildings. *Build Environ* 1998; 33(6): 385–96.

Chaplin GC, Randall JR, and Baker CJ. The turbulent ventilation of a single opening enclosure. *J Wind Eng Ind Aerodyn* 2000; 85: 145–61.

Chen Q. Ventilation performance prediction for buildings: A method overview and recent applications. *Build Environ* 2009; 44: 848–58.

Cheng V, Ng E, Chan C, and Givoni B. Outdoor thermal comfort study in a sub-tropical climate: A longitudinal study based in Hong Kong. *Int J Biometeorol* 2010; 56(1): 43–56.

Cheung JOP and Liu CH. CFD simulations of natural ventilation behavior in high-rise buildings in regular and staggered arrangements at various spacings. *Energy Build* 2011; 43: 1149–58.

Chiu YH and Etheridge DW. External flow effects on the discharge coefficients of two types of ventilation opening. *J Wind Eng Ind Aerodyn* 2007; 95: 225–52.

Choiniere Y, Tanaka H, Munroe JA, and Suchorski-Tremblay A. Prediction of wind induced ventilation for livestock housing. *J Wind Eng Ind Aerodyn* 1992; 41–44: 2563–74.

Chou PC, Chiang CM, Li YY, Lee CY, and Chang KF. Natural ventilation efficiency in a bedroom with a central-pivoting window. *Indoor Built Environ* 2008: 164–72.

Chu CR and Chiang BF. Wind-driven cross ventilation with internal obstacles. *Energy Build* 2013; 67: 201–9.

Chu CR and Chiang BF. Wind-driven cross ventilation in long buildings. *Build Environ* 2014; 80: 150–8.

Chu CR, Chen RH, and Chen JW. A laboratory experiment of shear-induced natural ventilation. *Energy Build* 2011; 43: 2631–7.

Chu CR, Chiu YH, Chen YJ, Wang YW, and Chou CP. Turbulence effects on the discharge coefficient and mean flow rate of wind-driven cross-ventilation. *Build Environ* 2009; 44: 2064–72.

CIBSE. *Natural Ventilation in Non-Domestic Buildings*. Norwich, UK: CIBSE Applications Manual AM10; 2005.

Cui S, Stabat P, and Marchio D. Numerical simulation of wind-driven natural ventilation: Effects of loggia and facade porosity on air change rate. *Build Environ* 2016; 106: 131–42.

Daish NC, Carrilho da Graca G, Linden PF, and Banks D. Impact of aperture separation on wind-driven single-sided natural ventilation. *Build Environ* 2016; 108: 122–34.

Dascalaki E, Santamouris M, Argiriou A, *et al.* Predicting single sided natural ventilation rates in buildings. *Solar Energy* 1995; 55: 327–41.

Dascalaki E, Santamouris M, Argiriou A, *et al.* On the combination of air velocity and flow measurements in single sided natural ventilation configurations. *Energy Build* 1996; 24: 155–65.

Eftekhari MM. Single-sided natural ventilation measurements. *Build Serv Eng Res Technol* 1995; 16(4): 221–5.

Eftekhari MM, Marjanovic LD, and Pinnock DJ. Air flow distribution in and around a single-sided naturally ventilated room. *Build Environ* 2003; 38: 389–97.

Escombe AR, Oeser CC, Gilman RH, *et al.* Natural ventilation for the prevention of airborne contagion. *PLoS Med* 2007; 4(2): e68.

Etheridge D. *Natural Ventilation of Buildings: Theory, Measurement and Design*. West Sussex: Wiley; 2011, Chapter 10.

Etheridge D. A perspective on fifty years of natural ventilation research. *Build Environ* 2015; 91: 51–60.

Etheridge D and Sandberg M. *Building Ventilation – Theory and Measurement*. West Sussex: Wiley; 1996.

Etheridge DW. Nondimensional methods for natural ventilation design. *Build Environ* 2002; 37: 1057–72.

Etheridge DW. Natural ventilation through large openings – Measurements at model scale and envelope flow theory. *Int J Vent* 2004; 2(4): 325–42.

Etheridge DW. Wind turbulence and multiple solutions for opposing wind and buoyancy. *Int J Vent* 2009; 7(4): 309–19.

Evola G and Popov V. Computational analysis of wind driven natural ventilation in buildings. *Energy Build* 2006; 38: 491–501.

Faggianelli GA, Brun A, Wurtz E, and Muselli M. Natural cross ventilation in buildings on Mediterranean coastal zones. *Energy Build* 2014; 77: 206–18.

Fanger PO and Toftum J. Extension of the PMV model to non-air-conditioned buildings in warm climates. *Energy Build* 2002; 34: 533–6.

Favarolo PA and Manz H. Temperature-driven single-sided ventilation through a large rectangular opening. *Build Environ* 2005; 40: 689–99.

Feriadi H and Wong NH. Thermal comfort for naturally ventilated houses in Indonesia. *Energy Build* 2004; 36(7): 614–26.

Franke J, Hellsten A, Schlnzen H, and Carissimo B. Best practice guideline for the CFD simulation of flows in the urban environment. Brussels, Belgium: COST Office; 2007.

Freire RZ, Abadie MO, and Mendes N. On the improvement of natural ventilation models. *Energy Build* 2013; 62: 222–9.

Fung Y and Lee W. Identifying the most influential parameter affecting natural ventilation performance in high-rise high-density residential buildings. *Indoor Built Environ* 2014; 24(6): 803–12.

Gan G. Effective depth of fresh air distribution in rooms with single-sided natural ventilation. *Energy Build* 2000; 31: 65–73.

Gao CF and Lee WL. Evaluating the influence of openings configuration on natural ventilation performance of residential units in Hong Kong. *Build Environ* 2011a; 46: 961–9.

Gao CF and Lee WL. Evaluating the influence of window types on the natural ventilation performance of residential buildings in Hong Kong. *Int J Vent* 2011b; 10(3): 227–38.

Gao CF and Lee WL. The influence of surrounding buildings on the natural ventilation performance of residential dwellings in Hong Kong. *Int J Vent* 2012; 11(3): 297–310.

Georgakis C and Santamouris M. Experimental investigation of air flow and temperature distribution in deep urban canyons for natural ventilation purposes. *Energy Build* 2006; 38: 367–76.

Ghiaus C, Allard F, Santamouris M, Georgakis C, and Nicol F. Urban environment influence on natural ventilation potential. *Build Environ* 2006; 41: 395–406.

Gilkeson CA, Camargo-Valero MA, Pickin LE, and Noakes CJ. Measurement of ventilation and airborne infection risk in large naturally ventilated hospital wards. *Build Environ* 2013; 65: 35–48.

Haghighat F, Rao J, and Fazio P. The influence of turbulent wind on air change rates—A modelling approach. *Build Environ* 1991; 26: 95–109.

Hasama T, Kato S, and Ooka R. Analysis of wind-induced inflow and outflow through a single opening using LES & DES. *J Wind Eng Ind Aerodyn* 2008; 96: 1678–91.

Heidari S and Sharples S. A comparative analysis of short-term and long-term thermal comfort surveys in Iran. *Energy Build* 2002; 34: 607–14.

Heiselberg P and Li Z. Buoyancy driven natural ventilation through horizontal openings. *Int J Vent* 2009; 8(3): 219–31.

Heiselberg P and Sandberg M. Evaluation of discharge coefficients for window openings in wind driven natural ventilation. *Int J Vent* 2006; 5(1): 43–52.

Heiselberg P, Bjorn E, and Nielsen PV. Impact of open windows on room air flow and thermal comfort. *Int J Vent* 2002; 1(2): 91–100.

Heiselberg P, Svidt K, and Nielsen PV. Characteristics of airflow from open windows. *Build Environ* 2001; 36: 859–69.

Hien WN and Istiadji AD. Effects of external shading devices on daylighting and natural ventilation. *8th International IBPSA Conference*, Eindhoven, the Netherlands, August 11-14, 2003.

HKPD. *Feasibility Study for Establishment of Air Ventilation Assessment System, Final Report, the Government of the Hong Kong Special Administrative Region*. Hong Kong Planning Department, http://www.pland.gov.hk/pland_en/p_study/comp_s/avas/paper&reports/final_report.pdf; 2005.

Horan JM and Finn DP. Sensitivity of air change rates in a naturally ventilated atrium space subject to variations in external wind speed and direction. *Energy Build* 2008; 40: 1577–85.

Hu CH, Kurabuchi T, and Ohba M. Applying the local dynamic similarity model and CFD for the study of cross-ventilation. *Int J Vent* 2006; 5(3): 301–12.

Hu CH, Ohba M, and Yoshie R. CFD modelling of unsteady cross ventilation flows using LES. *J Wind Eng Ind Aerodyn* 2008; 96: 1692–706.

Hughes BR, Chaudhry HN, and Ghani SA. A review of sustainable cooling technologies in buildings. *Renew Sustain Energy Rev* 2011; 15: 3112–20.

Hunt GR and Linden PF. Steady-state flows in an enclosure ventilated by buoyancy forces assisted by wind. *J Fluid Mech* 2001; 426: 355–86.

Indraganti M and Rao KD. Effect of age, gender, economic group and tenure on thermal comfort: A field study in residential buildings in hot and dry climate with seasonal variations. *Energy Build* 2010; 42(3): 273–81.

Indraganti M. Thermal comfort in naturally ventilated apartments in summer: Findings from a field study in Hyderabad, India. *Appl Energy* 2010; 87(3): 866–83.

Jensen JT, Heiselberg P, and Nielsen PV. Numerical prediction of natural ventilation by means of CFD. *Proceedings of ROOMVENT*, Copenhagen, Denmark, September 8-11, 2002a.

Jensen JT, Sandberg M, Heiselberg P, and Nielsen PV. Wind driven cross-flow analyzed as a catchment problem and as a pressure driven flow. *Int J Vent* 2002b; Hybrid Ventilation Special Edition: 88–101.

Ji L, Tan H, Kato S, Bu Z, and Tkahashi T. Wind tunnel investigation on influence o fluctuating wind direction on cross natural ventilation. *Build Environ* 2011; 46: 2490–9.

Jiang Y and Chen Q. Study of natural ventilation in buildings by large eddy simulation. *J Wind Eng Ind Aerodyn* 2001; 89: 1155–78.

Jiang Y and Chen Q. Effect of fluctuating wind direction on cross natural ventilation in buildings from large eddy simulation. *Build Environ* 2002; 37: 379–86.

Jiang Y and Chen Q. Buoyancy-driven single-sided natural ventilation in buildings with large openings. *Int J Heat Mass Transfer* 2003; 46: 973–88.

Jiang Y, Alexander D, Jenkins H, Arthur R, and Chen Q. Natural ventilation in buildings: Measurement in a wind tunnel and numerical simulation with large-eddy simulation. *J Wind Eng Ind Aerodyn* 2003; 91: 331–53.

Kao HM, Chang TJ, Hsieh YF, Wang CH, and Hsieh CI. Comparison of airflow and particulate matter transport in multi-room buildings for different natural ventilation patterns. *Energy Build* 2009; 41: 966–74.

Karava P, Stathopoulos T, and Athienitis AK. Impact of internal pressure coefficients on wind-driven ventilation analysis. *Int J Vent* 2006; 5: 255–66.

Karava P, Stathopoulos T, and Athienitis AK. Wind-induced natural ventilation analysis. *Solar Energy* 2007; 81: 20–30.

Karava P, Stathopoulos T, and Athienitis AK. Airflow assessment in cross-ventilated buildings with operable façade elements. *Build Environ* 2011; 46: 266–79.

Katayama T, Tsutsumi J, and Ishii A. Full-scale measurements and wind tunnel tests on cross-ventilation. *J Wind Eng Ind Aerodyn* 1992; 44(1–3): 2553–62.

Kato S. Flow network model based on power balance as applied to cross ventilation. *Int J Vent* 2004; 2(4): 395–408.

Kato S, Kono R, Hasama T, Tkahashi T, and Ooka R. A wind tunnel experimental analysis of ventilation characteristics of a room with single-sided opening in uniform flow. *Int J Vent* 2006; 5(1): 171–8.

Kato S, Murakami S, Mochida A, Akabashi S, and Tominaga Y. Velocity-pressure field of cross ventilation with open windows analyzed by wind tunnel and numerical simulation. *J Wind Eng Ind Aerodyn* 1992; 41–44: 2575–86.

Khan N, Su Y, and Riffat SB. A review on wind driven ventilation techniques. *Energy Build* 2008; 40: 1586–604.

King MF, Gough HL, Halios C, *et al.* Investigating the influence of neighbouring structures on natural ventilation potential of a full-scale cubical building using time-dependent CFD. *J Wind Eng Ind Aerodyn* 2017; 169: 265–79.

Kobayashi T, Sagara K, Yamanaka T, Kotani H, Takeda S, and Sandberg M. Stream tube based analysis of problems in prediction of cross-ventilation rate. *Int J Vent* 2009; 7: 321–34.

Kobayashi T, Sandberg M, Kotani H, and Claesson L. Experimental investigation and CFD analysis of cross-ventilated flow through single room detached house model. *Build Environ* 2010; 45: 2723–34.

Kurabuchi T, Ohba M, Endo T, Akamine Y, and Nakayama F. Local dynamic similarity model of cross-ventilation, Part 1 – Theoretical framework. *Int J Vent* 2004; 2(4): 371–82.

Larsen TS and Heiselberg P. Single-sided natural ventilation driven by wind pressure and temperature difference. *Energy Build* 2008; 40: 1031–40.

Larsen TS, Heiselberg P, and Sawachi T. Analysis and design of single-sided natural ventilation. *Proceedings of the 4th International Symposium on HVAC*, Beijing, China, 2003; pp. 159-163.

Larsen TS, Nikolopoulos N, Nikolopoulos A, Strotos G, and Nikas KS. Characterization and prediction of the volume flow rate aerating a cross ventilated building by means of experimental techniques and numerical approaches. *Energy Build* 2011; 43: 1371–81.

Laussmann D and Helm D. Air change measurements using tracer gases: Methods and results. Significance of air change for indoor air quality. *InTechOpen*, Rijeka, Croatia, 2011.

Li H, Li X, and Qi M. Field testing of natural ventilation in college student dormitories (Beijing, China). *Build Environ* 2014a; 78: 36–43.

Li Y. Buoyancy-driven natural ventilation in a thermally stratified one-zone building. *Build Environ* 2000; 35: 207–14.

Li Y and Delsante A. Natural ventilation induced by combined wind and thermal forces. *Build Environ* 2001; 36: 59–71.

Li Y, Delsante A, and Symons J. Prediction of natural ventilation in buildings with large openings. *Build Environ* 2000; 35: 191–206.

Li Y, Delsante A, Chen Z, *et al.* Some examples of solution multiplicity in natural ventilation. *Build Environ* 2001; 36: 851–58.

Li ZR, Ai ZT, Wang WJ, Xu ZR, Gao XZ, and Wang HS. Evaluation of airflow pattern in wind-driven naturally ventilated atrium buildings: Measurement and simulation. *Build Serv Eng Res Technol* 2014b; 35(2): 139–54.

Liddament MW. The applicability of natural ventilation – Technical Editorial. *Int J Vent* 2009; 8(3): 189–99.

Linden PF. The fluid mechanics of natural ventilation. *Annu Rev Fluid Mech* 1999; 31: 201–38.

Lishman B and Woods AW. On transitions in natural ventilation flow driven by changes in the wind. *Build Environ* 2009; 44: 666–73.

Liu H. *Wind Engineering: A Handbook for Structural Engineers*. Englewood Cliffs, NJ: Prentice Hall; 1991.

Lo LJ and Novoselac A. Cross ventilation with small openings: Measurements in a multi-zone test building. *Build Environ* 2012; 57: 377–86.

Lo LJ, Banks D, and Novoselac A. Combined wind tunnel and CFD analysis for indoor airflow prediction of wind-driven cross ventilation. *Build Environ* 2013; 60: 12–23.

Mak CM, Niu JL, Lee CT, and Chan KF. A numerical simulation of wing walls using computational fluid dynamics. *Energy Build* 2007; 39: 995–1002.

Martins NR and Carrilho da Graca G. Validation of numerical simulation tools for wind-driven natural ventilation design. *Int J Build Simul* 2016; 9(1): 75–87.

McWilliams J. *Review of Airflow Measurement Techniques*. Berkeley, CA, USA: Energy Performance of Buildings Group, Environmental Energy Technologies Division, Lawrence Berkeley National Laboratory; 2002.

Meroney RN. *Wind Tunnel and Numerical Simulation of Pollution Dispersion: A Hybrid Approach*. Invited Lecture, Croucher Advanced Study Institute, Hong Kong University of Science and Technology, 6–10 December, 2004.

Meroney RN. CFD prediction of airflow in buildings for natural ventilation. *11th Americas Conference on Wind Engineering*, San Juan, PR, USA, June 22-26, 2009.

Mochida A, Yoshino H, Takeda T, Kakegawa T, and Miyauchi S. Methods for controlling airflow in and around a building under cross-ventilation to improve indoor thermal comfort. *J Wind Eng Ind Aerodyn* 2008; 93: 437–49.

Montazeri H and Blocken B. CFD simulation of wind-induced pressure coefficients on buildings with and without balconies: Validation and sensitivity analysis. *Build Environ* 2013; 60: 137–49.

Murakami S, Kato S, Akabashi S, Mizutani K, and Kim YD. Wind tunnel test on velocity-pressure field of cross-ventilation with open windows. *ASHRAE Trans* 1991; 97(1): 525–38.

Narasaki M, Yamanaka T, and Higuchi M. Influence of turbulent wind on the ventilation of an enclosure with a single opening. *Environ Int* 1989; 15: 627–34.

Niachou K, Hassid S, Santamouris M, and Livada I. Comparative monitoring of natural, hybrid and mechanical ventilation systems in urban canyons. *Energy Build* 2005; 37: 503–13.

Niachou K, Hassid S, Santamouris M, and Livada I. Experimental performance investigation of natural, mechanical and hybrid ventilation in urban environment. *Build Environ* 2008; 43: 1373–82.

Nikas KS, Nikolopoulos N, and Nikolopoulos A. Numerical study of a naturally cross-ventilated building. *Energy Build* 2010; 42: 422–34.

Nikolopoulos N, Nikolopoulos A, Larsen TS, and Nikas KSP. Experimental and numerical investigation of the tracer gas methodology in the case of a naturally cross-ventilated building. *Build Environ* 2012; 56: 379–88.

Nishizawa S, Sawachi T, Narita K, Seto N, and Ishikawa Y. A wind tunnel full-scale building model comparison between experimental data and CFD results based on standard k-e turbulence representation. *Int J Vent* 2004; 2(4): 419–30.

O'Sullivan PD and Kolokotroni M. A field study of wind dominant single sided ventilation through a narrow slotted architectural Louvre system. *Energy Build* 2017; 138: 733–47.

Ohba M, Irie K, and Kurabuchi T. Study on airflow characteristics inside and outside a cross-ventilation model, and ventilation flow rates using wind tunnel experiments. *J Wind Eng Ind Aerodyn* 2001; 89: 1513–24.

Ohba M, Kurabuchi T, Endo T, Akamine Y, Kamata M, and Kurahashi A. Local dynamic similarity model of cross-ventilation, Part 2 – Application of local dynamic similarity model. *Int J Vent* 2004; 2(4): 383–94.

Omrani S, Garcia-Hansen V, Capra BR, and Drogemuller R. Effect of natural ventilation mode on thermal comfort and ventilation performance: Full-scale measurement. *Energy Build* 2017a; 156: 1–16.

Omrani S, Garcia-Hansen V, Capra BR, and Drogemuller R. On the effect of provision of balconies on natural ventilation and thermal comfort in high-rise residential buildings. *Build Environ* 2017b; 123: 504–16.

Oropeza-Perez I and Ostergaard PA. Energy saving potential of utilizing natural ventilation under warm conditions – A case study of Mexico. *Appl Energy* 2014a; 130: 20–32.

Oropeza-Perez I and Ostergaard PA. Potential of natural ventilation in temperate countries – A case study of Denmark. *Appl Energy* 2014b; 114: 520–30.

Pakakonstantinou KA, Kiranoudis CT, and Markatos NC. Numerical simulation of air flow field in single-sided ventilated buildings. *Energy Build* 2000; 33: 41–8.

Papamanolis N. An overview of the balcony contribution to the environmental behavior of buildings. *21st Conference on Passive and Low Energy Architecture*, Eindhoven, the Netherlands, September 19-22, 2004.

Park J, Sun X, Choi JI, and Rhee GH. Effect of wind and buoyancy interaction on single-sided ventilation in a building. *J Wind Eng Ind Aerodyn* 2017; 171: 380–9.

Park JS. Long-term field measurement on effects of wind speed and directional fluctuation on wind-driven cross ventilation in a mock-up building. *Build Environ* 2013; 62: 1–8.

Peren JI, van Hooff T, Leite BCC, and Blocken B. CFD simulation of wind-driven upward cross ventilation and its enhancement in long buildings: Impact of single-span versus double-span leeward sawtooth roof and opening ratio. *Build Environ* 2016; 96: 142–56.

Phaff H and De Gids W. Ventilation rates and energy consumption due to open windows: A brief overview of research in the Netherlands. *Air Infiltration Rev* 1982; 4(1): 4–5.

Prajongsan P and Sharples S. Enhancing natural ventilation, thermal comfort and energy savings in high-rise residential buildings in Bangkok through the use of ventilation shafts. *Build Environ* 2012; 50: 104–13.

Prianto E and Depecker P. Characteristic of airflow as the effect of balcony, opening design and internal division on indoor velocity: A case study of traditional dwelling in urban living quarter in tropical humid region. *Energy Build* 2002; 34: 401–9.

Prianto E and Depecker P. Optimization of architectural design elements in tropical humid region with thermal comfort approach. *Energy Build* 2003; 35: 273–80.

Qian H, Li Y, Seto WH, Ching P, Chiang WH, and Sun HQ. Natural ventilation for reducing airborne infection in hospitals. *Build Environ* 2010; 45: 559–65.

Ramponi R and Blocken B. CFD simulation of cross-ventilation for a generic isolated building: Impact of computational parameters. *Build Environ* 2012; 53: 34–48.

Roetzel A, Tsangrassoulis A, Dietrich U, and Busching S. A review of occupant control on natural ventilation. *Renew Sustain Energy Rev* 2010; 14: 1001–13.

Sandberg M. Airflow through large openings – A catchment problem. *Proceedings of ROOMVENT*, Copenhagen, Denmark, September 8-11, 2002.

Sandberg M. An alternative view on theory of cross-ventilation. *Int J Vent* 2004; 2 (4): 400–18.

Santamouris M, Synnefa A, Asssimakopoulos M, *et al.* Experimental investigation of the air flow and indoor carbon dioxide concentration in classrooms with intermittent natural ventilation. *Energy Build* 2008; 40: 1833–43.

Sawachi T. Detailed observation of cross ventilation and air flow through large openings by full scale building model in wind tunnel. *Proceedings of ROOMVENT*, Copenhagen, Denmark, September 8-11, 2002.

Sawachi T, Narita K, Kiyota N, Seto H, Nishizawa S, and Ishikawa Y. Wind pressure and air flow in a full-scale building model under cross ventilation. *Int J Vent* 2004; 2(4): 343–57.

Seifert J, Li YG, Axley J, and Rosler M. Calculation of wind-driven cross ventilation in buildings with large openings. *J Wind Eng Ind Aerodyn* 2006; 94: 925–47.

Sherman MH. Tracer-gas techniques for measuring ventilation in a single zone. *Build Environ* 1990; 25(4): 365–74.

Shetabivash H. Investigation of opening position and shape on the natural cross ventilation. *Energy Build* 2015; 93: 1–15.

Shirzadi M, Mirzaei PA, and Naghashzadegan M. Development of an adaptive discharge coefficient to improve the accuracy of cross-ventilation airflow calculation in building energy simulation tools. *Build Environ* 2018; 127: 277–90.

Simonson C. Energy consumption and ventilation performance of a naturally ventilated ecological house in a cold climate. *Energy Build* 2005; 37: 23–35.

Snyder WH. *Guideline for Fluid Modeling of Atmospheric Diffusion. Meteorology and Assessment*. Research Triangle Park, NC, USA: Division Environmental Sciences Research Laboratory, U.S. Environmental Protection Agency; 1981.

Stavrakakis GM, Koukou MK, Vrachopoulos MGr, and Markatos NC. Natural cross-ventilation in buildings: Building-scale experiments, numerical simulation and thermal comfort evaluation. *Energy Build* 2008; 40: 1666–81.

Stavrakakis GM, Zervas PL, Sarimveis H, and Markatos NC. Development of a computational tool to quantify architectural-design effects on thermal comfort in naturally ventilated rural houses. *Build Environ* 2010; 45: 65–80.

Stavrakakis GM, Zervas PL, Sarimveis H, and Markatos NC. Optimization of window-openings design for thermal comfort in naturally ventilated buildings. *Appl Math Model* 2012; 36: 193–211.

Stavridou AD and Prinos PE. Natural ventilation of buildings due to buoyancy assisted by wind: Investigating cross ventilation with computational and laboratory simulation. *Build Environ* 2013; 66: 104–19.

Straw MP, Baker CJ, and Robertson AP. Experimental measurements and computations of the wind-induced ventilation of a cubic structure. *J Wind Eng Ind Aerodyn* 2000; 88: 213–30.

Sun X, Park J, Choi JI, and Rhee GH. Uncertainty quantification of upstream wind effects on single-sided ventilation in a building using generalized polynomial chaos method. *Build Environ* 2017; 125: 153–67.

Tabor GR and Baba-Ahmadi MH. Inlet conditions for large eddy simulation: A review. *Comput Fluids* 2010; 39: 553–67.

Tecle A, Bitsuamlak GT, and Jiru TE. Wind-driven natural ventilation in a low-rise building: A boundary layer wind tunnel study. *Build Environ* 2013; 59: 275–89.

Tominaga Y and Blocken B. Wind tunnel experiments on cross-ventilation flow of a generic building with contaminant dispersion in unsheltered and sheltered. *Build Environ* 2015; 92: 452–61.

Tominaga Y, Mochida A, Yoshie R, *et al.* AIJ guidelines for practical applications of CFD to pedestrian wind environment around buildings. *J Wind Eng Ind Aerodyn* 2008; 96(10, 11): 1749–61.

True JJ, Sandberg M, Heiselberg P, and Nielsen PV. Wind driven cross-flow analyzed as a catchment problem and as a pressure driven flow. *Int J Vent* 2003; 1: 88–102.

van Hooff T and Blocken B. Coupled urban wind flow and indoor natural ventilation modelling on a high-resolution grid: A case study for the Amsterdam ArenA stadium. *Environ Model Softw* 2010a; 25: 51–65.

van Hooff T and Blocken B. On the effect of wind direction and urban surroundings on natural ventilation of a large semi-enclosed stadium. *Comput Fluids* 2010b; 39: 1146–55.

van Hooff T and Blocken B. Full-scale measurements of indoor environmental conditions and natural ventilation in a large semi-enclosed stadium: Possibilities and limitations for CFD validation. *J Wind Eng Ind Aerodyn* 2012; 104–106: 330–41.

van Hooff T and Blocken B. CFD evaluation of natural ventilation of indoor environments by the concentration decay method: CO_2 gas dispersion from a semi-enclosed stadium. *Build Environ* 2013; 61: 1–17.

van Hooff T, Blocken B, Aanen L, and Bronsema B. A venture-shaped roof for wind-induced natural ventilation of buildings: Wind tunnel and CFD evaluation of different design configurations. *Build Environ* 2011; 46: 1797–807.

van Hooff T, Blocken B, and Tominaga Y. On the accuracy of CFD simulations of cross-ventilation flows for a generic isolated building: Comparison of RANS, LES and experiments. *Build Environ* 2017; 114: 148–65.

van Moeseke G, Gratia E, Reiter S, and De Herde A. Wind pressure distribution influence on natural ventilation for different incidences and environment densities. *Energy Build* 2005; 37: 878–89.

Vickery BJ and Karakatstanis C. External wind pressure distributions and induced internal ventilation flow in low-rise industrial and domestic structures. *ASHRAE Trans* 1987; 93(2): 2198–213.

Visagavel K and Srinivasan PSS. Analysis of single side ventilated and cross ventilated rooms by varying the width of the window opening using CFD. *Solar Energy* 2009; 83: 2–5.

Wang H and Chen Q. A new empirical model for predicting single-sided, wind-driven natural ventilation in buildings. *Energy Build* 2012; 54: 386–94.

Wang H and Chen Q. Modeling of the impact of different window types on single-sided natural ventilation. *Energy Procedia* 2015; 78: 1549–55.

Wang H, Karava P, and Chen Q. Development of simple semiempirical models for calculating airflow through hopper, awning, and casement windows for single-sided natural ventilation. *Energy Build* 2015; 96: 373–84.

Wang J, Wang S, Zhang T, and Battaglia F. Assessment of single-sided natural ventilation driven by buoyancy forces through variable window configurations. *Energy Build* 2017; 139: 762–79.

Wang L and Wong NH. The impacts of ventilation strategies and fade on indoor thermal environment for naturally ventilated residential buildings in Singapore. *Build Environ* 2007; 42: 4006–15.

Wang L, Wong NH, and Li S. Facade design optimization for naturally ventilated residential buildings in Singapore. *Energy Build* 2007; 39: 954–61.

Wang Y, Zhao FY, Kuckelkorn J, Liu D, Liu J, and Zhang JL. Classroom energy efficiency and air environment with displacement natural ventilation in a passive public school building. *Energy Build* 2014; 70: 258–70.

Wang Z, Zhang L, Zhao J, and He Y. Thermal comfort for naturally ventilated residential buildings in Harbin. *Energy Build* 2010; 42: 2406–15.

Wargocki P, Wyon DP, Sundell J, Clausen G, and Fange, PO. The effects of outdoor air supply rate in an office on perceived air quality, sick building syndrome (SBS) symptoms and productivity. *Indoor Air* 2000; 10: 222–36.

Warren PR and Parkins LM. Single-sided ventilation through open windows. *Proceedings of Conference on Thermal Performance of the Exterior Envelopes of Buildings, ASHRAE SP 49*, Florida, 1985; pp. 209–228.

Warren PR. Ventilation through openings on one wall only. *Proceedings of International Centre for Heat and Mass Transfer Seminar Energy Conservation in Heating, Cooling, and Ventilating Buildings*, Washington, 1977.

WHO. *WHO Air Quality Guidelines for Particulate Matter, Ozone, Nitrogen Dioxide and Sulfur Dioxide – Summary of Risk Assessment*. Geneva, Switzerland: World Health Organization; 2006.

Wong NH, Feriadi H, Lim PY, Tham KW, Sekhar C, and Cheong KW. Thermal comfort evaluation of naturally ventilated public housing in Singapore. *Build Environ* 2002; 37: 1267–77.

Wong P, Prasad D, and Behnia M. A new type of double-skin fade configuration for the hot and humid climate. *Energy Build* 2008; 40(10): 1941–45.

Wright NG and Hargreaves DM. Unsteady CFD simulations for natural ventilation. *Int J Vent* 2006; 5(1): 13–20.

Yamanaka T, Kotani H, Iwamoto K, and Kato M. Natural, wind-forced ventilation caused by turbulence in a room with a single opening. *Int J Vent* 2006; 5(1): 179–87.

Yang L and Zhang G. Investigating potential of natural driving forces for ventilation in four major cities in China. *Build Environ* 2005; 40: 738–46.

Yang X, Zhong K, Zhu H, and Kang Y. Experimental investigation on transient natural ventilation driven by thermal buoyancy. *Build Environ* 2014; 77: 29–39.

Yim SHL, Fung JCH, Lau AKH, and Kot SC. Air ventilation impacts of the all effect – Resulting from the alignment of high-rise buildings. *Atmos Environ* 2009; 43: 4982–94.

Yoshie R, Mochida A, Tominaga Y, *et al.* Cooperative project for CFD prediction of pedestrian wind environment in the Architectural Institute of Japan. *J Wind Eng Ind Aerodyn* 2007; 95(9, 11): 1551–78.

Yuan S, Vallianos C, Athienitis A, and Rao J. A study of hybrid ventilation in an institutional building for predictive control. *Build Environ* 2018; 128: 1–11.

Zhou C, Wang Z, Chen Q, Jiang Y, and Pei J. Design optimization and field demonstration of natural ventilation for high-rise residential buildings. *Energy Build* 2014; 82: 457–65.

Chapter 10
Ventilation system design: numerical method

Wei Liu[1]

Ventilation system is a typical approach to supply air and construct air distribution for the desired indoor environment quality. A poorly designed ventilation system would lead to a variety of problems such as sick building syndrome, cross infections, and reduced productivity. This chapter focuses on the numerical methods for designing the ventilation system.

10.1 Introduction

A thermally comfortable, healthy, and productive indoor environment is basically created by appropriate air distribution, which regulates the air temperature, relative humidity, air speed, and chemical species concentrations (Melikov, 2004; Chen, 2009). Ventilation system is a typical approach to supply air and construct air distribution for the desired indoor environment quality. A poorly designed ventilation system would lead to a variety of problems such as sick building syndrome (Hedge *et al.*, 1996; Sun *et al.*, 2019), cross infections (Olsen *et al.*, 2003; Chen, 2020), and reduced productivity (Fisk, 2017; Geng *et al.*, 2017). Besides, the gradual deterioration of the outdoor environment and the occasionally pandemics pose new challenges to the performance of the ventilation system. Further, the requirements for indoor environments are changing with the improvement of our living standard. Therefore, it is critical to design the ventilation system for a desirable indoor environment.

This chapter focuses on the numerical methods for designing the ventilation system. The numerical methods here basically refer to the approaches of ventilation performance prediction for buildings, which connects the design variables and objectives. The design variables are parameters whose values can be varied in a certain range by the designers to determine the ventilation system. Sufficient freedom should be given to the design variables to create a desired indoor environment (Liu *et al.*, 2016). The design objectives are evaluation criterion for the ventilation performance such as indoor air quality (IAQ), thermal comfort, and

[1]Department of Civil and Architectural Engineering, KTH Royal Institute of Technology, Stockholm, Sweden

energy consumption. Besides, the design of a ventilation system generally involves a design algorithm for accelerating the speed of design process and ensuring the design to be optimal.

This chapter introduces the design variables, design objectives, numerical methods for ventilation performance prediction, and design algorithms in sequence. An application was then provided for demonstration.

10.2 Design variables

The ventilation system supplies air to maintain healthy IAQ by diluting and removing other pollutants emitted within a space, as well as to distribute heat for thermal comfort. Traditionally, the design variables for a ventilation system are the ventilation rate and air-supply temperature. The ventilation rate is basically determined for acceptable IAQ. For example, the ANSI/ASHRAE (American Society of Heating, Refrigerating and Air-Conditioning Engineers) Standard 62.1, Ventilation for Acceptable Indoor Air Quality (ANSI/ASHRAE, 2016) gives the minimum requirements on people/area outdoor air rate for different indoor environments. However, Persily (2015) pointed out "additional research into the health effects of contaminants and contaminant mixtures, source strengths in buildings, the performance of IAQ control technologies and new design approaches." The air-supply temperature is determined by the cooling/heating load and the desired comfort zone for thermal comfort. For example, the ANSI/ASHRAE Standard 55, Thermal Environmental Conditions for Human Occupancy (ANSI/ASHRAE, 2017) specifies conditions for acceptable thermal environments. However, even in a well-controlled environment such as an aircraft cabin, 25% of passengers were not satisfied with the thermal environment, as the upper body was too warm or the lower body was too cold (Park *et al.*, 2011). Actually, the IAQ and thermal comfort level is determined by the air distribution, which is created by all the thermo-fluid boundary conditions of the indoor environment. In considering the ventilation systems, the air-supply parameters and air-supply geometry, including air-supply locations, size, and shape of the air-supply inlets, all are possible design variables.

10.2.1 Air-supply parameters

The air-supply parameters for a ventilation system are air-supply velocity and temperature. The air-supply temperature is generally considered steady state, which is straightforward. For example, Zhai *et al.* (2014) designed the thermal environment in a confined space with air-supply temperature as a design variable. The air-supply velocity can be treated in two ways:

- velocity vector $\mathbf{U} = (U_1, U_2, U_3)$,
- velocity magnitude $|\mathbf{U}|$ and direction θ and φ,

where U_i are velocity components, θ is the angle from the polar direction, and φ is the azimuthal angle. Those two treatments are mathematically convertible:

$$U_1 = |\mathbf{U}| \sin \theta \cos \varphi \tag{10.1}$$

$$U_2 = |\mathbf{U}| \sin \theta \sin \varphi \tag{10.2}$$

$$U_3 = |\mathbf{U}| \cos \theta \tag{10.3}$$

In most researches on the design of ventilation systems, the air-supply velocity was assumed to be steady-state and uniform for a specific diffuser, which is the most common situation in engineering application. More specifically, a mean air-supply velocity was considered, and the turbulent features of the air-supply jet were generally considered by a presumed turbulence intensity or turbulent length ratio, etc. For example, Zhang *et al.* (2020) considered the ventilation rate (velocity magnitude) and angle as design variables in an optimal design of an office with displacement ventilation. Nabi and Grover (2017) used the velocity vectors of the air-supply jet as design variables for the identification of optimal buoyancy-driven ventilation flows. The treatment of the air-supply velocity is dependent on the design algorithm. If a genetic algorithm (GA) or artificial neural network (ANN) is adopted, either treatment works well. If an adjoint method is used, the air velocity in the vector form is recommended.

The air-supply velocity can also be transient. The purpose is to create unsteady air distribution for improving the effectiveness of indoor air pollutant removal. For example, Sattari and Sandberg (2013) and Fallenius *et al.* (2013) conducted the experimental investigation on a ventilation system with periodic air-supply velocity at 0.3, 0.4, and 0.5 Hz. van Hooff and Blocken (2017) numerically studied the ventilation performance of a room with air supplied periodically from two oppositely located diffusers. The air-supply velocity was defined as

$$U_i = U_0\left(1 \pm \sin\left(\frac{2\pi t}{P}\right)\right) \tag{10.4}$$

Then, the amplitude U_0 and period P are the actual design variables. Please note that the sine function in (10.4) is just an example for transient air-supply velocity. The reason for using sine function in the previous studies might be the feasibility in real engineering application.

In general, the air-supply velocity is not independent since it could be constrained by ventilation rate Q:

$$Q = \mathbf{U} \cdot \mathbf{A} \tag{10.5}$$

where $\mathbf{A}$ is the area vector of the diffuser. If the ventilation rate is a constant, the constraint defined by (10.5) should be considered, especially when the area of the diffuser is also a design variable (Liu *et al.*, 2015).

10.2.2 Air-supply geometry

The air-supply geometry includes the locations, size, and shape of the air-supply inlets because these two aspects are considered together. There are two ways to depict the air-supply geometry. The first way is to use the coordinates of boundary cells $\mathbf{X}_j$ for the air-supply inlet as the design variables, where the subscript

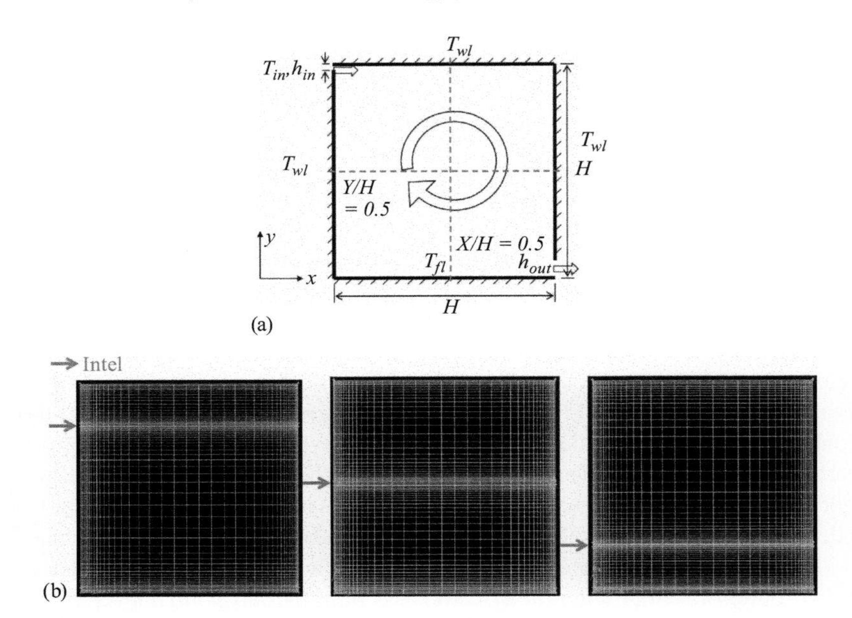

Figure 10.1 Design with coordinates of boundary cells: (a) geometry of a two-dimensional ventilated cavity (Liu et al., 2016) and (b) grids during the design

j represents the cell index. By adjusting the coordinates of boundary cells, Liu *et al.* (2016) designed the locations and size of air-supply inlets in a two-dimensional ventilated cavity as shown in Figure 10.1(a) and (b) that during the design, the grids were regenerated when the location of air-supply inlet varied. The grids close to the air-supply inlet were refined. This technique was further applied in designing the locations and size of air-supply inlets in an aircraft cabin for achieving the desired thermal environment (Liu *et al.*, 2015). One can notice that, once the coordinates of boundary cells changed, the computational grid for a numerical method needs to be regenerated. Thus in real engineering application, the designer needs to generate the grids for all possible air-supply inlets or apply automatic meshing tools such as script files for ICEM CFD (2012) or *BlockMesh* utility in OpenFOAM (Jasak *et al.*, 2007) for grid regeneration during the design process.

Another approach to consider the locations, size, and shape of the air-supply inlets is to use an area-constrained topology optimization method (Yamasaki *et al.*, 2010) and cluster analysis. As the study by Zhao *et al.* (2018) shows, this method involves three steps:

- Step 1, design with all the potential locations as air-supply inlets and the air-supply velocity is considered nonuniform. For example, Zhao *et al.* (2018) considered one side wall as the potential air-supply inlet and led to design air-supply velocities as shown in Figure 10.2(a). The design led to nonuniform air-supply velocity.

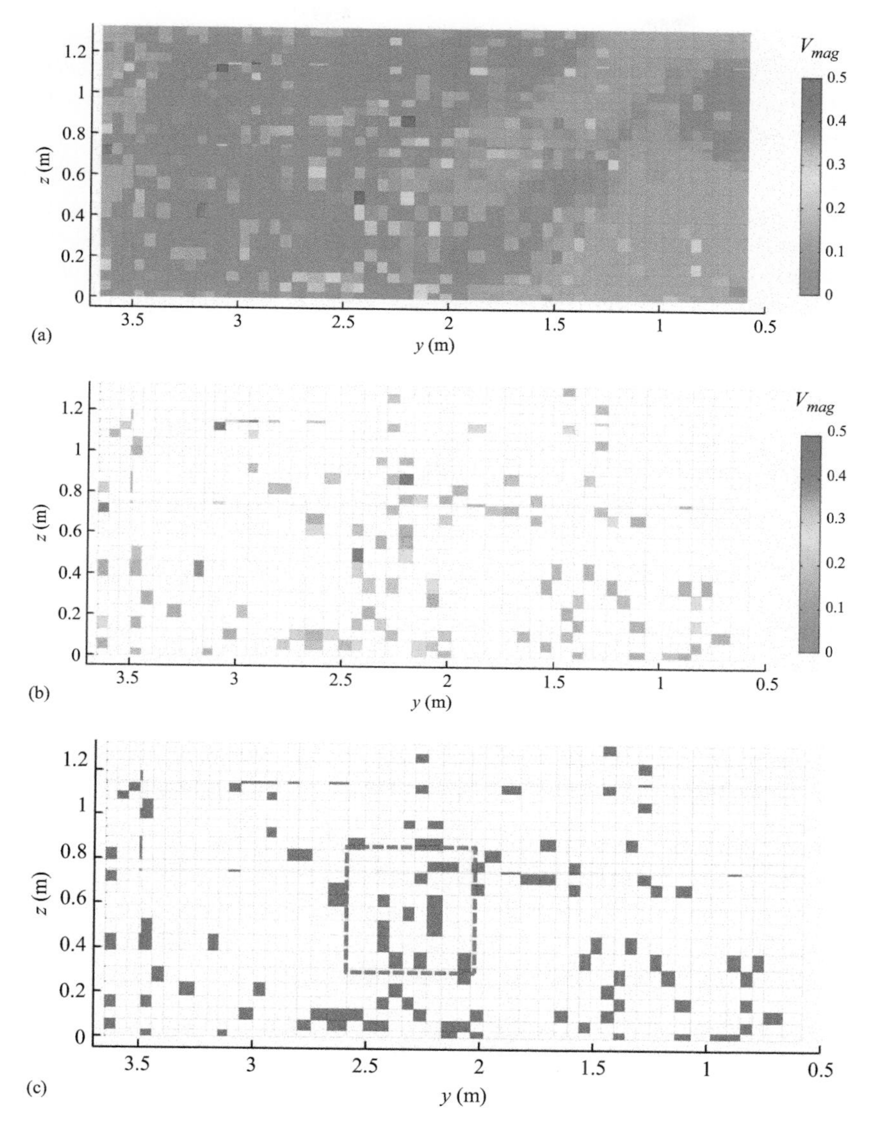

Figure 10.2 Design with area-constrained topology optimization method and cluster analysis (Zhao et al., 2018): (a) design with all the potential locations, (b) area-constrained topology applied, (c) cluster analysis for one square air-supply inlet

- Step 2, by assigning appropriate lower and upper bounds of the recommended range of the air changes per hour and the air-supply velocity, this step eliminates those cells with very low air velocity. The colored cells in Figure 10.2(b) are the open cells, and the other cells are closed. Theoretically, the air-supply inlet could take any shape. However, the example shows only rectangular openings because a structured grid was used. The design in Figure 10.2(b) is still impractical because of the scattered open cells and varying parameters of the air supply from the inlets.

- Step 3, consolidate the scattered open cells into the desired number of air-supply inlets by cluster analysis. In Figure 10.2(c), Zhao *et al.* (2018) applied a centroid-based hierarchical cluster analysis (Uppada, 2014) to consolidate the scattered cells into one square air-supply inlet. Please note that before the cluster analysis, the velocity for different grid cell is different to give sufficient freedom for the design. The cluster analysis consolidates the cells of air-supply inlet and gives only uniform air-supply parameters for each air-supply inlet in terms of being practical in design.

10.3 Design objective vs. objective function

The design objectives are used to evaluate the performance of the ventilation. In general, the design aims to achieve desired thermal comfort, draft sensation, IAQ, and minimum energy consumption. Those design objectives need to be quantitatively evaluated by establishing the corresponding objective functions. One can also construct an objective function for multiple design objectives.

10.3.1 Thermal comfort

To quantitatively evaluate the thermal comfort, one can use predicted mean vote (PMV), equivalent temperature (ET), or vertical temperature difference.

10.3.1.1 Predicted mean vote

PMV is an empirical expression to quantify the average vote of a large group of people on a seven-point thermal sensation scale where: +3 (hot), +2 (warm), +1 (slightly warm), 0 (neutral), −1 (slightly cool), −2 (cool), and −3 (cold) (Fanger et al., 1970). The value of PMV can be calculated by

$$\begin{aligned} PMV = (0.303e^{-0.036M} + 0.028)\{M - W - 3.05 \times 10^{-3} \times [5,733 \\ -6.99(M - W) - p_w] - 0.42 \times [(M - W) - 58.15] \\ -1.7 \times 10^{-5}M(5,867 - p) - 0.0014M(34 - T) \\ - 3.96 \times 10^{-8}f_{cl} \times [(T_{cl} + 273.15)^4 \\ - (T_r + 273.15)^4] - f_{cl}h_c(T - T_{cl})\} \end{aligned} \quad (10.6)$$

where

$$\begin{aligned} T_{cl} = \; 35.7 - 0.028(M - W) - I_{cl}\{3.96 \times 10^{-8}f_{cl} \times [(T_{cl} + 273.15)^4 \\ -(T_r + 273.15)^4] - f_{cl}h_c(T - T_{cl})\} \end{aligned} \quad (10.7)$$

$$h_c = \{ 2.38(T_{cl} - T)^{0.25} \text{ if } 2.38(T_{cl} - T)^{0.25} > 12.1\sqrt{|\mathbf{U}|}12.1\sqrt{|\mathbf{U}|} \text{ if } 2.38(T_{cl} - T)^{0.25} < 12.1\sqrt{|\mathbf{U}|} \tag{10.8}$$

$$f_{cl} = \begin{cases} 1.00 + 1.290 I_{cl} & \text{for } I_{cl} \leq 0.078 \ \text{m}^{2\text{o}}\text{C/W} \\ 1.05 + 0.645 I_{cl} & \text{for } I_{cl} > 0.078 \ \text{m}^{2\text{o}}\text{C/W} \end{cases} \tag{10.9}$$

In these equations, W (W) is the external work accomplished, M (W/m^2) the metabolic activity, f_{cl} the ratio of clothed surface area to DuBois surface area, I_{cl} (clo) the thermal resistance of the passengers' clothing, T $(^\circ\text{C})$ the air temperature, T_r $(^\circ\text{C})$ the mean radiant temperature, T_{cl} $(^\circ\text{C})$ the clothing temperature, $\mathbf{U}$ (m/s) the air velocity, and p_w (Pa) the relative humidity or partial water vapor pressure.

The PMV model can be applied to air-conditioned buildings (ANSI/ASHRAE, 2017). For a specific environment, such as a commercial aircraft cabin, the PMV needs to be tuned. An air cabin environment has a lower pressure, which will decrease human body heat loss through convection and increase that through evaporation (Cui *et al.*, 2014). Cui *et al.* (2014) modified PMV for air cabins (PMV_c) by considering the low pressure in a cruising aircraft cabin. PMV_c can be determined from PMV as shown in (10.1) and (10.2) for the summer and winter seasons, respectively:

$$PMV_c(\text{summer}) = -0.0758PMV^2 + 0.6757PMV - 0.1262 \tag{10.10}$$

$$PMV_c(\text{winter}) = -0.0696PMV^2 + 0.6906PMV - 0.1369 \tag{10.11}$$

Since the design objective is to achieve the most thermally comfortable indoor environment, PMV should approach zero. Therefore, the corresponding objective function could be constructed as

$$O_i(\xi) = \left(\frac{\int_\Omega (PMV/3)^2 d\Omega}{\int_\Omega d\Omega} \right)^{1/2} \tag{10.12}$$

where Ω is the design domain, which is normally the area around the occupants, and ξ is a vector that denotes a series of design variables. The denominator 3 in (10.12) is used for normalization. The minimum objective function leads to the most thermally comfortable indoor environment.

10.3.1.2 Equivalent temperature

ET was proposed to consider the thermally nonuniform microclimate encountered in automobiles (Wyon *et al.*, 1989) and applicable for indoor environment with displacement ventilation system or underfloor air distribution system. According to Zhou and Haghighat (2009), ET defines the temperature of a uniform enclosure in which a human body would experience the same rate of heat loss as in the actual thermally nonuniform environment. For each segment of a human body, ET can be calculated using

$$ET = T_s - I_{cl}Q_t \tag{10.13}$$

where Q_t is the local heat loss rate from skin surface (W/m^2). It is considered that there is no excessive thermal nonuniformity over the body if the variations of ET over the entire body are controlled in the range of −2 °C to 2 °C. Therefore, an objective function could be constructed as

$$O_i(\xi) = \left(\frac{\sum_{i=1}^{n} ET_i^2}{n}\right)^{1/2} \tag{10.14}$$

for a human body with n segments considered. The minimum objective function corresponds to the minimal thermal nonuniformity over the body.

10.3.1.3 Vertical temperature difference

According to the ANSI/ASHRAE Standard 55 (ANSI/ASHRAE, 2017), the temperature difference between the head and the ankle level should not exceed 3 °C, which corresponds to 5% of dissatisfaction. Then, an objective function could be constructed as

$$O_i(\xi) = |T_{head} - T_{ankle}| \tag{10.15}$$

where T_{head} and T_{ankle} could be the mean air temperature at the head and ankle level, respectively. The minimum objective function leads to the least vertical temperature difference. However, since $O(\xi) \leq 3$ °C is sufficient, the vertical temperature difference was sometimes used as a constraint in the design.

10.3.2 Draft sensation

The draft sensation for the occupants in an indoor environment could be quantified by the predicted percentage of people who are dissatisfied because of draft, PD, and local air speed. PD is an empirical expression to quantify the undesired local cooling of the human body or annoyance and distraction due to drafts. The inputs to PD are air temperature, air velocity, and turbulence intensity. According to the investigations by Fanger and Christensen (1986) and Fanger *et al.* (1989), PD could be expressed by

$$\mathrm{PD} = (34 - T)(|\mathbf{U}| - 0.05)^{0.62}(0.37|\mathbf{U}|TI + 3.14) \tag{10.16}$$

where T is the ambient air temperature (°C), TI is the turbulence intensity (%). In practice, if $|\mathbf{U}| < 0.05$ m/s, $|\mathbf{U}|$ is set to be 0.05 m/s. Besides, if the calculated PD is greater than 100%, it is set to be 100%. The expression for PD could be directly used to construct the objective function (Zhao *et al.*, 2018):

$$O_i(\xi) = \frac{\int_\Omega (PD/100)d\Omega}{\int_\Omega d\Omega} \tag{10.17}$$

The local air speed could also be used to quantify the draft sensation. According to the ANSI/ASHRAE Standard 55 (ANSI/ASHRAE, 2017), the air speed near an occupant in the office should be controlled less than or equal to 0.25 m/s to avoid dissatisfaction due to draft. The corresponding objective function could be constructed as

$$O_i(\xi) = \frac{\int_{\Omega} H(|\mathbf{U}| - 0.25)d\Omega}{\int_{\Omega} d\Omega} \tag{10.18}$$

where $H(x)$ is the Heaviside step function (Bracewell, 2000). The minimum objective function corresponds to the greatest percentage of the air speed on the design domain Ω being less than 0.25 m/s.

10.3.3 Indoor air quality

The IAQ could be quantified by age of air, ventilation effectiveness, and entransy theory for mass-transfer optimization.

10.3.3.1 Age of air

The age of air is mostly used to evaluate the "freshness" of air at certain locations in an indoor environment (Sandberg and Sjöberg, 1983). Numerically, the age of air could be predicted by solving the scalar transport equation of age of air $\overline{\tau}$ (Kato and Murakami, 1988):

$$(\mathbf{U} \cdot \nabla)\overline{\tau} = \Gamma_{\overline{\tau}} \nabla^2 \overline{\tau} + 1 \tag{10.19}$$

where $\Gamma_{\overline{\tau}} = \nu_t / \sigma_{\overline{\tau}} + \nu / \sigma$ is the diffusion coefficient with ν_t turbulent kinematic viscosity, $\sigma_{\overline{\tau}}$ turbulent Schmidt number, ν laminar kinematic viscosity, and σ laminar Schmidt number. One could use $\sigma_{\overline{\tau}} = 0.9$ in the simulation (Gan and Awbi, 1994). The boundary conditions for $\overline{\tau}$ is fixed zero value at the air-supply inlet and fixed zero gradient at the wall and outlet. The age of air is basically considered in the breathing zone. A small age of air leads to fresher local air. Therefore, the age of air could be directly used as the objective function.

10.3.3.2 Ventilation effectiveness

The ventilation effectiveness is also called contaminant removal effectiveness. It can be determined by the effectiveness in removing internally produced contaminants (Awbi, 2003). For example, Zhou and Haghighat (2009) used the CO_2 concentration to evaluate the ventilation effectiveness ε_v by

$$\varepsilon_v = \frac{c_{return} - c_{supply}}{c_{br} - c_{supply}} \tag{10.20}$$

where c_{return} is the CO_2 concentration in the return air, c_{supply} is the CO_2 concentration in the supply air, and c_{br} is the CO_2 concentration at breathing level of the occupants. Then an objective function could be constructed as

$$O_i(\xi) = \frac{\varepsilon_{V,\,\max}}{\varepsilon_V} \tag{10.21}$$

where $\varepsilon_{V,\,\max}$ is the observed maximal ventilation effectiveness during the design process.

10.3.3.3 Entransy theory for mass-transfer optimization

The objective of the design could be regarded as finding the ventilation system that can remove the contaminants with the highest efficiency. Such a mission could be accomplished by using the mass-transfer optimization method based on the entransy theory (Chen *et al.*, 2013). The mass entransy was defined as mass-transfer potential capacity. It has been proven theoretically that the mass entransy dissipation extremum would result in the best contaminant removal performance with the given boundary conditions (Chen *et al.*, 2009). Therefore, the optimal airflow field that has the highest contaminant removal efficiency can be found using the entransy theory. The objective functions for mass-transfer optimization is

$$O_i(\xi) = \int_\Theta \Gamma_\Phi |\nabla \Phi|^2 d\Theta \tag{10.22}$$

where Γ_Φ is the mass diffusion coefficient, Θ is the computational domain, and Φ is the contaminant concentration. By minimizing the objective function, the optimal airflow field for best removing the contaminant will be identified.

10.3.4 Energy consumption

The energy consumption of a ventilation system mainly includes the fan energy consumption and cooling/heating energy consumption. Another way to treat the energy consumption is the entransy theory for heat-transfer optimization.

10.3.4.1 Fan energy consumption

According to Zhou and Haghighat (2009), the fan energy consumption E_{fan} could be expressed by

$$E_{fan} = \frac{\Delta P \dot{V}_{air}}{\eta_{fan}} \tag{10.23}$$

where ΔP is the fan pressure rise, $\dot{V}_{air}$ is the ventilation rate, and η_{fan} is the fan efficiency. The ventilation rate is basically dependent on the air-supply velocity and the geometry of air-supply inlets. The fan pressure rise is related to ventilation rate by using the fan performance curve. Besides, the fan efficiency could be related to the ventilation rate, for example, by linear interpolation (Zhou and Haghighat, 2009). An objective function for the fan energy consumption could be

$$O_i(\xi) = \frac{E_{fan}}{E_{fan,\,\max}} \tag{10.24}$$

where $E_{fan,\max}$ is the maximum fan energy consumption observed during the design.

10.3.4.2 Cooling/heating energy consumption

The cooling/heating energy consumption is consisted of two parts (Xu and Niu, 2006; Nielsen and Drivsholm, 2010):

$$E_{cooling/heating} = E_{space} + E_{vent} \tag{10.25}$$

where E_{space} is the cooling/heating energy used to remove/add sensible heat load and E_{vent} is the cooling/heating energy used to condition the fresh air to return air states. Therefore,

$$E_{space} = \rho_{air}\dot{V}_{air}c_p|T_{return} - T_{supply}| \tag{10.26}$$

$$E_{vent} = \rho_{air}\dot{V}_{fresh}|h_{out} - h_{return}| \tag{10.27}$$

where ρ_{air} is the air density, $\dot{V}_{air}$ the ventilation rate, c_p the specific heat of air, T_{return} the return air temperature, T_{supply} the return air temperature, $\dot{V}_{fresh}$ the airflow rate of fresh air, h_{out} the specific enthalpy of outdoor air, and h_{return} the specific enthalpy of return air. An objective function for the cooling/heating energy consumption could be

$$O_i(\xi) = \frac{E_{cooling/heating}}{E_{cooling/heating,\max}} \tag{10.28}$$

where $E_{cooling/heating,\max}$ is the maximum cooling/heating energy consumption observed during the design.

10.3.4.3 Entransy theory for heat-transfer optimization

The objective of the design could be regarded as finding the ventilation system that can remove/add heat load with the highest efficiency. Similar with the entransy theory for mass-transfer optimization, it has been proven theoretically that the heat entransy dissipation extremum would result in the best heat removal performance with the given boundary conditions (Chen *et al.*, 2009). Therefore, the optimal airflow field that has the highest heat removal efficiency can be found using the entransy theory. The objective functions for heat-transfer optimization is

$$O_i(\xi) = \int_{\Theta} \kappa|\nabla T|^2 d\Theta \tag{10.29}$$

where κ is the heat diffusion coefficient. By minimizing the objective function, the optimal airflow field for best removing the heat load will be identified.

10.3.5 Multiple design objectives

The design of a ventilation system normally involves multiple design objectives, including the abovementioned thermal comfort, draft sensation, IAQ, and energy consumption. In order to consider the multiple design objectives, one can use a single-objective function that is constructed by aggregating numbers of objective functions using predefined weighting factors (Laverge and Janssens, 2013; Li *et al.*, 2013). For example, to consider the thermal comfort, IAQ, and energy consumption, one can construct an objective function like

$$O(\xi) = \sum \omega_i O_i(\xi) = \omega_1 \left(\int_{\Omega} (PMV/3)^2 d\Omega / \int_{\Omega} d\Omega \right)^{1/2} + \omega_2 \frac{\varepsilon_{v,\,\max}}{\varepsilon_v} + \omega_3 E_{cooling/heating} / E_{cooling/heating,\,\max} \tag{10.30}$$

and

$$\sum \omega_i = 1 \tag{10.31}$$

The values for weights ω_i could be determined by the designer's experience or questionnaire survey. For example, Zhao *et al.* (2018) designed the thermal environment in an office by considering the PMV and PD. The weighting factors used the values from a questionnaire survey in offices made by Ncube and Riffat (2012).

Such an objective function is simple to construct, but its disadvantage is straightforward. The design could be sensitive to the weighting factors and there is no formal protocol to determine those weighting factors. Besides, the design with such an objective function would give only one near optimal solution, then the designer has no flexibility to strike a balance or "trade-off" of the conflicting parameters (Li *et al.*, 2017).

Another approach to resolve the multiobjective design is the design algorithms. A vector-evaluated GA was proposed by Schaffer (1985) and was further improved by Zitzler and Thiele (1999) and Lu and Yen (2003) to find optimal global multi-solutions for multiple objectives. Zhai *et al.* (2014) developed two system design methods: the constraint and the optimization for a multiobjective GA. Li (2003) proposed a nondominated sorting method to extend the original particle swarm optimization to multiobjective optimization problems. Li *et al.* (2017, 2019) integrated this algorithm with a Kriging method (Forrester *et al.*, 2008) to perform optimization for the ventilation system design of a typical office room and high speed train. These multiobjective design algorithms are able to overcome the disadvantages of using a single objective function with weighting factors.

10.4 Numerical methods for ventilation performance prediction

With given design variables, one needs to predict the ventilation performance for calculating the objective functions. The ventilation performance is basically dependent on the airflow and heat and mass transfer in the indoor environment, which is governed by the Navier–Stokes (NS) equation. The airflow and heat and mass transfer can be obtained by means of analytical and empirical models, computer simulations, or experimental measurements. This chapter focuses on the numerical methods, the computational fluid dynamics (CFD), for solving the NS equations. Based on the CFD simulations, a few surrogate modeling techniques are also introduced.

10.4.1 Computational fluid dynamics

To evaluate the ventilation performance, the air distribution needs to be predicted. This chapter considers the NS equations for incompressible Newtonian fluid:

$$\nabla \cdot \mathbf{U} = 0 \tag{10.32}$$

$$\frac{\partial \mathbf{U}}{\partial t} + (\mathbf{U} \cdot \nabla)\mathbf{U} = -\nabla p + \nu \nabla^2 \mathbf{U} + \mathbf{F} \tag{10.33}$$

where t is the time, p is the air pressure, ν is the effective viscosity, $\mathbf{F}$ is the body force. For indoor airflow, the heat transfer and contaminant transport are generally considered by solving extra scalar equations:

$$\frac{\partial T}{\partial t} + (\mathbf{U} \cdot \nabla)T = \nabla \cdot (\kappa \nabla T) + S_T \tag{10.34}$$

$$\frac{\partial \Phi}{\partial t} + (\mathbf{U} \cdot \nabla)\Phi = \nabla \cdot (\Gamma_\Phi \nabla \Phi) + S_\Phi \tag{10.35}$$

where κ is the effective conductivity, S_T is the source term for T, Φ is the species concentration, Γ_Φ is the effective mass diffusion coefficient, and S_Φ is the source term for Φ.

If a steady-state air distribution is needed in the design, the conventional way to solve the abovementioned equations is using the semi-implicit method for pressure linked equations (SIMPLE) algorithm (Caretto *et al.*, 1973) with the temporal terms $\partial \mathbf{U}/\partial t$, $\partial T/\partial t$, and $\partial \Phi/\partial t$ eliminated. For transient indoor airflow, one can solve (10.32)–(10.35) with SIMPLE, pressure implicit with splitting of operators (PISO) (Issa, 1986), PIMPLE (a merged PISO and SIMPLE algorithm) (Holzmann, 2016), or fast fluid dynamics (FFD, a pressure-correction scheme) (Chorin, 1968; Zuo and Chen, 2009). A transient solver with SIMPLE algorithm is regarded to be robust if sufficient iteration in each time step is used. The calculation with the PISO algorithm is more efficient than the SIMPLE algorithm but limited in the time step based on the Courant number (*Co*). The PIMPLE algorithm combines both algorithms and allows one to use bigger time step size ($Co \gg 1$). FFD uses a two-step (Liu *et al.*, 2016) or three-step, time splitting scheme (Mortezazadeh and Wang, 2020) to solve the momentum equations of the time-dependent NS equations. This algorithm is generally more than ten times faster than the other algorithms with similar accuracy maintained. One can run the transient solver with sufficient physical flow time to obtain a steady-state solution.

The accuracy and efficiency of CFD is highly dependent on the turbulence model. Zhai *et al.* (2007) conducted systematic investigations on the evaluation of various turbulence models in predicting airflow and turbulence in indoor environments. These models cover a wide range of CFD approaches, including Reynolds averaged NS (RANS) modeling, hybrid RANS and large eddy simulation (LES) (or detached eddy simulation), and LES. The RANS turbulence models tested include the indoor zero equation model, three two-equation models (the RNG $k-\varepsilon$, low Reynolds number $k-\varepsilon$, and SST $k-\omega$ models), a three-equation model (v^2-f model), and a Reynolds stress model. The evaluations with experimental data on typical indoor airflow found that the LES provided the most detailed flow features while it was much more computational demanding than RANS models and the accuracy might not always be the best. The RNG $k-\varepsilon$ and the v^2-f model had the best overall performance among the RANS models.

10.4.2 Surrogate models

A surrogate model is an engineering method used when an outcome of interest cannot be easily directly obtained, so a model of the outcome is used instead (Queipo *et al.*, 2005). Since the CFD simulations could be computational demanding, surrogate models were investigated to achieve fast evaluation of ventilation performance. A surrogate model generally first use CFD to create a few training samples since CFD with a proper turbulence model could be accurate. Then a surrogate model reconstructs the overall mapping between the design variables and design objectives by using those training samples. The common surrogate models used for ventilation system designs are proper orthogonal decomposition (POD) method (Lumley, 1967) and machine learning (Gorissen *et al.*, 2010).

10.4.2.1 Proper orthogonal decomposition method

The POD method provides a quick approach to describing forward mapping of a thermo-fluid distribution from various boundary setting conditions (Chen *et al.*, 2017). POD establishes a basis for the modal decomposition of an ensemble of data, such as thermo-flow fields obtained from CFD simulations or measurements. A thermo-flow field can be expressed as a combination of the orthogonal spatial modes with their amplitudes or coefficients (Holmes *et al.*, 2012):

$$\Phi_c = \sum_{k=1}^{K} c_k \phi_k \tag{10.36}$$

where Φ_c is the constructed field, c_k is the amplitude or coefficient of the spatial mode ϕ_k, k is the index of the spatial mode, and K is the number of required spatial modes. To reduce the effort required to solve for the spatial modes, Sirovich (1987) proposed the snapshot method for extracting the spatial modes as samples. Generally, the number of snapshots is far less than the number of spatial grid points, so considerable computing expense can be saved. The spatial mode is calculated as a combination of each snapshot field with its coefficient as

$$\phi = \sum_{s=1}^{S} a_s \Phi_s \tag{10.37}$$

where a_s is the coefficient of a snapshot field Φ_s, and S is the number of snapshots. Once the coefficients of the spatial modes are provided, a field can be constructed on the basis of (10.37). According to the orthogonality principles, the coefficient of a spatial mode corresponding to a snapshot field is

$$c_k = (\Phi_s \cdot \phi_k) \tag{10.38}$$

The coefficients of the POD modes can correspond to a number of causal variables, for example, the air-supply parameters. By interpolating the coefficients for the rest air-supply parameters that are not used to extract the POD modes, the rest thermo-flow distribution can be instantly constructed.

For ventilation system design, POD is mainly used for a fast prediction of indoor thermo-flow and pollutant concentration and an optimization of air-supply parameters. Elhadidi and Khalifa (Sirovich, 1987) applied the POD analysis to efficiently predict the velocity and temperature distributions inside an empty office. Sempey *et al.* (2009) performed a POD-based prediction of temperature distribution in air-conditioned rooms in a fixed-flow context. The POD prediction of thermo-fluid flow could be integrated with optimization algorithm for the design of air-supply velocity and temperature (Li *et al.*, 2013). However, since the reconstruction of thermo-flow distribution by POD for evaluating ventilation performance could be very efficient, the POD prediction could be used by exhaust searching. For example, Wang *et al.* (2018) applied POD and exhausted searching to determine appropriate air-supply parameters in an aircraft cabin environment. The major weakness of POD prediction is accuracy. To overcome this shortcoming, one can further apply fine-tuning after the design by POD. Wei *et al.* (2019) applied a CFD-based adjoint method with the optimal design by POD analysis for fine-tuning the air-supply conditions in an aircraft cabin.

10.4.2.2 Machine learning

Machine learning is a subset of artificial intelligence that improves automatically through experience by the study of computer algorithms (Mitchell, 1997). Machine learning algorithms build a model based on sample data, known as "training data," in order to make predictions or decisions without being explicitly programmed to do so. For the ventilation system design, the training data could be generated by CFD with an appropriate turbulence model. Besides, performing machine learning involves creating a model, which is trained by the training data and then can process additional data to make predictions. The most common type used in built environment is ANN.

ANN is a computing system vaguely inspired by the biological neural networks (NNs) that constitute animal brains. Such systems "learn" to perform tasks by considering examples, generally without being programmed with any task-specific rules. ANN can be used as a surrogate model of the CFD to predict indoor velocity, temperature, thermal comfort, energy cost, and contaminant source. There are mainly three types of ANN: standard NN, convolutional NN, and recurrent NN. The convolutional NN is basically applied in image recognition (Simard *et al.*, 2003) and recurrent NN is a sequence model that can be applied to natural language processing problems (Zaremba *et al.*, 2014). For the design of ventilation system, a

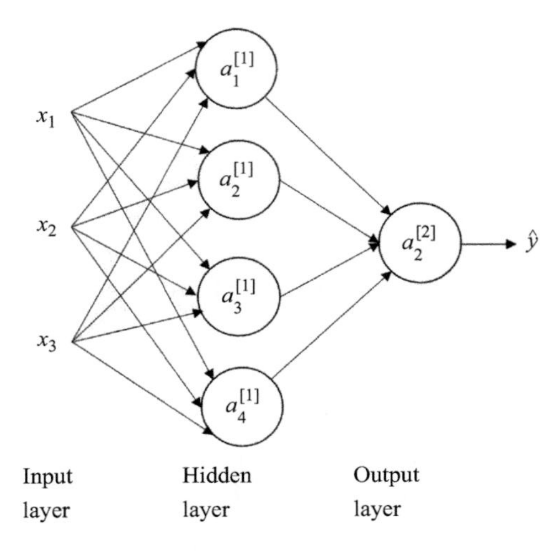

Figure 10.3 A standard NN with two layers

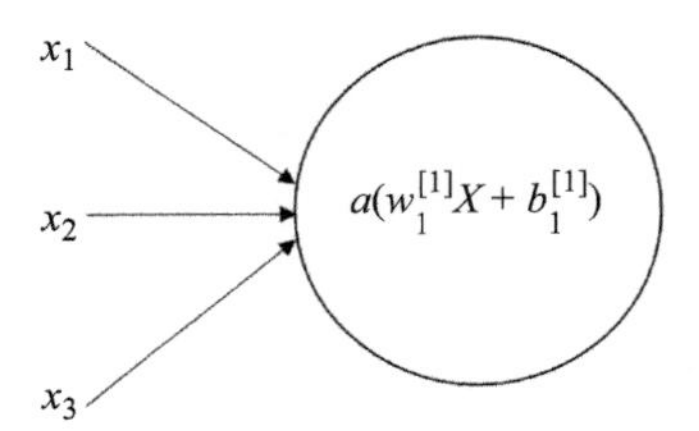

Figure 10.4 Computation of a neuron

standard NN is generally applied. Figure 10.3 shows a two-layer standard NN, since the input layer does not count. The ξ_i represents the ith input (design variable) and $a_i^{[j]}$ represents the ith neuron of the jth layer. The $\widehat{y}$ is the predicted value such as air distribution or objective function in this chapter. An ANN with more than two layers is called deep NN. The machine learning with a deep NN is deep learning.

If we have a close look to a neuron, for example, $a_1^{[1]}$, Figure 10.4 shows the computation. The computation gives

$$a_1^{[1]} = \alpha(\mathbf{w}_1^{[1]T}\mathbf{X} + b_1^{[1]}) \tag{10.39}$$

where α is an activation function, $\mathbf{w}_1^{[1]}$ is the weight and its dimension is 3×1, $\mathbf{X} = [x_1, x_2, x_3]^T$ is the three inputs and its dimension is again 3×1, and $b_1^{[1]}$ is the bias. Since the air distribution is governed by the NS equations that are nonlinear, the nonlinearity is introduced in an ANN by the activation function. The popular activation functions include *LogSig*, *TanSig*, *purelin*, and *ReLU* (rectified linear units) activations (Chen *et al.*, 2017; Glorot *et al.*, 2011).

The values for weight $\mathbf{w}$ and bias $\mathbf{b}$ are determined by training and testing. In general, from a data pool (numbers of CFD simulations in this chapter), one would randomly choose part of the data as training samples and the rest as testing sample. With a set of training data $(\mathbf{X}, y)$, the training aims to identify the weight $\mathbf{w}$ and bias $\mathbf{b}$ that lead to least difference between y and $\widehat{y}$. Then a loss function is established to quantify the agreement between a sample and the prediction, such as

$$L(\widehat{y}, y) = (y - \widehat{y})^2 \tag{10.40}$$

For m training samples, a cost function could be constructed as

$$J(\mathbf{w}, \mathbf{b}) = \frac{1}{m}\sum_{i=1}^{m} L(\widehat{y}^{(i)}, y^{(i)}) \tag{10.41}$$

where the superscript (i) is the index of samples. The computations of an NN are organized in terms of a forward propagation step and commonly followed by a back-propagation step, which is used to compute gradients of cost function J over $\mathbf{w}$ and $\mathbf{b}$. The gradients are further used to update the weights and bias that lead to minimal cost function. With the gradient, one also needs to determine the learning rate in updating the weights and bias. Various training algorithms based on back-propagation have been used to train ANNs, such as the steepest descent algorithm, Newton's method, the Gauss–Newton algorithm, and the Levenberg–Marquardt and Bayesian regularization algorithms (Wilamowski and Irwin, 2011). The training algorithm could be divided into two types: incremental and batch (Pérez-Sánchez *et al.*, 2013). The incremental training updates the weights and bias whenever an input is introduced and batch training updates those after all the inputs are introduced.

Thomas and Soleimani-Mohseni (2007) compared the prediction accuracy of ANN with other models and found that the ANN model gave more accurate temperature predictions. Zhang and You (2014) used ANN to identify the indoor boundary conditions by the data of velocity and temperature measured at observation locations. The Bayesian regularization training algorithm was used to train ANN. The ANN could also be integrated with other methods such as GA, multivariate regression analysis, and fuzzy logic controller for the optimal design of ventilation system (Zhang *et al.*, 2020).

10.5 Optimization algorithm

The traditional design of ventilation system by CFD uses a trial-and-error process. A designer first assumes values to design variables according to experience. Then the resulting distributions of the air temperature, air velocity, and relative humidity in the indoor space will be compared with the design objectives. If the air distributions do not meet the standards, the values for the design variables will be changed until the design objectives are met. The trial-and-error process is robust and easy to use. However, the design process can take days and the obtained air distributions may not be optimal. If one uses a fast model such POD or ANN for evaluating the ventilation performance, an exhaust searching could be used. But the design by POD or ANN might not be as accurate as that by CFD. In order to design the ventilation system by CFD in an efficient way, one can integrate CFD with optimization algorithms. According to Liu *et al.* (2015), the adjoint method and GA can be integrated with CFD simulations for an optimal design of ventilation system.

10.5.1 CFD-based adjoint method

The CFD-based adjoint method calculates the gradient $dO/d\xi$ for finding the direction to adjust ξ to minimize O(ξ). Since the relationship between O and ξ is constrained by the NS equations, the adjoint method introduces a Lagrangian multiplier $\chi_a = (p_a, \mathbf{U}_a, T_a, \Phi_a)$ to produce an augmented objective function as

$$L(\xi) = O(\xi) + \int_{\Theta} \chi_a \cdot \mathbf{N} d\Theta \tag{10.42}$$

where χ_a represents the adjoint variables and $\mathbf{N}(\chi) = (N_1, N_2, N_3, N_4, N_5, N_6)$ with N_1 the continuity equation (Equation (10.32)), N_2, N_3, and N_4 the momentum equations (Equation (10.33)), N_5 the energy equation (Equation (10.34)), and N_6 the scalar transport equation (Equation (10.35)). $\chi = (p, \mathbf{U}, T, \Phi)$ are the state variables. This mathematical manipulation transforms min $O(\xi)$ to min $L(\xi)$. This is because the minimum of $L(\xi)$ should satisfy

$$\frac{\partial L}{\partial \chi_a} = 0 \rightarrow \mathbf{N} = 0 \tag{10.43}$$

$$\frac{\partial L}{\partial \chi} = 0 \rightarrow \frac{\partial O}{\partial \chi} + \int_{\Theta} \chi_a \cdot \frac{\partial \mathbf{N}}{\partial \chi} d\Theta = 0 \tag{10.44}$$

$$\frac{\partial L}{\partial \xi} = 0 \rightarrow \frac{\partial O}{\partial \xi} + \int_{\Theta} \chi_a \cdot \frac{\partial \mathbf{N}}{\partial \xi} d\Theta = 0 \tag{10.45}$$

Equation (10.43) is the NS equations. Further, (10.43)–(10.45) could not be solved analytically. The adjoint method solved them iteratively. With an initialized ξ, (10.43) is solved first to obtain ξ and calculate O. Then the variables in (10.44) other than ξ_a can be obtained. In the following step, adjoint equation (Equation (10.44)) is solved to obtain ξ_a. Afterwards, (10.45) is used to compute the gradient $\partial L/\partial \xi$. By using the simple steepest descent algorithm (Bryson, 1975), the design variables can be adjusted by

$$\xi_{new} = \xi_{old} - \lambda \left[\frac{\partial L}{\partial \xi}\right]^T \tag{10.46}$$

where λ is a positive constant. With the adjusted ξ, (10.43)–(10.46) are solved again to adjust the ξ. This procedure is repeated until the O is minimized. The procedure for obtaining a new ξ is called one design cycle.

10.5.2 CFD-based genetic algorithm

The GA uses evolution operations to transform a population of design variables into a new population with higher average fitness values (Koza and Koza, 1992). The higher fitness value an individual has, the smaller value the objective function is. The GA is capable of resolving nonlinear and multi-solution problems (Sakamoto *et al.*, 1999). The GA procedure used in this study is as follows:

- Step 1: Generate a random initial population of potential solutions to the design variables ξ, which contains several individuals.
- Step 2: Evaluate the fitness value of each individual with the objective function $O(\xi)$ by solving the NS equations.
- Step 3: Check whether the population meets the prescribed optimization criterion; if not, apply genetic operations such as selection, crossover, and mutation to the population to create a new generation of potential solutions to ξ.

After step 1, the GA method repeats steps 2 and 3 until the optimization criterion is satisfactory. The criterion in this study was that the fitness value of the best individual in each generation remains unchanged for five generations. This study used the tournament selection method (Miller *et al.*, 1995) for gene selection operation, multipoint-crossover method for the crossover process, and constant mutation rate for the mutation process.

10.6 Applications

To demonstrate the application of numerical methods for the design of ventilation system, this section compares the accuracy and computing efficacy of the CFD-based GA and adjoint methods by using them for the identification of the air-supply location (y_{in}, the y coordinate of the middle point of the inlet) and parameters ($\mathbf{U}_{in}$, T_{in}) ($\xi = (y_{in}, \mathbf{U}_{in}, T_{in})$) for a two-dimensional, non-isothermal case with the experimental data from Blay (1992). In the experiment, Figure 10.1(a) shows that the inlet location was at $x_{in} = 0$ and $y_{in}/H = 0.991$, the inlet height $h_{in} = 18$ mm, the inlet air velocity $U_{in,x} = 0.57$ m/s and $U_{in,y} = 0.0$ m/s, the inlet air temperature (T_{in}) 15 °C, and the outlet height $h_{out} = 24$ mm. The temperature of the walls (T_{wl}) was 15 °C and that of the floor (T_{fl}) 35.5 °C. The case contained representative flow characteristics in an indoor environment, such as jets and thermal plumes. The experiment measured the air velocity and temperature on the red dash lines as shown in Figure 10.1(a). Our design set the measured air velocity and temperature as the design objective. Then the objective function for this case was

$$O(\xi) = \int_{\Omega} \left[(U_x^* - U_{exp,x}^*)^2 + (U_y^* - U_{exp,y}^*)^2 + (T^* - T_{exp}^*)^2 \right] d\Omega \qquad (10.47)$$

where $*$ represents the normalized value by the experimental data. This design objective was clearly attainable because it measured data from the experiment with the corresponding boundary conditions used. Figure 10.1(b) shows the resolution of

Table 10.1 Prescribed ξ for the trial-and-error process

Resolution	Inlet location y_{in}/H	$\|\mathbf{U}_{in}\|$ (m/s)	α (°)	T_{in} (°C)
2 × 2 × 2 × 2	0.25; 0.75	0.25; 0.75	−22.5; 22.5	12.5; 17.5
3 × 3 × 3 × 3	0.166; 0.5; 0.834	0.16; 0.5; 0.83	−30; 0; 30	11.6; 15; 18.3
4 × 4 × 4 × 4	0.125; 0.375	0.125; 0.375	−33.75; −11.25	11.25; 3.75
	0.625; 0.875	0.625; 0.875	11.25; 33.75	16.25; 18.75

the mesh when the inlet changed during the design. The mesh resolution at the inlet, outlet, and near wall region was finer because of the high gradient of velocity and temperature there. This principle was applied to regenerate the mesh when the air-supply location changed in the design.

The design requires the initialization of the design variables ξ and this study assumes the range of ξ. This study assumed that the left wall in Figure 10.1(b) was a possible location for the inlet. For the air-supply velocity, this study assumed that its magnitude was between 0 and 1 m/s and its direction was $\alpha \in [-45°,+45°]$ (0° means the air velocity direction was normal to the inlet face). The range of the air-supply temperature was 10 °C–20 °C.

For the adjoint method, the initialized air-supply conditions were $y_{in}/H = 0.5$, $|\mathbf{U}_{in}| = 0.5$ m/s, $\alpha = 0°$, and $T_{in} = 12.5$ °C. For the GA method, the initialized ξ in the first generation with eight individuals were generated stochastically within the ranges. As a comparison, this investigation used the trial-and-error process with prescribed design variables as shown in Table 10.1. The trial-and-error process uses different sets of design variables ξ to obtain the objective function and selects the best one as the final design. Within the presumed range for each design variable, this study divides the range equally to generate different sets of ξ. For instance, if a design variable can vary between 0 and 1.0, the value used for the trail-and-error process is 0.25 and 0.75 under the resolution of 2. The CFD-based trail-and-error process for designing indoor environment is as follows:

- Step 1: Generate the values for the design variables under a certain resolution.
- Step 2: Obtain a set of $O(\xi)$ under this sampling resolution by CFD calculation.
- Step 3: Check any of the $O(\xi)$ can meet the desired objective. If yes, select the one with the best fit to the design objective as the final design. If not, generate new values of the design variables with a higher resolution.
- Step 4: Repeat steps 2 and 3 until an acceptable design is found.

Figure 10.5 shows the objective function vs. the number of cases calculated for each method. The objective function for the adjoint method decreases very fast and the local optima could be identified within five design cycles (in this study, the calculation stops when $|O_{new} - O_{old}| < 0.01$), which means the number of cases calculated was ten. This is because the adjoint method calculated both the NS equations and adjoint equations in one design cycle and the time spent on solving the adjoint equations was almost the same as that for solving the NS equations.

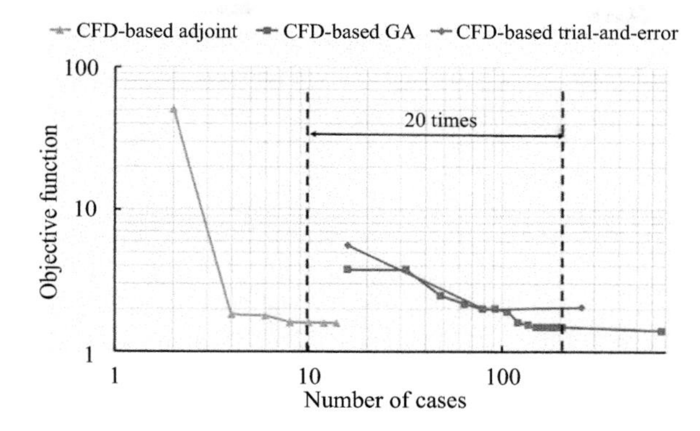

Figure 10.5 Objective function vs. number of cases for different methods

The GA method can lead to the smallest objective function. This is reasonable because the GA method has higher chance to identify the global optima. However, both the GA and trial-and-error approaches require the calculation of about 200 cases to obtain the converged solutions and the computing effort was about 20 times greater than that of the adjoint method. For the GA method, this study continued the calculation after the converged results were obtained. It was possible to have a smaller objective function after more than 600 cases. It implied that GA method cannot guarantee the identification of global optima. The optimal solution of the adjoint method was $|\mathbf{U}_{in}| = 0.561$ m/s, $\alpha = -2°$, $T_{in} = 286.5$K, and $y_{in} = 1.031$ m and that for GA method was $|\mathbf{U}_{in}| = 0.91$ m/s, $\alpha = 45°$, $T_{in} = 287.4$K, and $y_{in} = 0.977$ m.

For the trial-and-error process, the computing effort increased dramatically with the refined sampling resolution for design variables. However, Figure 10.5 shows that the increased resolution did not necessarily improve the designed results. Besides, the method cannot lead to an objective function to be as small as that of the adjoint or GA methods.

10.7 Summary

This chapter systematically introduces the components of ventilation system design by a numerical method. These components include the design variables, design objectives, numerical methods for evaluating ventilation performance, and optimization algorithm. The mathematical treatments of the design variables and design objectives were generally provided. The design variables should be given sufficient freedom to achieve a desired design objective. The numerical methods in this chapter are CFD simulations and CFD-based surrogate models. The optimization algorithm introduced includes the adjoint method and GA. The application of optimization algorithm is dependent on the design variables, design objectives, and the numerical methods. If a fast numerical tool is adopted, the optimization algorithm is not a necessity. In the end, this chapter compares the accuracy and computing efficacy of the CFD-based GA and adjoint methods by using them for the identification of air-supply conditions in a ventilated cavity for demonstration.

References

ANSI/ASHRAE. *ASHRAE Standard 55–2017. Thermal Environmental Conditions for Human Occupancy*. American Society of Heating, Refrigerating and Air-Conditioning Engineers, Inc, Atlanta, GA; 2017.

ANSI/ASHRAE. *Standard 62.1-2016. Ventilation for Acceptable Indoor Air Quality*. American Society of Heating, Refrigerating and Air-Conditioning Engineers, Inc, Atlanta, GA; 2016.

Awbi HB. *Ventilation of Buildings*. New York, USA, Taylor& Francis; 2003.

Blay D. Confined turbulent mixed convection in the presence of horizontal buoyant wall jet. HTD Vol. 213, Fundermentals of Mixed Convection. 1992.

Bracewell R. Heaviside'sunit step function. In: *The Fourier Transform and Its Applications*, McGraw-Hill, New York, USA; 2000. p. 61–65.

Bryson AE and Ho YC. *Applied Optimal Control: Optimization, Estimation and Control*. New York, USA, Hempshire Publishrng Co.; 1975.

Caretto L, Gosman A, Patankar S, *et al.* Two calculation procedures for steady, three-dimensional flows with recirculation. In: *Proceedings of the third international conference on numerical methods in fluid mechanics*. Springer; 1973. 60–68.

Chen Q, Liang XG, and Guo ZY. Entransy theory for the optimization of heat transfer—a review and update. *International Journal of Heat and Mass Transfer*. 2013;63:65–81.

Chen Q, Ren J, and Guo Z. The extremum principle of mass entransy dissipation and its application to decontamination ventilation designs in space station cabins. *Chinese Science Bulletin*. 2009;54(16):2862–2870.

Chen Q, Zhai Z, You X, et al. *Inverse Design Methods for the Built Environment*. Oxford, England, Routledge; 2017.

Chen Q. Can we migrate COVID-19 spreading risk?. *Frontiers of Environmental Science & Engineering*. 2020;15(3):1–4.

Chen Q. Ventilation performance prediction for buildings: a method overview and recent applications. *Building and Environment*. 2009;44(4):848–858.

Chorin AJ. Numerical solution of the Navier-Stokes equations. *Mathematics of Computation*. 1968;22(104):745–762.

Cui W, Ouyang Q, and Zhu Y. Field study of thermal environment spatial distribution and passenger local thermal comfort in aircraft cabin. *Building and Environment*. 2014;80:213–220.

Cui W, Ouyang Q, Zhu Y, et al. Prediction model of human thermal sensation under low-air-pressure environment. In: *Proceedings of the 8th International Symposium on Heating, Ventilation and Air Conditioning*. Springer; 2014. p. 329–336.

Fallenius BE, Sattari A, Fransson JH, et al. Experimental study on the effect of pulsating inflow to an enclosure for improved mixing. *International Journal of Heat and Fluid Flow*. 2013;44:108–119.

Fanger PO and Christensen N. Perception of draught in ventilated spaces. *Ergonomics*. 1986;29(2):215–235.

Fanger PO. *Thermal Comfort. Analysis and Applications in Environmental Engineering*, Copenhagen: Danish Technical Press; 1970.

Fanger PO, Melikov AK, Hanzawa H, et al. Turbulence and draft. The turbulence of airflow has a significant impact on the sensation of draft. *ASHRAE Journal*. 1989;31(4):18–25.

Fisk WJ. The ventilation problem in schools: literature review. *Indoor Air*. 2017;27 (6):1039–1051.

Forrester A, Sobester A, and Keane A. *Engineering Design Via Surrogate Modelling: A Practical Guide*. John Wiley & Sons, Chichester, England; 2008.

Gan G and Awbi H. Numerical prediction of the age of air in ventilated rooms. In: *Proceedings of 4th International Conference on Air Distribution in Rooms-Roomvent*. vol. 94; 1994. p. 15–27.

Geng Y, Ji W, Lin B, et al. The impact of thermal environment on occupant IEQ perception and productivity. *Building and Environment*. 2017;121:158–167.

Glorot X, Bordes A and Bengio Y.Deep sparse rectifier neural networks. In: Proceedings of the Fourteenth International Conference on Artificial Intelligence and Statistics; 2011. p. 315–323.

Gorissen D, Couckuyt I, Demeester P, et al. A surrogate modeling and adaptive sampling toolbox for computer based design. *The Journal of Machine Learning Research*. 2010;11:2051–2055.

Hedge A, Erickson WA, and Rubin G. Predicting sick building syndrome at the individual and aggregate levels. *Environment International*. 1996;22 (1):3–19.

Holmes P, Lumley JL, Berkooz G, et al. *Turbulence, Coherent Structures, Dynamical Systems and Symmetry*. Cambridge University Press, New York, USA; 2012.

Holzmann T. *Mathematics, Numerics, Derivations and OpenFOAM*. Holzmann CFD, Loeben, Germany; 2016.

ANSYS Inc. ANSYS ICEM CFD 14.0 User Manual. Southpointe, 2012.

Issa RI. Solution of the implicitly discretised fluid flow equations by operator-splitting. *Journal of Computational Physics*. 1986;62(1):40–65.

Jasak H, Jemcov A, Tukovic Z, *et al.* OpenFOAM: A C++ library for complex physics simulations. In: International Workshop on Coupled Methods in Numerical Dynamics. vol. 1000. IUC Dubrovnik Croatia; 2007. p. 1–20.

Kato S and Murakami S. New ventilation efficiency scales based on spatial distribution of contaminant concentration aided by numerical simulation. *ASHRAE Transactions*. 1988;94:309–330.

Koza JR and Koza JR. *Genetic Programming: On the Programming of Computers by Means of Natural Selection*. vol. 1. MIT Press, Cambridge, MA, USA; 1992.

Laverge J and Janssens A. Optimization of design flow rates and component sizing for residential ventilation. *Building and Environment*. 2013;65:81–89.

Li X. A non-dominated sorting particle swarm optimizer for multiobjective optimization. In: *Genetic and Evolutionary Computation Conference*. Springer, Berlin, Germany; 2003. p. 37–48.

Li K, Xue W, Xu C, et al. Optimization of ventilation system operation in office environment using POD model reduction and genetic algorithm. *Energy and Buildings*. 2013;67:34–43.

Li N, Cheung SC, Li X, et al. Multi-objective optimization of HVAC system using NSPSO and Kriging algorithms—a case study. In: *Building Simulation*. vol. 10. Springer, Tsinghua University Press; 2017. p. 769–781.

Li N, Yang L, Li X, et al. Multi-objective optimization for designing of high-speed train cabin ventilation system using particle swarm optimization and multi-fidelity Kriging. *Building and Environment*. 2019;155:161–174.

Liu W, Duan R, Chen C, et al. Inverse design of the thermal environment in an airliner cabin by use of the CFD-based adjoint method. *Energy and Buildings*. 2015;104:147–155.

Liu W, Jin M, Chen C, et al. Implementation of a fast fluid dynamics model in OpenFOAM for simulating indoor airflow. *Numerical Heat Transfer, Part A: Applications*. 2016;69(7):748–762.

Liu W, Jin M, Chen C, et al. Optimization of air supply location, size, and parameters in enclosed environments using a computational fluid dynamics-based adjoint method. *Journal of Building Performance Simulation*. 2016;9 (2):149–161.

Liu W, Zhang T, Xue Y, et al. State-of-the-art methods for inverse design of an enclosed environment. *Building and Environment*. 2015;91:91–100.

Lu H and Yen GG. Rank-density-based multiobjective genetic algorithm and benchmark test function study. *IEEE Transactions on Evolutionary Computation*. 2003;7(4):325–343.

Lumley JL. The structure of inhomogeneous turbulent flows. In: *Atmospheric Turbulence and Radio Wave Propagation*, Moscow: Nauka; 1967.

Melikov AK. Personalized ventilation. *Indoor Air*. 2004;14:157–167.

Miller BL and Goldberg DE. Genetic algorithms, tournament selection, and the effects of noise. *Complex Systems*. 1995;9(3):193–212.

Mitchell T. Introduction to machine learning. *Machine Learning*. 1997;7:2–5.

Mortezazadeh M and Wang LL. Solving city and building microclimates by fast fluid dynamics with large timesteps and coarse meshes. *Building and Environment*. 2020:106955.

Nabi S, Grover P, and Caulfield CP.Adjoint-based optimization of displacement ventilation flow. *Building and Environment*. 2017;124:342–356.

Ncube M and Riffat S. Developing an indoor environment quality tool for assessment of mechanically ventilated office buildings in the UK—a preliminary study. *Building and Environment*. 2012;53:26–33.

Nielsen TR and Drivsholm C. Energy efficient demand controlled ventilation in single family houses. *Energy and Buildings*. 2010;42(11):1995–1998.

Olsen SJ, Chang HL, Cheung TYY, et al. Transmission of the severe acute respiratory syndrome on aircraft. *New England Journal of Medicine*. 2003;349(25):2416–2422.

Park S, Hellwig RT, Grün G, et al. Local and overall thermal comfort in an aircraft cabin and their interrelations. *Building and Environment*. 2011;46(5): 1056–1064.

Pérez-Sánchez B, Fontenla-Romero O, Guijarro-Berdiñas B, et al. An online learning algorithm for adaptable topologies of neural networks. *Expert Systems with Applications*. 2013;40(18):7294–7304.

Persily A. Challenges in developing ventilation and indoor air quality standards: the story of ASHRAE Standard 62. *Building and Environment*. 2015;91:61–69.

Queipo NV, Haftka RT, Shyy W, et al. Surrogate-based analysis and optimization. *Progress in Aerospace Sciences*. 2005;41(1):1–28.

Sakamoto Y, Nagaiwa A, Kobayasi S, et al. An optimization method of district heating and cooling plant operation based on genetic algorithm. *ASHRAE Transactions*. 1999;105:104.

Sandberg M and Sjöberg M. The use of moments for assessing air quality in ventilated rooms. *Building and Environment*. 1983;18(4):181–197.

Sattari A and Sandberg M. PIV study of ventilation quality in certain occupied regions of a two-dimensional room model with rapidly varying flow rates. *International Journal of Ventilation*. 2013;12(2):187–194.

Schaffer JD. Multiple objective optimization with vector evaluated genetic algorithms. In: *Proceedings of the First International Conference on Genetic Algorithms and Their Applications*. Lawrence Erlbaum Associates. Inc.; 1985.

Sempey A, Inard C, Ghiaus C, et al. Fast simulation of temperature distribution in air conditioned rooms by using proper orthogonal decomposition. *Building and Environment*. 2009;44(2):280–289.

Simard PY, Steinkraus D, Platt JC, et al. Best practices for convolutional neural networks applied to visual document analysis. In: *ICDAR*. vol. 3; 2003.

Sirovich L. Turbulence and the dynamics of coherent structures. I. Coherent structures. *Quarterly of Applied Mathematics*. 1987;45(3):561–571.

Sun Y, Hou J, Cheng R, et al. Indoor air quality, ventilation and their associations with sick building syndrome in Chinese homes. *Energy and Buildings*. 2019;197:112–119.

Thomas B and Soleimani-Mohseni M. Artificial neural network models for indoor temperature prediction: investigations in two buildings. *Neural Computing and Applications*. 2007;16(1):81–89.

Uppada SK. Centroid based clustering algorithms—a clarion study. *International Journal of Computer Science and Information Technologies*. 2014;5 (6):7309–7313.

van Hooff T and Blocken B. Mixing ventilation flows driven by two opposite out-of-phase wall jets: comparison of URANS and LES. In: *Proceedings of the 10th International Symposium on Heating Ventilation and Air Conditioning (ISHVAC)*; 2017. p. 1–6.

Wang J, Zhang T, Zhou H, et al. Inverse design of aircraft cabin environment using computational fluid dynamics-based proper orthogonal decomposition method. *Indoor and Built Environment*. 2018;27(10):1379–1391.

Wei Y, Liu W, Xue Y, et al. Inverse design of aircraft cabin ventilation by integrating three methods. *Building and Environment*. 2019;150:33–43.

Wilamowski BM and Irwin JD. *The Industrial Electronics Handbook-Five Volume Set*. CRC Press, Boca Raton, FL, USA; 2011.

Wyon DP, Larsson S, Forsgren B, et al. Standard procedures for assessing vehicle climate with a thermal manikin. *SAE Transactions*. 1989:46–56.

Xu H and Niu J. Numerical procedure for predicting annual energy consumption of the under-floor air distribution system. *Energy and Buildings*. 2006;38 (6):641–647.

Yamasaki S, Nomura T, Kawamoto A, et al. A level set based topology optimization method using the discretized signed distance function as the design variables. *Structural and Multidisciplinary Optimization*. 2010;41(5):685–698.

Zaremba W, Sutskever I, and Vinyals O. Recurrent neural network regularization. arXiv preprint arXiv:14092329; 2014.

Zhai ZJ, Xue Y, and Chen Q. Inverse design methods for indoor ventilation systems using CFD-based multi-objective genetic algorithm. In: *Building Simulation*. vol. 7. Springer, Tsinghua University Press; 2014. p. 661–669.

Zhai ZJ, Zhang Z, Zhang W, et al. Evaluation of various turbulence models in predicting airflow and turbulence in enclosed environments by CFD: Part 1—Summary of prevalent turbulence models. *HVAC&R Research*. 2007;13 (6):853–870.

Zhang T, Liu Y, Rao Y, et al. Optimal design of building environment with hybrid genetic algorithm, artificial neural network, multivariate regression analysis and fuzzy logic controller. *Building and Environment*. 2020:106810.

Zhang T-h and You X-y. Applying neural networks to solve the inverse problem of indoor environment. *Indoor and Built Environment*. 2014;23(8):1187–1195.

Zhao X, Liu W, Lai D, *et al.* Optimal design of an indoor environment by the CFD-based adjoint method with area-constrained topology and cluster analysis. *Building and Environment*. 2018;138:171–180.

Zhou L and Haghighat F. Optimization of ventilation system design and operation in office environment, Part I: Methodology. *Building and Environment*. 2009;44(4):651–656.

Zitzler E and Thiele L. Multiobjective evolutionary algorithms: a comparative case study and the strength Pareto approach. *IEEE transactions on Evolutionary Computation*. 1999;3(4):257–271.

Zuo W and Chen Q. Real-time or faster-than-real-time simulation of airflow in buildings. *Indoor Air*. 2009;19(1):33.

Chapter 11
Ventilation system design: fast prediction

Zhiqiang (John) Zhai[1] *and Haidong Wang*[2]

Accurate and rapid prediction of ventilation in built environment is highly desired for the design, construction and operation of energy-efficient, comfortable, and healthy buildings. Developing fast prediction techniques can advocate and accelerate a broader and better application of modeling tools for engineering practices, such as building emergency management, early-stage building and system design, and real-time system control and continuous optimization. This chapter first reviews and introduces the prevalent modeling techniques for built environment ranging from the simplest mixing model to the sophisticated field (computational fluid dynamics—CFD) model. The focus is put on those fast and accurate modeling methods, including zonal models, reduced-order models (ROMs), zero-equation turbulence models, coarse-grid CFD methods, and pressure–velocity decoupling algorithms. The chapter then demonstrates the application of some of these fast simulation techniques and methods for ventilation studies through several engineering and research case studies.

11.1 Motivation

Computational modeling and prediction of built environment started from the 1970s with the development of computer technologies. It has been experiencing an exponential growth since the 1990s in parallel to active commercialization efforts of in-house research modeling programs. In the past three decades, computational simulation tools have been playing an important role in built environment study and have been applied to various areas and stages of building environment research and practice, ranging from schematic building design (Hensen *et al.*, 2004; Augenbroe and Hensen, 2004) to pollutant dispersion prediction (Sun *et al.*, 2002; Gousseau *et al.*, 2011) to real-time building control and monitoring (Zerihun Desta *et al.*, 2004). Integration of different modeling tools (with different modeling principles,

[1]Department of Civil, Environmental and Architectural Engineering, University of Colorado, Boulder, Colorado, USA
[2]School of Environment and Architecture, University of Shanghai for Science and Technology, Shanghai, China

scales, and capabilities) also shows popularity, such as integrating sophisticated CFD techniques with whole-year whole-building energy and thermal simulation tools (Zhai and Chen, 2005; Djunaedy, 2005), and coupling CFD with whole-building zonal airflow models (Wang and Chen, 2007), to obtain more informative results and improve accuracy of simulations.

Significant computing cost of most built environment simulation tools such as CFD, however, has become the main hurdle that restricts a broader and better application of these tools for engineering practices, such as building emergency management, early-stage architectural design, and real-time system control and continuous optimization. In such applications, a large variety of parameters and their dynamic interactions need to be evaluated in a timely manner and require real-time or faster-than-real-time predictions and responses. A rapid evolution of computer hardware and scientific computing technologies, such as parallel computing, cloud computing, and compute unified device architecture programming with graphical processing unit (Cohen and Molemake, 2009), provides good alternative solutions for this challenge. However, the high cost of equipment, maintenance, and operation prohibits the wide deployment of these advanced technologies for building environment-related applications and does not fundamentally deliver a practical solution upon the current affordability of computer resources. Research on fast simulation techniques for built environment study, therefore, deserves attentions and investigations in order to address current engineering challenges timely and accurately.

11.2 Overview of fast simulation techniques for built environment study

Accurate prediction of built environment conditions is crucial for designing an energy-efficient, comfortable, and healthy building. In particular, the spatial and temporal distributions of the following critical parameters in a built environment can be directly or indirectly utilized to quantitatively analyze the built environment quality and judge the system performance:

1. air velocity in three directions (and air speed),
2. air pressure,
3. air temperature,
4. relative humidity,
5. turbulent intensity, and
6. contaminant concentrations.

Air velocity, temperature, relative humidity, and surface temperatures of building enclosures are the four most important parameters influencing thermal comfort of indoor space. The concentration level of different pollutants such as CO_2 and volatile organic compound can be employed to evaluate building indoor air quality (IAQ). In addition, Fanger *et al.* (1989) indicated that turbulence intensity has a significant impact on draft and thus thermal comfort.

Airflow model is dominant in built environment simulation, which determines all the key parameters, including air velocity, pressure, turbulence intensity, temperature, humidity, and pollutant concentration. Airflow model was first developed to estimate wind- and buoyancy-driven infiltration effects on buildings. After that, different types of airflow models were created to address all kinds of airflow-related problems, such as outdoor wind pressure on buildings, indoor air pollutant distributions, zone-to-zone air exchanges, natural ventilation, envelope interior surface convection, efficiency of air handling, and distribution systems. Modeling airflow in a confined environment can assist the assessment of indoor thermal comfort and air quality as well as the prediction of space energy need/cost. Figure 11.1 presents a general classification of indoor airflow models, evolving from the simplest mixing model to the most sophisticated field (CFD) model.

11.3 Mixing, nodal, and zonal airflow models

The mixing model is the simplest airflow model, which assumes that the air in each single space is completely mixed and has uniform properties (e.g., density, viscosity, velocity, temperature, humidity, and contaminant concentration). It is a special case of the nodal model and is thus also called single-nodal model in some literatures. Nodal airflow models were developed in the 1970s for simulating both infiltration and internal airflow between spaces. Nodal models treat the building and room air as an idealized network of nodes connected with flow paths. In a nodal model, a specific airflow configuration is assumed, and the mass and energy balances are written for each node of the nodal network. Lebrun (1970) was apparently the first who proposes a nodal model to provide a rough estimate of thermal stratification in the context of building energy use. The work focused on

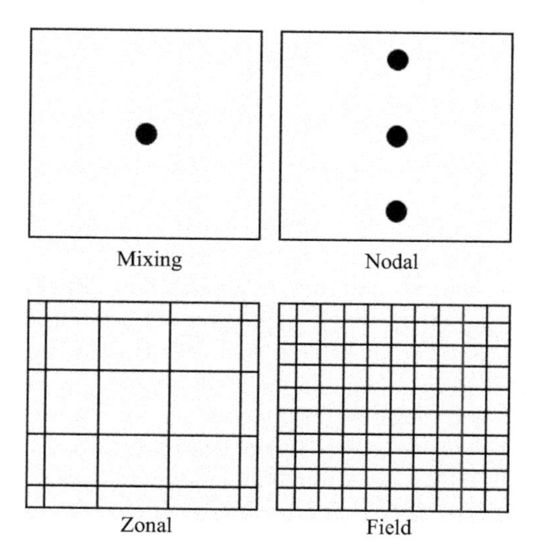

Figure 11.1 Classification of indoor airflow models

modeling how heat is convected within a room by a baseboard heater under a cold window. Lebrun's pioneering work has led to the development of nodal models. Allard and Inard (1992) reviewed various levels of nodal models used in the prediction of the thermally coupled behavior of a room and its heating system. Other nodal models (e.g., Mundt, 1996; Yuan *et al.*, 1998; Rees and Haves, 2001) can predict the vertical temperature gradient in rooms with displacement ventilation. Harrington (2001) developed a nodal model for natural ventilation where the model selects among five different airflow patterns based on the Archimedes number. The primary drawback of nodal models is that prior knowledge of the airflow pattern is required to specify mass flow in the thermal network, which makes the models difficult to use for most designers.

Zonal models introduce more flow dynamics into the prediction of mean airflows compared to nodal models. Depending on the physically valid hypotheses and experimental experience, zonal models divide the space of a room into several subspaces (zones). These subzones are combined together with the conservation of mass and energy equations. In a zonal model, airflow rates are usually solved based on temperature differences, length scales, and initial momentum. Take a certain zone, for example. Assume that there is airflow inlet and outlet, with the mass and mechanical energy conservation expressed as

1. Mass conservation:

$$\frac{dm}{dt} = \rho_i A_i v_i - \rho_o A_o v_o \tag{11.1}$$

2. Mechanical energy conservation:

$$\left(P_i + \frac{1}{2}\rho_i v_i^2 + \rho_i g z_i\right) = \int_{in}^{out} \frac{\partial v}{\partial t} ds + \left(P_o + \frac{1}{2}\rho_o v_o^2 + \rho_o g z_o\right) + \Delta P \tag{11.2}$$

3. Concentration conservation:

$$V\frac{dC}{dt} = C_i A_i v_i - C_o A_o v_o \tag{11.3}$$

The mass conservation equation applies to air mass as well as any species that transport with airflow, so the concentration equation has the same format as mass. The mechanical energy conservation expression is the transient version of the Bernoulli equation, where ΔP denotes the pressure loss/gain of the connection. For example, a duct–fan system enhancing the airflow will give positive ΔP, while a crack with resistance will exert negative ΔP to the equation. $\int_{in}^{out} (\partial v/\partial t) ds$ term accounts for the energy dissipation in the control volumes. The temperature is usually not calculated (Dols, 2001) so the thermal energy equation is not included. Temperature is often specified by schedules to each zone.

The equation sets of interactive zones eventually form linear algebraic system equations. Most multizone programs use the Newton–Raphson method (Feustel and

Rayner-Hooson, 1990) or its variants to solve the system equations. It is fundamentally an iterative method. The convergence of this iterative process is sensitive to the initial guess.

Bouia and Dalicieux (1996) were apparently the first to publish what is termed a pressure-zonal model that uses pressure as a state variable and solves energy and mass balance equations in the context of building room air modeling. Inard *et al.* (1996) demonstrated a functional three-dimensional pressure zonal model with special cells (handlings) for walls, jets, and plumes. A difficulty in applying pressure-zonal models to most building simulations is the requirement of using special laws to describe flows in certain regions. Togari *et al.* (1993) presented a temperature-zonal model that was intended for use in HVAC applications for large vertical spaces such as atriums. The temperature-zonal models use empirical correlations based on temperature differences in combination with special laws for flows like jets and plumes. The main problem with temperature-zonal models is that it is difficult to obtain general-purpose convective heat transport terms that are usually developed from a small set of experiments.

Zonal models provide increased generality compared to nodal models. Most zonal models can also be used in a personal computer and require relatively small computing power (Griffith, 2006). The most attractive reason for design engineers is that limited investment and training are needed to apply these models. However, zonal models are highly experiment dependent. The existing zonal models are limited to the prediction of mean temperature and mass flow rate in each zone. More specifically, zonal models employ two main assumptions: (1) the primary driving flows (boundary layer, jet or thermal plume) can be pre-predicted and (2) users have a good knowledge of the entire flow structures so that the whole flow field can be divided into zones with distinct features. These two assumptions largely limit the application of these methods in a prediction process. It is not always possible to make a clear decision about the main flow pattern. Much more work is needed to get a better knowledge of the heat and mass transfer in buoyancy-driven flows in a non-isothermal and nonadiabatic environment. Further limitations of using these models include

1. the need to quantify the driving forces and to account accurately for all openings in the room;
2. the assumption of uniformly mixed air and pollutant in each zone; and
3. uncertainty in the results obtained with zonal models.

Beyond these, Griffith (2006) found that many zonal models are not numerically stable. Hence, it is necessary to develop and utilize more general, accurate, detailed, and stable air movement models.

Nevertheless, zonal models can efficiently simulate the whole building air change and mass transfer with very low computing intensity, plus the wide availability of prevalent zonal simulation software, such as AIRNET (Walton, 1989), CONTAM (Walton *et al.*, 2005), and COMMIS (Feustel *et al.*, 1990), makes it popular in various aspects of building environment research such as evaluation of ventilation systems (Blomsterberg and Johansson, 2005), epidemic disease transmission (Li *et al.*, 2005a, 2005b), and contaminant source tracking (Liu and Zhai, 2009). Zonal models are a valuable way to promptly predict air movement in built environments with many compartments and complex mechanical systems and can be integrated with more sophisticated airflow models such as CFD.

11.4 Field airflow models (CFD)

With the development of fluid mechanics theories and numerical methodologies, CFD techniques become popular for predicting engineering flows, including building airflows since the early 1970s. CFD offers the potential of much richer details, a higher degree of flexibility, and lower cost than the traditional laboratory study. As the most sophisticated airflow model, CFD provides the detailed spatial (field) distributions of variables such as air pressure, velocity, temperature, humidity, contaminants, and turbulence intensity by solving the conservation equations of mass, momentum, energy, and species concentrations.

Nielsen (1974) was probably the first one who applied CFD techniques to room air motion. Applications of CFD for room airflow were mushrooming in the 1980s. The International Energy Agency Annex 20 was particularly devoted to room airflow prediction with participants from 13 countries (Moser, 1994). Ladeinde and Nearon (1997) reviewed CFD applications in the HVAC industry. These reviews concluded that CFD is powerful in predicting building air movement, although users' knowledge, experience, and skills with CFD are essential to the accuracy of CFD results.

CFD numerically solves the flow-governing equations (i.e., the Navier–Stokes equations), which are a set of partial differential equations (PDEs) used to mathematically describe flow phenomena. Most flows in built environment are turbulent, and turbulence modeling method is thus critical for building airflow simulation. Among various turbulence modeling methods, the Reynolds-averaged Navier–Stokes (RANS)-based approach becomes most popular in modeling airflows in built environment due to its reasonable accuracy and computing speed as well as significantly small requirements on computer resources and user skills (Zhao *et al.*, 2007).

CFD models are computationally expensive to solve compared to nodal and zonal models due to the strong nonlinearity of the flow-governing equations that typically requires iterative numerical solutions. CFD solves a large-scale computational system resulting from discretization of the PDEs for fluid flow. For a two-equation turbulence model in three dimensions, the finite-volume method produces seven degrees of freedom (DOF) per grid cell (P, u, v, w, k, ε, and T). All these DOF solved for are necessary to model the airflow and heat transfer in built environment. Millions of grid cells (subzone) are required for CFD to produce grid-independent solutions and predict results with adequate spatial resolution for result analysis. This will demand hours to days computing time on a personal computer with large memory capacity.

Most CFD models being applied to building environment study focus on the steady-state flows, although transient or unsteady-state flow characteristics broadly exist in built environment with profound engineering implications. Unsteady CFD modeling requires more boundary and initial conditions as inputs and imposes further burden on computing power. Well-validated CFD models for steady-state calculation, although providing a decent prediction on the steady-state flow field, are not necessarily accurate on the intermediate result for the flow transiting from one state to another. The commonly used CFD models for steady-state airflow simulation, once adapted to a transient problem, will require hundreds of iterations within each time step, making the acquisition of CFD even more computationally intensive. However, the potential applications of a fast CFD technique, such as for building emergency management and integrated building energy simulation, primarily depend on the accuracy of a time-dependent solution. Hence, research on the transient features of CFD models and investigating the advanced techniques of fast transient simulation are also of great interest.

The voice for fast simulation has been leading to various means and efforts to make CFD models less computationally expensive. Some of these efforts include developing simplified turbulence models such as zero-equation models (Chen and Xu, 1998), utilization of coarse grid (Andrews *et al.*, 2005; Mora *et al.*, 2003), reforming solution algorithms for pressure–velocity decoupling such as PISO (Issa, 1986) and projection methods (Chorin, 1997); creating ROMs (Burkardt *et al.*, 2006) to quantitatively describe original dynamics of flow systems with simplified numerical models. The following paragraphs will introduce these techniques in detail.

11.4.1 Turbulence model simplification

Turbulence treatment is an important factor that determines the speed of CFD. Different turbulence modeling strategies have different requirements on the resolutions of grid and time step. RANS-based turbulence models provide an economical way to reasonably simulate the primary flow characteristics that are of great interest to building environment analysis.

There is no single universally acceptable turbulence model that works for all flows and all regimes; therefore, numerous turbulence models have been developed over the past decades. Studies show that for enclosed environment modeling, the RNG k–ε model has the best overall performance (Zhai *et al.*, 2007; Zhao *et al.*, 2007). However, these types of two-equation models add two more PDEs to the flow-governing equations to calculate turbulence properties, which increases the computing cost. The calculated turbulence properties, specifically the turbulence viscosity, are mostly used for the closure of the PDE system and are not a necessity of studying environment quality and system performance under most circumstances.

This computing cost can thus be reduced by using simplified, empirical algebraic equation based turbulence models to predict turbulent viscosity, which are called zero-equation turbulence model. Zero-equation models are the simplest eddy viscosity models based on a mixed length hypothesis (Zhai *et al.*, 2007) that solves zero PDE in addition to the RANS equations. The empirical formulation of the zero-equation model developed specifically for building environment modeling

(Chen and Xu, 1998) has been well validated and widely used for different environment studies with acceptable accuracy and significant reduction in computing time.

The earliest zero-equation model was developed by Prandtl (1925) with the mixing-length hypothesis. Although the mixing-length model is not theoretically sound and the mixing-length needs calibrations for each specific type of flow, the model has yielded good results in predicting simple turbulent flows. Some simple zero-equation models, once calibrated, may even provide surprisingly good results for mean-flow quantities of some complex flows. For instance, the study of Nielsen (1998) revealed that the constant eddy-viscosity model provides results closer to the measured data than the standard k–ε model for the prediction of smoke movement in a tunnel. Nilsson (2007) also used the constant eddy-viscosity model to study the comfort conditions around a thermal manikin, which provided acceptable accuracy with significantly less computing effort. A common constant eddy-viscosity model for indoor modeling assumes

$$v_t = 100v \tag{11.4}$$

where v_t is the turbulent kinematic viscosity and v is the physical kinematic viscosity. The constant coefficient (i.e., 100) can be adjusted according to specific flow conditions.

One important development in zero-equation models for modeling airflows in enclosed environments is the zero-equation model developed by Chen and Xu (1998). By using the assumption of uniform turbulence intensity, they derived an algebraic formula to express turbulent viscosity v_t as a function of local mean airflow speed, U_D, and the normal distance to the nearest wall, L:

$$v_t = 0.03874|U_D|L \tag{11.5}$$

The previous equation has an empirical constant of 0.03874 for different flows. The validations conducted by Chen and Xu (1998), Srebric *et al.* (1999), and Morrison (2000) have demonstrated the feasibility of this model in predicting general room airflows. The model has been widely used for simulating airflows in different indoor environments with acceptable accuracy and significant computational saving (e.g., Kameel and Khalil, 2003; Chen *et al.*, 2005). Li *et al.* (2005a, 2005b) further applied this zero-equation model for outdoor thermal environment simulations, which also provided reasonable predictions when compared with the measured data. Some commercial CFD software for HVAC applications has adopted this model as its default.

11.4.2 Coarse grid simulation

Grid resolution directly determines the CFD modeling speed because the number of foating-point operations is proportional to the number of grid points in CFD (Barnard *et al.*, 1999). Although a fine grid shall be employed in most CFD studies to obtain grid-independent solutions that can be analyzed further (Wang and Zhai, 2012a, 2012b), information from all grid points used in a CFD (usually in the magnitude of millions) may not be necessary for the analyses. Computational results on a much coarser grid resolution, in most scenarios, may be adequate for macro- or mesoscale flow pattern analysis.

Coarse-grid CFD prediction, although less accurate than those with fine grids that level out most discretization-caused numerical error, has been applied to various engineering problems (Andrews *et al.*, 2005; Weathers and Spitler, 1993; Moulton and Steinhoff, 2000) where accuracy is not the main concern. Not only can coarse grid decrease the number of floating point operation, but also it cuts the necessary iteration numbers to achieve convergence comparing to fine grids.

However, the credibility of CFD simulation with coarse grids has always been questioned due to the unknown numerical errors that coarse grids have brought. This leads to a necessity of understanding and quantifying the trade-off between CFD grid resolution and simulation accuracy. Guidelines for selecting appropriate CFD grids based on the accuracy requirements of a simulation for realistic building airflow conditions are crucial. In addition, grid distribution, assuming total grid points are the same, does not have significant impact on CFD speed but will affect the accuracy of predicted results. Wang and Zhai (2012a, 2012b), Wang *et al.* (2014) provided detailed analysis on trade-off between grid resolution and simulation accuracy and presented guidelines on selecting appropriate coarse grids for fast CFD modeling of indoor environments.

The following paragraphs demonstrate the theory analysis approach to deriving the coarse-grid-induced numerical viscosity with upwind scheme. The same approach can be applied to analyze the numerical viscosity caused by other numerical schemes (e.g., central, hybrid, and quick schemes). To simplify the theory analysis and highlight the role of numerical viscosity from convection term discretization, the following flow-governing equation is analyzed upon a simple case that assumes

1. 2D flow;
2. grid cell size $\Delta x = \Delta y = h$; and
3. first-order forward scheme for time discretization:

$$\frac{\partial u}{\partial t} = f - (\vec{u} \cdot \nabla) u - \frac{1}{\rho} \Delta p \tag{11.6}$$

The momentum transport equation of velocity u can be discretized by using the upwind scheme for the convection term and thus yields

$$\frac{u_{i,j}^{n+1} - u_{i,j}^{n}}{\Delta t} = f_{x;i,j} - \left(U \frac{u_{i,j}^{n} - u_{i-1,j}^{n}}{\Delta x} + V \frac{u_{i,j}^{n} - u_{i,j-1}^{n}}{\Delta y} \right) - \frac{1}{\rho} \frac{p_{i+1,j}^{n} - p_{i,j}^{n}}{\Delta x} \quad \text{if } U > 0, V > 0 \tag{11.7}$$

$$\frac{u_{i,j}^{n+1} - u_{i,j}^{n}}{\Delta t} = f_{x;i,j} - \left(U \frac{u_{i+1,j}^{n} - u_{i,j}^{n}}{\Delta x} + V \frac{u_{i,j+1}^{n} - u_{i,j}^{n}}{\Delta y} \right) - \frac{1}{\rho} \frac{p_{i+1,j}^{n} - p_{i,j}^{n}}{\Delta x} \quad \text{if } U < 0, \ V < 0 \tag{11.8}$$

where (i, j) is the grid cell numbers in the x and y directions, U and V are the u and v velocity at cell (i, j) that is used to simplify the expression.

Taking (11.7) as an example, (11.7) is what is numerically solved in CFD that represents a numerical (or modified) form of (11.6). Additional information was introduced to (11.7) due to the numerical discretization, which does not exist in physics. Hence, it is necessary to find what (11.7) really stands for in mathematics and what unphysical information has been brought into the simulation results.

Since

$$u_{i,j}^{n+1} = u_{i,j}^{n} + \Delta t \cdot \frac{\partial u}{\partial t}\Big|_{i,j}^{n} + \frac{\Delta t^2}{2}\frac{\partial^2 u}{\partial t^2}\Big|_{i,j}^{n} + O(\Delta t^3) \tag{11.9}$$

$$u_{i-1,j}^{n} = u_{i,j}^{n} - \Delta x \cdot \frac{\partial u}{\partial x}\Big|_{i,j}^{n} + \frac{\Delta x^2}{2}\frac{\partial^2 u}{\partial x^2}\Big|_{i,j}^{n} + O(\Delta x^3) \tag{11.10}$$

$$u_{i,j-1}^{n} = u_{i,j}^{n} - \Delta y \cdot \frac{\partial u}{\partial y}\Big|_{i,j}^{n} + \frac{\Delta y^2}{2}\frac{\partial^2 u}{\partial y^2}\Big|_{i,j}^{n} + O(\Delta y^3) \tag{11.11}$$

This yields

$$\frac{u_{i,j}^{n+1} - u_{i,j}^{n}}{\Delta t} = \frac{\partial u}{\partial t}\Big|_{i,j}^{n} + \frac{\Delta t}{2}\frac{\partial^2 u}{\partial t^2}\Big|_{i,j}^{n} + O(\Delta t^2) \tag{11.12}$$

$$\frac{u_{i,j}^{n} - u_{i-1,j}^{n}}{\Delta x} = \frac{\partial u}{\partial x}\Big|_{i,j}^{n} - \frac{\Delta x}{2}\frac{\partial^2 u}{\partial x^2}\Big|_{i,j}^{n} + O(\Delta x^2) = \frac{\partial u}{\partial x}\Big|_{i,j}^{n} - \frac{h}{2}\frac{\partial^2 u}{\partial x^2}\Big|_{i,j}^{n} + O(h^2) \tag{11.13}$$

$$\frac{u_{i,j}^{n} - u_{i,j-1}^{n}}{\Delta y} = \frac{\partial u}{\partial y}\Big|_{i,j}^{n} - \frac{\Delta y}{2}\frac{\partial^2 u}{\partial y^2}\Big|_{i,j}^{n} + O(\Delta y^2) = \frac{\partial u}{\partial y}\Big|_{i,j}^{n} - \frac{h}{2}\frac{\partial^2 u}{\partial y^2}\Big|_{i,j}^{n} + O(h^2) \tag{11.14}$$

and

$$\frac{p_{i+1,j}^{n} - p_{i,j}^{n}}{\Delta x} = \frac{\partial p}{\partial x}\Big|_{i,j}^{n} + \frac{\Delta x}{2}\frac{\partial^2 p}{\partial x^2}\Big|_{i,j}^{n} + O(\Delta x^2) = \frac{\partial p}{\partial x}\Big|_{i,j}^{n} + \frac{h}{2}\frac{\partial^2 p}{\partial x^2}\Big|_{i,j}^{n} + O(h^2) \tag{11.15}$$

Therefore, (11.7) can be rewritten as

$$\frac{\partial u}{\partial t}\Big|_{i,j}^{n} + \frac{\Delta t}{2}\frac{\partial^2 u}{\partial t^2}\Big|_{i,j}^{n} + O(\Delta t^2) = f_{x;i,j} - \begin{pmatrix} \left(U\frac{\partial u}{\partial x}\Big|_{i,j}^{n} - \frac{Uh}{2}\frac{\partial^2 u}{\partial x^2}\Big|_{i,j}^{n} + U \cdot O(h^2)\right) + \\ \left(V\frac{\partial u}{\partial y}\Big|_{i,j}^{n} - \frac{Vh}{2}\frac{\partial^2 u}{\partial y^2}\Big|_{i,j}^{n} + V \cdot O(h^2)\right) \end{pmatrix}$$

$$-\frac{1}{\rho}\left(\frac{\partial p}{\partial x}\Big|_{i,j}^{n} + \frac{h}{2}\frac{\partial^2 p}{\partial x^2}\Big|_{i,j}^{n} + O(h^2)\right) \tag{11.16}$$

Further as

$$\frac{\partial u}{\partial t}\Big|_{i,j}^{n} = f_{x;i,j} - \left(U\frac{\partial u}{\partial x}\Big|_{i,j}^{n} + V\frac{\partial u}{\partial y}\Big|_{i,j}^{n}\right) + \frac{h}{2}\left(U\frac{\partial^2 u}{\partial x^2}\Big|_{i,j}^{n} + V\frac{\partial^2 u}{\partial y^2}\Big|_{i,j}^{n}\right) - \frac{1}{\rho}\left(\frac{\partial p}{\partial x}\Big|_{i,j}^{n}\right)$$
$$+\left(-\frac{h}{2\rho}\frac{\partial^2 p}{\partial x^2}\Big|_{i,j}^{n} - \frac{\Delta t}{2}\frac{\partial^2 u}{\partial t^2}\Big|_{i,j}^{n} + O(h^2) + O(\Delta t^2)\right) \tag{11.17}$$

To eliminate $\partial^2 u/\partial t^2$ term in (11.17), (11.6) can be converted to

$$\frac{\partial^2 u}{\partial t^2} = \frac{\partial f}{\partial t} - \left(U\frac{\partial}{\partial x}\left(\frac{\partial u}{\partial t}\right) + V\frac{\partial}{\partial y}\left(\frac{\partial u}{\partial t}\right)\right) - \frac{1}{\rho}\frac{\partial}{\partial x}\left(\frac{\partial p}{\partial t}\right) \tag{11.18}$$

Substituting (11.6) into (11.18) yields

$$\frac{\partial^2 u}{\partial t^2} = \frac{\partial f}{\partial t} - \left(U\left(\frac{\partial f}{\partial x} - U\frac{\partial^2 u}{\partial x^2} - V\frac{\partial^2 u}{\partial x \partial y} - \frac{1}{\rho}\frac{\partial^2 p}{\partial x^2}\right)\right.$$
$$\left. + V\left(\frac{\partial f}{\partial y} - U\frac{\partial^2 u}{\partial x \partial y} - V\frac{\partial^2 u}{\partial y^2} - \frac{1}{\rho}\frac{\partial^2 p}{\partial y^2}\right)\right) - \frac{1}{\rho}\frac{\partial^2 p}{\partial x \partial t}$$
$$= \left(U^2\frac{\partial^2 u}{\partial x^2} + V^2\frac{\partial^2 u}{\partial y^2}\right) + \left(2UV\frac{\partial^2 u}{\partial x \partial y} + \frac{\partial f}{\partial t} - U\frac{\partial f}{\partial x} - V\frac{\partial f}{\partial y}\right) + \frac{U}{\rho}\frac{\partial^2 p}{\partial x^2}$$
$$+ \frac{V}{\rho}\frac{\partial^2 p}{\partial y^2} - \frac{1}{\rho}\frac{\partial^2 p}{\partial x \partial t} \tag{11.19}$$

Substituting (11.19) into (11.17) to eliminate $\partial^2 u/\partial t^2$ will provide

$$\frac{\partial u}{\partial t}\Big|_{i,j}^{n} = f_{x;i,j} - \left(U\frac{\partial u}{\partial x}\Big|_{i,j}^{n} + V\frac{\partial u}{\partial y}\Big|_{i,j}^{n}\right)$$
$$+\left(\frac{U}{2}(h - U\Delta t)\frac{\partial^2 u}{\partial x^2}\Big|_{i,j}^{n} + \frac{V}{2}(h - V\Delta t)\frac{\partial^2 u}{\partial y^2}\Big|_{i,j}^{n}\right) - \frac{1}{\rho}\left(\frac{\partial p}{\partial x}\Big|_{i,j}^{n}\right)$$
$$+\left(\frac{(-h - U\Delta t)}{2\rho}\frac{\partial^2 p}{\partial x^2}\Big|_{i,j}^{n} - \frac{V\Delta t}{2\rho}\frac{\partial^2 p}{\partial y^2}\Big|_{i,j}^{n} + \frac{\Delta t}{2\rho}\frac{\partial^2 p}{\partial x \partial t} - \frac{\Delta t}{2}\right.$$
$$\left.\times\left(\frac{\partial f}{\partial t} - U\frac{\partial f}{\partial x} - V\frac{\partial f}{\partial y}\right) + O(h^2) + O(\Delta t^2)\right) \tag{11.20}$$

Thus, by solving (11.7) in CFD, the following differential equation is really solved:

$$\frac{\partial u}{\partial t} = f_x - \left(U\frac{\partial u}{\partial x} + V\frac{\partial u}{\partial y}\right) + \left(\frac{U}{2}(h - U\Delta t)\frac{\partial^2 u}{\partial x^2} + \frac{V}{2}(h - V\Delta t)\frac{\partial^2 u}{\partial y^2}\right) - \frac{1}{\rho}\left(\frac{\partial p}{\partial x}\right)$$
$$+\left(\frac{(-h - U\Delta t)}{2\rho}\frac{\partial^2 p}{\partial x^2} - \frac{V\Delta t}{2\rho}\frac{\partial^2 p}{\partial y^2} + \frac{\Delta t}{2\rho}\frac{\partial^2 p}{\partial x \partial t} - \frac{\Delta t}{2} \times \left(\frac{\partial f}{\partial t} - U\frac{\partial f}{\partial x} - V\frac{\partial f}{\partial y}\right)\right.$$
$$\left. + O(h^2) + O(\Delta t^2)\right) \tag{11.21}$$

This brings in the unphysical viscosity that does not exist in the original Euler equation (Equation (11.6)). The numerical/artificial viscosities for x and y directions are, respectively,

$$\nu_x = \frac{U}{2}(h - U\Delta t) = \frac{Uh}{2}\left(1 - \frac{U\Delta t}{h}\right) \tag{11.22}$$

$$\nu_y = \frac{V}{2}(h - V\Delta t) = \frac{Vh}{2}\left(1 - \frac{V\Delta t}{h}\right) \tag{11.23}$$

The same analysis can be performed for Equation (11.8) with negative velocity. Wang and Zhai (2012a, 2012b) conducted comprehensive theoretical analysis for other numerical schemes to understand the fundamentals and mechanisms of the influence of coarse grids on numerical simulation. They further investigated the feasibility of utilizing inherent numerical viscosity induced by coarse CFD grid, coupled with simplest turbulence model, to greatly reduce the computational cost while maintaining reasonable modeling accuracy of CFD (Wang and Zhai, 2012a, 2012b; Wang *et al.*, 2014). Their studies provided guidelines on coarse-grid CFD simulation. As a rule of thumb, assuming that the characteristic length of geometry under investigation is L, the grid size of 1 percent L is recommended for the local refinement areas, while for other areas and directions, 10 percent L is suggested. The geometry height is usually a good representative of characteristic length as indicated by different indoor cases tested. With at most around 5 percent of the original computing cost, the optimized coarse grid according to the guidelines can have comparable numerical results as grid-independent solutions. The improved computing speed at about 100 times faster than a fine-grid CFD makes it possible to simulate a complicated three-dimensional built environment in real time (or near real time) with a personal computer.

11.4.3 Velocity–pressure decoupling algorithm

Velocity–pressure decoupling algorithm is critical for solving highly coupled Navier–Stokes equations. A semi-implicit method for pressure-linked equations (SIMPLE) (Patankar and Spalding, 1972) algorithm and different modified versions of such algorithms, such as SIMPLER (SIMPLE Revised) (Patankar, 1980), SIMPLEC (SIMPLE Consistent) (Van Doormaal and Raithby, 1984), SIMPLEM (SIMPLE modified) (Acharya and Moukalled, 1989), and SIMPLEX (SIMPLE extrapolation) (Van Doormaal and Raithby, 1985), are mainly used for steady-state simulation and have been successfully applied to popular commercial CFD codes such as Fluent, Phoenics, Star-CD, and open-source code such as OpenFOAM.

They are applicable to transient simulation as well by splitting a simulation into each time step and executing iterations in each time step to obtain converged solutions. Within each iteration, numeric operation is conducted for a large data set that contains the information for every grid node. This makes the SIMPLE algorithm and its variants not computationally economical.

Other algorithms designed for transient CFD simulation applications include the PISO (pressure-implicit with splitting of operators) (Issa, 1986) algorithm, projection method (Harlow and Welch, 1965), and fractional time step methods (Kim and Moin, 1985), which require no iteration within the selected time step.

PISO is a non-iterative algorithm developed for transient flow simulation and thus provides great potential for decreasing computing cost. It literally reaches the exact solution of the discretized equation through two stages of predictor–corrector and is much faster than any iteration-based algorithm such as SIMPLE. For steady-state simulation, PISO has no advantage over SIMPLE in computing cost when temperature is closely linked to velocity (Jang *et al.*, 1986). PISO can be much faster for calculating isothermal flow (Audi, 2009); as for a velocity–temperature strongly coupled problem, PISO can be four times slower and gives correct solutions only when the time step is small. An improved PISO for buoyancy-driven flow (Oliveira and Issa, 2001) may stress this problem.

Besides the coupling between velocity and pressure, the velocity components along each direction are also highly coupled in the momentum equations. Foster and Metaxas (1996, 1997) implemented the projection method (Chorin, 1997) to simulate the 3D motion of hot, turbulent gas using a relatively coarse grid. Stam (1999) proposed using semi-Lagrangian advection and fast Fourier transformation to speed up the computation to a real-time or faster-than-real-time level. Zuo and Chen (2009) gave this algorithm a name Fast Fluid Dynamics (FFD), and first applied this operator splitting algorithm to indoor environment modeling. They improved the sequence of operators, tested higher orders of differencing schemes, and evaluated the accuracy levels. Semi-Lagrangian scheme (Courant *et al.*, 1952) used in FFD not only overcomes such coupling but also escapes the restriction of CFL condition (Courant *et al.*, 1928) on a time step, which is especially beneficial for transient flow simulation.

Xue *et al.* (2016) proposed a new semi-Lagrangian-based PISO (SLPISO) method for fast and accurate indoor environment modeling. Figure 11.2 demonstrates the general principle and procedure of the SLPISO algorithm. The proposed SLPISO algorithm, without the corrector steps, is similar to FFD except that it takes into consideration the pressure field from the previous time step. The FFD algorithm neglects the influence of pressure from the previous time step and assumes that pressure is solely determined by the velocity field under the continuity restriction. In FFD, the advection term is completely separated from the rest of the momentum equation and is solved by using the semi-Lagrangian algorithm, which is faster and more stable compared to the conventional method of directly solving the advection equation. But the accuracy of PISO, theoretically and practically, has more advantages over FFD. The integrated algorithm (SLPISO) is expected to improve the accuracy of FFD without sacrificing much computing speed. The

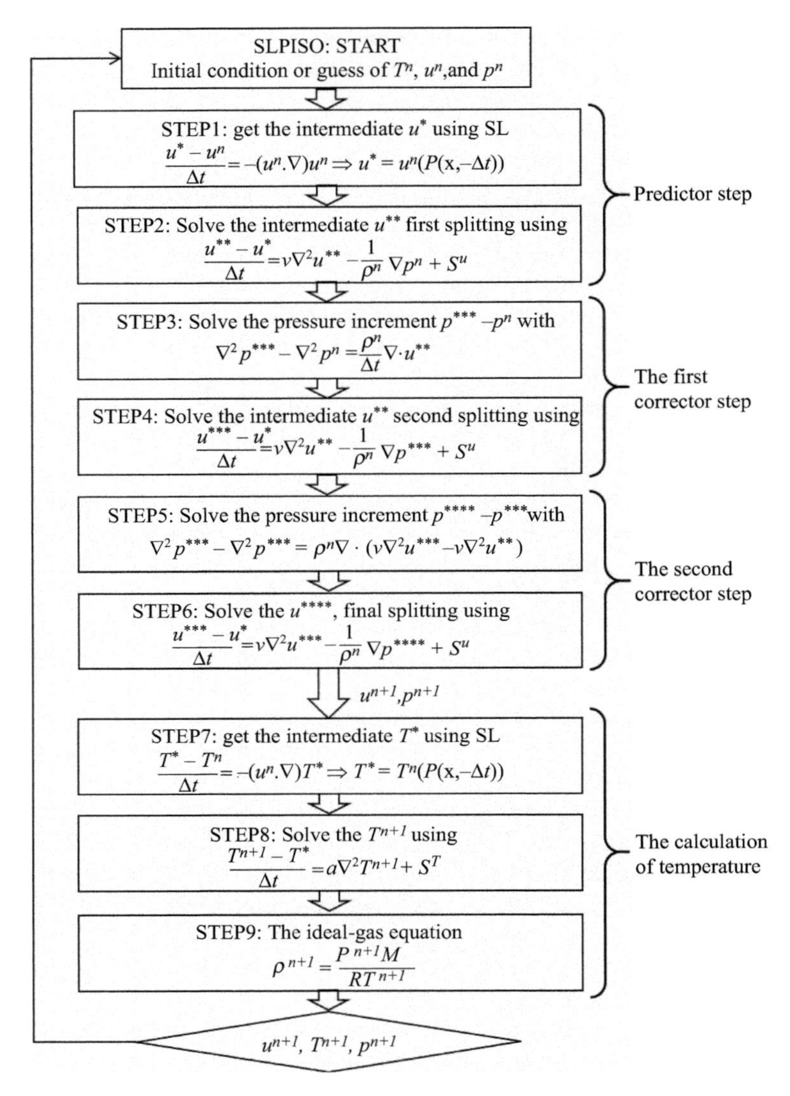

Figure 11.2 Operation sequence of semi-Lagrangian PISO

semi-Lagrangian advection algorithm is anticipated to largely reduce the computing cost of the direct solving of the advection term in the original PISO algorithm.

11.4.4 Reduced-order model

ROM has the potential of quantitatively describing the dynamics of systems at a computational cost much lower than the original numerical model. ROMs thus provide a means by which system dynamics can be readily interpreted (Lucia *et al.*, 2004).

ROMs generate the systematic cost-efficient representations of large-scale systems. In many unsteady CFD applications, a small number of inputs and outputs of interest can be identifed, and computationally efficient ROMs can be obtained that preserve the desired input–output mapping (Bui-Thanh *et al.*, 2008).

For nonlinear, time-dependent problems in CFD, ROMs are typically based on the proper orthogonal decomposition (POD) (Borggaard *et al.*, 2007) combined with the Galerkin projection (Antoulas, 2005). The POD uses principle component analysis to decompose large systems into a series of fundamental modes; the Galerkin projection method is to project the governing equations onto modal subspace. There are many successful applications of this model in related areas such as building contaminant transport (Surana *et al.*, 2008) and incompressible jet flow (Hou, 2010). A successful application of this method on data center thermal management (Rambo, 2007), as shown in Figure 11.3, was able to reduce numerical models containing 103–105 DOF down to less than 20 DOF while still retaining greater than 90 percent accuracy over the domain.

There are two other widely used ROM approaches, called the Volterra theory of nonlinear systems (Schetzen, 1980) and harmonic balance formulation (Hall *et al.*, 2002), which are prevalent approaches with many successful applications to fluid dynamics problems (Lucia *et al.*, 2004).

Substantial challenges, however, need to be overcome before ROM methods can be routinely applied to practical problems, in the three categories: construction, generality, and accuracy assessment (Lucia *et al.*, 2004). While most ROMs can operate in near real time, their construction can however be computationally expensive as it requires accumulating a large number of system responses to input excitations. ROMs usually lack robustness with respect to parameter changes and therefore must often be rebuilt for each parameter variation (Amsallem and Farhat, 2008).

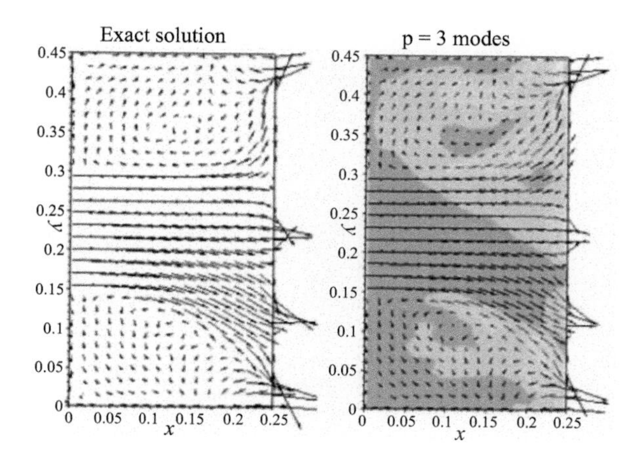

Figure 11.3 An example of successful application of reduced-order modeling with POD to a flow system. The contour indicates an error range of 0–0.08

Accuracy assessment is also an important topic dealing with the ability to evaluate the quality, inherently or posteriori, of ROM solutions.

Specifically, in indoor environment-related fields, the ROMs serve building environment control purposes and building control systems well (Borggaard *et al.*, 2012), but the character of ROMs and these challenges constitute a great barrier for its application in general built environment simulation. A simple input–output relationship model does not provide informative data required for thermal and airflow environment analysis.

11.5 Demonstration of fast simulation techniques for ventilation study

11.5.1 Zonal model simulation

IAQ and ventilation in built environments can be readily studied by using a zonal model simulation tool such as CONTAM that is developed by the US National Institute of Standards and Technology. The following example demonstrates the simulation and analysis of ventilation and filtering performance in a typical residential single-family house, built according to the 2012 International Energy Conservation Code (IECC) residential prototype building model, with associated building characteristics corresponding to ASHRAE climate zone 4A (the test location).

11.5.1.1 Building model

The two-story residential building has one lump-sum window on each wall and one door on the south and north wall of the first floor, respectively (Figure 11.4). Table 11.1 summarizes the building dimensions and the designed mechanical system airflow rates. A whole-building mechanical system supplies 0.2063 m^3/s of air to each

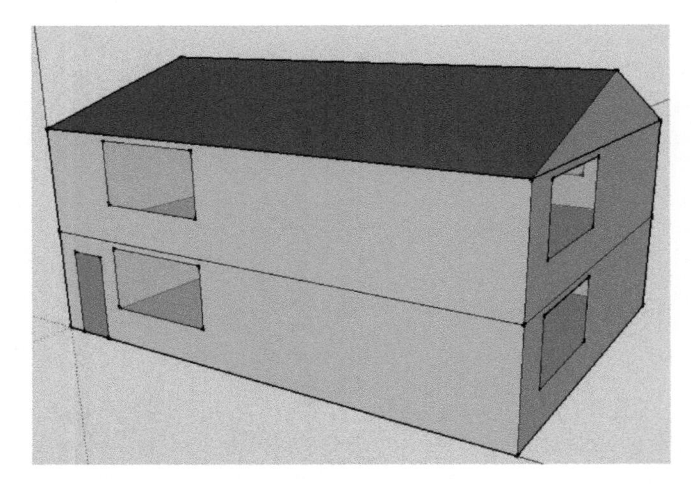

Figure 11.4 Graphic view of the 2012 IECC compliant residential prototype building model

Table 11.1 Dimensions and ventilation flow rates of 2012 IECC compliant residential prototype building

Dimension	**9.144 × 12.192 m**	**Floor to ceiling height: 2.6 m each floor**
Windows	2.7 × 1.5 m	One on each wall
Door	0.9 × 2.1 m	One on north wall and one on south wall
System air-flow	0.413 m^3/s	Mechanical system supplies 0.2063 m^3/s of air to each floor
OA intake	0.0283 m^3/s	

Table 11.2 2012 IECC residential building model air-tightness parameters

	Leakage settings	**Note**
Wall	1.2 cm^2/m^2 (total 265.33 cm^2)	
Window	0.28 cm^2/m (total 19 cm^2)	Each wall has a same-sized window each floor has 4 windows (total 8 windows)
Door	0.45 cm^2/each (total 0.9 cm^2)	The first floor has two doors on the north and south wall
Stairs	75 cm^2/each (total 75 cm^2)	One indoor stair connects the first and second floor
Ceiling	1.8 cm^2/m^2 (total 200.67 cm^2)	

floor with a total of 0.0283 m^3/s of outdoor air (OA) (i.e., 6.87 percent of the total air-supply rate) (International Code Council, 2012). The study also simulates 15 and 30.02 percent OA ratio as the comparison. Building air tightness condition is critical for analyzing IAQ and ventilation performance. The 2012 IECC requires that residential buildings in the ASHRAE climate zone 4A can allow infiltration or exfiltration up to 3 air change rate per hour (ACH) at 50 Pa (International Code Council, 2012). This study uses 2.4 ACH at 50 Pa for demonstration and Table 11.2 lists the leakage area settings for the CONTAM model.

Figure 11.4 shows the single-zone CONTAM model, where walls are represented with solid lines and leakage areas with hollow circles. The first and second floor, respectively, has ten and eight leakages. Indoor air can move between the first and second floors through the indoor stairs and the ceiling leakage. Ventilation ductworks, which are shown as blue lines in Figure 11.5, supply OA to the indoors, as well as circulating air to both the first and second floors.

11.5.1.2 Ventilation system model

The building is pressurized by a mechanical ventilation system as sketched in Figure 11.6. Most of the indoor air is recirculated, and the rest of the air is exfiltrated through leakages at the building envelopes. Figure 11.7 shows the mechanical system in the CONTAM model. The OA fan introduces OA through the

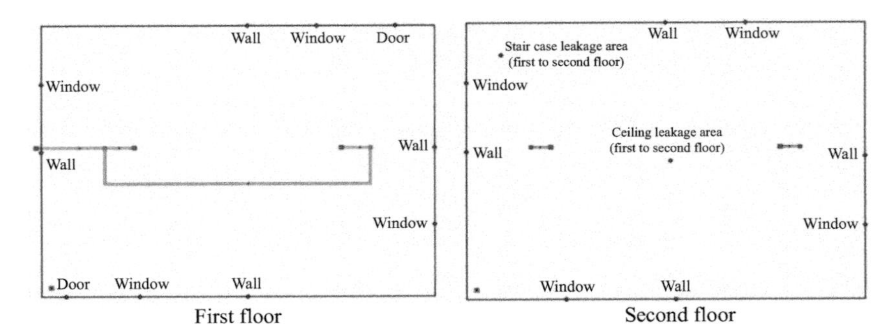

Figure 11.5 CONTAM model with leakage area descriptions for each floor

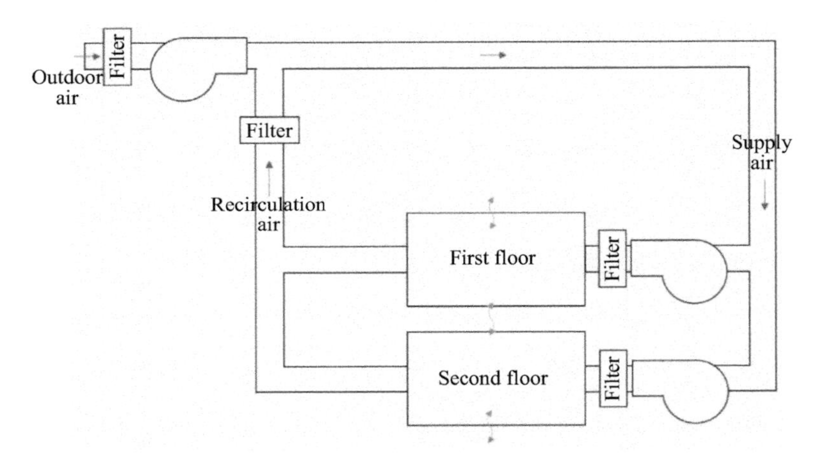

Figure 11.6 Mechanical system and airflow in the building

opening, "OA intake," on the first floor. The amount of OA can be adjusted by the OA fan that is located just next to the opening. The introduced OA is mixed with the recirculated air and distributed to the first and second floors. The amount of air supplied to each floor is controlled by supply fans that are located just before the first and second floors.

Filter rating, minimum efficiency reporting value (MERV), affects the quality of filtered air and thus IAQ. ASHRAE (2004) recommends using MERV 6 or higher rated filters for an effective removal of PM10 and using MERV 11 or higher rated filters for an effective removal of PM2.5. This study focuses on the removal of PM2.5 particles and therefore MERV 11 filters are tested in the simulation. In addition to the MERV 11 filters, MERV 6 and 15 filters are also simulated for comparison. Table 11.3 presents the PM2.5 removal efficiency of the three MERV filters from the MERV Curve Filters library retrieved from the CONTAM website

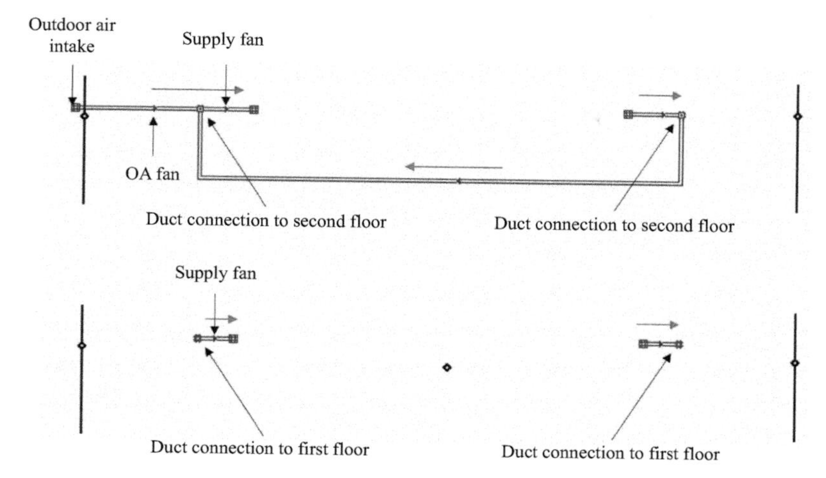

Figure 11.7 *Whole-building mechanical ventilation system in the CONTAM model*

Table 11.3 *PM2.5 removal efficiency of MERV-rated filters*

	PM2.5 removal efficiency (%)
MERV 6	39.0
MERV 11	87.8
MERV 15	99.9

(link: http://www.bfrl.nist.gov/IAQanalysis/CONTAM/libraries.htm). Three filter installation locations are tested in the simulation:

- OA filter: to filter outdoor particles;
- recirculation filter: to filter room particles;
- supply filter: to filter conditioned air from heating/cooling coils.

As a demonstration, this study only tests the filtering performance of single-filter installation; without considering serial filter installations that can install air filters at two or three locations in the system as commonly occur in commercial buildings.

The study also compares the performance of room air purifiers. Air purifiers with high-efficiency particulate air filters at 99.97 percent removal efficiency are tested. The required air purifier size of the tested residential building is calculated based on China GB/T 18801-2015 Standard (Lu, 2015). The standard recommends using 928.916–1,592.428 m^3/h of air purifiers for each floor. This study utilizes 1,019.406 m^3/h of room air purifiers for each floor.

For simplification, the follow assumptions are made to the simulation:

Three occupants: including two parents and one child, with two bedrooms in the house.

Density of particle: 1,200 kg/m^3 for all particles in the air.

Density of air: 1.2042 kg/m^3.

No particle generation in the house: particles only from outdoor environment.

11.5.1.3 Weather and pollution data

Local weather and pollutant data (i.e., from Beijing) are obtained and applied to investigate the potential influence of PM2.5 on IAQ through ventilation and filtration. This study simulates January and July as the peak heating and cooling season. Figure 11.8 shows the PM2.5 concentration measured at Dongsi, Beijing in January and July of 2015, which are obviously above the 12 μg/m^3 threshold as regulated by US EPA and 35 μg/m^3 by China.

11.5.1.4 Results and analysis

Heating season

Figure 11.9(a) presents and compares the whole-building monthly average PM2.5 removal efficiency for different filtering scenarios. Figure 11.10(a) shows the total number of hours that meet EPA Primary Air Quality Standard in the heating season. ASHRAE 62.2 recommends the use of MERV 11 or higher filters for effective removal of PM2.5. The results verify the noticeable PM2.5 concentration difference between the cases using MERV 11 and MERV 6 filters. However, there is a minor improvement between the cases using MERV 11 and MERV 15 filters, and thus a marginal gain of IAQ improvement with higher MERV than 11.

When the amount of OA delivered through the whole-building mechanical system increases, using supply air (Sup) filters or OA filters is a better option than using recirculation (Rec) filters or indoor air purifiers. As expected, OA filters and Sup filters are more effective as OA supply increases, while indoor air purifiers or Rec filters are more effective as indoor/recirculated air is dominant. A Sup filter shows an overall better performance than the others, attributed to the good airtightness of the space as regulated.

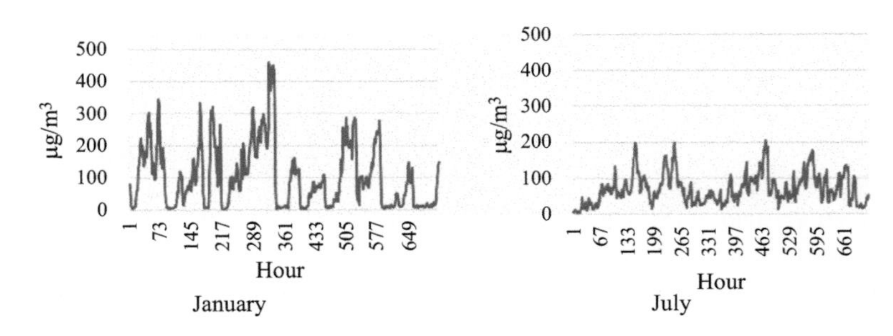

Figure 11.8 January (peak heating) and July (peak cooling) ambient PM2.5 concentration of 2015 in Beijing

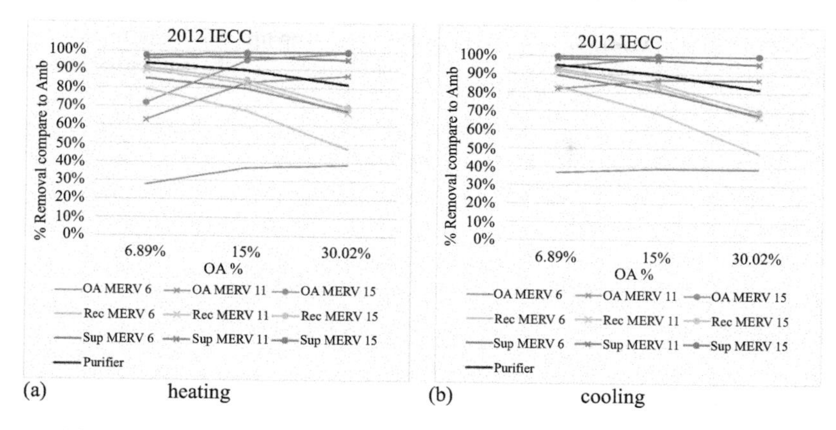

Figure 11.9 Average PM2.5 removal efficiency under different filtering scenarios in (a) heating and (b) cooling season

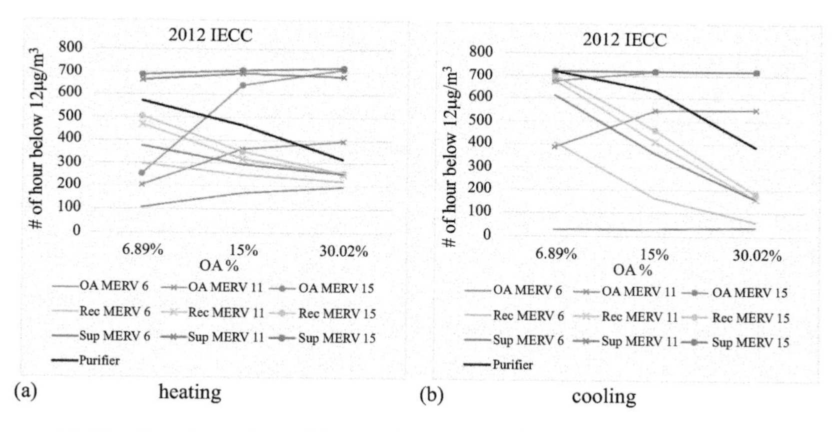

Figure 11.10 Total number of hours that meet EPA primary air quality standard in (a) heating and (b) cooling season

The following design and operation implications can thus be obtained to achieve a healthful IAQ during the polluted heating days: (1) installing a high MERV filter at supply location; (2) improving building airtightness; (3) using OA or a supply air filter when delivering a large amount of OA through whole-building mechanical system; and (4) using an indoor air purifier or a recirculated air filter for well-sealed buildings.

Cooling season

The performance of air filters and indoor air purifiers in the cooling season shows similar trends to those in the heating season (Figures 11.9(b) and 11.10(b)). Filters installed at the supply air location demonstrate the best overall performance. The average PM2.5 removal efficiency in the cooling season is generally better than the heating season. For example, MERV 15 filters at the supply air location remove

98.92 percent of PM2.5 during the heating season, while 99.83 percent of PM2.5 removed during the cooling season. Compared to the EPA standard, significant differences are observed between the heating and cooling seasons, because ambient PM2.5 concentrations in the cooling and heating season are different as shown in Figure 11.8. MERV 15 filters at the supply air location can maintain the building within the EPA standard for 704.83 and 720.00 h, respectively, during the tested heating and cooling period.

11.5.2 Zero-equation turbulence model

Displacement ventilation is an advanced indoor ventilation approach. Unlike the conventional mixing ventilation, displacement ventilation provides a cleaner indoor environment with less energy consumption. A typical displacement ventilation system supplies fresh air at or near floor level at a very low velocity and a temperature slightly below room temperature. Exhausts are located at or near the ceiling. The supply air spreads across the floor and rises as it is heated by sources such as people and equipment (Figure 11.11), removing indoor heat and contaminants directly from the occupied zone to the upper zone without mixing. Since only the occupied zone must be maintained at the room set-point temperature while the upper zone may be warmer, the supply airflow rate can be significantly reduced due to the vertical temperature gradient, resulting in reduced fan energy.

However, the complexity added by this nonuniform air temperature distribution makes displacement ventilation systems more difficult to model and design than mixing ventilation systems. Yuan *et al.* (1998) provide a review of many issues and models associated with displacement ventilation. This study simulates a typical displacement ventilation case by using two different turbulence models: the standard k–ε model and the zero-equation model in (11.5), compared with the laboratory test results (Yuan *et al.*, 1999). Figure 11.12, Tables 11.4 and 11.5 provide detailed information about the geometrical, thermal, and flow conditions of the case.

The computational grid is 55 × 37 × 29, which is sufficient for obtaining the grid-independent solution according to the investigation of Srebric (2000). Figure 11.13(a) shows the calculated air velocity and temperature distributions in the middle section of

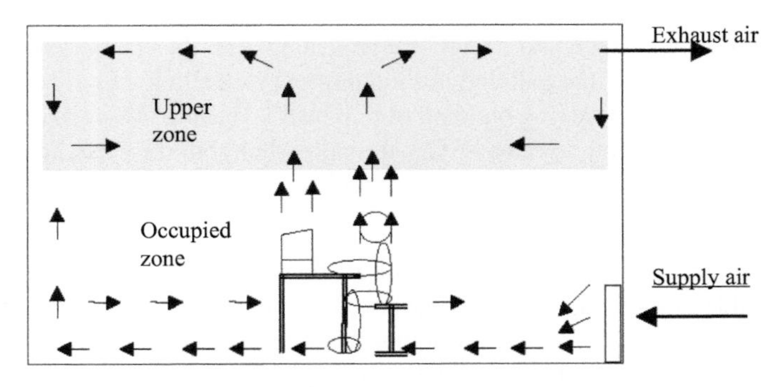

Figure 11.11 Schematic of conventional displacement ventilation

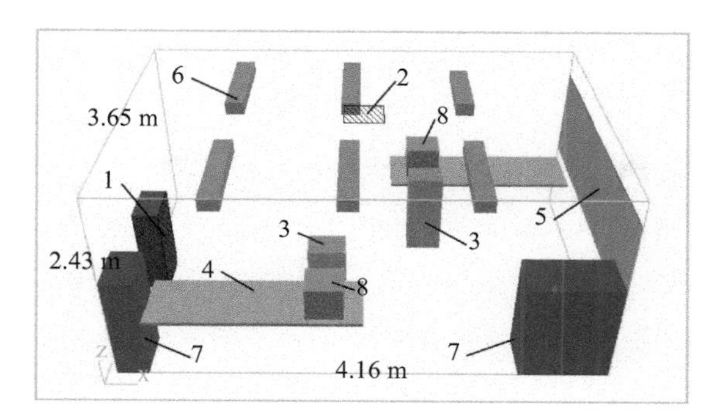

Figure 11.12 The layout of the displacement ventilation case (inlet—1, outlet—2, person—3, table—4, window—5, fluorescent lamps—6, cabinet—7, computer—8)

Table 11.4 The geometrical, thermal, and flow conditions for the diffuser and window

Displacement ventilation case			
Inlet diffuser	Size: 0.53 × 1.1 m	Temperature: 17.0 °C	Velocity: 0.086 m/s
Window	Size: 3.65 × 1.16 m	Temperature: 27.7 °C	Closed

Table 11.5 The size and capacity of the heat sources

Heat source	**Size (m^3)**	**Power (W)**
Person	0.4 × 0.35 × 1.1	75
Computer 1	0.4 × 0.4 × 0.4	108
Computer 2	0.4 × 0.4 × 0.4	173
Overhead lighting	0.2 × 1.2 × 0.15	34

the room with the zero-equation model. The solutions with the standard k–ε model are fairly similar. The computed results are in very good agreement with the flow pattern observed by smoke visualization, as illustrated in Figure 11.13(b). The large recirculation in the lower part of the room, which is known as a typical flow characteristic of displacement ventilation, is well captured by the CFD simulation. The airflow and temperature patterns in the respective sections across a person and a computer, as shown in Figure 11.13(c) and (d), clearly exhibit the upward thermal plumes due to the positive buoyancy from the heat sources.

The study further compares the measured and calculated velocity and air temperature profiles at nine pole locations where detailed measurements were carried out. The pole locations are illustrated in the lower right pictures of Figures 11.14 and 11.15. Figure 11.14 shows the comparison of measured and calculated air velocities

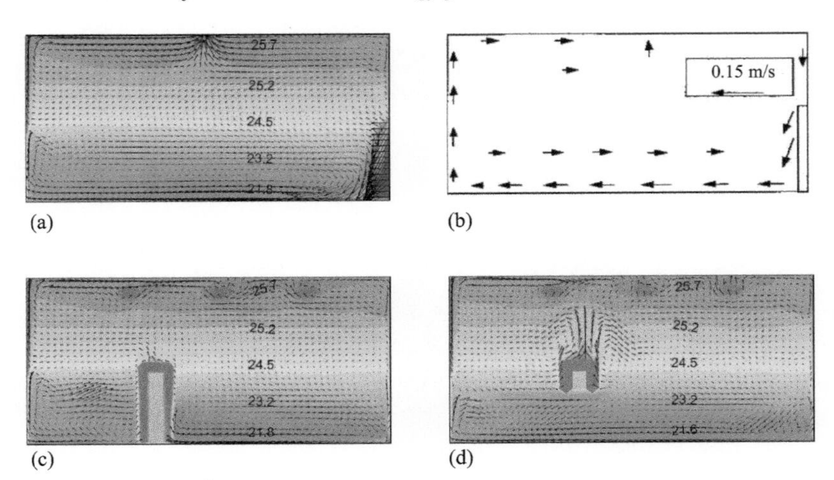

Figure 11.13 *Velocity and temperature distributions for the displacement ventilation case: (a) calculated results in the middle section, (b) observed airflow pattern with smoke visualization in the middle section, (c) calculated results in the section across a computer, and (d) calculated results in the section across an occupant*

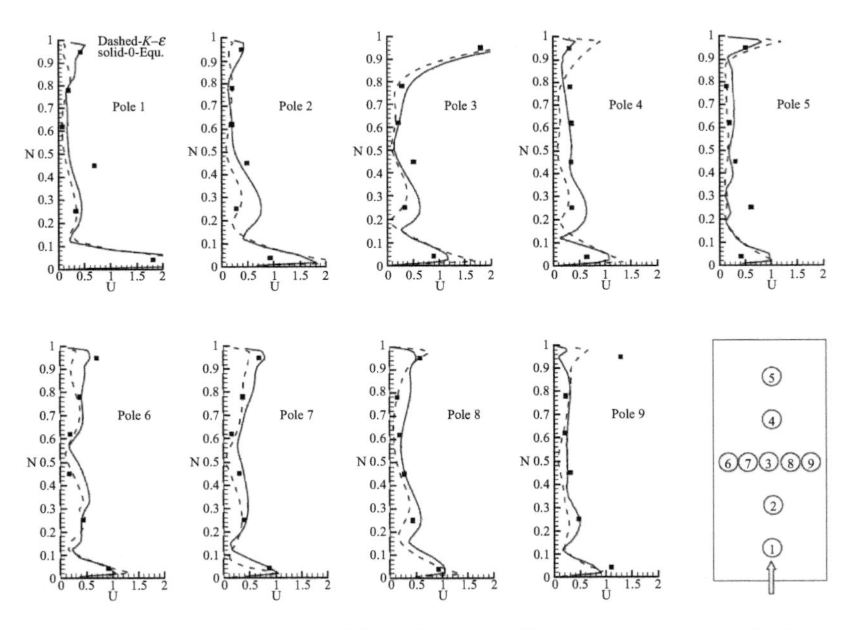

Figure 11.14 *The comparison of the velocity profiles at nine positions in the room between the calculated and measured data for the displacement ventilation case: Z = height/total room height (H), V = velocity/ inlet velocity (V_{in}), H = 2.43 m, V_{in} = 0.086 m/s*

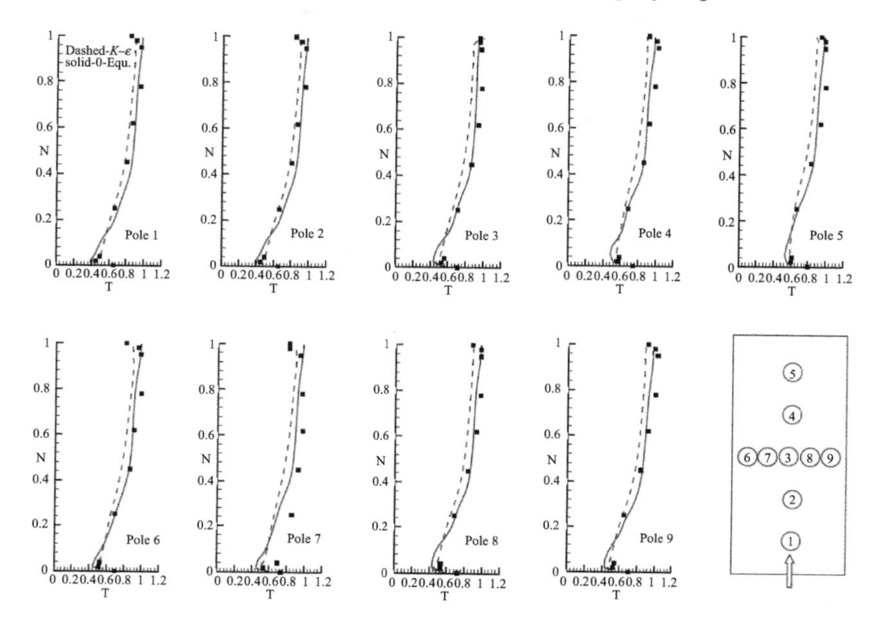

Figure 11.15 The comparison of the temperature profiles at nine positions in the room between the calculated and measured data for the displacement ventilation case: Z = height/total room height (H), $T = (T_{air} - T_{in}/T_{out} - T_{in})$, H = 2.43 m, T_{in} = 17.0 °C, T_{out} = 26.7 °C

along the nine vertical poles and Figure 11.15 shows the comparison of air temperatures. Both the figures have a very good agreement between the computed and measured results, indicating that the turbulence models have good capability to predict this kind of indoor airflow. The computing time with the zero-equation model is significantly less than that with the standard *k*–ε model due to (1) two less PDEs solved and (2) faster convergence (fewer iterations).

11.5.3 Coarse-grid simulation

A buoyancy-driven natural ventilation room (Jiang and Chen, 2003) is demonstrated as the first coarse-grid simulation case. The detailed configuration of this experiment is in Figure 11.16. The test chamber adjacent to the environmental chamber had a heater inside. The chamber system was put inside a larger room. The chambers are well insulated, and the wall is supposed to be adiabatic. Air velocity and temperature were measured along five vertical poles marked as P1–P5. In the CFD model, only the test chamber is modeled, so the data for comparison in this study is on P2 to P5.

According to the general rule of coarse-grid specification, local refinement applies to heat source/sink for this buoyancy-driven natural ventilation. Figure 11.17 shows the cell distribution of an optimized coarse grid for this case

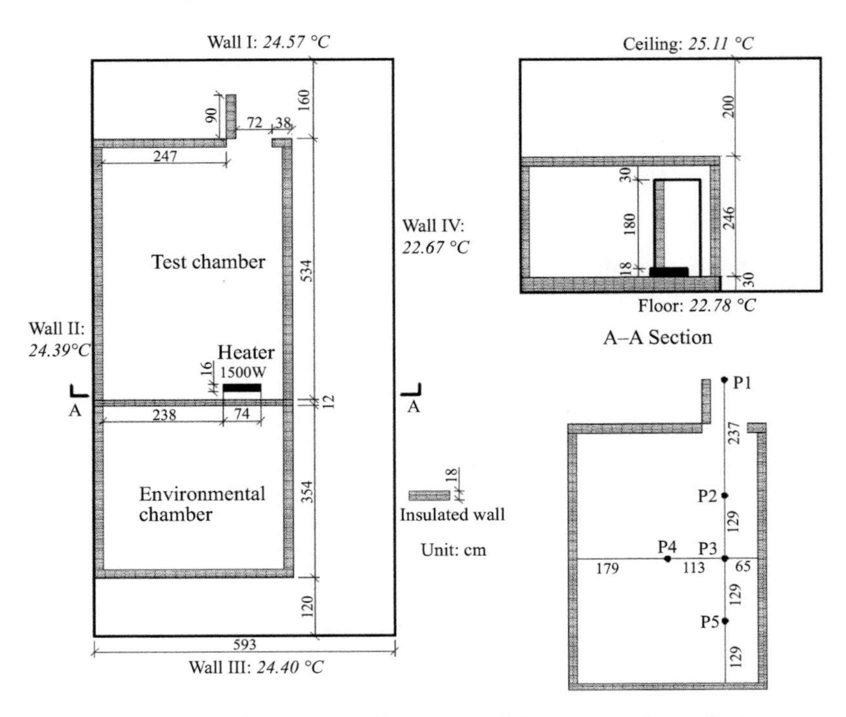

Figure 11.16 Configuration of buoyancy-driven natural ventilation case

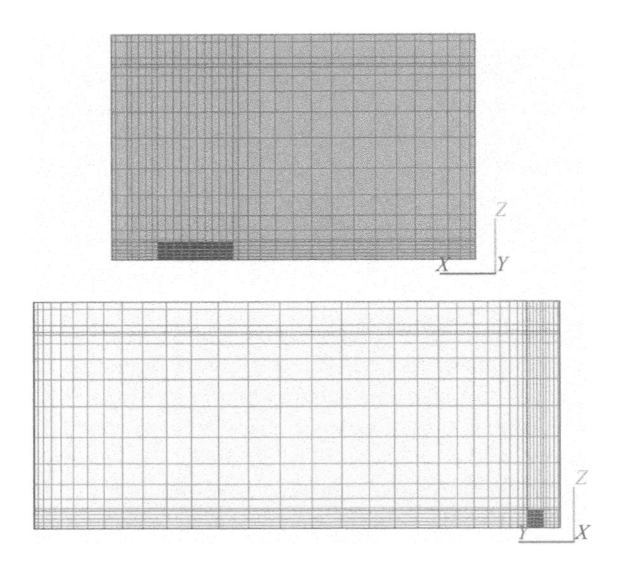

Figure 11.17 Distribution of optimized coarse grid for buoyancy-driven natural ventilation case

Table 11.6 Grid resolutions and normalized RMSE results compared to experimental data for natural ventilation case

Grid index	**Grid number ($X \times Y \times Z$)**	**Computing cost (%)**	**Normalized RMSE to experimental data**	
			V	T
Grid independent	$80 \times 78 \times 50$	$t = 100$	0.3989	0.1400
Optimized coarse	$33 \times 40 \times 19$	$t = 5.0$	0.4033	0.2043

from the view of two vertical intersections. Because of radiation, all inner surfaces act as heat sources, and local refinement normal to each surface thus applies.

The measured and predicted temperatures are normalized by exhaust air temperature and surrounding (environment) air temperature as

$$T = \frac{t - T_{environ}}{T_{exhaust} - T_{environ}} = \frac{t - 25}{33 - 25} \tag{11.24}$$

Table 11.6 summarizes the grid resolution and prediction performance of two grid resolutions in terms of computing cost and normalized root mean square error (RMSE) value compared to experimental data. With about 5 percent of the original computing cost, the coarse grid gives a comparable prediction to the grid-independent one. Figure 11.18 confirms the closeness of the predictions with fine and coarse grids.

The side-wall supply displacement ventilation case shown in Figure 11.12 is also used for verification and demonstration. The coarse grid is complex for this case due to the number and positions of objects. For the main heat source surfaces such as lamp, computer, and person, local refinement with about 1 percent the height of the room is applied. The same local refinement applies to the region surrounding the inlet diffuser and exhaust. The detailed distribution of the coarse grid is shown in Figure 11.19.

The measured and predicted data are normalized for further comparison. Temperature is normalized by inlet and outlet temperatures as (11.25), while velocity is normalized by equivalent inlet velocity as (11.26):

$$T = \frac{t - T_{in}}{T_{out} - T_{in}} = \frac{t - 17.0}{26.7 - 17.0} \tag{11.25}$$

$$V = \frac{vel.}{u_{inlet}} = \frac{vel.}{0.086} \tag{11.26}$$

Table 11.7 summarizes the prediction performance of the two grid resolutions in terms of computing cost and normalized RMSE. The optimized coarse grid consumes 5.9 percent of the uniform fine grid computing cost based on the same computer platform. The prediction results are close to each other. The profile

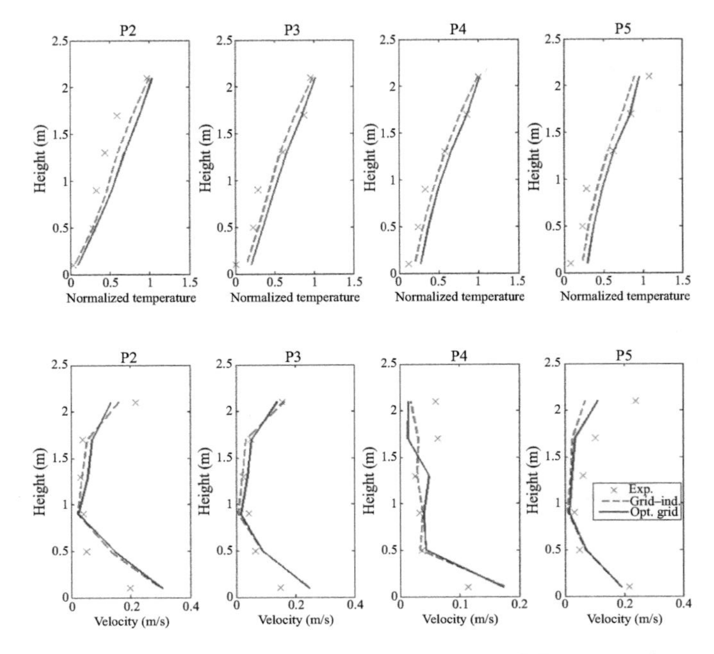

Figure 11.18 Profile comparison of predictions with different grids against experimental data for natural ventilation case

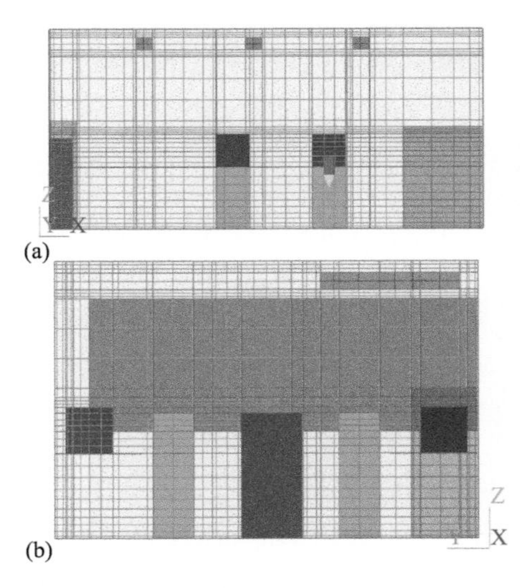

Figure 11.19 Distribution of optimized coarse grid for displacement ventilation case

Table 11.7 Grid resolutions and normalized RMSE results for displacement ventilation case

Grid index	Grid number ($X \times Y \times Z$)	Computing cost (%)	Normalized RMSE compared to experimental data			
			V-streamwise	V-cross section	T-streamwise	T-cross section
Grid independent	123 × 86 × 54	$t = 100$	0.7320	0.4301	0.0259	0.0288
Optimized coarse	48 × 44 × 32	$t = 5.9$	0.7777	0.5140	0.0207	0.0291

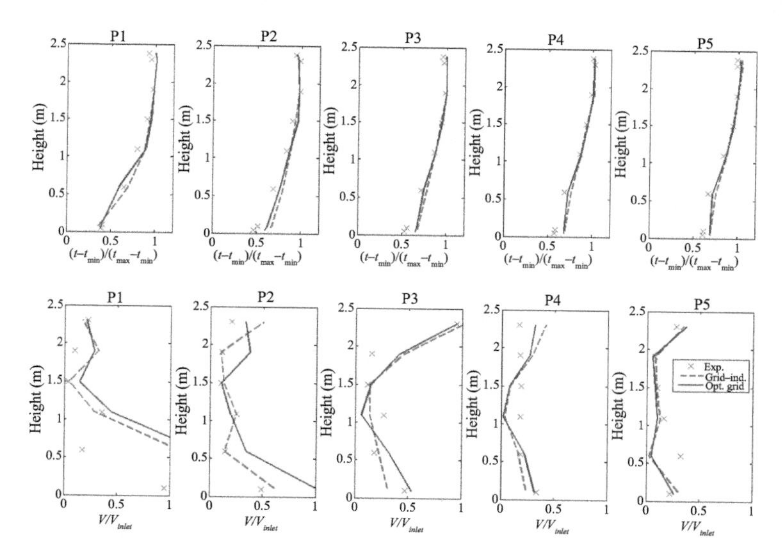

Figure 11.20 Profile comparison of predictions with different grids against experimental data for displacement ventilation case

comparison against experimental data is presented in Figure 11.20, which shows the results along the vertical poles 1–5 as demonstration. The temperature predictions are almost identical except few points.

11.5.4 Velocity–pressure decoupling algorithm

A two-dimensional mixing convection case as shown in Figure 11.21 is simulated to demonstrate the influences of different velocity–pressure decoupling algorithms on simulation accuracy and speed. Experimental results were obtained from the literature (Blay *et al.*, 1992), which were measured in a laboratory chamber of 1.04 × 1.04 × 0.7 m ($x \times y \times z$) equipped with an 18-mm wide inlet slot and a 24-mm wide outlet slot. The experiment produced a fairly good 2D flow at the

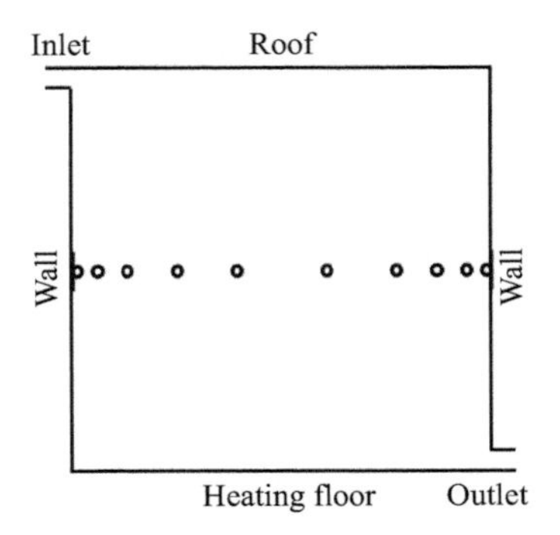

Figure 11.21 Model of the mixing convection case

central plate. The experiment measured wall temperatures and supply air conditions, respectively, as $T_{roof} = T_{walls} = 15\,°C$, $T_{floor} = 35.5\,°C$, $T_{inlet} = 15\,°C$, $V_{inlet} = 0.57$ m/s (normal to the inlet slot), as well as temperature, V_y at the ten points along the middle line on the central plate (Figure 11.21). The CFD study uses the zero-equation turbulence model in (11.4) to consider the turbulence impact. The mesh size is 80 × 80 and the time step size is 0.005 s. Results show that the SLPISO algorithm provides similar results as the PISO method. FFD has a large disparity in temperature prediction (Figure 11.22). SLPISO has similar computational speed as FFD, while they are still slower than PISO (Figure 11.23).

When the simulation increases the grid number from 80 × 80 to 300 × 300, and further to 1,000 × 1,000, the computational cost performances for these algorithms change as shown in Figure 11.24(a) and (b). As the number of grid increases, SLPISO and FFD are faster than PISO. The reason for this is the inherent characteristic of the semi-Lagrangian scheme. As the grid number increases, the computing cost of the traditional solvers, such as SIMPLE and PISO, demonstrates exponential growth trend, while the semi-Lagrangian scheme shows an almost linear growth as revealed in Figure 11.24(c). The influence of correction steps makes the calculation cost growth of FFD and SLPISO not exactly the linear though.

The previous comparison of simulation speed is under the situation of using the same time step. However, the stability analysis shows that SLPISO can tolerate a larger time step than PISO. The study uses the case with a mesh resolution of 1,000 × 1,000 to check the actual calculation speed of different solvers with different time steps. Figure 11.25 shows computing time for the largest time steps that the different solvers can handle. To reach stable and acceptable results for this case, the largest time steps are 0.02, 0.005, 0.08, and 0.1 s, for SIMPLE, PISO, SLPISO, and FFD, respectively. The gray columns in Figure 11.25 show the relative computing cost with the time step size of 0.005 s for all the solvers, using SIMPLE as the benchmark. The black columns show the relative computing cost using their own largest time steps. While the predicted results for velocity and temperature are similar to those in Figure 11.22, the modeling speeds of SLPISO and FFD with larger time steps significantly increase.

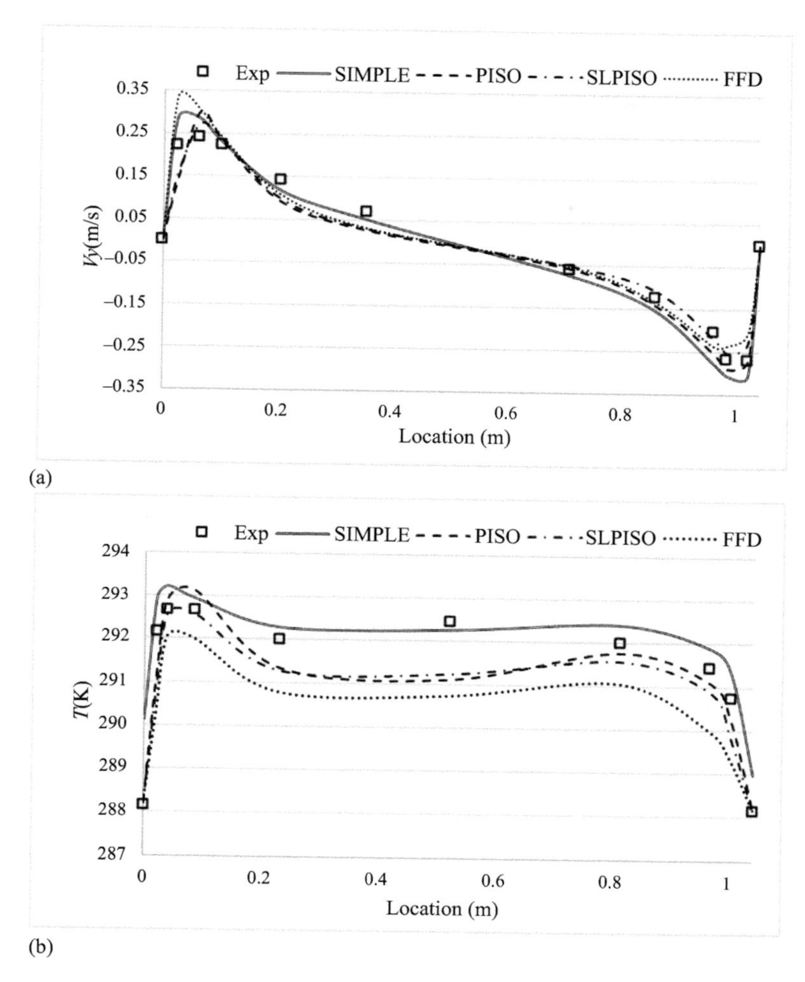

Figure 11.22 Measured and simulated results for the mixing convection case: (a) velocity comparison and (b) temperature comparison

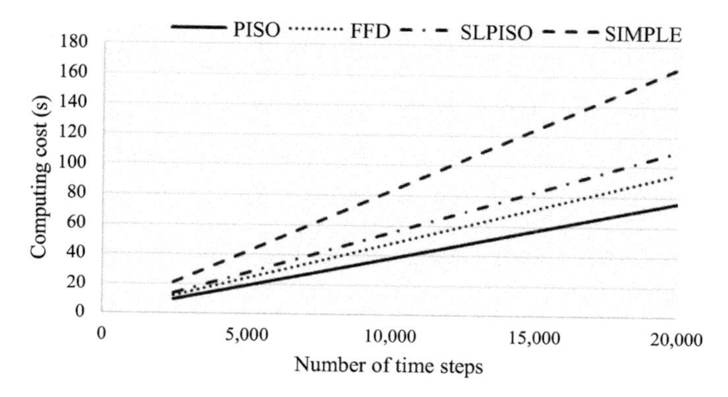

Figure 11.23 Computational cost comparison of different velocity-pressure decoupling algorithms

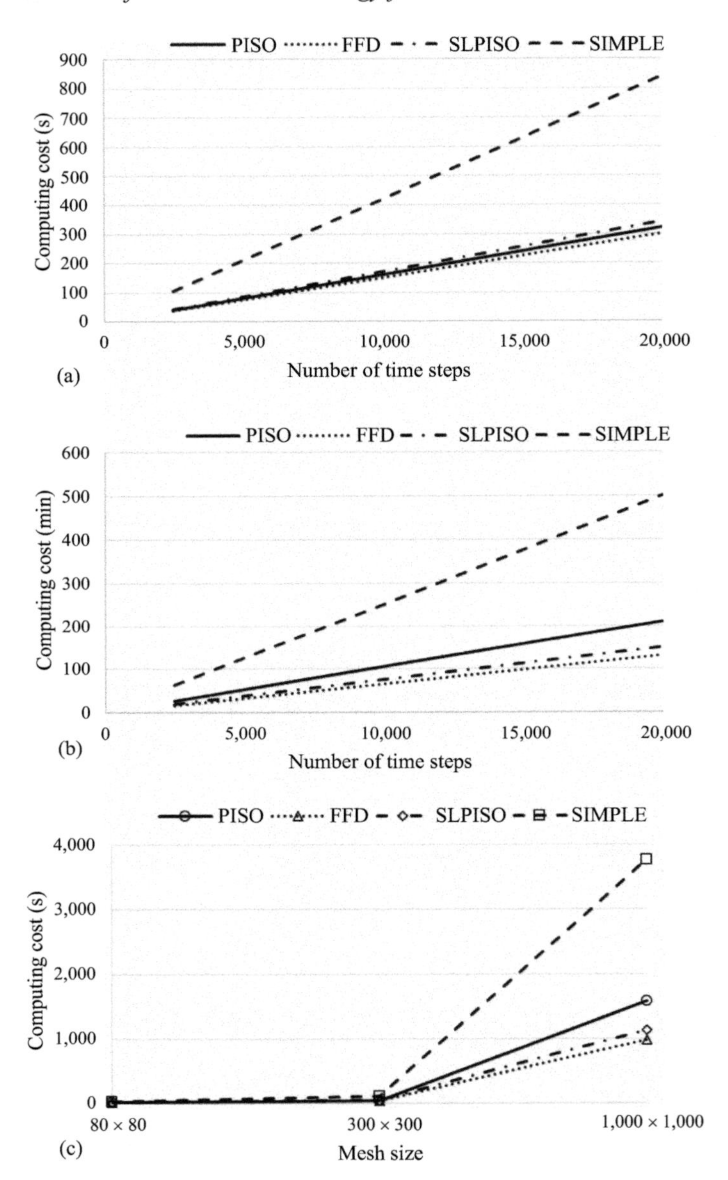

Figure 11.24 Computational cost comparison of different algorithms with different grids: (a) grid resolution of 300 × 300, (b) grid resolution of 1,000 × 1,000, (c) time steps = 2,500

To evaluate the transient simulation accuracy of SLPISO and FFD, a transient flow in the 2D mixing convection case as shown in Figure 11.21 is simulated. Since no transient experiment results exist for this case, the SIMPLE algorithm results are

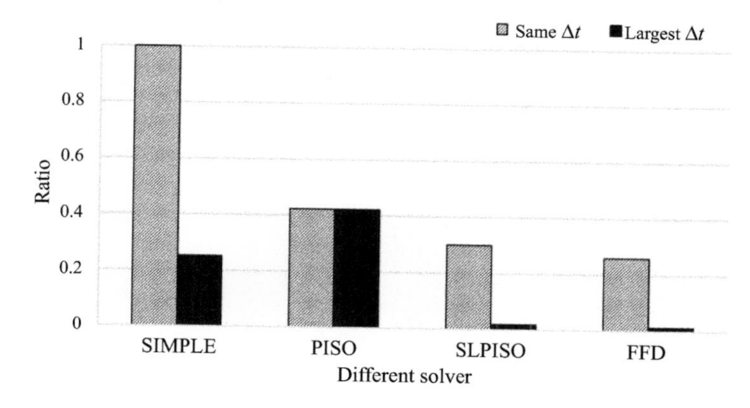

Figure 11.25 Computational cost comparison with different time steps

used as the reference. These "experimental" data are taken every 4 s from the SIMPLE prediction at the middle point of the test chamber. The study again uses the effective kinematic viscosity and heat-transfer coefficient, namely, 100 times the physical values (11.4), to consider the turbulence impact. The mesh resolution used is 80×80. The time step is varied from 0.005 to 0.08 s in the test. It is noted that when the time step increases, the predicted transient and steady-state results of SLPISO and FFD deviate, and SLPISO performs overall better than FFD (Figure 11.26). FFD must use smaller time steps to obtain similar results as SLPISO. The increased deviation of the SLPISO results at a larger time step is attributed to the induced false diffusion of the time term, which can be improved by adjusting the effective kinematic viscosity to a lower value, thus compensating the increased false diffusion due to the larger time step.

11.6 Conclusions

Different simulation techniques for modeling indoor airflow are reviewed in this chapter with the focus to identify proper strategies for fast indoor environment modeling. Engineering practice shows that CFD is one of the most popular methods applied to indoor environment quality study. Various other types of models such as nodal and zonal models are able to predict average flow conditions for a building with many compartments and sophisticated systems under transient indoor and outdoor environments. These models, however, are dependent on CFD to provide higher levels of details in order to make superior predictions.

CFD provides the most informative data that are useful for various applications in built environment study. However, the intensive computational cost is the main reason that restricts its broader and better usages for design and engineering applications. Potentials for decreasing the computing cost of CFD simulation are explored. It is identified that the grid resolution, turbulence model, and equation-solving algorithm are closely relevant to computing speed. Computing operation is

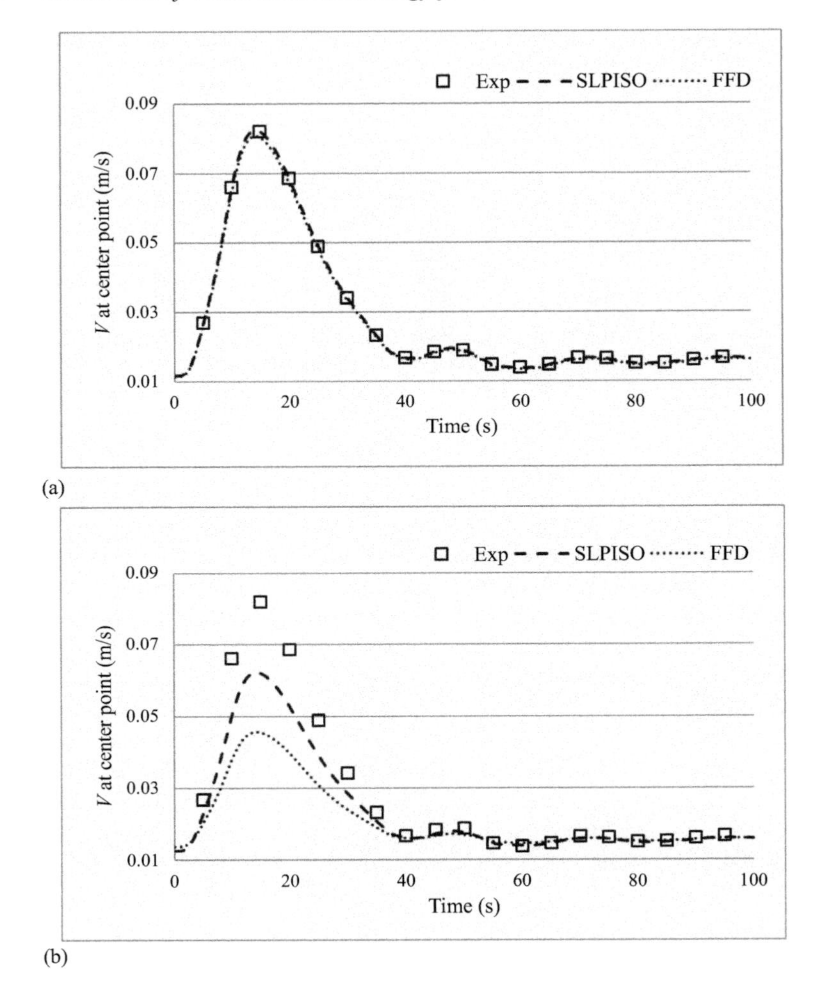

Figure 11.26 Accuracy comparison with different time steps for transient simulation with SLPISO and FFD: (a) time step of 0.005 s and (b) time step of 0.05 s

directly related to the total grid number in a CFD model; therefore, coarse grid, although it brings in numerical truncation error and has impact on the accuracy of CFD solution, can significantly reduce computing cost. Both theoretical analyses and numerical tests reveal that properly designed coarse grid can provide reasonable simulation while keeping the benefit of computational saving.

Turbulence modeling is another important factor that influences CFD speed. Simplified zero-equation turbulence models, once calibrated and validated, can perform great prediction for flow and heat-transfer phenomena in built environments that share the same or similar flow mechanisms.

Numerical iteration impacts the computational and converging efficiency of CFD. PISO and projection methods present noniterative alternatives to solving the flow governing equations with different velocity–pressure decoupling algorithms. Semi-Lagrangian advection scheme overcomes the restriction of the stability condition and weakens the coupling between the equations. The combination of these algorithms and schemes deliver great potential for developing a fast and accurate version of CFD model.

References

Acharya, S. and Moukalled, F. (1989) Improvements to Incompressible flow calculation on a nonstaggered curvilinear grid. *Numerical Heat Transfer*, 15, 131–52.

Allard, F. and Inard, C. (1992) Natural and mixed convection in rooms: Prediction of thermal stratification and heat transfer by zonal models. In *International Symposium of Room Air Convection and Ventilation Effectiveness*. Japan.

Amsallem, D. and Farhat, C. (2008) Interpolation method for adaptation reduced-order models and application to aeroelasticity. *AIAA Journal*, 1803–13.

Andrews, A.T., Loezos, P.N. and Sundaresan, S. (2005) Coarse-Grid Simulation of Gas-Particle Flows in Vertical Risers. *Industrial and Engineering Chemistry Research*, 44(16), 6022-37

Antoulas, A. C. (2005) *Approximation of Large-Scale Dynamical Systems (Advances in Design and Control)*. Philadelphia, PA, USA: Society for Industrial and Applied Mathematics.

ASHRAE. (2004) *ANSI/ASHRAE Standard 62.2-2004: Ventilation and Acceptable Indoor Air Quality in Low-Rise Residential Buildings*. Atlanta, GA: ASHRAE.

Audi, E. (2009) Comparison of pressure-velocity coupling schemes for 2D flow problems. *International conference on advances in computational tools for engineering applications*. Zouk Mosbeh. pp. 245–48.

Augenbroe, G. and Hensen, J. (2004) Simulation for better building design. *Building and Environment*, 39(8), 875–7.

Barnard, S., Biswas, R., Saini, S., *et al.* (1999) Large-scale distributed computational fluid dynamics on the information power grid using Globus. *Proceeding fronties 99. The seventh symposium on the frontiers of massively parallel computation*; MD, USA, pp. 60–7.

Blay, D., Mergui, S. and Niculae, C. (1992) Confined turbulent mixed convection in the presence of a horizontal buoyant wall jet. *Fundamentals of Mixed Convection*, 213, 65–72.

Blomsterberg, A. and Johansson, T. (2005) Use of multi-zone air flow simulations to evaluate a hybrid ventilation system. In *9th International IBPSA Conference*. Montreal, Canada.

Borggaard, J., Cliff, E. and Gugercin, S. (2012). *Model Reduction for Indoor-Air Behavior in Control Design for Energy-Efficient Buildings. American Control Conference (ACC)*, Montreal, QC, June 2012. pp. 2283–8

Borggaard, J., Hay, A. and Pelletier, D. (2007) Interval-based reduced-order models for unsteady fluid flow. *International Journal of Numerical Analysis and Modeling*, 4(3-4), 353–67.

Bouia, H. and Dalicieux, P. (1996) Simplified modeling of air movements inside dwelling room. In *Proceedings of the Building Simulation Conference*, 106–10.

Bui-Thanh, T. Willcox, K. and Ghattas, O. (2008) Parametric reduced-order models for probabilistic analysis of unsteady aerodynamic applications. AIAA journal, 46(10), 2520–9.

Burkardt, J., Gunzburger, M. and Lee, H.-C. (2006) POD and CVT-based reduced-order modeling of Navier–Stokes flows. *Computer Methods in Applied Mechanics and Engineering*, 196(1-3), 337–55.

Chen, Q. and Xu, W. (1998) *A Zero-Equation Turbulence Model for Indoor Airflow Simulation*. Energy and buildings, 28(2), 137–144.

Chen, X., Zhao, B. and Li, X. (2005) *Numerical Investigation on the Influence of Contaminant Source Location, Occupant Distribution and Air Distribution on Emergency Ventilation Strategy*. Indoor and Built Environment, 14(6), 455–67.

Chorin, A. J. (1997) *A Numerical Method for Solving Incompressible Viscous Flow Problems. Journal of Computational Physics*, 135(2):118–125.

Cohen, J. and Molemake, J. (2009) A fast double precision CFD code using CUDA. In *Proceedings of Parallel Computational Fluid Dynamics (ParCFD2009)*, 414–29. California, USA.

Courant, R., Friedrichs, K. and Lewy, H. (1928) Über die partiellen Differenzengleichungen der mathematischen Physik. *Mathematische Annalen*, 100, 32–74.

Courant, R., Isaacson, E. and Rees, M. (1952) On the solution of nonlinear hyperbolic differential equations by finite differences. Communications on pure and applied mathematics, 5(3), 243–55.

Djunaedy, E. (2005) *External Coupling Between Building Energy Simulation and Computational Fluid Dynamics*. Eindhoven, the Netherlands: Technische Universiteit Eindhoven.

Dols, W.S. (2001) A Tool for modeling airflow and contaminant transport. *ASHRAE Journal*, 43(3), 35–42.

Fanger, P. O., Melikov, A. K., Hanzawa, H. and Ring, J. (1989) Turbulence and draft. *ASHRAE Journal*, 18–25.

Feustal, H., Allard, F., Dorer, V., Garcia, E., Herrlin, M., Grosso, M., *et al.* (1990). COMIS Fundamentals. Lawrence Berkeley National Laboratory. LBNL Report: LBL-28560.

Foster, N. and Metaxas, D. N. (1996) Realistic animation of liquids. Graphical models and image processing, 58(5), 471–83.

Foster, N. and Metaxas, D. N. (1997) Modeling the motion of a hot, turbulent gas. In *International Conference on Computer Graphics and Interactive Technique*, 181–8.

Gousseau, P., Blocken, B., Stathopoulos, T. and van Heijst, G. J. F. (2011) CFD simulation of near-field pollutant dispersion on a high-resolution grid: A case study by LES and RANS for a building group in downtown Montreal. *Atmospheric Environment*, 45(2), 428–38.

Griffith, B. T. (2006) Incorporating nodal and zonal room air models into building energy calculation procedures. M.Sc Dissertation, Massachusetts Institute of Technology, USA.

Hall, K. C., Thomas, T. J. and Clark, W. S. (2002) Computation of unsteady non-linear flows in cascades using a harmonic balance technique. *AIAA Journal*, 40(5), 879–86.

Harlow, F. H. and Welch, J. E. (1965) Numerical calculation of time-dependent viscous incompressible flow of fluid with free surface. *The Physics of Fluids*, 8(12), 2182–9.

Harrington, L. (2001) Computer modelling of night-time natural ventilation. Doctoral Thesis, Loughborough University, UK.

Hensen, J., Djunaedy, E., Trcka, M. and Yahiaoui, A. (2004) Building performance simulation for better design: Some issues and solutions. In *Proceedings of the 21st International Conference on Passive and Low Energy Architecture*, 1185–90. Eindhoven, the Netherlands.

Hou, Y. (2010) Reduced-order modeling of incompressible jet flow using proper orthogonal decomposition and Galerkin projection. Doctoral dissertation, Northeastern University, USA.

Inard, C., Bouia, H. and Dalicieux, P. (1996) Prediction of air temperature distribution in buildings with a zonal model. *Energy and Buildings*, 24(2), 125–32.

International Code Council. (2012) International Energy Conservation Code. Falls Church, VA.

Issa, R. I. (1986) Solution of the implicitly discretised fluid flow equations by operator-splitting. *Journal of Computational Physics*, 62(1), 40–65.

Jang, D. S., Jetli, R. and Acharya, S. (1986) Comparison of the PISO, SIMPLER, and SIMPLEC algorithms for the treatment of the pressure-velocity coupling in steady flow problems. *Numerical Heat Transfer*, Part A: Applications, 10 (3), 209–28.

Jiang, Y. and Chen, Q. (2003) Buoyancy-driven single-sided natural ventilation in buildings with large openings. *International Journal of Heat and Mass Transfer*, 46(6), 973–88.

Kameel, R. and Khalil, E. (2003) The prediction of airflow regimes in surgical operating theatres: A comparison of different turbulence models. In *41st Aerospace Sciences Meeting and Exhibit*.

Kim, J. and Moin, P. (1985) Application of a fractional-step method to incompressible Navier-Stokes equations. *Journal of Computational Physics*, 59, 308–23.

Ladeinde, F. and Nearon, M. (1997) CFD applications in the HVAC&R industry. *ASHRAE Journal*, 39(1), 44–8

Lebrun, J. (1970) *Physiological Requirements and Physical Conditions for Air Conditioning by Concentrated Static Source*. Belgium: Universite de Liege.

Li, X., Yu, Z., Zhao, B. and Li, Y. (2005a) Numerical analysis of outdoor thermal environment around buildings. *Building and Environment*, 40(6), 853–66.

Li, Y., Duan, S., Yu, I. T. and Wong, T. W. (2005b) Multi-zone modeling of probable SARS virus transmission by airflow between flats in Block E, Amoy Gardens. *Indoor Air*, 15(2), 96–111.

Liu, X. and Zhai, Z. J. (2009) Prompt tracking of indoor airborne contaminant source location with probability-based inverse multi-zone modeling. *Building and Environment*, 44(6), 1135–43.

Lu, J. (2015) Analysis of new national standard (GB/T 18801-2015) technical requirements of household air purifier. *Journal of Appliance Technology*, 11, 12-3. (in Chinese).

Lucia, D. J., Beran, P. S. and Silva, W. A. (2004) Reduced-order modeling: New approaches for computational physics. *Progress in Aerospace Sciences*, 40(1-2), 51–117.

Mora, L., Gadgil, A. J. and Wurtz, E. (2003) *Comparing Zonal and CFD Model Predictions of Isothermal Indoor Airflows to Experimental Data*. Indoor Air, 13(2), 77–85.

Morrison, B. I. (2000) *The Adaptive Coupling of Heat and Air Flow Modeling Within Dynamic Whole-Building Simulation*. Glasgow: University of Strathclyde.

Moser, A. (1994) The IEA works on guidelines for ventilation of large enclosures. In *Building Research Establishment, BEPAC Conference, Building Environmental Performance – Facing the Future. Proc*, UK.

Moulton, M. and Steinhoff, J. (2000) A technique for the simulation of stall with coarse-grid CFD methods. In *38th Aerospace Sciences Meeting and Exhibit*.

Mundt, E. (1996) The performance of displacement ventilation systems: Experimental and theoretical studies. Royal Institute Of Technology, S-100 44 Stockholm 70, Sweden.

Nielsen, P. V. (1974) *Flow in Air-Conditioned Rooms*. Copenhagen: Technical University of Denmark.

Nielsen, P.V. (1998) The Selection of Turbulence Models for Prediction of Room Airflow. Department of Building Technology and Structural Engineering. Indoor Environmental Engineering Vol. R9828 No. 86.

Nilsson, H. O. (2007) Thermal comfort evaluation with virtual manikin methods. *Building and Environment*, 42, 4000–5.

Oliveira, P. J. and Issa, R. I. (2001) *An Improved PISO Algorithm for the Computation of Buoyancy-Driven Flows. Numerical Heat Transfer Part B: fundamentals*, 40(6), 473–93.

Patankar, S. V. (1980) *Numerical Heat Transfer and Fluid Flow*. Washington, DC: Hemisphere Publishing Corporation.

Patankar, S. V. and Spalding, D. B. (1972) A calculation procedure for heat, mass and momentum transfer in three-dimensional parabolic flows. *International Journal of Heat and Mass Transfer*, 15(10), 1787–806.

Prandtl, L. (1925) Über die ausgebildete turbulenz. *ZAMM*, 5, 136–9.

Rambo, J. D. (2007) Reduced-Order Modeling of Multiscale Turbulent Convection: Application to Data Center Thermal Management. Doctoral dissertation, Georgia Institute of Technology, USA.

Rees, S. J. and Haves, P. (2001) *A Nodal Model for Displacement Ventilation and Chilled Ceiling Systems in Office Spaces. Building and Environment*, 36(6), 753–62.

Schetzen, M. (1980) *The Volterra and Wiener Theories of Nonlinear Systems*. New York, NY: Wiley.

Srebric, J. (2000) Simplified methodology for indoor environment designs. Doctoral dissertation, Massachusetts Institute of Technology, USA.

Srebric, J., Chen, Q. and Glicksman, L. R. (1999) *Validation of a Zero-Equation Turbulence Model for Complex Indoor Airflow Simulation.* ASHRAE Transactions, 105(2), 414–27.

Stam, J. (1999) Stable fluid. In *International Conference on Computer Graphics and Interactive Techniques*, 121–8.

Sun, H., Stowell, R. R., Keener, H. M., and Michel, F. C. Jr. (2002) Two-dimensional computational fluid dynamics (CFD) modeling of air velocity and ammonia distribution in a high-rise (TM) hog building. *Transactions of the ASAE*, 45(5), 1559–68.

Surana, A., Hariharan, N., Narayanan, S. and Banaszuk, A. (2008) *Reduced Order Modeling for Contaminant Transport and Mixing in Building Systems: A Case Study Using Dynamical Systems Techniques. American Control Conference*, 902–7.

Togari, S., Arai, Y. and Miura, K. (1993) A simplified model for predicting vertical temperature distribution in a large space. *ASHRAE Transactions*, 99, 84–99.

Van Doormaal, J. P. and Raithby, G. D. (1984) Enhancements of the SIMPL E method for predicting incompressible fluid flows. *Numerical Heat Transfer*, 7(2), 147–63.

Van Doormaal, J. P. and Raithby, G. D. (1985) An evaluation of the segregated approach for predicting incompressible fluid flows. In *National Heat Transfer Conference*. Denver, CO.

Walton G.N. (1989) *AIRNET: A computer program for building airflow network modeling*. NISTIR report 89–4072.

Walton G.N., and Dols W.S. (2005) *CONTAM user guide and program documentation Version 2.4*. NISTIR report 7251.

Wang, H. and Zhai, Z. (2012a) Analyzing grid independency and numerical viscosity of computational fluid dynamics for indoor environment applications. *Building and Environment*, 52, 107–18.

Wang, H. and Zhai, Z. (2012b) Application of coarse grid CFD on indoor environment modeling: Optimizing the trade-off between grid resolution and simulation accuracy. *HVAC&R Research*, 18(5), 915–33.

Wang, H., Zhai, Z. J. and Liu, X. (2014) Feasibility of utilizing numerical viscosity from coarse grid CFD for fast turbulence modeling of indoor environments. *Building Simulation*, 7(2), 155–64.

Wang, L. and Chen Q. (2007) Validation of a coupled multizone-CFD program for building airflow and contaminant transport Simulations. *HVAC&R Research*, 13(2), 267–81.

Weathers, J. W. and Spitler, J. D. (1993) A comparative study of room airflow: Numerical prediction using computational fluid dynamics and full-scale experimental measurements. *ASHRAE Transactions*, 100, 114–57.

Xue, Y., Liu, W. and Zhai, Z. (2016) New semi-Lagrangian-based PISO method for fast and accurate indoor environment modeling. *Building and Environment*, 105, 236–44.

Yuan, X., Chen, Q. and Glicksman, L. R. (1998) *(RP-949) A Critical Review of Displacement Ventilation. ASHRAE Transactions*,104(1), 78–90.

Yuan, X., Chen, Q. and Glicksman, L. R. (1999) *(RP-949)Models for Prediction of Temperature Difference and Ventilation Effectiveness with Displacement Ventilation. ASHRAE Transactions*, 105(1), 353–67.

Zerihun Desta, T., Janssens, K., Van Brecht, A., Meyers, J., Baelmans, M. and Berckmans, D. (2004) CFD for model-based controller development. *Building and Environment*, 39(6), 621–33.

Zhai, Z. J. and Chen, Q. Y. (2005) Performance of coupled building energy and CFD simulations. Energy and Buildings, 37(4), 333–44.

Zhai, Z., Wei, Z., Zhao, Z. and Chen, Q. (2007) Evaluation of various turbulence models in predicting airflow and turbulence in enclosed environments by CFD: Part 1—Summary of prevalent turbulence models. *HVAC&R Research*, 13(6), 853–70.

Zhao, Z., Zhai, Z., Wei, Z. and Chen, Q. (2007) Evaluation of various turbulence models in predicting airflow and turbulence in enclosed environments by CFD: Part 2—Comparison with experimental data from literature. *HVAC&R Research*, 13(6), 871–86.

Zuo, W. and Chen, Q. (2009) Real-time or faster-than-real-time simulation of airflow in buildings. *Indoor Air*, 19(1), 33–44.

Chapter 12
HVAC online monitoring and control strategy

Shi-Jie Cao[1]

12.1 What is HVAC online monitoring and control strategy?

HVAC systems play an important role in improving air quality (Chen *et al.*, 2018) and thermal comfort (Bluyssen *et al.*, 2011) in the building environment and can also cause building energy waste and even air pollution problems (Kim and Yu, 2018). The main reason is that the traditional HVAC system cannot be based on dynamic changes of indoor environmental parameters and nonuniform distribution characteristics to achieve optimal real-time adjustment of air supply parameters, resulting in an excessive growth of energy consumption for running the air-conditioning system (Ren and Cao, 2020). To further save energy and improve indoor air quality, this chapter introduces an HVAC online monitoring and control strategy for unsteady, nonlinear, nonuniform and real-time response to multiple environmental parameters (e.g., CO_2 concentration, temperature, and humidity) to create a healthy, comfortable, and energy-efficient building environment.

The HVAC online monitoring strategy mainly includes three key components: (1) online monitoring, (2) "faster-than-real-time" prediction, and (3) optimal evaluation and control. The online monitoring module mainly uses limited sensors to obtain online monitoring data and provide effective input for "faster-than-real-time" prediction. The "faster-than-real-time" prediction module builds input–output prediction database by combining the limited monitoring data and the output for environmental parameter distribution data, to achieve a rapid prediction of global environmental parameters under the effect of any source and air supply parameter. "Faster-than-real-time" is defined as the prediction time of environmental parameter that is much smaller than the diffusion time of environmental parameter. The optimal evaluation and control module mainly combine the rapid prediction results and the optimal evaluation indexes to realize the evaluation and control of the optimal air supply parameters. Figure 12.1 is a basic flowchart of the HVAC online control system.

[1]School of Architecture, Southeast University, Nanjing, China

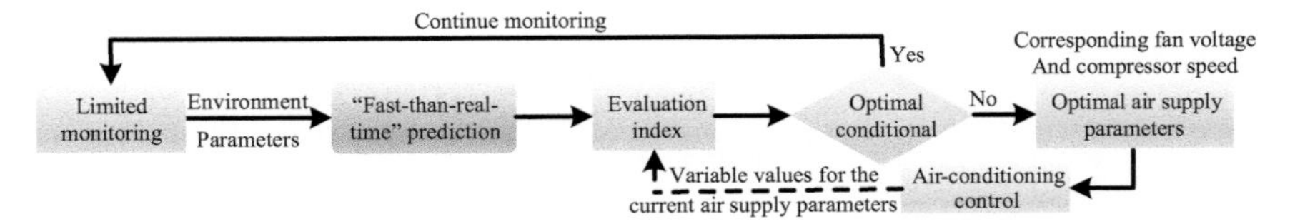

Figure 12.1 Flowchart of the HVAC online control system

12.2 Problem that still needs to be solved

Considering the unsteady, nonlinear and nonuniform distribution characteristics of the indoor environment, an efficient HVAC online monitoring strategy should first address the following core issues.

The first question is how to achieve the "faster-than-real-time" prediction of indoor environmental distribution. Computational fluid dynamics (CFD) can basically satisfy the needs of indoor environmental parameter prediction (Chen *et al.*, 2014). However, the method is generally limited by the number of grids and computational cost, which makes it difficult to meet the real-time demand for prediction results. As a result, rapid prediction methods such as multi-zone models and fast fluid dynamics have been developed (Cao, 2019). The multi-zone model effectively reduces the cost of predicting indoor environmental parameters by simplifying the computational grid, and the fast fluid dynamics method also solves the control equations quickly using a special numerical format, increasing the prediction speed by a factor of about 50 (compared to CFD methods) (Feng *et al.*, 2019). However, the multi-zone models use uniform distribution of indoor airflow as a fundamental assumption, and the fast-hydrodynamic methods also ignore important turbulence effects. As a result, the previous fast prediction models may have large prediction errors, i.e., they cannot meet the prediction accuracy requirements (for practical engineering applications). In order to better balance prediction speed and accuracy, machine learning models (e.g., artificial neural network (ANN) models) and the contribution ratio of indoor environment (CRI) have become a hot research topic in recent years and provide a variety of effective ways for an "faster-than-real-time" prediction of indoor environmental parameters. Many researches have also demonstrated the reliability of these models (Cao and Ren, 2018; Ren and Cao, 2019a; Zhu *et al.*, 2019). Thus, both the ANN model and CRI method provide an effective guarantee for online HVAC monitoring.

The second question is how to combine limited monitoring for optimal air supply parameter evaluation and control. It is common for HVAC online monitoring to use sensors near the return air or at single-point indoors to obtain monitoring data to evaluate and control the air supply parameters. However, a limited number of sensors cannot effectively characterize the global (or indoor human activity areas) environmental parameters and therefore cannot ensure that the system obtains the optimal air delivery parameter evaluation results. In view of sensor cost constraints, the development of "faster-than-real-time" predictive models (e.g., ANN) based on limited monitoring data can further improve the effectiveness of online HVAC monitoring strategies.

Thus, an efficient online HVAC monitoring strategy requires the coupling of "faster-than-real-time" prediction models, such as ANN model and CRI method, with limited monitoring techniques to achieve rapid prediction of global environmental parameters and evaluation and control of optimal air supply parameters. To further improve prediction efficiency, coupling multiple downscaled linear models (including low-dimensional linear ventilation model (LLVM), low-dimensional linear temperature model (LLTM), and low-dimensional linear humidity model (LLHM)) with the "faster-than-real-time" prediction model will effectively reduce the cost of building the prediction database.

12.3 How to achieve "faster-than-real-time" prediction?

The construction of "faster-than-real-time" prediction models is mainly based on various low-dimensional linear models (ventilation, temperature, humidity, etc.), ANN model, indoor environmental contribution prediction method and coupled with limited indoor environmental monitoring data. Specifically, it takes limited monitoring data as inputs and uses "faster-than-real-time" prediction methods to reconstruct and predict the indoor environment under different air conditionings, evaluate and make optimal ventilation decisions, and issue control commands to the system.

12.3.1 Low-dimensional linear model

The main function of low-dimensional linear model is to achieve the rapid expansion of initial prediction database (built by CFD method) and the reduction of data dimensionality, effectively reducing the database construction cost. The model consists of low-dimensional model and linear model. By discrete Green's function approach, starting from the Navier-Stokes equations, the transport equations and energy equation for indoor pollutant concentrations, a low-dimensional model can be derived that can accurately estimate the distribution of indoor pollutant concentrations (Cao and Meyers, 2012). Figure 12.2 shows the basic principle of the low-dimensional model (Cao and Ren, 2018). The grid discretization method (including linear and non-linear discretization) will directly affect the accuracy of low-dimensional models. in this context, a self-adaptive low-dimensional model was proposed to rapidly determine the optimal mesh grid discretization approach corresponding to any given indoor environment space, which can effectively improve the discretization efficiency and accuracy of low-dimensional models (Ren and Cao2020a). Linear (scalar) models can be divided into linear ventilation (LVM), temperature (LTM), and humidity (LHM) models, which are mainly used to rapidly reconstruct the fields of multiple environmental scalars (e.g., concentration, temperature, and humidity) produced by any type of sources (e.g., pollution, heat and humidity sources) (Cao and Meyers, 2014). Generally, these linear models are based on the precondition that indoor velocity field is unchanged and indoor environmental scalars have less influence on airflow distribution. For a real and complex indoor environment potentially with sharply-varying sources, the feasibility of linear models should be validated and considered for two important

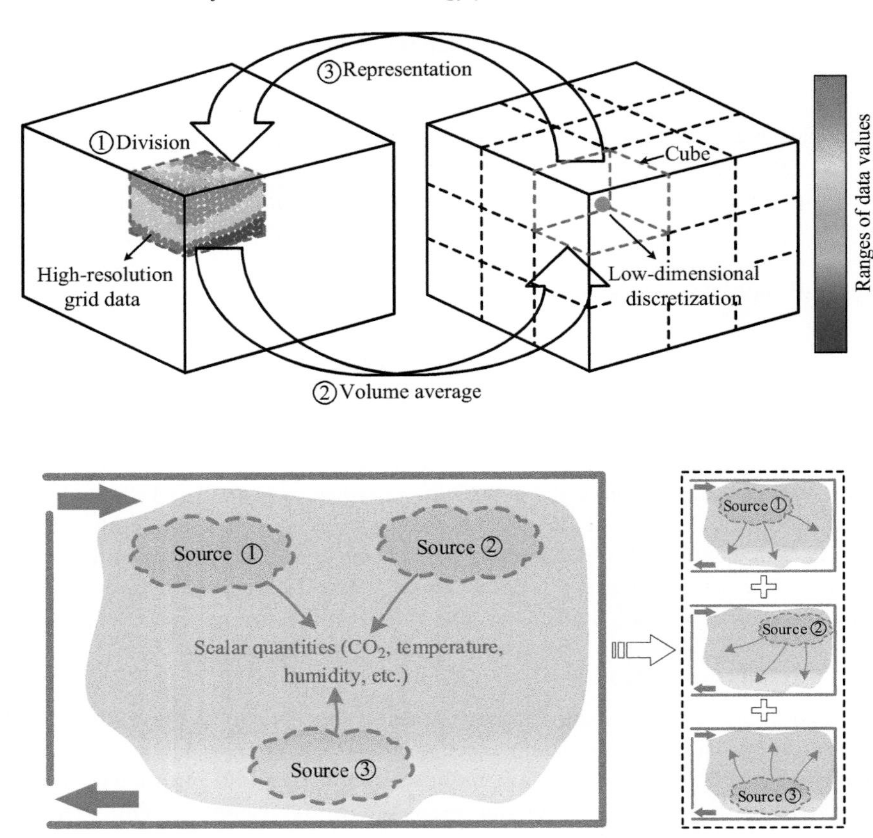

Figure 12.2 Schematic diagram of low-dimensional model (Ren and Cao, 2019a)

issues: (1) when to adopt the linear assumption? (2) how to construct a valid linear model? In this regard, a dimensionless ratio of air-mass-flow rate to pollutant-mass-flow rate was proposed to determine when and how to use a well-behaved LVM for rapid prediction of pollutant fields (Cao and Meyers, 2014). Similarly, a dimensionless heat exchange ratio was put forward for the judgment of the decoupling of momentum and energy equations, further employed for the construction of LTM (Ren and Cao, 2021).

12.3.2 Artificial neural network

ANN is a neural system-like model, which is used to construct the database and train the mapping between input and output vectors to obtain fast prediction results of target variables. Radial basis function belongs to the basic ANN model, which consists of input, hidden, and output layers, with good prediction performance for indoor environmental parameters (e.g., CO_2 concentration) (Cao and Ren, 2018).

12.3.3 Contribution ratio of indoor environment (CRI)

The function of contribution ratio of indoor climate ($CRI_{(T)}$) was to quantize the effect of each single heat source on indoor temperature distribution. The definition of $CRI_{(T)}$ was the ratio of the temperature rise (or decrease) caused by one independent heat source at any position to the absolute value of the temperature rise (or decrease) caused by the same heat source with its increasing (or decreasing) heat distributed uniformly (Zhang *et al.*, 2013). The $CRI_{(T)}$ expressed the diffusion distance of the generated heat in space, i.e., the scope and extent of the impact of each heat source. The $CRI_{(T)}$ of the heat source m at the location x_i was defined as

$$\mathrm{CRI_m}(x_i) = \frac{\Delta T_{(x_i)}}{\Delta T_{\mathrm{m},0}} = \frac{T_\mathrm{m}(x_i) - T_\mathrm{n}}{T_{\mathrm{m},0} - T_\mathrm{n}} = \frac{T_\mathrm{m}(x_i) - T_\mathrm{n}}{(Q_\mathrm{m}/C_p\rho V)} \tag{12.1}$$

where x_i is the spatial coordinates (i=1, 2, 3 for x, y, z); Q_m is the convective heat transfer from source m; C_p is the specific heat of indoor air; ρ is the air density; V is the volume of supply air; T_n is the neutral temperature that is only used for calculating $CRI_{(T)}$; $T_{\mathrm{m},0}$ is the temperature of the room when the heat transfer Q_m from heat source m is diffused uniformly; $T_\mathrm{m}(x_i)$ is the temperature at position x_i caused by heat source m as calculated by CFD; $\Delta T_{\mathrm{m},0}$ is the temperature rise of the room from T_n; $\Delta T_\mathrm{m}(x_i)$ is the temperature rise from T_n at the position x_i caused by heat source m (Ren and Cao, 2019a).

The contribution rate of indoor humidity $CRI_{(H)}$ was utilized to calculate the influence of various moisture sources on indoor humidity distribution. $CRI_{(H)}$ indicated the ratio of the rise (or fall) in humidity at a point from an individual moisture source to the rise (or fall) in humidity with the perfect mixing conditions for the same moisture source (Huang *et al.*, 2011). The equation of $CRI_{(H)}$ can be defined as follows:

$$\mathrm{CRI_{(H)}}(x,n) = \frac{\delta X(x,n)}{X_n}, X_n = \frac{q_n}{\rho \cdot Q} \tag{12.2}$$

where $\delta X(x,n)$ was the rise (or fall) in humidity value from standard status at a point x due to the nth moisture source [kg/kg′]; X_n was the rise (or fall) in humidity value under the perfect mixing conditions due to the nth moisture source [kg/kg′]; q_n was the moisture flux generated by the nth moisture source [kg/s]; ρ was the air density [kg′/m^3]; and Q was the airflow rate [m^3/s] (Huang *et al.*, 2011).

12.3.4 Coupled prediction method

Combining the low-dimensional linear model and fast prediction model can construct coupled prediction model, which includes LLVM-based ANN, LLTM-based contribution ratio of indoor temperature (LLTM-based $CRI_{(T)}$), and LLHM-based contribution ratio of indoor humidity (LLHM-based $CRI_{(H)}$). In LLVM-based ANN model, LLVM is mainly used to reduce the construction cost of the prediction database. The limited monitoring data and the prediction database are coupled to build an input–output ANN prediction database, with the limited monitoring data as inputs and the low-dimensional CO_2 concentration field as outputs. Based on the

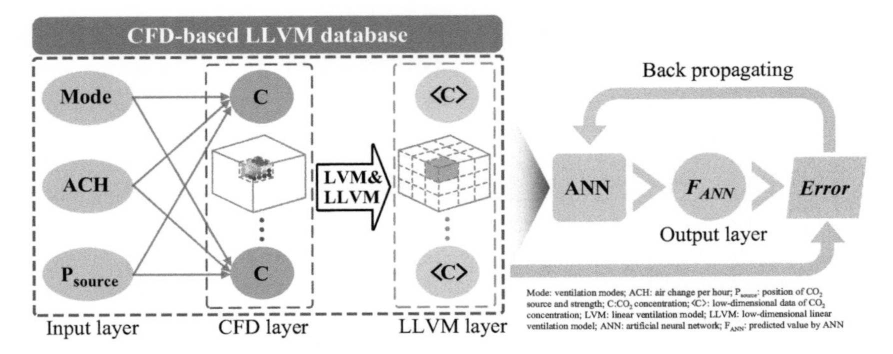

Figure 12.3 Structure of LLVM-based ANN model coupled with limited monitoring data (Cao and Ren, 2018)

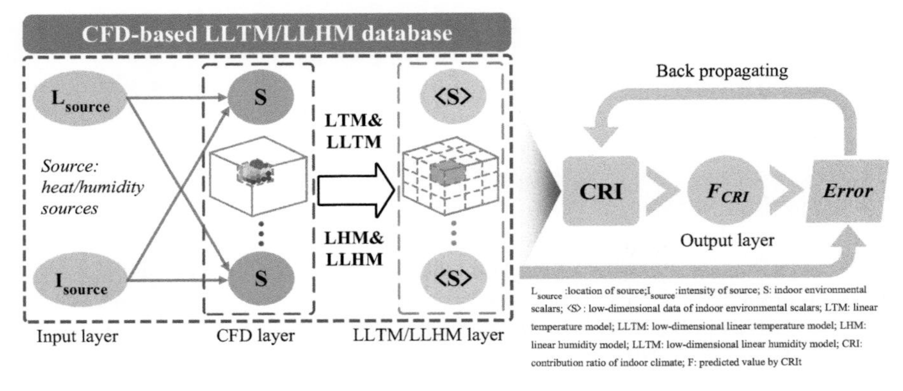

Figure 12.4 Basic structure of LLTM-based $CRI_{(T)}$ and LLHM-based $CRI_{(H)}$ coupled with the limited monitoring data (Ren and Cao, 2019a)

database, the LLVM-based ANN model coupled with limited monitoring data is constructed to realize a "faster-than-real-time" prediction of low-dimensional concentration field. Figure 12.3 shows the basic structure of LLVM-based ANN model coupled with the limited monitoring data.

Similarly, LLTM-based $CRI_{(T)}$ and LLHM-based $CRI_{(H)}$ have similar steps for a rapid prediction of indoor temperature and humidity. Combining the LLTM and LLHM, respectively, with contribution ratio of indoor climate and humidity (i.e., CRI $_{(T)}$ and $CRI_{(H)}$), we get LLTM and LLHM-based contribution ratio of indoor climate (i.e., LLTM-based $CRI_{(T)}$ and LLHM-based $CRI_{(H)}$) to realize the rapid prediction of temperature and humidity distribution. Among them, the main function of LLTM and LLHM is also to effectively reduce the cost of constructing the prediction database. Combined with limited monitoring data, the input–output CRI prediction database was built, and the output was the low-dimensional temperature or humidity distribution results. By coupling the limited monitoring data and LLTM-based $CRI_{(T)}$ and LLHM-based $CRI_{(H)}$, the "faster-than-real-time" prediction of low-dimensional temperature field or humidity field can be effectively realized. Figure 12.4 shows the basic structure of LLTM-based $CRI_{(T)}$ and LLHM-based $CRI_{(H)}$ coupled with the limited monitoring data (Ren and Cao, 2019a).

12.4 How to achieve optimal evaluation and control?

Indoor environmental parameters have "unsteady,nonlinear and nonuniform" dynamic distribution characteristics; thus the "optimal evaluation and control" strategy of ventilation online monitoring system needs to integrate the multi-dimensional and multifactor roles to regulate the HVAC system. The HVAC evaluation strategy needs to evaluate indoor environmental quality (IEQ) and energy consumption according to the comprehensive evaluation criteria (such as the balance between indoor air quality and ventilation energy consumption), so as to judge whether to execute the optimal environmental control instructions, in order to create a better indoor environment and optimize the energy structure. Based on the previous principles, the comprehensive evaluation of the optimal balance between IEQ (expressed as CO_2 concentration, predicted mean vote (PMV), and thermal sensation (TS)) and energy consumption (expressed as ventilation rate, air supply temperature, and humidity) was carried out. Customized evaluation indicators (i.e., E_V, E_T, E_H) were obtained to comprehensively evaluate the indoor environmental conditions and obtain the optimal indoor environmental parameters. After that, the HVAC system can be controlled by the control module.

12.4.1 Evaluation index

With the gradual increase of the air change rate per hour (ACH), the proportion of ventilation energy consumption is also increasing (Cao and Deng, 2019). Therefore, the evaluation of optimal ACH is mainly to realize the optimal assessment of CO_2 concentration and ventilation energy consumption, that is, to effectively reduce the ventilation energy consumption within the acceptable concentration range. The CO_2 concentration is used to characterize the quality of the ventilation environment, while the ventilation energy consumption is represented by the ACHs. The CO_2 concentration can also be represented by the average CO_2 concentration in the respiratory zone, considering that the main activity range of indoor personnel is located near the respiratory zone. The optimal ACHs evaluation index E_V is shown in the following equation:

$$E_V = W_{V1} \frac{\text{ACH}}{\text{Max(ACH)}} + W_{V2} \frac{C_{\text{mean}}}{\text{Max}(C_{\text{mean}})} \tag{12.3}$$

where W_{V1} and W_{V2} are the weights for ACH and C_{mean}, respectively. On the basis of index E_V, the ACH for the minimum E_V value corresponds to the optimal ACH.

Based on the optimal selection of ACHs, the evaluation index of optimal air supply temperature mainly realizes the comprehensive assessment of personnel thermal comfort and air-conditioning energy consumption. The optimal air supply temperature not only meets the human comfort needs but also reduces the energy consumption of air-conditioning operation. Personnel thermal comfort can be expressed quantitatively by the PMV, and air-conditioning energy consumption can be replaced by the air supply temperature. Optimal air supply temperature evaluation index E_T is shown in the following equation:

$$E_T = W_{T1}\frac{|\mathrm{PMV}|}{\mathrm{Max}(|\mathrm{PMV}|)} + W_{T2}\frac{T}{\mathrm{Max}(T)} \quad (12.4)$$

where W_{T1} and W_{T2} are the weightings for supply air temperature and PMV (absolute value). The temperature with the minimum E_T value corresponds to the optimal supply air temperature. Besides, PMV was calculated by the average room air velocity (V_R) and temperature (T_R) (Deng *et al.*, 2018), as shown in the following equation:

$$\mathrm{PMV} = \frac{7}{83}T_R V_R + \frac{28}{75}T_R \frac{689}{74} \quad (12.5)$$

On the basis of optimal evaluations of ACH and supplied air temperature, the evaluation index of optimal air supply humidity is used to realize the integrated evaluation of personnel perception of the humidity environment and humidity load. Among them, the human perception of the humidity environment can be replaced by the thermal sensory index equivalent, and humidity load can be indicated by the relative humidity (RH) of supply air. The optimal air supply humidity evaluation index E_H was shown in the following formula:

$$E_H = W_{H1}\frac{\mathrm{TS}}{\mathrm{Max}(\mathrm{TS})} + W_{H2}\frac{\mathrm{H}}{\mathrm{Max}(\mathrm{H})} \quad (12.6)$$

where W_{H1} and W_{H2} are the weightings for supply air RH (H) and thermal sensation (TS). The RH with the minimum E_H value corresponds to the optimal supply air RH. Besides, the expression of index TS was presented as follows (Paula Xavier and Roberto, 2000):

$$\mathrm{TS} = 0.219T_0 + 0.012\mathrm{RH} - 0.547V_a - 5.83 \quad (12.7)$$

where T_0 was the temperature (°C); RH was the relative humidity (%); and V_a was the air velocity (m/s). The scales of TS were indicated in Table 12.1.

Since the results of the rapid prediction of indoor humidity parameters are absolute humidity values, the evaluation of the optimal air supply humidity needs to be converted from absolute humidity to RH values, as shown in the following three equations (Feng *et al.*, 2018):

$$h_a = 1.006t_a + w(2{,}501 + 1.86t_a) \quad (12.8)$$

Table 12.1 Assessment scales of thermal sensation (TS)

TS sensations scale						
Cold	Cool	Slightly cool	Neutral	Slight warm	Warm	Hot
−3	−2	−1	0	1	2	3

Table 12.2 The coefficients C_1–C_6 in (12.7) (Zhu et al., 2021)

Coefficient	C_1	C_2	C_3	C_4	C_5	C_6
Constant	−5.8e+03	1.391	−4.864e−02	4.176e−05	−1.445e−08	6.546

$$w = 0.622\frac{\varphi P_{\text{sw}}}{P_{\text{b}} - \varphi P_{\text{sw}}} \tag{12.9}$$

$$\ln P_{\text{sw}} = C_1/T_{\text{a}} + C_2 + C_3 T_{\text{a}} + C_4 T_{\text{a}}^2 + C_5 T_{\text{a}}^3 + C_6 \ln T_{\text{a}} \tag{12.10}$$

where t_{a} was the air dry-bulb temperature (°C); w was the moisture content (kg/kg for dry air); φ was the RH; P_{sw} was the saturation vapor pressure over liquid water (Pa); P_{b} was the atmosphere pressure (Pa); C_1–C_6 were the regression coefficients indicated in Table 12.2; and T_{a} was the absolute air temperature (K) (Feng *et al.*, 2018).

12.4.2 Indoor environment monitoring and control module

Based on the optimal air supply parameters evaluation strategy and evaluation results, the control system mainly combines the limited monitoring and control module to achieve the optimal air supply parameters online monitoring. The main function of the limited monitoring module is to use limited sensors to obtain real-time monitoring data of environmental parameters (such as pollutant concentration, temperature, and humidity) to provide reliable inputs for the rapid global environmental prediction and evaluation of the optimal air supply parameters. For this purpose, the number of sensors and their placement strategies should be researched to improve the accuracy and efficiency of prediction. Ren and Cao (2019b) analyzed the effect of sensor placement strategy (including number and location) on the prediction performance of environmental parameters in a full-scale environmental space. This study showed that when the number of sensors is greater than or equal to 5, the prediction performance is better. When the number of sensors is equal to 3 or 4, the prediction results have large deviations.

However, by adjusting the sensor location, the prediction performance can be improved. Through placing three hypothetic sensors in low-dimension zones (cf. Figure 12.5), further research concluded the strategies of sensor deployment: (a) sensors should be avoided to be along the same or parallel with the main stream plane; (b) sensors should be better placed in or near the 'well-mix zone (close to outlet region)' but avoiding close to inlet. This deployment strategy provides some reference value for the initial planning of sensor regions.

In order to obtain a more systematic and optimal sensor deployment scheme, Cao *et al.* (2020) used clustering method of Fuzzy C-means (FCM) algorithm to partition the pollutant distribution regions and determine the cluster center of each region as the optimal sensor location, which enables the prediction of concentration field. As is shown in Figure 12.6, the FCM algorithm realized the identification of regions with different characteristics and the cluster center of respective regions.

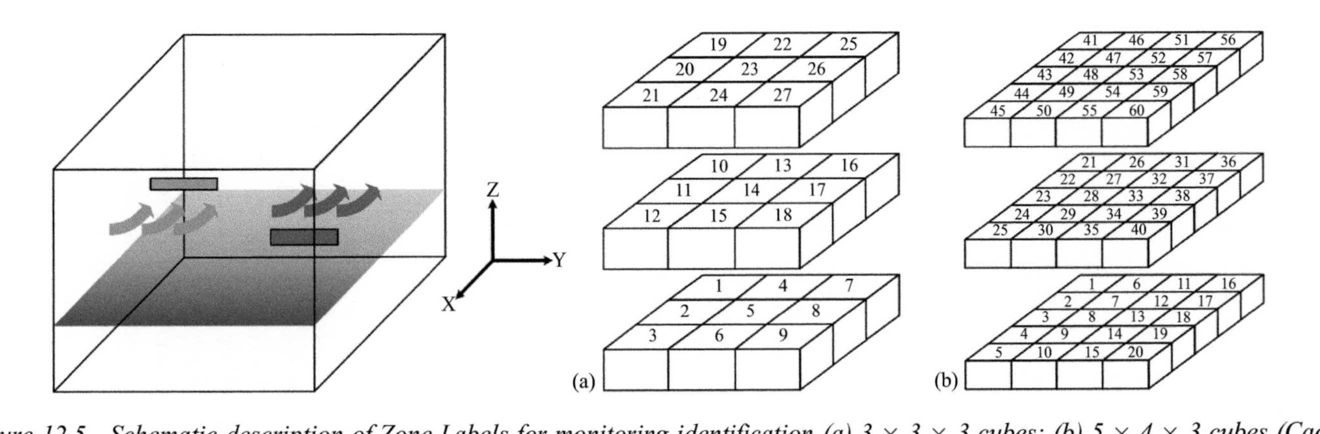

Figure 12.5 Schematic description of Zone Labels for monitoring identification (a) 3 × 3 × 3 cubes; (b) 5 × 4 × 3 cubes (Cao et al., 2020)

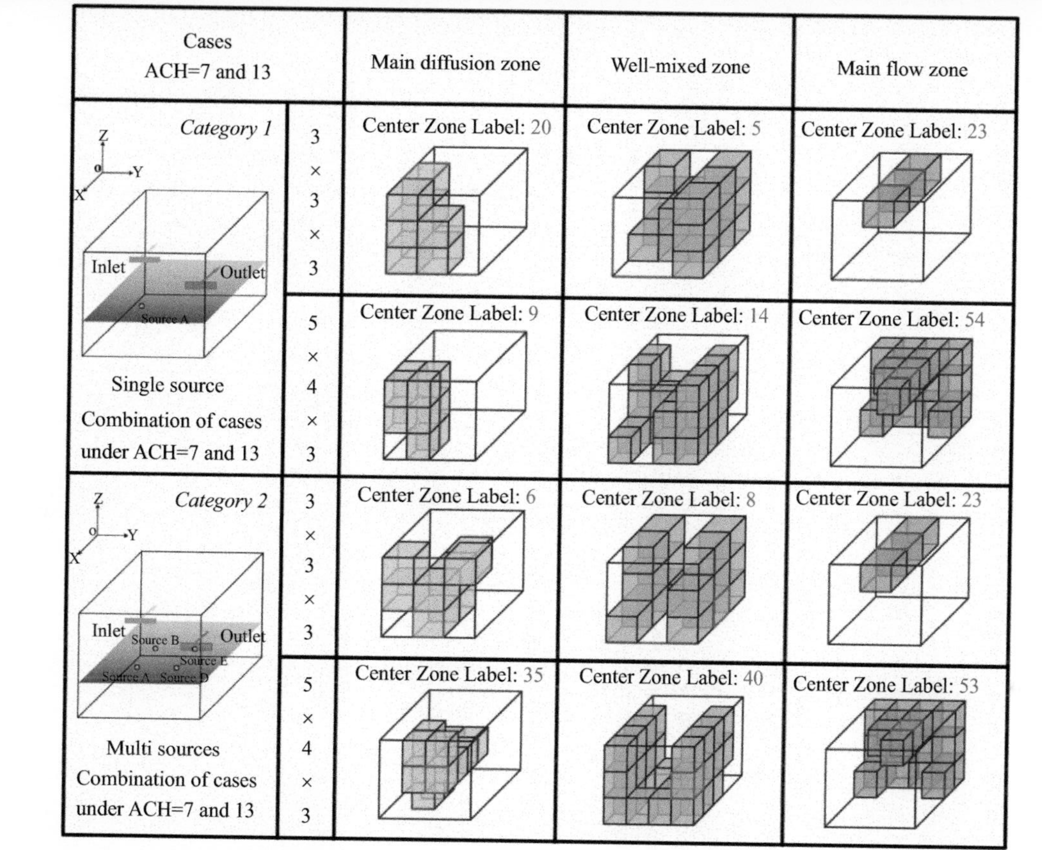

Figure 12.6 *The clustering results of hypothetic monitoring points and two categorized cases with the pollutant source position of A and ABDE, divided by (a) 3 × 3 × 3 cubes; (b) 5 × 4 × 3 cubes (Cluster center is lined by red and zone label indicated above) (Cao et al., 2020)*

The ventilated area of building environments can be mainly clustered as three main zones: (a) main flow zone with higher airflow velocity and lower concentration level; (b) diffusion zone with higher pollutant concentration and lower airflow velocity; (c) well-mixed zone with uniformly distributed lower airflow velocity and concentration level. The results show that the clustering algorithm has good performance with a maximum prediction error of 6.5% for CO_2 concentration. Based on the previous research, the online monitoring module can acquire real-time monitoring data of environmental parameters (such as CO_2 concentration, temperature, and humidity) using a limited number of sensors to provide reliable inputs for rapid global environmental prediction.

Control module mainly combines the evaluation results of the optimal air supply parameters (such as ACH, temperature, and humidity) to achieve real-time control of the HVAC system's air supply parameters. For different air supply parameters, the HVAC online monitoring and control system can select fan voltage, compressor speed, and other variables, respectively, as real-time control instructions for ACH, air supply temperature, and humidity. For example, for optimal ACH's control, based on LLVM-based ANN model, the ventilation online monitoring system is actually constructed on the strength of the small-scale environmental chamber (combined with ZigBee wireless communication technology and visualization device) to further demonstrate the visual control effect of the optimal ACH (based on the optimal ACH evaluation index E_V to obtain the evaluation results), as shown in Figures 12.7 and 12.8. The locations of the pollution source are A and B, and the initial number of ACH is 7, corresponding to the optimal ACH of 9 (Ren and Cao, 2020). Similarly, for temperature and humidity, the previous same steps can also be used to achieve indoor temperature and humidity control, thus realizing the online control of HVAC systems.

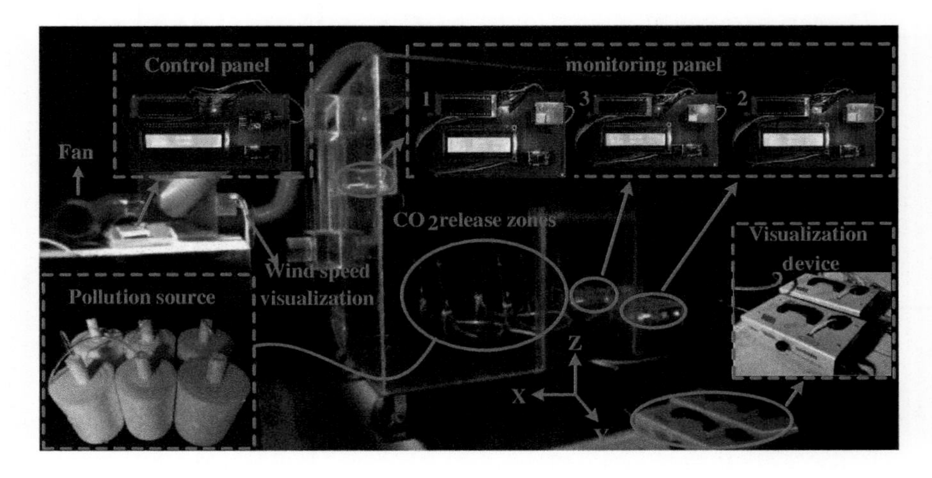

Figure 12.7 Online ventilation monitoring system based on a small size environmental chamber (Ren and Cao, 2020)

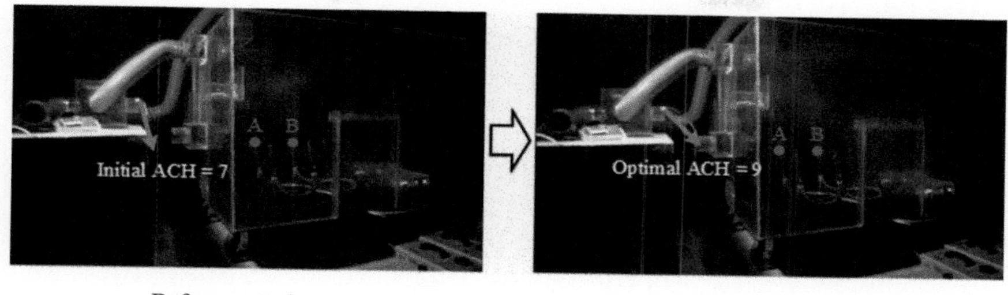

Figure 12.8 Visualization and control effect of optimal ACH based on small size environmental chamber (source location AB) (Ren and Cao, 2020)

12.5 Brief description of a case study

12.5.1 Experimental setting

The dimensions of the full-scale experimental model are 3.5 (length)×3.4 (width)× 2.5 m (height) with a window of 1.5×1.0 m, as shown in Figure 12.10(a). The source locations correspond to A (0.875, 2.55, 1.1), B (2.625, 2.55, 1.1), C (0.875, 0.85, 1.1), and D (2.625, 0.85, 1.1) m, respectively, as shown in Figure 12.10(b). In this study, we use the side-supply and side-return air supply mode. Of this mode, inlet-1 and inlet-2 are located on the sidewall with the size of 18 (length)×18 cm (width). The outlet is located on the same sidewall with a size of 18 (length)×18 cm (width). The CFD simulation grid number is 2123820. The turbulence model is selected as renormalization group k–ε model to simulate CO_2 concentration and temperature distribution (Cao and Ren, 2018; Ren and Cao, 2019a), and the low Reynolds number k–ε model is used to simulate the humidity distribution (Huang *et al.*, 2011).

For the simulation cases of temperature and humidity, inlet-1 is responsible for room temperature control and inlet-2 is responsible for humidity control. Therefore, the supply air temperature of the inlet-1 is set in the range of 16 °C–22 °C (289 K–295 K), and the supply air temperature of inlet-2 is set at 26 °C. It has shown that using 26 °C as the standard indoor temperature, there is little difference in thermal comfort and perceived air quality under high and moderate humidity conditions (Razjouyan *et al.*, 2019). The RH of 30%–60% can achieve an acceptable thermal comfort according to ASHRAE 55-1989 (ASHRAE, 2009). Therefore, the supply air RH for inlet-2 was set at 30%–60%. In addition, a related study showed that RH value around 45% may have a negligible effect on subjective perceptions. Therefore, in this study, a constant RH value of 45% was set for the inlet-1. Figure 12.9 shows the flowchart of this study, and Figure 12.10 shows the case's simulation model.

12.5.1.1 Pollutant case setting

In this case study, four pollutant source locations are set, as shown in Figure 12.10 (b). For a single source, the release rate is set as 5×10^{-6} kg/s and the background

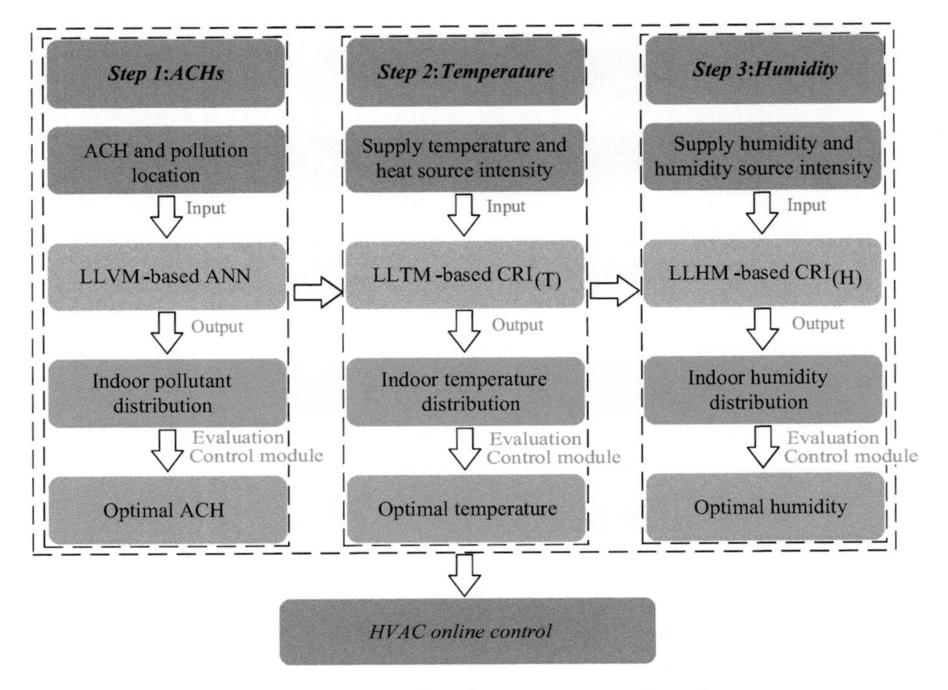

Figure 12.9 HVAC online monitoring flowchart

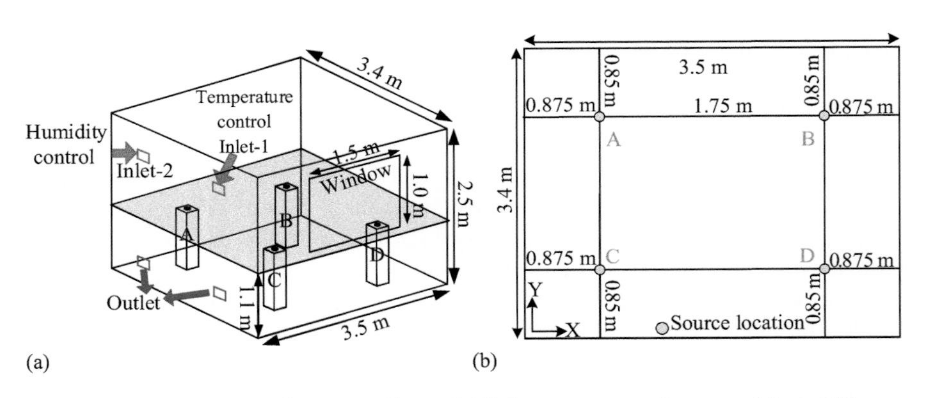

Figure 12.10 (a) Ventilation mode and (b) location coordinates of four CO_2 sources, i.e., A, B, C, and D (Zhu et al., 2021)

concentration value (molar fraction) is initialized to 5×10^{-4} (500 ppm) (Cao and Ren, 2018). ACH values of 2, 4, 6, 8, and 10 were used as initial conditions for construction of CFD database.

12.5.1.2 Temperature case setting

In this case study, based on the location of the pollutant source (as shown in Figure 12.10(b)), four heating sources with dimensions of 0.2 (length)×0.2 (width)×1.1 m (height) were set up in this case. One heating source can represent the combination of a human body and a personal computer. The intensity settings of the heat source are shown in Table 12.3 (Ren and Cao, 2019a), and all heat sources are defined by user-defined function (UDF). In addition, another heat source for window is set on the east wall of the room (with wall thickness equal to 0.2 m) (Ren and Cao, 2019a). In this study, the outdoor temperature (i.e., window surface temperature) was set at 30.7 °C (303.7 K).

12.5.1.3 Humidity case setting

In this case study, four humidity sources with dimensions of 0.45 (length)× 0.33 (width)×0.88 m (height) (Huang *et al.*, 2011) were designed, and their locations correspond to pollution sources A, B, C, and D (Figure 12.10(b)). A humidity source can represent the release of humidity from a human body and an object (e.g., a water bottle as a humidity generator). In practice, a certain amount of humidity can be transferred through the walls. In this case, we are mainly concerned with the humidity variation of indoor people, objects, and air intakes. Therefore, we did not consider the walls with low humidity release as an independent source of moisture. We continue to use the UDF method to define the humidity source for CFD simulations. More details of the simulation conditions are shown in Table 12.4.

12.5.2 Rapid prediction and optimal control of ACH

LVM is used to predict the concentration distribution of indoor pollutants (e.g., CO_2) based on the CFD simulation results. Specifically, the concentration field of pollutants resulting from multiple sources is equivalent to the superposition of the concentration field resulting from a single source, which is used to rapidly expand the database based on the initial simulation results. On the basis of LVM, LLVM is further proposed, which can reduce the data storage of the CO_2 concentration field to a large extent. Besides, an ANN is able to rapidly predict the relationship between inputs (e.g., location and intensity of CO_2 source) and outputs (e.g., CO_2 concentration). In combination with LLVM, we use an LLVM-based ANN for the rapid prediction of indoor CO_2 concentrations. It is evident from Figure 12.11 that

Table 12.3 Thermal parameters of heating sources, including human body, personal computer, and window (Ren and Cao, 2019a)

	Heat source (human body)				Window	Personal computer
	A	B	C	D		
Source intensity	85 W	85 W	85 W	85 W	6.17 W/(m^2 K)	50 W

Table 12.4 Conditions and settings for humidity simulation (Zhu et al., 2021)

CFD model	**Low Reynolds number *k*–*ε* model**
Mesh	2123820
Differencing scheme	Convection term: second-order upwind
Air inlet	• Velocity: based on the optimal ACH value • Humidity: inlet-1: 45% relative humidity • Inlet-2: 30%–60% relative humidity
Moisture generation	Moisture (human):60 g/h; moisture (kettle): 40 g/h
Wall	Materials: no moisture generation; others: no-slip

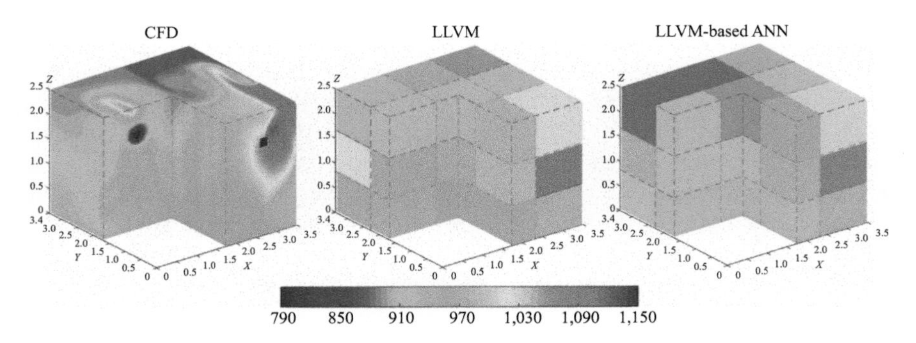

Figure 12.11 Comparisons of CO_2 concentrations (ppm) between CFD, LLVM, and LLVM-based ANN when ACH, equal to 4 (Zhu et al., 2021)

the prediction result is within the acceptable error range. Therefore, LLVM-based ANN approach can be used to efficiently predict indoor environmental fields (e.g., indoor CO_2 concentration distribution).

Then, with the assistance of fast prediction results of indoor CO_2 concentration and self-defined evaluation index E_V, we comprehensively assessed the balance between indoor CO_2 concentration levels and energy consumption (ventilation rate) by calculating the E_V values, as shown in Table 12.5, the minimum E_V value corresponded to the optimal selection of the targeted ACH value, i.e., 4 in this case.

12.5.3 Rapid prediction and optimal control of supply air temperature

Based on the optimal ACH value, the next step will derive the optimization of cooling load (i.e., supply air temperature) of the HVAC system. The validity of LLTM has already been validated (Ren and Cao, 2019a). By combing $CRI_{(T)}$ model, we can further adopt the well-behaved LLTM-based $CRI_{(T)}$ method for a rapid prediction of indoor temperature field.

Table 12.5 The calculated results of E_V index for different inlet ACHs (Zhu et al., 2021)

ACH	2	4	6	8	10
E_V	0.54	0.51	0.62	0.74	0.80

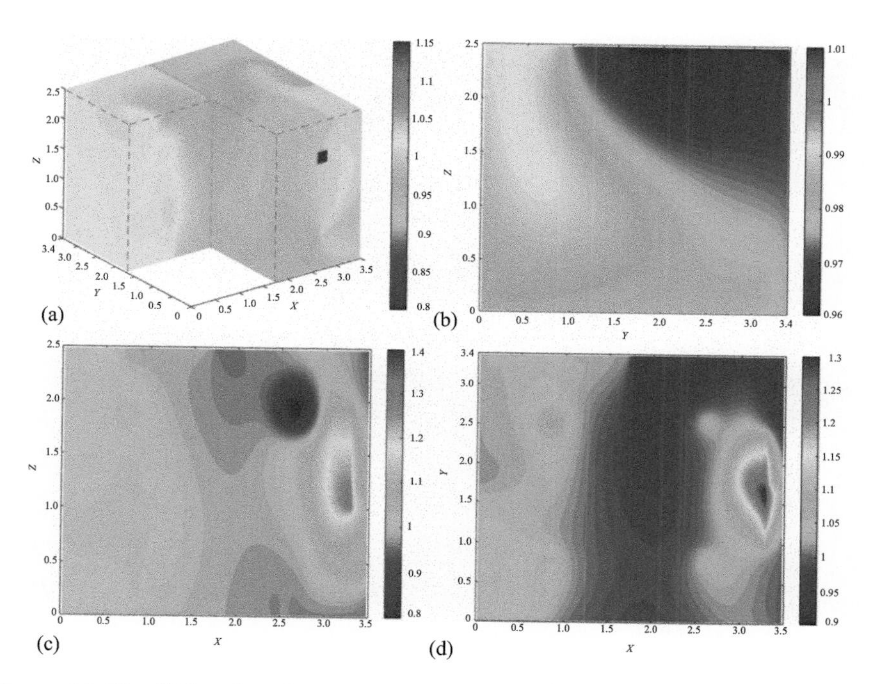

Figure 12.12 $CRI_{(T)}$ distribution with inlet temperature of 16 °C (Zhu et al., 2021): (a) stereogram; (b) plane of x = 1.75 m; (c) plane of y = 1.7 m; (d) plane of z = 1.1 m

According to the $CRI_{(T)}$ equation, the distributions of $CRI_{(T)}$ were calculated for the case with the temperature of inlet-1 being equal to 16 °C (as shown in Figure 12.12).

Then, the rapid prediction efficiency of indoor temperature distribution is improved by the combination of LLTM and $CRI_{(T)}$. Figure 12.13 shows the comparisons of the temperature fields obtained by the LLTM-based $CRI_{(T)}$ model and CFD simulation, respectively (with the temperature of inlet-1 set at 20 °C). It is found that the LLTM-based $CRI_{(T)}$ temperature distribution results are consistent with the CFD results. Therefore, it is believed that the LLTM-based $CRI_{(T)}$ model can be more effective for a rapid prediction of temperature distribution with less data storage.

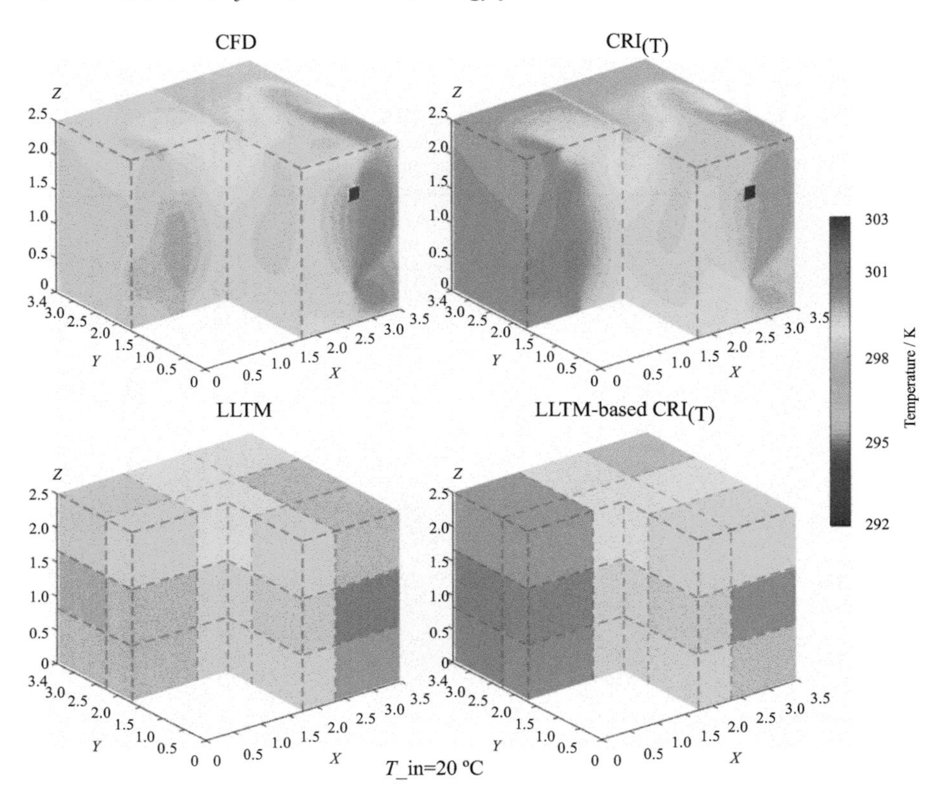

Figure 12.13 Comparisons of temperature distributions from CFD and LLTM-based $CRI_{(T)}$ prediction with inlet-1 temperate set as 20 °C (Zhu et al., 2021)

On the basis of indoor temperature prediction results, the balance between indoor thermal comfort (expressed as PMV value) and energy consumption (i.e., air supply temperature) is evaluated using the evaluation index E_T (as shown in (12.4)). The minimum E_T value for different inlet temperature conditions corresponds to the optimal supply air temperature selection at inlet-1. Considering the adjustment range of the supply air temperature of 16 °C–22 °C (in summer conditions), the optimal supply air temperature of 21 °C was obtained in this study. To show the control effect of the optimum temperature, Table 12.6 shows the corresponding absolute PMV values for different supply air temperatures. From the table, it can be seen that when the temperature of the inlet-1 is equal to 21 °C, the PMV is closest to the ideal value for optimal thermal comfort (i.e., PMV = 0). Furthermore, it has shown that the energy consumption of HVAC systems can increase by approximately 8% when the average indoor temperature or the supply air temperature of the air conditioning system is decreased by 1 °C (Cao and Deng, 2019). Therefore, by optimizing the inlet air temperature, the HVAC system has the

Table 12.6 The calculated results of PMV for different inlet-1's temperatures (Zhu et al., 2019)

	16 °C	17 °C	18 °C	19 °C	20 °C	21 °C	22 °C
|PMV|	0.821	0.682	0.421	0.302	0.218	0.0102	0.015

potential to reduce the energy consumption of the cooling load by about 32% in this case study.

12.5.4 *Rapid prediction and optimal control of supply air humidity*

Based on the optimal ACH and temperature, we start to predict and control the optimal indoor humidity in this case. The LHM and LLHM models have already been validated by CFD simulation in the previous research (with acceptable errors for engineering application) (Zhu *et al.*, 2021).

Then, we calculated the $CRI_{(H)}$ values, when the inlet temperature and RH of inlet-1 were equal to 21 °C and 45%, respectively, and the inlet temperature and RH of inlet-2 were equal to 26 °C and 25%, respectively. Figure 12.14 shows the stereogram viewpoint and the planar $CRI_{(H)}$ distribution for $x = 1.75$ m, $y = 1.7$ m, and $z = 1.1$ m. The $CRI_{(H)}$ values were then used to predict the indoor humidity distribution.

Combining $CRI_{(H)}$ and LLHM, LLHM-based $CRI_{(H)}$ model can further improve the prediction efficiency of indoor humidity distribution. Figure 12.15 shows the comparative results of CFD and LLHM-based $CRI_{(H)}$ for inlet-2 with supply air RH equal to 40%. It can be seen that the results of the LLHM-based $CRI_{(H)}$ are well acceptable compared to the CFD results.

Then, the thermal comfort indicator TS and HVAC energy consumption (indicated by RH of inlet-2) was combined in the evaluation index E_H to assess the humidity environment for obtaining the optimal supplied air RH (corresponding to the minimum value of E_H). Regarding the TS indicator, when the value of TS was close to zero, the comfortable state of indoor personnel should be ideal. It was analyzed by the co-effects of temperature and humidity on the total energy consumption of integrated HVAC system, which could be more closely matching the real energy consumption of HVAC system during daily regulation. According to (12.8), the enthalpy required for the control process is calculated from the outdoor air parameters before control and the indoor air parameters after control are as a reference to the energy calculation. In standard GB50736-2012 (2012), it can be obtained that the temperature and humidity parameters for outdoor calculation (taking Guangzhou city as example) are respectively corresponding to 30.7 °C and 68%. Besides, the temperature and humidity parameters at the inlet processed by the traditional HVAC system are corresponding to 16 °C and 50%. On these bases, the selections of inlet-2's supplied air RH for the optimal control solutions were shown in Table 12.7, by considering five different conditions of humidity sources

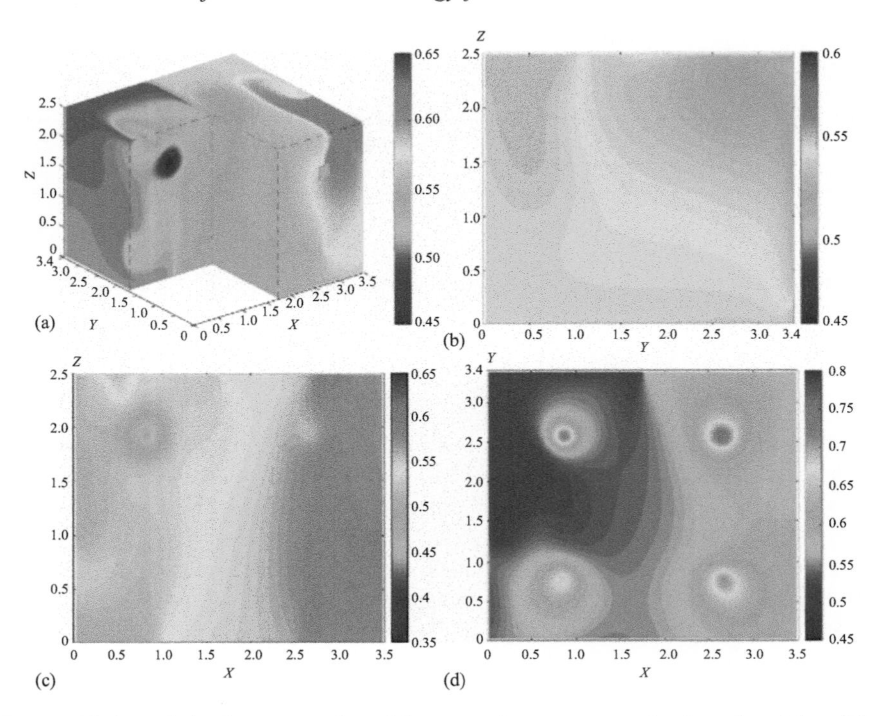

Figure 12.14 Calculation results of the contribution ratio of indoor humidity (Zhu et al., 2021): (a) stereogram; (b) plane of x = 1.75 m; (c) plane of y = 1.7 m; (d) plane of z = 1.1 m

(including human body and kettle), and the RH control ranges of inlet-2 were from 30% to 60% (with the optimal inlet-1's supplied air temperature equal to 21 °C). It can be seen that the optimal supplied air RH at inlet-2 varied between 30% and 40%. Table 12.7 also lists the TS values in the occupied areas and energy savings of HVAC system corresponding to the optimal strategies. According to (12.8), the enthalpy value required for the operation of air-conditioning system could be rapidly calculated in order to further obtain the energy-saving results. Based on the optimal selections of supplied air RH, the total energy consumption of HVAC system (considering cooling and humidity loads) could be maximally decreased by 35% (with inlet-2's RH and supplied air temperature corresponding to 40% and 26 °C).

12.6 Occupant based ventilation control for infection prevention

COVID-19 is a respiratory disease caused by a coronavirus (SARS-COV-2), which has a high-level of infection and morbidity. Ventilation plays a vital role in

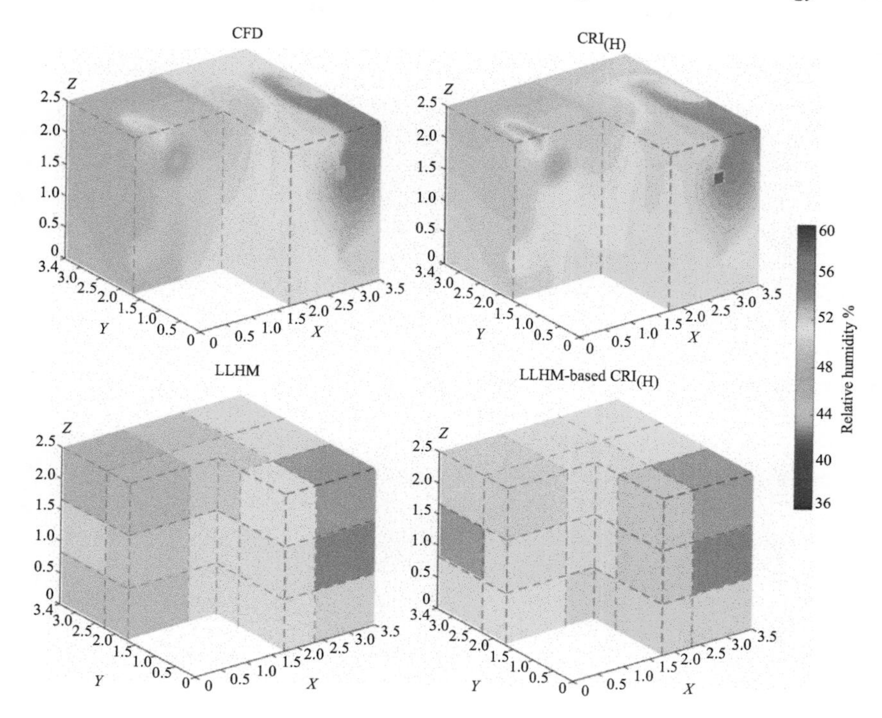

Figure 12.15 Comparisons of relative humidity distribution between CFD and LLHM-based $CRI_{(H)}$ with the inlet-2's supplied air relative humidity being equal to 40% (Zhu et al., 2021)

Table 12.7 Selection results of supplied air relative humidity for the optimal control strategy with corresponding TS values in the occupied areas and energy savings of HVAC system (considering different humidity conditions of humidity source including human body and kettle) (Zhu et al., 2021)

Indoor humidity conditions (g/h)				Control strategies corresponding to the minimum E_H (control range of relative humidity at inlet-2: 30%–60%)		
A	**B**	**C**	**D**	**Optimal humidity (%)**	**TS**	**Energy saving (%)**
60	60	60	60	40	−0.008	35
100	60	60	60	30	−0.05	24
100	100	60	60	35	−0.009	29
100	100	100	60	35	0.003	29
100	100	100	100	30	0.007	24

prevention and control of COVID-19 and alike diseases in enclosed indoor environment and specially in high-occupant-density spaces. To reduce the infection risk, ventilation rate should be manipulated based on occupant situation. Thus, occupant detection is fundamental. Currently, environmental monitoring (Jin *et al.*, 2018), information and communication technology (Wang *et al.*, 2019) and cameras (Wang *et al.*, 2018) can be used in occupant detection. Here, a camera-based occupant detection method is proposed because it can only provide quantity and spatial distribution of occupants, but also detecte or track the crowd behaviors. Besides, surveillance cameras are usually installed in buildings, which enhances the data accessibility.

Figure 12.16 shows the proposed occupant based ventilation control, control schematic and hardware prototype. A deep-learning based object detection algorithm, YOLO (You Only Look Once), was adopted, which achieved a 92.5% of the detection accuracy (acceptable for engineering applications). In the case study (Wang *et al.*, 2021), the traditional fixed ventilation mode (15% outdoor air ratio) achieved an infection probability of 12.5%. In comparison, the proposed occupant based ventilation control mode can reduce the infection probability to 2% and save 11.7% of energy consumption at the same time, which provides a feasible and promising solution to fight against COVID-19.

12.7 Summary

This chapter introduces a systematic online monitoring strategy for HVAC, including online monitoring, "faster-than-real-time" prediction, and optimal evaluation and control module to adapt to the unsteady, nonlinear and nonuniform dynamic distribution of indoor environmental parameters (such as CO_2 concentration, temperature, and humidity). The strategy aims to effectively address the "faster-than-real-time" prediction issues of indoor environmental parameters by combining limited monitoring data with various "faster-than-real-time" prediction models, including low-dimensional linear models, ANN, contribution ratio of indoor climate ($CRI_{(T)}$), contribution ratio of indoor humidity ($CRI_{(H)}$), and coupled prediction models (i.e., LLVM-based ANN, LLTM-based $CRI_{(T)}$, and LLHM-based $CRI_{(H)}$). Then, this strategy, respectively, achieved the accurate evaluation and control of optimal air supply parameters (such as ACH, air supply temperature, and humidity) so as to achieve good control effects (e.g., effective improvement of IEQ, including the quality of ventilation, thermal and humidity environments, and significant reduction in energy consumption for HVAC system operation). This chapter can provide a theoretical basis for the practical construction of online monitoring systems for HVAC.

For the online monitoring of HVAC systems, it mainly focuses on separative monitoring strategies for multiple environmental parameters, i.e., CO_2 concentration, temperature, and humidity. However, for actual HVAC systems, the coupled evaluation and control strategies of indoor environment (including ventilation, temperature and humidity environment) need to be studied, especially for the

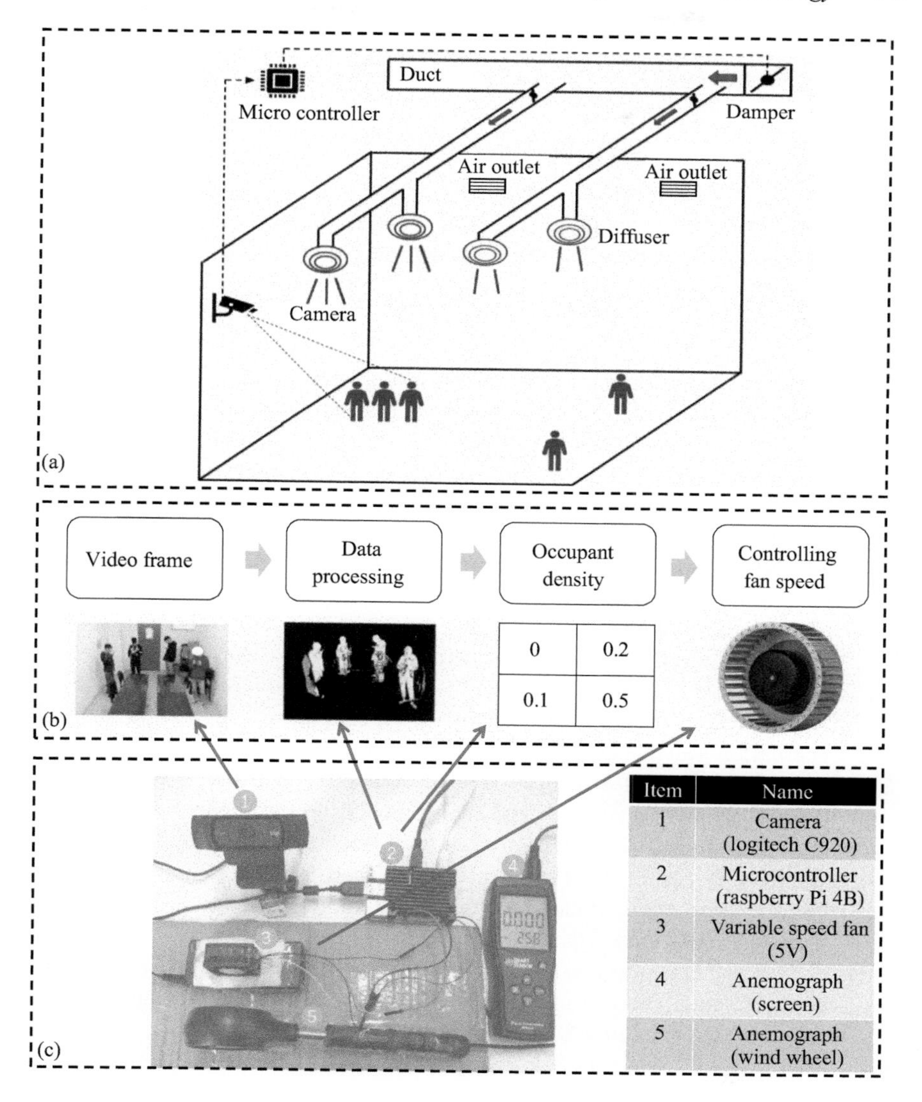

Item	Name
1	Camera (logitech C920)
2	Microcontroller (raspberry Pi 4B)
3	Variable speed fan (5V)
4	Anemograph (screen)
5	Anemograph (wind wheel)

Figure 12.16 (a) occupant based ventilation control; (b) the control schematic; (c) hardware prototype (Wang et al., 2021)

densely populated built environment (e.g., public and industrial buildings). To this end, it is important to develop “faster-than-real-time” prediction models coupled with limited monitoring data and nonlinear dimensionality reduction models. In addition, exploring the coupled evaluation strategy for multiple air supply parameters (e.g., ACH, air supply temperature, and humidity) will further improve the feasibility of HVAC online control strategy and provide reference for constructing the actual HVAC online monitoring system.

References

ASHRAE. *ASHRAE Handbook – Fundamentals*. Atlanta, GA, USA: American Society of Heating. Refrigerating and Air-Conditioning Engineers, 2009.

Bluyssen P M, Aries M, and Dommelen P V. Comfort of workers in office buildings: The European HOPE project. *Building and Environment*, 2011, 46(1): 280–288.

Cao S J, Ding J W, and Ren C. Sensor deployment strategy using cluster analysis of Fuzzy C-means algorithm: Towards online control of indoor environment safety and health. *Sustainable Cities and Society*, 2020, 59(8): 102190.

Cao S J and Meyers J. Asymptotic conditions for the use of linear ventilation models in the presence of buoyancy forces. *Building Simulation*, 2014, 7(2): 131–136.

Cao S J and Meyers J. On the construction and use of linear low-dimensional ventilation models. *Indoor Air*, 2012, 22(5): 427–441.

Cao S J and Ren C. Ventilation control strategy using low-dimensional linear ventilation models and artificial neural network. *Building and Environment*, 2018, 144(10): 316–333.

Cao S J. Challenges of using CFD simulation for the design and online control of ventilation systems. *Indoor and Built Environment*, 2019, 28(1): 3–6.

Cao S-J and Deng H Y. Investigation of temperature regulation effects on indoor thermal comfort, air quality, and energy savings toward green residential buildings. *Science and Technology for the Built Environment*, 2019, 25: 309–321.

Chen C, Zhao B, Lai D Y, *et al.* A simple method for differentiating direct and indirect exposure to exhaled contaminants in mechanically ventilated rooms. *Building Simulation*, 2018, 11(3): 1039–1051.

Deng Y L, Feng Z B, Fang J, *et al.* Impact of ventilation rates on indoor thermal comfort and energy efficiency of ground-source heat pump system. *Sustainable Cities and Society*, 2018, 37(2): 154–163.

Feng Z B, Yu C W, and Cao S J. Fast prediction for indoor environment: Models assessment. *Indoor and Built Environment*, 2019, 28(6): 727–730.

Feng Z B, Zhou X Q, Xu S H, *et al.* Impacts of humidification process on indoor thermal comfort and air quality using portable ultrasonic humidifier. *Building and Environment*, 2018, 133(4): 62–72.

GB50736-2012. Design Code for Heating, Ventilation and Air Conditioning in Residential Buildings. Ministry of Housing and Urban-Rural Development of the People's Republic of China (MOHURD), 2012.

Huang H, Kato S, Hu R, and Ishida Y. Development of new indices to assess the contribution of moisture sources to indoor humidity and application to optimization design: Proposal of $CRI_{(H)}$ and a transient simulation for the prediction of indoor humidity. *Building and Environment*, 2011, 46: 1817–1826.

Kim J T and Yu C W F. Sustainable development and requirements for energy efficiency in buildings – The Korean perspectives. *Indoor and Built Environment*, 2018, 27: 734–751.

Paula Xavier A A de and Roberto L. Indices of thermal comfort developed from field survey in Brazil. U.S. Department of Energy Office of Scientific and Technical Information. *ASHRAE Winter Meeting*. United States: ASHRAE Transactions, 2000: 929.

Razjouyan J, Lee H, Gilligan B, Lindberg C, and Najafi B. Wellbuilt for wellbeing: Controlling relative humidity in the workplace matters for our health. *Indoor Air*, 2019, 30(1).

Ren C and Cao S J. Development and application of linear ventilation and temperature models for indoor environmental prediction and HVAC systems control. *Sustainable Cities and Society*, 2019a, 51(11): 101673.

Ren J and Cao S J. Incorporating online monitoring data into fast prediction models towards the development of artificial intelligent ventilation systems. *Sustainable Cities and Society*, 2019b, 47(5): 101498.

Ren C and Cao S J. Implementation and visualization of artificial intelligent ventilation control system using fast prediction models and limited monitoring data. *Sustainable Cities and Society*, 2020, 52(1): 101860.

Ren J and Cao S J. Development of self-adaptive low-dimension ventilation models using OpenFOAM: Towards the application of AI based on CFD data. Building and Environment, 2020a, 171:106671.

Ren C and Cao S J. Construction of linear temperature model using non-dimensional heat exchange ratio: Towards fast prediction of indoor temperature and heating, ventilation and air conditioning systems control. *Energy and Buildings*, 2021, 251:111351.

Chen X, Li A G, and Gao R. Numerical investigation on particle deposition in a chamber with an attached-wall heat source. *Indoor and Built Environment*, 2014, 23: 640–652.

Zhang W R, Hiyama K, Kato S, and Ishida Y. Building energy simulation considering spatial temperature distribution for nonuniform indoor environment. *Building and Environment*, 2013, 63, 89–86.

Zhu H C, Yu C W, and Cao S J. Ventilation online monitoring and control system from the perspectives of technology application. *Indoor and Built Environment*, 2019, 29(4): 587–602.

Zhu H C, Ren C, and Cao S J. Fast prediction for multi-parameters (concentration, temperature and humidity) of indoor environment towards the online control of HVAC system. *Building Simulation*, 2021, 14(3): 1–17.

Chapter 13
Ventilation and health

Chan Lu[1], *Jing Li*[2] *and Qihong Deng*[1,2,3]

13.1 Indoor ventilation and human health

13.1.1 Indoor ventilation type

Ventilation, normally characterized by outdoor and indoor, is closely associated with indoor air quality (IAQ). Indoor air pollution has a greater impact than outdoor air pollution as most of the people spend up to 90% of their time in indoor environment (Nazaroff and Goldstein, 2015; Zhang and Smith, 2003). Thus, IAQ is greatly important for the quality of human life, including the productivity (Wargocki and Wyon, 2013) and performance (Theodosiou and Ordoumpozanis, 2008), and health condition (Sundell *et al.*, 2011). Both natural and mechanical ventilations (MVs) determine the level of indoor air pollution and air quality. However, indoor air pollution and IAQ mainly depend on outdoor ventilation or air quality especially in the buildings with both natural and MV systems (Kukadia and Palmer, 1998).

There are two major types of indoor ventilation which could improve IAQ in residential buildings: natural ventilation (NV) and MV (Ye *et al.*, 2017; Zhao *et al.*, 2018a) (Figure 13.1). NV needs low maintenance with zero-energy requirement, which can increase the level of thermal comfort and produce a green indoor environment via continuously supplying fresh air indoors, mainly by opening windows and doors (Omrani *et al.*, 2017). However, NV is not always safe as it may bring outdoor polluted air to indoors which could elevate the exposure level of indoor air pollution from traffic-related air pollution, industrial air pollution, and outdoor particulate matter (PM) (Xu *et al.*, 2017). Furthermore, the increased outdoor air pollution can interplay with the indoor environmental pollution, including renovation (Jiang *et al.*, 2018), mould/dampness (Lu *et al.*, 2019), and environmental tobacco smoke (ETS) (Lu *et al.*, 2020a), which largely increased the risk of health outcomes (Lu *et al.*, 2020b, 2020c). On the other hand, MV system with relatively high energy requirement is commonly characterized as the kitchen

[1]XiangYa School of Public Health, Central South University, Changsha, China
[2]School of Energy Science and Engineering, Central South University, Changsha, China
[3]School of Public Health, Zhengzhou University, Zhengzhou, China

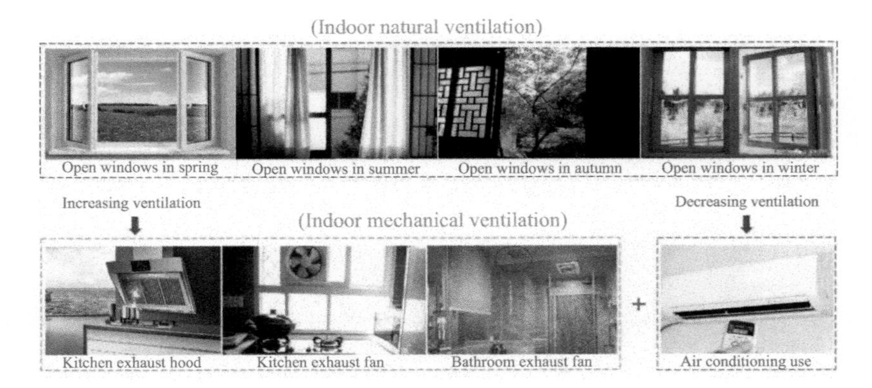

Figure 13.1 Two major types of indoor ventilation in residential houses

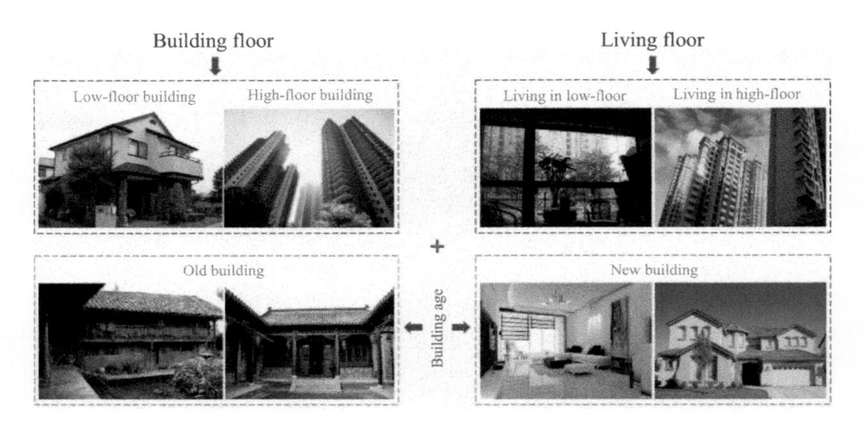

Figure 13.2 Different types of the living buildings in China

exhaust hood or fan, bathroom exhaust fan, and air conditioning, as shown in Figure 13.1. Mounting evidence studies have demonstrated that the use of MV system, especially containing the air filter, could effectively reduce the concentration of indoor air pollution (Park *et al.*, 2014; Quang *et al.*, 2013; Stephens and Siegel, 2012). Moreover, the building characteristics or house condition (Figure 13.2) can also influence the ventilation rate and IAQ. However, the distribution and trend of different ventilation types on indoor air pollution and human health are rarely reported, particularly in developing countries, such as China (Figure 13.3).

NV is not always predictable or sufficient in indoor environment (Ye *et al.*, 2017). Thus, living area, building characteristic, and house condition need to be carefully considered and designed to improve indoor ventilation level. It is well known that NV in multi-storey or high-floor buildings has become an emerging topic in China (Zhou *et al.*, 2014). China is now producing the largest quality of

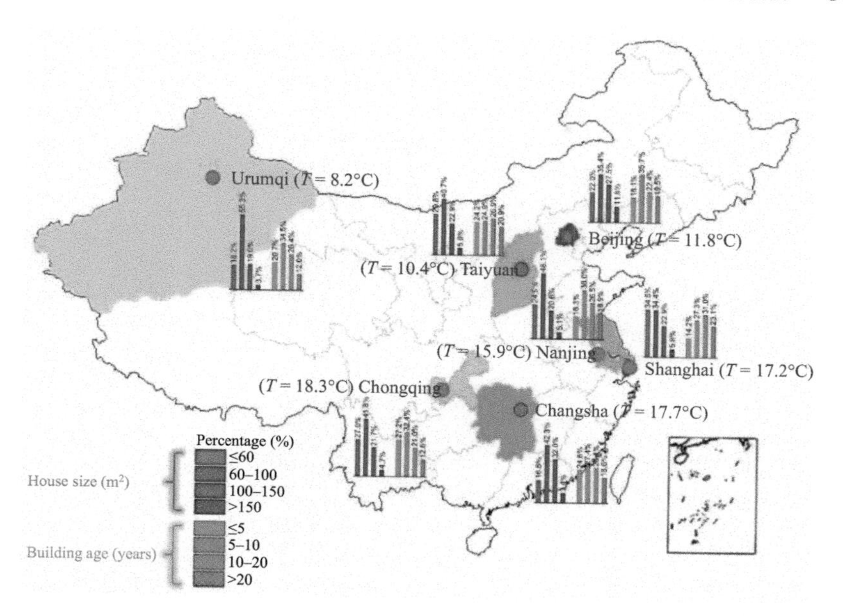

Figure 13.3 Map of seven cities in China from the "China-Children-Home-Health (CCHH)" survey during 2010–12 (annual mean air temperature, and the percentage of the different house sizes and building ages among surveyed family with children aged 3–6 in each city)

new furniture, wood-based panels, and coatings worldwide (Liu *et al.*, 2012), and thereby many chemical materials for new furnishing, painting, and decoration in new buildings have emerged in our house and life. Exposure to these toxic chemicals, such as volatile organic compounds (VOCs) and semi-VOCs (SVOCs) derived from new furniture and redecoration in the homes, has become a major concern and leads to serious diseases burden, including allergies (asthma, allergic rhinitis, and eczema) (Lu *et al.*, 2020b, 2020c), infectious diseases (pneumonia, ear infection, and common cold) (Deng *et al.*, 2017a; Jiang *et al.*, 2018; Norbäck *et al.*, 2018), birth outcomes (Lu *et al.*, 2019, 2020a), and sick building syndrome (SBS) (Lu *et al.*, 2016) during the past decades, which suggests that poor ventilation in the indoor environment (i.e., homes, office, and schools) can be a significant problem (Zhang *et al.*, 2013b).

With a rapid development of economy and industrialization, China has witnessed a serious increase in ambient air pollution over past years (Deng *et al.*, 2015a). Outdoor air pollution in China is characterized by high concentration level and mixed pollution sources, which is significantly different compared to the developed countries (Deng *et al.*, 2015a). Theoretically, indoor NV can be unhealthy to use in a condition of severe outdoor air pollution unless air has been properly purified or treated (Ye *et al.*, 2017). Although using air cleaners could effectively reduce indoor air pollution even under haze or extreme polluted days meanwhile with NV via opening windows, only using air cleaners is unable to

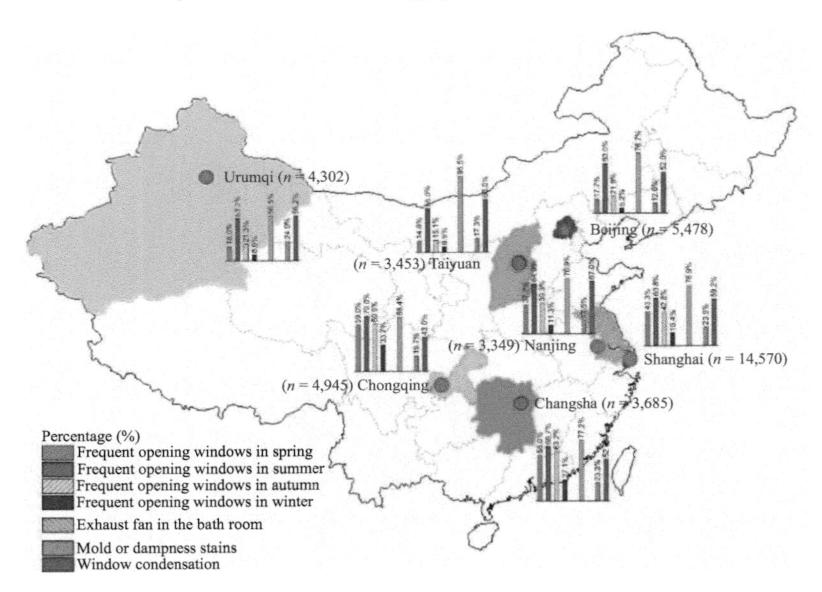

Figure 13.4 Map of seven cities in China from the "China-Children-Home-Health (CCHH)" survey during 2010–12 (percentage of the natural ventilation via opening windows in the four seasons, mechanical ventilation via exhaust fan in the bathroom, and poor ventilation-related indicators, including mould/damp stains and window condensation among surveyed family with children aged 3–6 in each city)

remove all indoor air pollution, such as particles and VOCs, and some may generate some other pollutants from by-products during the utilization (Zhang *et al.*, 2011). Hence, the role of NV on IAQ in dwellings can be in both positive and negative ways, which indicates that NV cannot be the only or the best way to choose particularly in developing countries with serious air pollution both indoors and outdoors. In a way, the MV, mainly including kitchen exhaust hood and bath room exhaust fan, is also necessary in the residential homes. However, a majority of residences in developing countries mainly use NV via opening window except under extreme weather, such as China (Figure 13.4).

Low ventilation rate due to both poor natural and MV indoors can lead to an increase in various dampness-related indicators including mould/damp stains, mouldy odour, and window pane condensation (Liu *et al.*, 2015b; Wargocki, 2013). In China, the frequency of NV via opening windows changed greatly between different seasons with the highest in summer while the lowest in winter; the MV indicated as exhaust fan in the bathroom was widely used with the percentage ranging from 56.5% to 95.5%; and the percentages of families with poor ventilation-related indicators, including mould/damp stains and window condensation, were 12.6%–24.9% and 43.0%–67.0%, respectively (Figure 13.4). These poor ventilation indicators such as mould and dampness not only aggravate pre-existing respiratory conditions but also cause new onset symptoms and allergic diseases (D'Amato *et al.*, 2015; Mendell *et al.*, 2011). Also, it could contribute to the depression (Shenassa *et al.*, 2007) and respiratory infections (Fisk *et al.*, 2010) which may in turn increase the risk of asthma and respiratory symptoms (Atkinson, 2013; Douwes *et al.*, 2011). A recent review suggested that visible mould and mouldy odour are associated with the development and exacerbations of asthma and allergies in children, indicating a causal relationship (Caillaud *et al.*, 2018). Some other review studies have indicated that the exposure to the poor ventilation indicators, mainly for mould and dampness, during postnatal or current period, is strongly related with childhood allergies (i.e., asthma, allergic rhinitis, and eczema) and infections (i.e., common cold) (Mendell *et al.*, 2011). However, most available evidence focused on allergy and infection risk of postnatal or current exposure to mould and dampness, but the prenatal or perinatal effect has been rarely addressed (Norbäck *et al.*, 2017, 2018).

13.1.2 Indoor air pollution and health effect

Air pollution is considered to be the main influencing factor of global burden of diseases (GBD 2015 Risk Factors Collaborators, 2016), among which indoor and outdoor air pollutions rank the fourth and ninth, respectively (GBD 2017 Risk Factor Collaborators, 2017; Lelieveld *et al.*, 2015). Recently, the World Health Organization (WHO) estimates that 7 million people die of air pollution annually in the world, and the air pollution concentration of most people's living environment exceeds the limit standard of WHO (Brauer *et al.*, 2012). China is one of the countries with the most serious air pollution in the world (Guan *et al.*, 2016; Zhao *et al.*, 2018b), and the number of death caused by air pollution is the highest in the world (GBD 2017 Risk Factor Collaborators, 2017; Zhou *et al.*, 2019).

Indoor air pollution is mainly caused by burning solid fuel sources, including firewood, crop waste, biomass combustion for cooking and heating, ETS, and VOCs emitted from new furniture and redecoration, as well as poor ventilation-related indicators such as mould/dampness. These indoor air pollutants, particularly in poor households, lead to respiratory allergic and infectious diseases which can result in premature death. Therefore, the WHO calls indoor air pollution 'the world's largest single environmental health risk (WHO, 2014a)'.

13.1.2.1 Indoor air pollution and premature death

Indoor air pollution is one of the world's largest environmental problems, particularly for the poorest in the world who often do not have access to clean fuels and great building design. The Global Burden of Disease estimates that indoor air pollution is a risk factor for several of the world's leading causes of death, including heart disease, pneumonia, stroke, diabetes, and lung cancer globally (WHO, 2014b). According to the Global Burden of Disease study (GBD 2017 Risk Factor Collaborators, 2017) 1.6 million people died prematurely in 2017 as a result of indoor air pollution, while the WHO estimates 4.3 million deaths derived from indoor air pollution in 2012 (the latest available data) (WHO, 2014a) (Figure 13.5).

Indoor air pollution is one of the leading causes for deaths globally, while it ranks the leading risk factor for death in low-income countries (Ritchie and Roser, 2013).

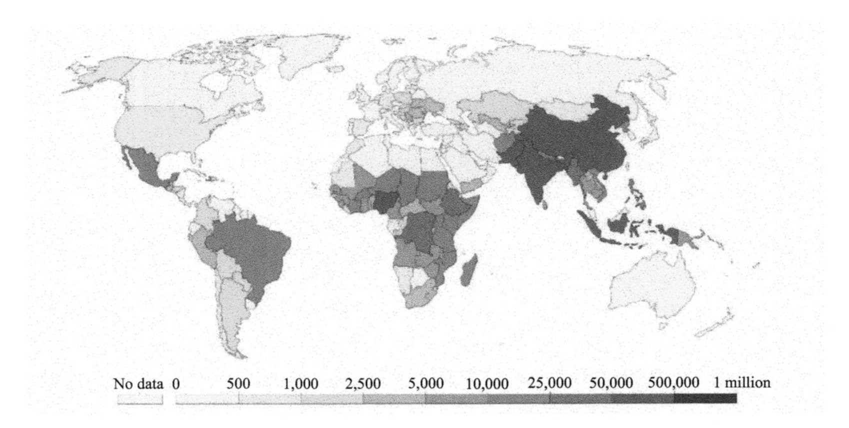

Figure 13.5 Number of deaths from indoor air pollution, 2017 (GBD 2017 Risk Factor Collaborators, 2017). Source: IHME, Global Burden of Disease. OurWorldInData.org/indoor-air-pollution/

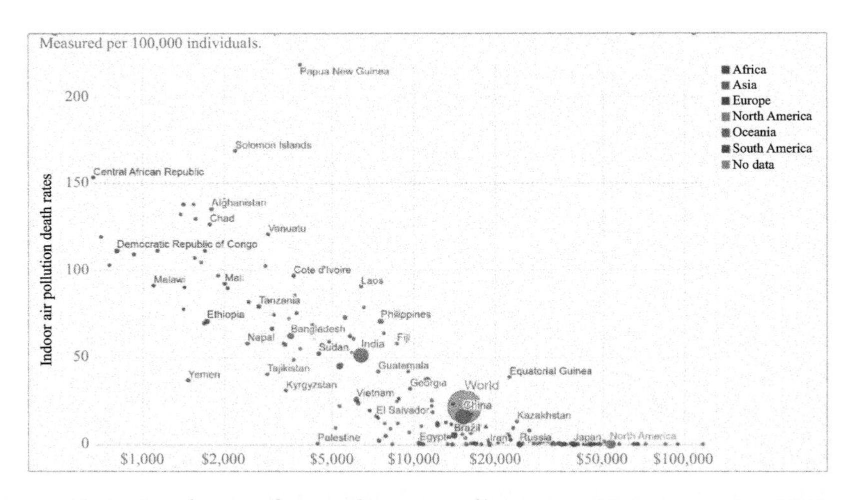

Figure 13.6 Death rates from indoor air pollution vs. GDP per capita, 2017. Source: Institute for Health Metrics and Evaluation (IHMW); World Bank. OurWorldInData.org/indoor-air-pollution/

The Institute for Health Metrics and Evaluation (IHME) estimates that 1.6 million people died prematurely as a result of indoor air pollution in 2017 which accounts for 3% of global deaths, while it accounts for 6% of deaths in low-income countries. The issue of indoor air pollution therefore has a clear economic split: it is a problem that has almost been entirely eliminated across high-income countries but remains a large environmental and health problem at lower incomes. There is a strong negative relationship: death rates decline as countries get richer, as shown in Figure 13.6.

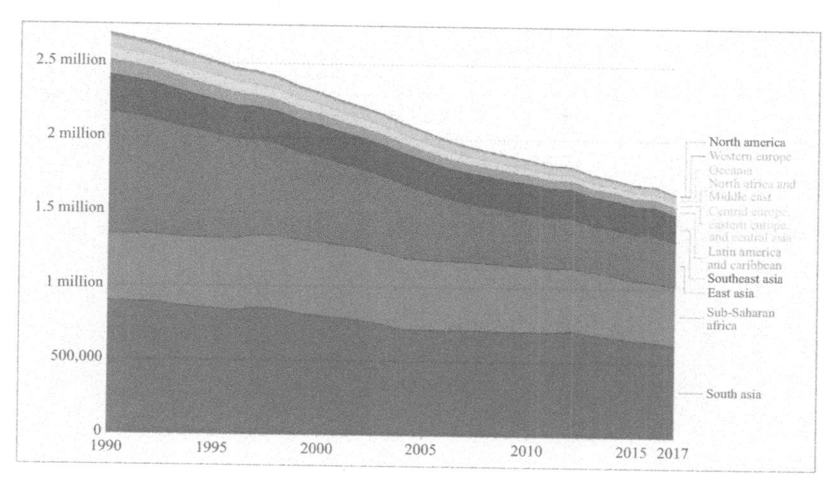

Figure 13.7 Annual number of premature deaths attributed to household air pollution from the use of solid fuels for cooking and heating, including crop water, dung, charcoal, and coal by region, 1990–2017

Whilst indoor air pollution still remains one of the leading risk factors for mortality, particularly the largest risk factor at low incomes, the world has also witnessed a significant progress across many countries over recent decades. Globally, the number of annual deaths from indoor air pollution has fallen by more than 1 million since 1990. As shown in Figure 13.7, an estimated 2.7 million died prematurely as a result in 1990 and then had fallen to 1.6 million by 2017. This means that despite continued population growth in recent decades, the total deaths from indoor air pollution have still been declining. In 2017, 1.6 million people died prematurely as a result of indoor air pollution – 45% were aged 70 and older while the young children under 5-year old had the fastest declining rate in premature death since 1990 (www.childmortality.org), suggesting an age-specific breakdown of deaths attributed to indoor air pollution. Although the total number of child deaths has more than halved from 11.8 million in 1990 to 5.4 million in 2017, the major causes of child deaths have largely remained the same.

13.1.2.2 Indoor air pollution increases the risk of infectious diseases

About 15% of all children died from lower respiratory infection (LRI) in 2017, which has remained the leading cause of mortality over the past three decades (Figure 13.8). Pneumonia as the leading LRI is primarily caused by bacterial infections (Troeger *et al.*, 2018), which is also associated with air pollution from both indoor and outdoor (Jiang *et al.*, 2018; Norbäck *et al.*, 2018). Indoor and outdoor air pollution was responsible for 29% and 18% of pneumonia deaths in 2017 (Troeger *et al.*, 2018).

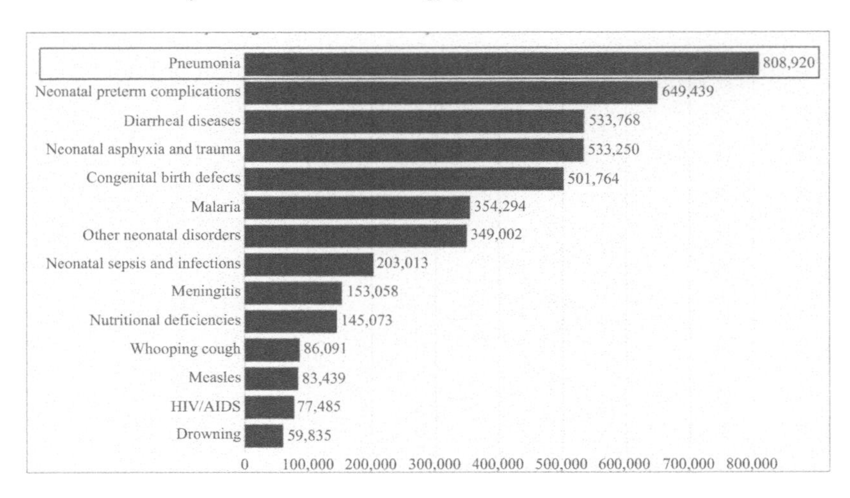

Figure 13.8 Annual number of deaths by leading causes in in children under 5-year old in the world, 2017. Source: IHME, Global Burden of Disease (GBD)

With the rapid economic growth and urbanization development in developing countries such as China during past decades, a huge amount of people have moved into new buildings with an installation of a lot of new furniture and heavy redecoration (Deng *et al.*, 2015b). The new furniture and house renovation have become a major source of indoor air pollution and their impact on the health of children who was born or will be born is an increasing concern in China (Gao *et al.*, 2014). However, there is a lack of investigation for the potential impact of indoor air pollution on infectious diseases.

13.1.2.3 Indoor environmental pollution and birth outcomes

On the other hand, neonatal preterm complications contributed ranked No. 2 of the causes of death in children under 5-year old worldwide in 2017, and the numbers of preterm birth (PTB) in low- and middle-income countries are extremely high (Figure 13.9). As pregnant women spend more time indoors, indoor environment, such as poor ventilation-related air pollution, should be an important risk factor for birth outcomes such as PTB (Lu *et al.*, 2019) and low birth weight (Lu *et al.*, 2020a) which is a major public health concern worldwide (Blencowe *et al.*, 2012). These birth outcomes are not only the leading cause of neonatal death (Bukowski *et al.*, 2007; Goldenberg *et al.*, 2008) and infant mortality under age 5 (Liu *et al.*, 2015a) but also has long-term adverse consequences in survived children (Moster *et al.*, 2008; Goldenberg and Culhane, 2007; Lakshmanan *et al.*, 2015). PTB rate is rising globally and China ranks No. 2 in the number (Blencowe *et al.*, 2012). Children with LBW increased worldwide over the past two decades (Kana *et al.*, 2017; Shan *et al.*, 2014), most of which occurred in developing countries (Chen *et al.*, 2013).

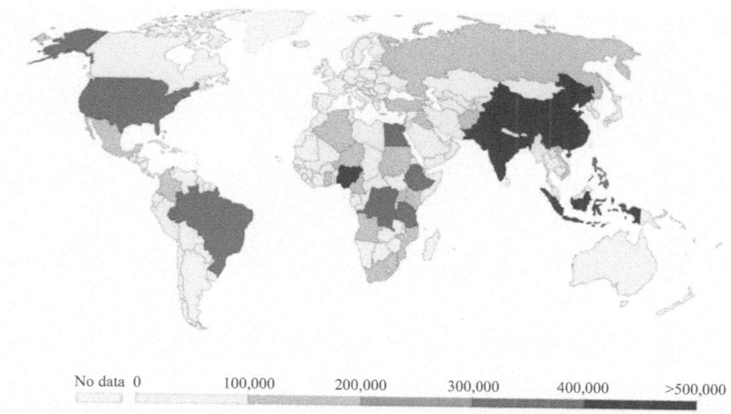

Figure 13.9 Distribution of the number of preterm births worldwide in 2014 (Chawanpaiboon et al., 2019)

Although numerous studies have suggested that exposure to indoor environment during pregnancy was significant associated with childhood diseases (Deng *et al.*, 2015a, 2016a; Gordon *et al.*, 2014; Sun *et al.*, 2018), only several studies examined the risk of birth outcomes due to the exposure to ventilation-related indoor environmental factors, including house renovation (Liu *et al.*, 2018), VOC emitting household products (Sørensen *et al.*, 2010), and mould or dampness (Harville and Rabito, 2018). These scarce results are not consistent and debatable. A majority of pregnant women are heavily exposed to indoor air pollution in the low/middle income countries (Kadir *et al.*, 2010), such as China (Deng *et al.*, 2016b). Thus, evidence in the home environment from Asian areas like China would provide further insights on the potential effect of indoor air pollution on birth outcomes.

13.1.2.4 Effect of indoor air pollution on allergic diseases

Indoor air pollution is an important environmental factor related to the development of childhood allergies (Ahn, 2014; Dick *et al.*, 2014a). Due to the rapid urbanization in China over the past few decades, a large number of people, especially new couples and expecting parents, moved into new buildings in urban areas (Deng *et al.*, 2015b; Franklin, 2007). New building materials, decoration materials, and new furniture may greatly increase the levels of indoor chemicals, including VOCs and SVOCs, which are associated with allergic diseases (Cavaleiro Rufo *et al.*, 2016; Choo and Jalaludin, 2015; Huang *et al.*, 2020). Due to a lack of central air-conditioning and heating systems or the adverse building characteristics, some poor ventilation-related indicators, including mould/damp stains and window condensation in dwellings, are common, particularly in the areas with a subtropical

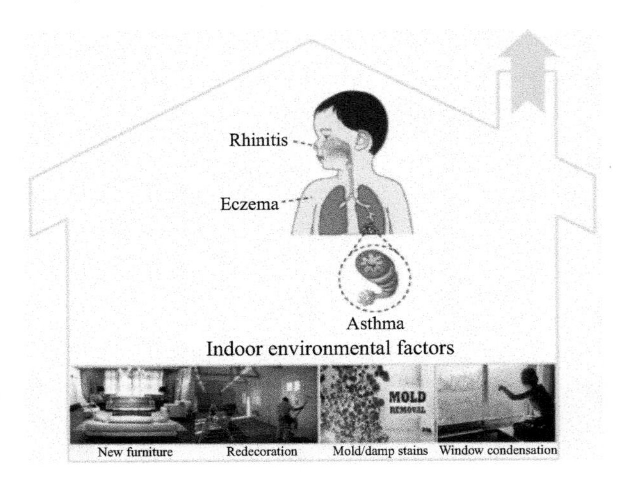

Figure 13.10 Combined effects of ventilation-related air pollution from both outdoor and indoor on childhood allergic and infectious diseases

climate. Indoor mould and dampness are associated with allergy risk (Dick *et al.*, 2014b; Fisk *et al.*, 2007, 2010; Tischer *et al.*, 2011) as shown in Figure 13.10. Thus, indoor environmental pollution plays an important role in the development of childhood allergies (Castro-Rodriguez *et al.*, 2016; Dick *et al.*, 2014a), especially in developing countries. Due to the serious exposure to air pollution from both indoor and outdoor, China has witnessed a rapid increase in the prevalence of allergic diseases among preschool children during recent years (Zhang *et al.*, 2013b). However, only a few recent studies have addressed the allergy risk of indoor exposure to air pollution, such as new furniture, redecoration, mould/damp stains, and window condensation (Lu *et al.*, 2020b, 2020c).

13.1.3 Effect of ventilation on indoor air pollutants

Indoor air pollutants mostly come from indoor sources, including occupants and their activities, building materials, furnishings and carpeting, cleaning products, and personal care products. In addition, other pollutants often emanate from the infiltration of outdoor air pollutants through doors, windows, and their gaps (Salvador *et al.*, 2019). Indoor pollutant concentrations are related to three major variables: indoor pollutant generation rate, outdoor pollutant concentration, and ventilation rate (Maroni *et al.*, 1995). Ventilation is one of the effective technical means to reduce indoor air pollution. However, when contaminant concentration levels are higher outdoors than indoors, contaminant will migrate to buildings. Therefore, the rational use of ventilation is the key to reduce indoor air pollution. The physico-chemical nature, the occurrence and sources of each indoor air pollutant are different, and the dilution effect of ventilation is also different. Therefore, this section will focus on PM and VOCs, the two most common indoor air pollutants, and explore the effect of ventilation on its concentration levels.

13.1.3.1 Particulate matter

PM refers to a mixture of tiny solid particles and liquid droplets suspended in the air, which are subdivided into three modes according to formation processes: ultrafine ($d_p \leq 0.1$ μm), accumulation ($0.1 < d_p \leq 2.5$ μm), and coarse ($d_p > 2.5$ μm). An alternative classification scheme uses two modes: fine ($d_p < 2.5$ μm) and coarse ($d_p > 2.5$ μm).

Figure 13.11 schematically depicts some of the potentially important factors that can affect indoor concentrations (Thatcher *et al.*, 2003). There were two main sources of indoor particles: one was from outdoor air, which contains some particle attribute at a concentration C_{out} (μg/m^3), entered the building through infiltration from the building envelope (Nazaroff, 2004); the other was indoor pollution, such as human activities, combustion sources (combustion appliances and tobacco smoking), and other indoor sources of particles. If we assumed to ignore these factors: coagulation between indoor particles, gas-particle condensation, filtration by air-conditioning systems, or filters, and resuspension caused by indoor human activities, the equation was established according to mass conservation as follows:

$$\frac{dC_{in}}{dt} = apC_{out} - (a+k)C_{in} + \dot{S} \tag{13.1}$$

where C_{in} and C_{out} were indoor and outdoor mass concentrations (μg/m^3) or number concentrations (particles m^{-3}); $\dot{S}$ was the emission rate of indoor particles (μg/m^3 h or particles/m^3 h); a was the air exchange rate (h^{-1}); k was the particle deposition loss-rate coefficient; p was the particle penetration factor, $0 \leq p \leq 1$.

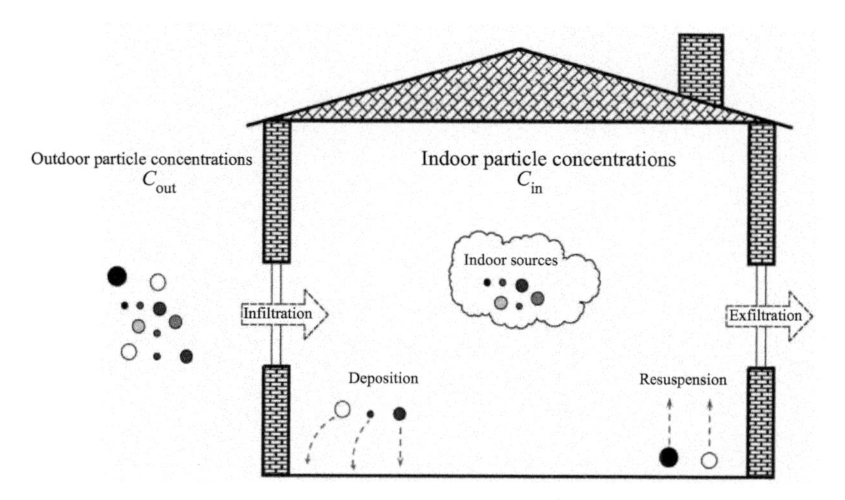

Figure 13.11 Schematic of particle transport, transformation, and removal processes in the indoor environment (Thatcher et al., 2003)

Under steady-state conditions ($dC_{in}/dt = 0$), (13.1) is further simplified as

$$C_{in} = \frac{ap}{\alpha + k}\alpha C_{out} + \frac{\dot{S}}{\alpha + k} \quad (13.2)$$

Equation (13.2) visually described that indoor particulate concentration was mainly composed of outdoor and indoor source contributions. The air exchange rate (α) has an important influence on outdoor and indoor source contributions. Table 13.1 listed some recent studies on the effect of ventilation on indoor-to-outdoor (*I*/*O*) particle concentrations ratios. The influence of ventilation on the *I*/*O* ratio was affected by the presence of indoor sources.

Table 13.1 Literature statistics of I/O ratios of particulate matter

Reference	**Country/ Area**	**PM**	***I/O***	**Test conditions**
Banerjee and Annesi-Maesano (2012)	France	PM2.5	0.87	Classroom: without air-conditioning/mechanical ventilation
Ben-David and Waring (2016)	USA	PM2.5	0.679	Office: mechanical ventilation with filters
			0.811	Office: natural ventilation with filters
Chao and Wong (2002)	Hong Kong	PM10	0.88	Residence: high air-change rate ($ACH > 3.5\ h^{-1}$)
		PM2.5	0.92	
		PM10	1.04	Residence: low air-change rate ($ACH < 3.5\ h^{-1}$)
		PM2.5	1.09	
Chiesa *et al.* (2019)	Italy	PM10	0.69	Residence: winter, windows closed
		PM2.5	0.58	
		PM10	0.87	Residence: winter, windows opened
		PM2.5	0.79	
Cyrys *et al.* (2004)	Germany	PM2.5	0.63	Residence: window closed
			0.83	Residence: ventilation twice a day for 15 min each time
Dai *et al.* (2018)	China	PM2.5	0.88–0.97	Residence: natural ventilation
Meng *et al.* (2005)	USA	PM2.5	0.85	Air exchange rates for California residences were 1.22 h^{-1}
			0.99	Air exchange rates for New Jersey residences were 1.22 h^{-1}
			1.16	Air exchange rates for Texas residences were 0.71 h^{-1}
Mohammadyan and Shabankhani (2013)	Iran	PM2.5	1.26	Classroom: without air-conditioning/mechanical ventilation
Othman *et al.* (2019)	Malaysia	PM2.5	1.01±0.03	Classroom: weekdays, natural ventilation
Pallarés *et al.* (2019)	Spain	PM10	2.6	Classroom: urban, natural ventilation
			2.1	Classroom: rural, natural ventilation
Wang *et al.* (2016)	China	PM2.5	0.876	Residence: natural ventilation
			0.463	Residence: with centralized HVAC

Without indoor pollution sources

Under without indoor sources conditions, we compared the influence of air exchange rate on *I/O* ratios in various documents and found that the *I/O* ratios gradually increased with the increase of indoor air exchange rate, that is, the concentration levels of indoor particles gradually approached the outdoor concentration levels (Figure 13.12). In particular, for higher air exchange rates (α>3 h^{-1}), the *I/O* ratio remains unchanged, while for lower air exchange rates (α<3 h^{-1}), the *I/O* ratios varied significantly.

It should be noted that indoor air pollution would be aggravated when outdoor pollution was more serious due to the increased impact of outdoor particles by ventilation. Therefore, when the outdoor air quality was better, strengthening the ventilation was conducive to improving the IAQ; but the ventilation should be avoided when the outdoor air pollution was serious, and the door and window closing at this time was conducive to reduce indoor air pollution.

Figure 13.13 illustrates the real-time change of indoor and outdoor PM2.5 concentrations at different air exchange rates, in which Figure 13.13(a) data were measured on 20–23 June 1998, with air exchange rate of 4.9; Figure 13.13(b) data were measured on 20–23 June 1998, with an air exchange rate of 0.62. Such data clearly demonstrated that the concentrations of the indoor particles followed the outdoor concentrations change closely with little time lag between them if the indoor air exchange rate was very high, such as windows and doors were left open in summer. It responded quickly to the variation of outdoor particle concentrations, and the indoor concentration levels were basically the same as the outdoor levels, *I/O*≈1. In contrast, the windows and doors were left close in winter, the indoor air exchange rate was consistently low, and the response of indoor particle concentrations to outdoor concentrations variation became slower. The large separation between indoor and outdoor concentrations was a distinct indoor–outdoor time lag. The indoor particle concentrations were substantially lower than outdoor levels, *I/O*<1.

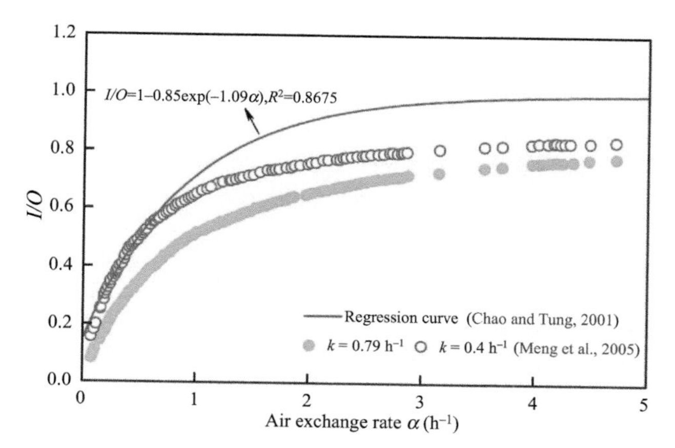

Figure 13.12 Variation of the I/O ratio of PM2.5 with air exchange rate (Chao and Tung, 2001; Nazaroff, 2004)

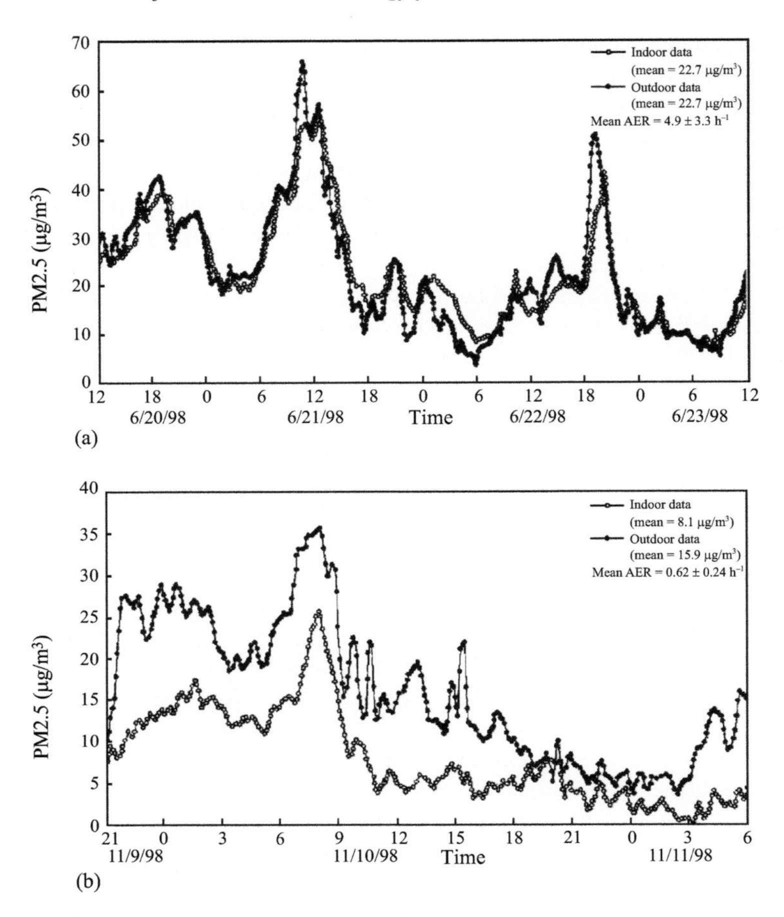

Figure 13.13 Effect of air exchange rate on indoor particulate concentration level (Long et al., 2001): (a) 20–23 June 1998 and (b) 9–11 November 1998

The aforementioned results enunciated that the outdoor particle concentrations can be approximated to judge the indoor particle concentrations only when the room air exchange rates were high, so it was also used for health evaluation. However, a decrease in the indoor ventilation resulted in a decreased outdoor ambient contribution, and the concentrations of the outdoor particles cannot represent concentration levels and variation regularity of indoor particle, so it cannot be used for health evaluation.

Indoor pollution sources

In order to illustrate the influence of indoor sources on the concentration levels of PM, we analysed the changes in fine (particle diameter less than 0.5 μm, measured with SMPS) and coarse (particle diameter greater than 0.5 μm, measured with APS) particle number concentrations, as well as PM2.5 mass concentrations (measured with DustTrak) with time during the cooking test, as shown in Figure 13.14 (He *et al.*, 2004). The fine particles reacted very quickly to the event, the cooking started, the number concentrations increased rapidly, while the cooking stopped, and the number concentrations decreased immediately. But there was a time lag in the increase or decrease in PM2.5 mass concentrations and coarse particles number concentrations compared with fine particles. This was mainly due to the coagulation of PM, which resulted in the shift in particle size distribution towards larger sizes over time. The indoor particle concentrations were still higher than background levels 50 min after the end of cooking under minimum ventilation conditions, as depicted in Figure 13.14(a). But the decay rate under normal ventilation condition was clearly higher than that under minimum ventilation condition. The indoor particle concentration levels close to background levels about 10 min after the end of cooking in Figure 13.14(b). Therefore, indoor particle concentrations

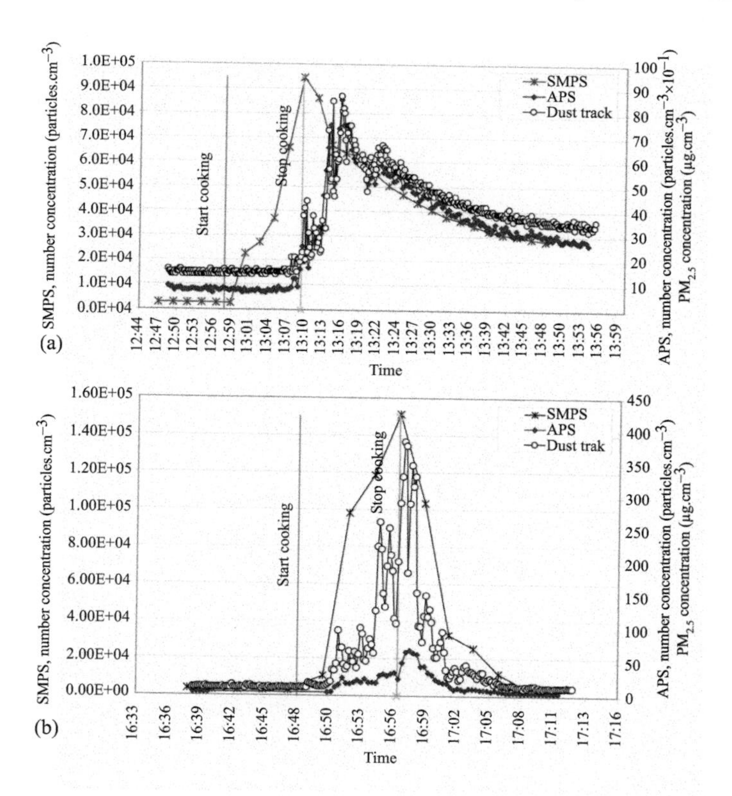

Figure 13.14 Emission characteristics of indoor sources (He et al., 2004): (a) Cooking test under minimum ventilation condition (α=0.62) and (b) Cooking test under normal ventilation condition (α=4.9)

decayed slowly when the air exchange rate was low (slow settlement and slow discharge); when the air exchange rate was high, the concentrations of indoor particle decreased rapidly (fast settlement and discharge), which basically had no influence on the total indoor concentration levels, so the concentrations were low.

In conclusion, we found that when the indoor air exchange rate was large, the effect of indoor sources was small and momentary as well as can be discharged rapidly under the high ventilation condition, so the concentration levels of indoor particle were mainly affected by outside, and $I/O \leq 1$. On the other hand, indoor particles cannot be discharged in time and gradually coagulation under the low air exchange rate, which led to the increase of indoor particle concentration, $I/O>1$. Consequently, strengthening indoor ventilation was an important means to reduce the impact of indoor sources on the concentrations of indoor particle. It is particularly noteworthy that this is the opposite conclusion between the existence of indoor sources and no indoor sources.

Although ventilation was the most effective and economical means to control indoor air pollution, it could not directly open the window for ventilation in the case of heavy outdoor air conditions. The IAQ could be significantly improved by sending the filtered outdoor air into the indoor. Figure 13.15 presented the effect of

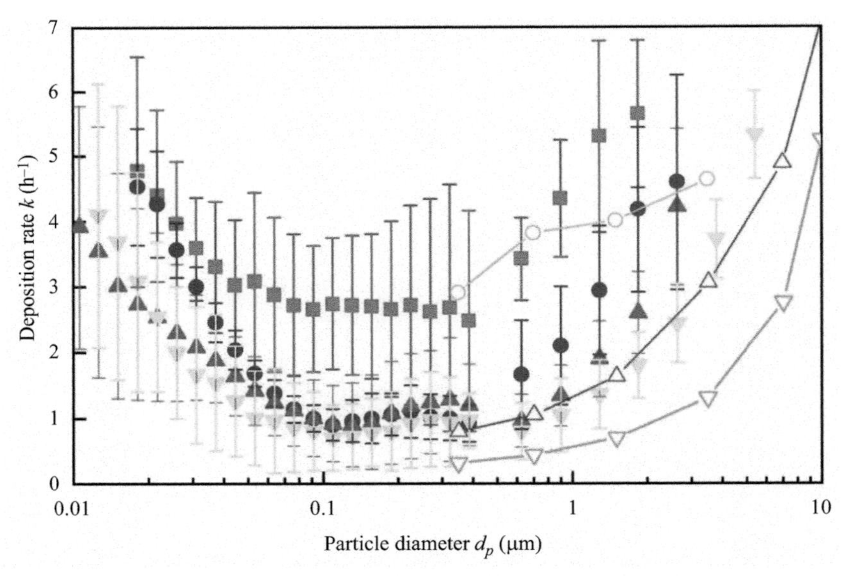

■ Wallace *et al.* (2004) ESP(high efficiency filter)	○ Howard-Reed *et al.* (2003) fan on with air cleaner
● Wallace *et al.* (2004) MECH(low efficiency filter)	▽ Howard-Reed *et al.* (2003) fan off
▲ Wallace *et al.* (2004) fan on without filter	△ Howard-Reed *et al.* (2003) fan on
▼ Wallace *et al.* (2004) fan off	

Figure 13.15 Effect of filtration and ventilation on deposition rates

the combination of ventilation and air filters on the deposition rate of indoor particles, indicating that high-performance filtration materials and MV could significantly improve the deposition rate of particles. In addition, the deposition rate varied with particle diameter in a V-shaped distribution under different filtration efficiencies or ventilation types. It illustrated that filter and ventilation efficiencies vary greatly with particle size. Filtration and ventilation had higher efficiency in treating ultrafine and coarse particles (higher k value), but lower efficiency in treating the accumulation mode particles (lower k value). This result revealed that the accumulation mode particles were the most difficult part to solve in indoor particles, and its control strategy should be enriched.

13.1.3.2 Volatile organic compounds

VOCs are a kind of low boiling point organic compounds. The WHO defines VOCs as organic compounds with melting point lower than room temperature and boiling point between 50 °C and 260 °C measured at a standard atmospheric pressure. VOCs mainly include benzene series, organic chlorides, freon series, organic ketones, amines, alcohols, ethers, esters, acids, and petroleum hydrocarbons. VOCs were ubiquitous in the indoor environment. Almost all materials, consumer goods, furniture, pesticides, fuels, and outdoor sources released VOCs into the indoor air, among which building materials usually contribute the most (Maroni *et al.*, 1995).

Ventilation was the most direct and effective method to remove indoor VOCs when the interior source of decoration materials and furniture could not be changed. There were two purposes to increase the indoor ventilation rates: one was to accelerate the VOC emission rate of materials so as to consume emission source more quickly; the other was to reduce the resulting airborne concentrations of emitted pollutants to acceptable levels (Holøs *et al.*, 2019). Although the ventilation rate is key to controlling airborne concentrations, the emission rates for seven VOCs were not significantly affected by changes to the air exchange rates as shown in Figure 13.16 (Caron *et al.*, 2020). Therefore, the emission rates were not affected by the ventilation rate, and the key to depleting VOCs source was time. For newly decorated houses, delaying occupancy was the best way to avoid VOCs damage to health.

According to the reference (Xu and Zhang, 2011), there is no exact definition for the significance of (B) in Figure 13.17 and here significant impact is not based on a statistical analysis but a general word. Thus, we did not provide a definition here. Figures 13.17 and 13.18, respectively, show the effect of air-change rates on VOCs concentrations in air through the experimental test and numerical simulation. The trends of the simulation results were consistent with the experimental results. The application of ventilation with an increase in air exchange rate had an effect of reducing the VOCs concentration at a much faster rate to reach a stable in a shorter period (Wang *et al.*, 2012). Besides that, the steady-state concentrations of VOCs were reduced even lower by adding air cleaner under the same air exchange rate. Thus, the advantages of the combination of air cleaner and ventilation which complement each other provide the most effective and efficient method to remove VOCs.

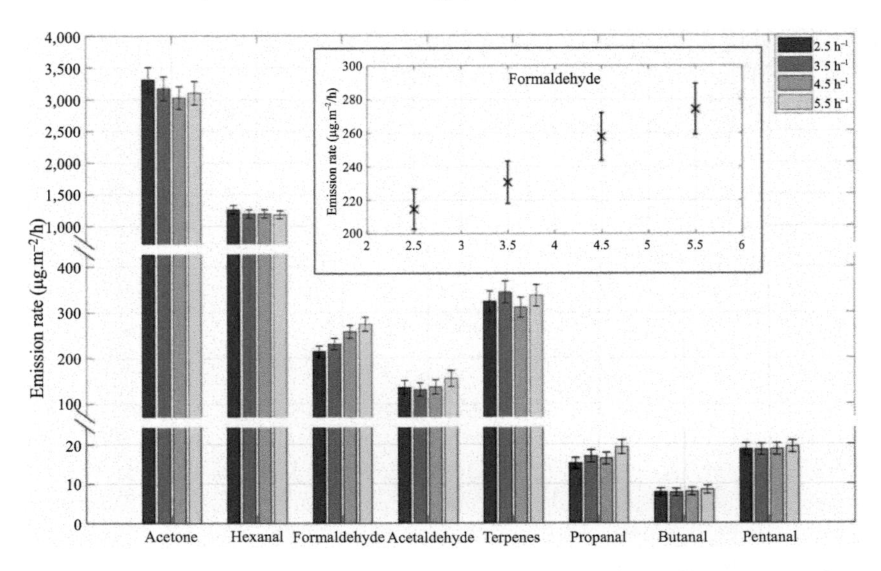

Figure 13.16 Emission rate for the eight main VOCs under different air exchange rate conditions (Caron et al., 2020)

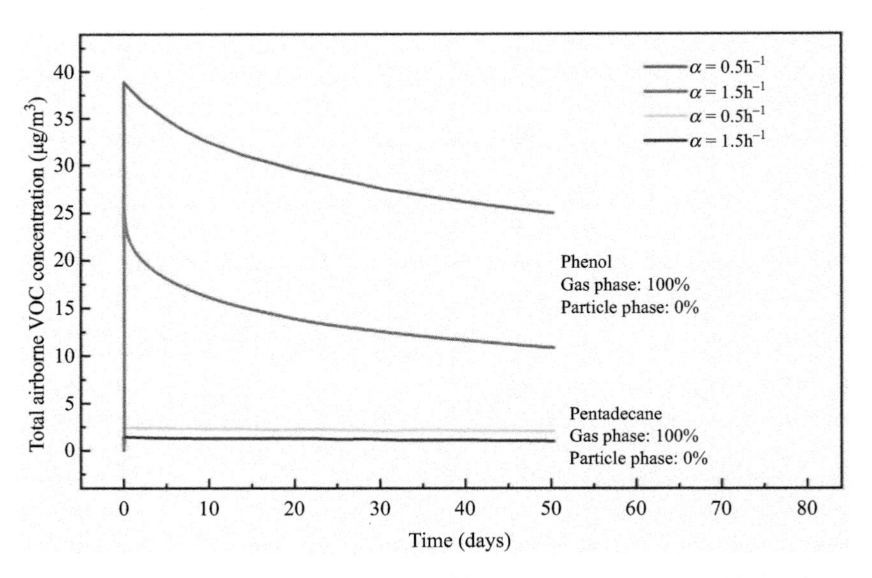

Figure 13.17 The influence of air exchange rate on total airborne VOC (B) concentrations (Xu and Zhang, 2011)

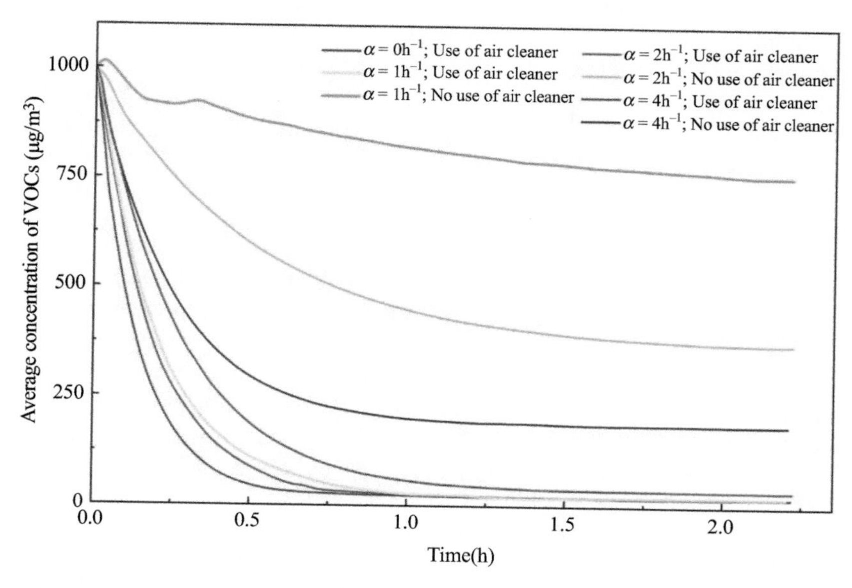

Figure 13.18 The change trend of VOCs average concentrations with time for numerical simulation cases (Wang et al., 2012)

13.2 Early-life ventilation and children's health

13.2.1 Introduction and significance

Climatic factors, such as temperature, relative humidity, wind velocity, and precipitation were indicated to be related with the epidemics of allergic diseases (D'Amato *et al.*, 2010) and respiratory infections (Jendritzky *et al.*, 2012; Mirsaeidi *et al.*, 2016). Of these stimuli, temperature and wind speed are determined to be key variables to influence RSV infection (Walton *et al.*, 2010). Although the association between temperature and allergic diseases (Xu *et al.*, 2012) and respiratory infections have been well addressed (Lam and Chan, 2019), the wind speed has rarely been considered till now. Many studies have suggested that changes in weather conditions, especially cold weather, may trigger respiratory problems especially in younger children (Yousif and Al Muhyi, 2019), whereas wind indicated as the most important urban ventilation variable has dual effects on disease vectors/hosts, which may influence the development of allergic diseases (Hu *et al.*, 2020) and affect the transmission of respiratory diseases (Wu *et al.*, 2016). Although increasing studies have analysed the effects of postnatal and current exposure to climatic factors, including wind speed on childhood allergic and infectious diseases (D'Amato *et al.*, 2014, 2015), no study has so far examined the prenatal effect. Our current work strikingly demonstrated that prenatal exposure to both indoor and outdoor environmental factors such as air pollution, closely related with meteorological conditions including wind speed, was associated with the development of both allergic (Deng *et al.*, 2015a, 2016b, 2016c; Lu *et al.*, 2017) and infectious diseases in childhood (Deng *et al.*, 2017a, 2017b; Jiang *et al.*, 2018). Recent evidence even indicated that the climatic factors played a more important role in childhood respiratory diseases compared to air pollution (Hu *et al.*, 2020). Therefore, the hypothesis with aspect of the effect of wind speed, an important indicator for outdoor ventilation, during pregnancy and early postnatal period warrants investigations.

Indoor ventilation, as an important determinant factor for the IAQ (Sarigiannis, 2013; WHO, 2011), has been suggested to be related to both childhood allergic diseases (Sundell *et al.*, 2011; Wargocki, 2013) and respiratory infections (Mendell *et al.*, 2011; Ucci *et al.*, 2004). A growing number of evidence from China have associated both indoor NV (Sun *et al.*, 2017) and MV (Sun and Sundell, 2011) with an increase in prevalence of allergy and respiratory tract infections (RTIs). On the other hand, a low ventilation rate due to both poor natural and MV indoors may contribute to an increase in dampness-related indicators, including mould/damp stains, mouldy odour, and window pane condensation (Liu *et al.*, 2015b; Wargocki, 2013). Recent review studies have indicated that the exposure to the poor ventilation indicators, mainly for mould and dampness, during postnatal or current period is strongly related with childhood allergies (i.e., asthma, allergic rhinitis, and eczema) and infections (i.e., common cold) (Mendell *et al.*, 2011). However, very few studies have been addressed for prenatal effect (Norbäck *et al.*, 2017, 2018), and thus no clear critical timing window has been identified as relative importance in poor ventilation indicators exposure for the development of childhood allergy and infection.

In the present study, we assumed that childhood allergy and infection are associated with early-life ventilation during both prenatal and early postnatal periods. To test our hypothesis, a retrospective cohort study was conducted in Changsha of China, which belongs to a part of nationwide the ‘China-Children-Homes-Health (CCHH)’ study (Deng *et al.*, 2015a; Zhang *et al.*, 2013b). Our study aimed to examine the role of early-life exposure to both indoor and outdoor ventilation variables during pregnancy and postnatal period on allergic and infectious diseases among preschool children in Changsha, China. The flow chart of the questionnaire survey and retrospective cohort protocol on early-life ventilation and childhood allergy and infection in CCHH study is shown in Figure 13.19.

13.2.2 Effect of early-life exposure to ventilation on childhood allergy and infection

Table 13.2 provides the estimate effects of outdoor ventilation indicated as WS exposure during pregnancy and postnatal period on childhood allergic and infectious diseases. We found that prenatal exposure to an increase in WS particularly in the first trimester of pregnancy was negatively associated with childhood asthma, pneumonia, and frequent common cold, respectively, with ORs (95% CI) of 0.67 (0.48–0.94), 0.84 (0.71–1.00), and 0.75 (0.56–0.99) for an IQR increase in WS after adjustment for all the covariates and outdoor air pollutants (PM10, SO_2, and NO_2).

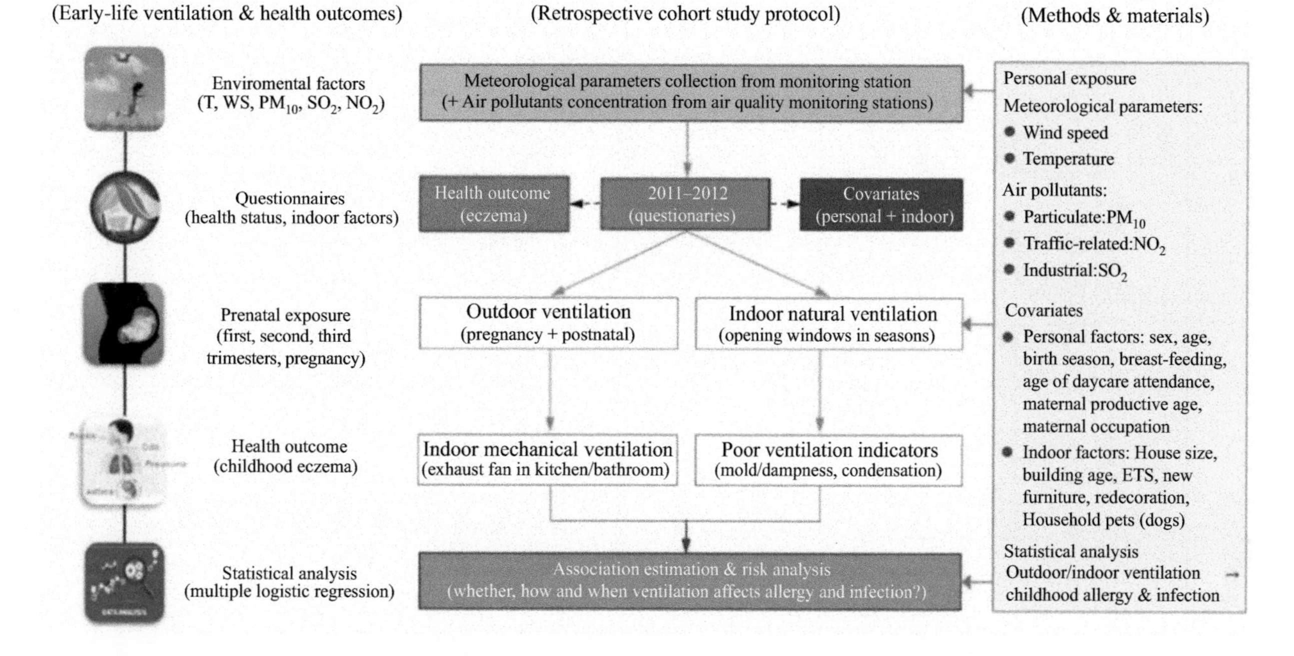

Figure 13.19 Flow chart of the questionnaire survey and retrospective cohort protocol on early-life ventilation and childhood allergy and infection in CCHH study

Table 13.2 Associations between prenatal and postnatal exposure to outdoor wind speed (WS) and childhood allergies and infections among children aged 3–6 years in Changsha, China, with OR (95% CI) (n = 2,598)

WS (m/s)	Allergy		Infection	
	Asthma	**Rhinitis**	**Pneumonia**	**≥ 5 colds/year**
First trimester	0.67 (0.48, 0.94)*	0.85 (0.62, 1.15)	0.84 (0.71, 1.00)*	0.75 (0.56, 0.99)*
Second trimester	1.01 (0.74, 1.37)	0.93 (0.70, 1.25)	0.82 (0.70, 0.96)*	0.82 (0.63, 1.08)
Third trimester	1.10 (0.82, 1.48)	1.03 (0.76, 1.39)	1.09 (0.93, 1.27)	0.88 (0.67, 1.15)
Entire pregnancy	0.79 (0.57, 1.10)	0.87 (0.64, 1.18)	0.90 (0.76, 1.06)	0.70 (0.53, 0.93)*
Entire postnatal	1.25 (0.87, 1.80)	1.18 (0.83, 1.66)	1.29 (1.07, 1.56)**	1.44 (1.05, 1.98)*

OR (95% CI) was estimated for per IQR increase in wind speed (WS) during different time windows. Models were adjusted for all the covariates in Table 13.1, and outdoor air pollutants (PM10, SO_2, and NO_2). * $p \leq 0.05$; ** $p \leq 0.01$.

A negative association was also observed for pneumonia and frequent common cold with WS exposure during the second trimester and entire pregnancy, respectively, whereas we found that postnatal exposure to an increase in WS was positively associated with infectious diseases, including pneumonia and frequent common cold with ORs (95% CI) = 1.29 (1.07–1.56) and 1.44 (1.05–1.98), respectively.

Table 13.3 provides the summary of personal exposure to indoor NV and its effects on childhood allergies and infections. There was a great difference in the prevalence of NV indicated as opening windows, with the highest prevalence of frequent opening windows in summer while the lowest in winter. We found that opening windows in winter significantly decreased the risk of both allergies (asthma) and infections (pneumonia and frequent common cold) compared to never opening windows, respectively, with ORs (95% CI) of 0.61 (0.38–0.98), 0.73 (0.57–0.94), and 0.47 (0.31–0.70). A negative association was also observed between opening windows in spring and frequent common cold. However, no relationship was observed for allergic rhinitis with opening windows in any season.

Table 13.4 presents the summary of personal exposure to indoor MV and poor ventilation indicator, and their impact on childhood allergic and infectious diseases. We observed that using a fan in the kitchen/bathroom was associated with a decreased risk of infection for pneumonia with OR (95% CI) = 0.73 (0.54–0.98) but was not related with allergy for asthma or allergic rhinitis. On the other hand, our analysis showed that exposure to mould/damp stains due to poor ventilation was positively associated with infections for pneumonia and frequent common cold

Table 13.3 Associations between home natural ventilation (opening windows) and childhood allergies and infections among children aged 3–6 years, with OR (95% CI) (n = 2,598)

Opening windows	***n*** **(%)**	**Allergy**		**Infection**	
		Asthma	**Rhinitis**	**Pneumonia**	**≥ 5 colds/year**
Spring					
Never	124 (5)	1.00	1.00	1.00	1.00
Sometimes	916 (35)	1.10 (0.49, 2.51)	0.61 (0.31, 1.18)	0.78 (0.52, 1.17)	0.66 (0.36, 1.18)
Frequently	1,529 (59)	1.07 (0.48, 2.40)	0.58 (0.30, 1.11)	0.77 (0.52, 1.14)	0.48 (0.27, 0.86)*
Summer					
Never	109 (4)	1.00	1.00	1.00	1.00
Sometimes	626 (24)	0.76 (0.32, 1.80)	1.45 (0.55, 3.81)	1.04 (0.67, 1.62)	0.72 (0.35, 1.45)
Frequently	1,828 (70)	0.97 (0.43, 2.16)	1.43 (0.56, 3.62)	1.05 (0.69, 1.60)	0.76 (0.39, 1.46)
Autumn					
Never	99 (4)	1.00	1.00	1.00	1.00
Sometimes	797 (31)	0.93 (0.38, 2.27)	0.68 (0.30, 1.50)	0.79 (0.50, 1.24)	0.92 (0.44, 1.93)
Frequently	1,667 (64)	0.92 (0.39, 2.19)	0.68 (0.32, 1.48)	0.79 (0.51, 1.23)	0.70 (0.34, 1.45)
Winter					
Never	481 (19)	1.00	1.00	1.00	1.00
Sometimes	1,354 (52)	0.78 (0.52, 1.16)	0.86 (0.57, 1.29)	0.82 (0.66, 1.03)	0.54 (0.39, 0.77)***
Frequently	708 (27)	0.61 (0.38,0.98)*	0.80 (0.50, 1.27)	0.73 (0.57,0.94)*	0.47 (0.31, 0.70)***

Models were adjusted for all the covariates in Table 13.1, and outdoor temperature and air pollutants (PM10, SO_2, and NO_2). * $p \leq 0.05$; *** $p \leq 0.001$.

with ORs (95% CI) of 1.23 (1.01–1.51) and 1.46 (1.06–2.01) but was not associated with allergic diseases. Furthermore, exposure to the other poor ventilation indicator, window condensation, significantly increased the risk of both allergic diseases, including asthma (1.64, 1.16–2.31) and allergic rhinitis (1.88, 1.34–2.66) and infectious diseases, including pneumonia (1.19, 1.00–1.42) and frequent common cold (1.43, 1.06–1.93).

Figure 13.20 shows the association between outdoor ventilation indicated as WS and allergies and infections stratified by children's sex. We found a significantly negative association between prenatal exposure to WS and both allergy (asthma) and infection (pneumonia and frequent common cold) in girls but not in boys. We further observed a positive association between postnatal exposure to WS and allergy and infection only in girls.

Table 13.4 Associations between indoor mechanical ventilation and poor ventilation indicators and childhood allergies and infections among children aged 3–6 years, with OR (95% CI) (n = 2,598)

	n	(%)	Allergy		Infection	
			Asthma	**Rhinitis**	**Pneumonia**	**≥ 5 colds/year**
Exhaust fan[a]	2,354	(91)	0.68 (0.40, 1.15)	1.46 (0.75, 2.84)	0.73 (0.54, 0.98)*	0.79 (0.49, 1.28)
Mould/damp stains	606	(23)	1.37 (0.95, 1.97)	0.92 (0.62, 1.36)	1.23 (1.01, 1.51)*	1.46 (1.06, 2.01)*
Window condensation	1,369	(53)	1.64 (1.16, 2.31)**	1.88 (1.34, 2.66)***	1.19 (1.00, 1.42)*	1.43 (1.06, 1.93)*

Models were adjusted for all the covariates in Table 13.1, and outdoor temperature and air pollutants (PM_{10}, SO_2, and NO_2). * $p \leq 0.05$; ** $p \leq 0.01$; *** $p \leq 0.001$.

[a] Exhaust fan in the kitchen and/or bathroom.

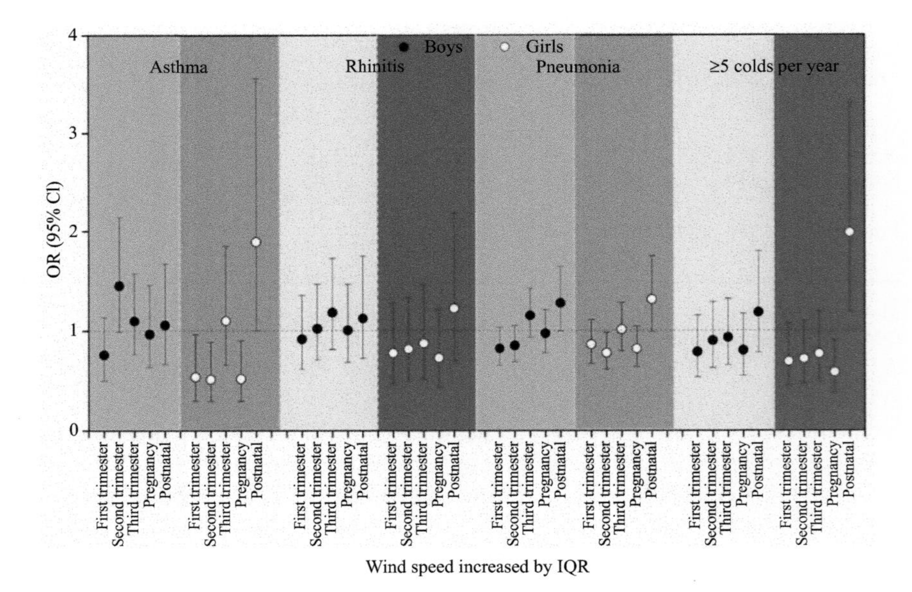

Figure 13.20 Associations between outdoor wind speed, indoor ventilation, and poor ventilation indicators and childhood allergies and infections among boys and girls

Figure 13.21 presents the association between indoor NV indicated as opening windows and allergies and infections stratified by children's sex. We found that boys were more susceptible to the protective effect of frequent opening windows especially in winter and spring on both allergy (asthma) and infection (pneumonia and frequent common cold) than girls.

Figure 13.22 shows the association between indoor MV and poor ventilation indicators and allergies and infections stratified by children's sex. We found that

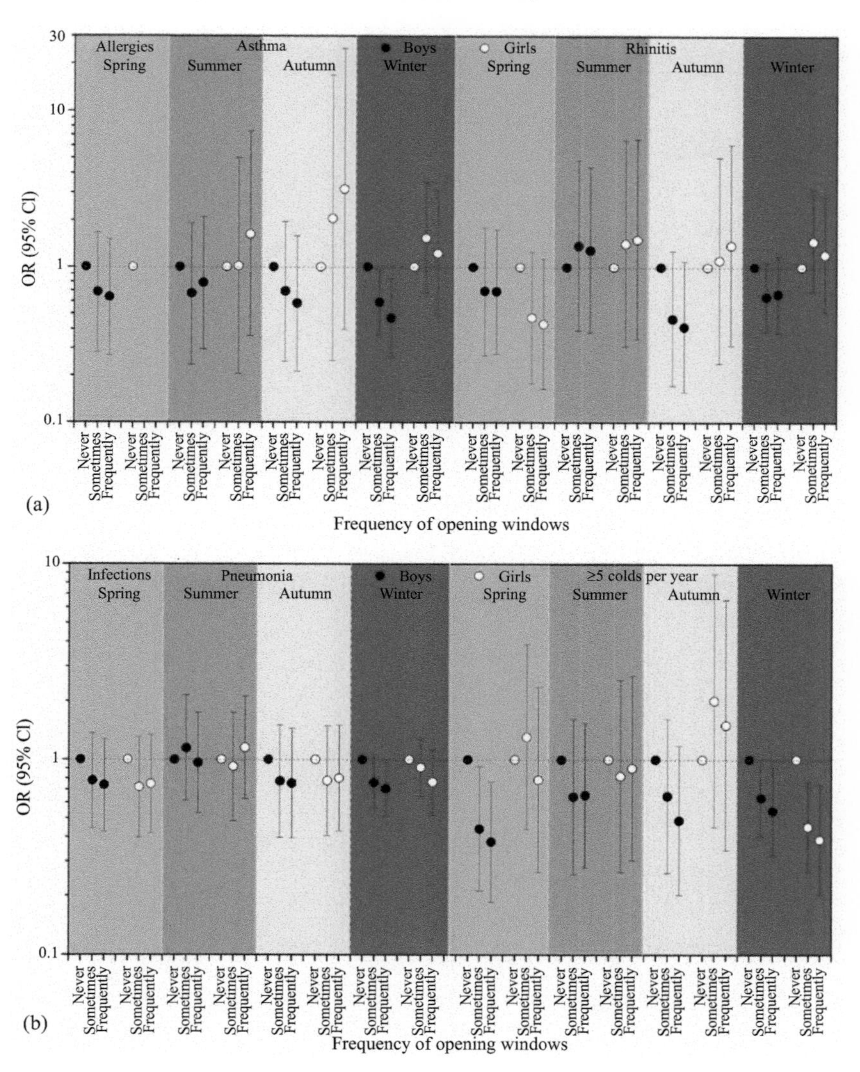

Figure 13.21 Associations between indoor natural ventilation by frequency of opening windows and childhood allergies (a) and infections (b) among boys and girls

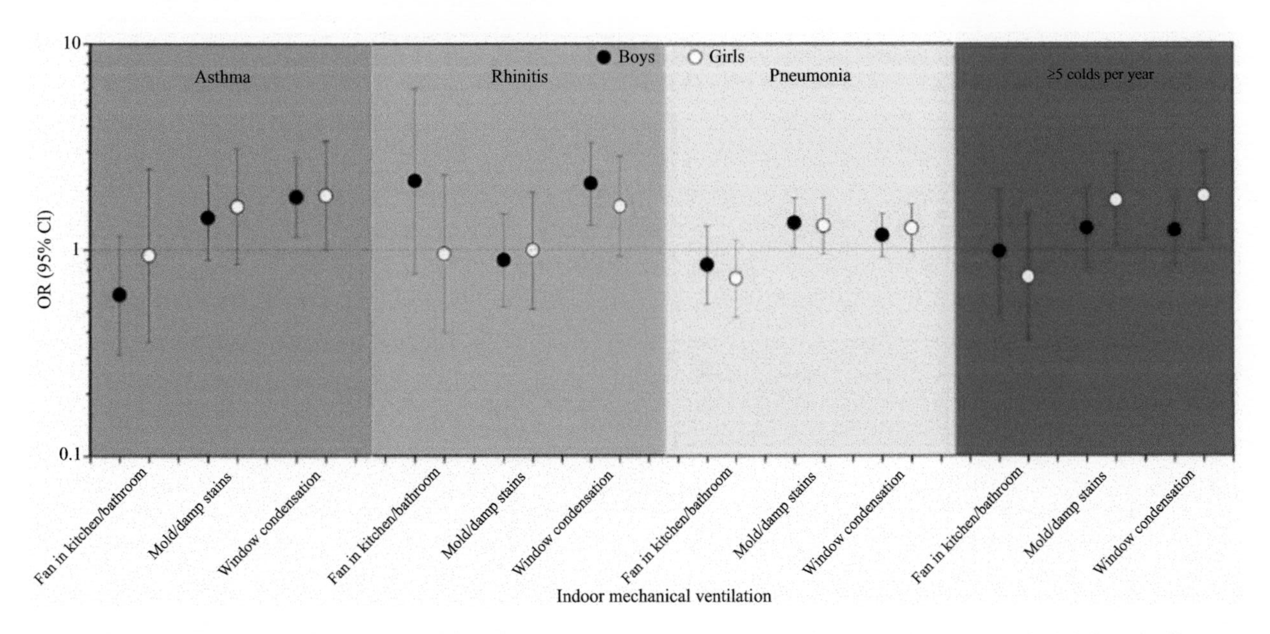

Figure 13.22 Associations between indoor mechanical ventilation and poor ventilation indicators and childhood allergies and infections among boys and girls

boys were more susceptible to the risk of allergy (asthma and allergic rhinitis) due to poor ventilation indicators, including mould/damp stains and window condensation, while girls were more sensitive to the risk of infection (pneumonia) attributed to these poor ventilation indicators.

13.3 Discussions and implications

Our study is the first to explore the association between early-life exposure to both outdoor and indoor ventilation during prenatal and postnatal periods and allergic and infectious diseases among preschool children. In our analysis, prenatal exposure to outdoor ventilation indicated as wind speed during pregnancy particularly in the early stage was negatively associated with both allergy and infection, while a positive association was observed between postnatal exposure to WS and infection. On the other hand, indoor NV indicated as frequent opening windows mainly in winter decreased the risk of both allergies and infections. Furthermore, early life exposure to indoor MV indicated as fan in the kitchen/bathroom decreased pneumonia risk, while poor ventilation indicators elevated the risk of allergic and infectious diseases. Sensitivity analysis indicated that girls were more susceptible to the allergic and infectious risk of outdoor ventilation in early life, while boys were more sensitive to the effect of indoor ventilation. Our findings further supported our hypothesis of 'Fetal origin of childhood diseases' (Deng *et al.*, 2016a, 2017a; Norbäck *et al.*, 2017).

Strikingly, we found that the outdoor ventilation indicated as wind speed in early life displays significant association with childhood allergic and infectious diseases in a negative way for prenatal exposure particularly in the first trimester and a positive way for postnatal exposure. Many studies have examined the association between climatic factors and respiratory diseases (Chan *et al.*, 2002; de Souza *et al.*, 2012; du Prel *et al.*, 2009; Freitas *et al.*, 2010; Gurgel *et al.*, 2016; Kim *et al.*, 2016; Zhan *et al.*, 2017) but only few found a relationship with WS (D'Amato *et al.*, 2015; Freitas *et al.*, 2010; Hossain *et al.*, 2019; Hu *et al.*, 2020; Matyasovszky *et al.*, 2011). To the best acknowledge, no study has been addressed on the association between prenatal outdoor ventilation or WS and childhood allergy and infection, and therefore it is difficult to compare our result with those from previous literatures. Some of our recent work found positive relationships between meteorological parameters, including ambient temperature and diurnal temperature variation during prenatal and childhood common cold (Lu *et al.*, 2018) and pneumonia (Miao *et al.*, 2017; Zeng *et al.*, 2017), suggesting a potential role of climatic factors in utero on the development of infection. On the other hand, positive associations of postnatal exposure to WS with respiratory diseases were also reported in Nigeria (Omonijo and Matzarakis, 2014) and China (Zhang *et al.*, 2011), which is in-line with our study. By contrast, a recent study in China found that daily mean wind speed was inversely associated with the clinical visits for childhood allergic diseases, including asthma, allergic rhinitis, and atopic dermatitis (Hu *et al.*, 2020). Some other recent studies in China also found an inverse correlation between mean monthly wind velocity and RSV detection rates among children with acute respiratory infection (Ji *et al.*, 2011; Zhang *et al.*, 2013a). Similarly, a study in Brazil indicated that a decrease in wind speed favoured hospitalization due to pneumonia (de Souza *et al.*, 2012). Recent study has indicated that meteorological factors such as wind speed could promote the diffusion of PM and general air pollutants (SO_2, NO_2, etc.), which could help to reduce the incidence of acute respiratory diseases (Chen and Wu, 2020). However, some recent studies found no significant relationship between WS and allergic diseases, including asthma (Yamazaki *et al.*, 2015) and RTIs in Greece (Falagas *et al.*, 2008), Turkey (Tasci *et al.*, 2018), and China (Liu *et al.*, 2016), which may be due to a relatively short-term effect of WS on the microorganism or individual (Tasci *et al.*, 2018). The WS exhibits a dual character because strong winds facilitate decreasing hospital admissions by reducing the levels of the pollutants throughout the year, but they also help to desiccate the air and hence encourage respiratory problems (Matyasovszky *et al.*, 2011). The former effect prevails in the pollen season due to the pollution-diluting effect of the wind, while the latter effect seems to be higher in the pollen-free season since the wind speed varies proportionally with the susceptible populations such as children with weak immune systems and thus their repeated exposition to air desiccated by winds may lead to an inflammation of the respiratory tracts (Strausz, 2003).

We further found that improved indoor NV characterized by frequent opening windows especially in winter showed a protective effect on both allergy and infection. We found that opening windows were associated with asthma, pneumonia, and frequent common cold, even after adjusting for outdoor air pollution, temperature, and other potential confounders. A recent Chinese study suggested crude negative associations between ventilation/cleaning habits, including frequent opening windows and childhood asthma and allergies (Lin *et al.*, 2015). Another recent study from China observed no significance between frequent opening windows in winter and current rhinitis or history of allergic rhinitis among children aged 1–8 years (Wang *et al.*, 2014). Some studies from China also showed an association between frequency of opening windows and high frequency and long duration of common cold (Sun *et al.*, 2011, 2017). The previous findings are consistent with our study. Furthermore, a study in Canada reported that respiratory infection was significantly associated with mean CO_2 level which is a key factor indicating the indoor ventilation rate (Kovesi *et al.*, 2007). Some other studies observed a positive association between low indoor ventilation rate indicated by high CO_2 level in the office environment and upper respiratory problems (Chao *et al.*, 2003) and SBS symptoms (Apte *et al.*, 2000; Erdmann and Apte, 2004). These results are in agreement with our studies. The ventilation was expected to be poorest when the doors and windows are normally kept shut especially during winter and generally coincides with the peak rates of RTI (Kovesi *et al.*, 2007). Though relatively few data exist, airflow and ventilation seem to play a role in respiratory virus infectivity and transmission (Pica and Bouvier, 2012). Inadequate ventilation has been implicated in the airborne transmission of respiratory viruses and thus may increase the concentration of airborne virus (Kovesi *et al.*, 2007;

Moser *et al.*, 1979; Myatt *et al.*, 2004; Rudnick and Milton, 2003). A previous study observed the effect of airflow on the transmissibility of rhinovirus in the mouse model, which indicates an inverse association between ventilation with fresh outside air and rhinovirus transmission (Myatt *et al.*, 2004). However, most existing data are inadequate for conclusions to be drawn whether ventilation rates directly cause respiratory allergies and infections after adjusting for personal factors, indoor environmental factors, and outdoor environmental and meteorological factors.

We found that indoor MV indicated as the use of exhaust fan in the kitchen or bathroom was negatively associated with infectious disease. A recent Chinese study showed that the use of exhaust fan in the bathroom at home was negatively associated with ever and current wheeze in preschool children, but not with doctor-diagnosed asthma (Liu *et al.*, 2014). Some other evidence also suggested that improving cooking devices and indoor MV could reduce the fraction of acute respiratory infection (ARI) in children under 5 years attributable to exposure to biomass smoke (Akunne *et al.*, 2006). A previous study in the United States found that the rates of febrile acute respiratory disease were significantly higher among persons living in the new barracks with low rates of mechanical outdoor air supply compared to those living in the older barracks with higher rates of mechanical air supply (Brundage *et al.*, 1988), whereas a Swedish study in homes found a dose–response relationship between ventilation rate in the homes and childhood asthma and allergy (Bornehag *et al.*, 2005). There are other evidence indicated that the use of MV at home resulted in reduction in numbers of house dust mites (Warner *et al.*, 2000) and childhood asthma and allergy (Choi *et al.*, 2014; Sundell *et al.*, 2011), which are inconsistent with our study.

We found strong associations between poor ventilation indicator in the homes and childhood allergy and infection. Our study is consistent with mounting evidence showing a significant association between dampness variables in the home and increased allergic and infectious diseases and symptoms (Fisk *et al.*, 2010; Gunnbjornsdottir *et al.*, 2006; Mendell *et al.*, 2011; Peat *et al.*, 1998; Stark *et al.*, 2003; Tischer *et al.*, 2011; Vardoulakis *et al.*, 2015). On the one hand, several studies found a higher prevalence of frequent common cold in occupants with dampness in their home (du Prel *et al.*, 2006; Kilpeläinen *et al.*, 2001; Spengler *et al.*, 2004) and dormitory rooms (Sun *et al.*, 2011). Some of our recent studies also indicated that perinatal and current exposure to damp stains was significantly associated with an increase in childhood pneumonia (Norbäck *et al.*, 2018) and common cold (Norbäck *et al.*, 2017). Furthermore, our finding is consistent with one longitudinal study showing that self-reported home dampness at birth was associated with a higher incidence of URTIs in the first year of life (Biagini *et al.*, 2006). Inconsistently, one study from Finland did not find any association between home dampness and common cold (Karvonen *et al.*, 2009). Moreover, some previous studies reported associations between early-life exposure to visible mould and mouldy odour and childhood asthma and allergic symptoms in young children, and the earlier exposures may trigger the earlier onset (Chen *et al.*, 2011; Hagerhed-Engman *et al.*, 2009). The inconsistency with our study may be due to that we used lifetime mould/damp stains, including both perinatal and current to exposure, but the early-life exposure particularly during prenatal or perinatal periods, as a more critical timing window, played a key role in the development of asthma and allergies among early childhood diseases (Lu *et al.*, 2020b, 2020c).

On the other hand, several Chinese studies found that both perinatal and current exposure to window condensation increased the risk of both diagnosis and onset of childhood asthma and allergies and decreased their remission (Lu *et al.*, 2020b, 2020c; Norbäck *et al.*, 2019; Wang *et al.*, 2015), which is consistent with our study. Furthermore, there are some previous studies supporting our study regarding on the association between window pane condensation in winter and respiratory infections. In one study among children in Japan, window pane condensation in winter time was associated with more respiratory infections needing antibiotic treatment (Takaoka *et al.*, 2016). Our recent work also found a consistent positive relationship between exposure to window pane condensation in the homes during both perinatal and current periods and childhood common cold (Norbäck *et al.*, 2017) and pneumonia (Norbäck *et al.*, 2018). Mould/dampness was indicated as an important risk factor in our study and it can be assumed that families with a poor ventilation level due to less natural or MV would have more dampness or condensation at home (Seppänen and Kurnitski, 2009; Ucci *et al.*, 2004). The built environment can also play an important role in the airborne transmission of infections through poor ventilation (Vardoulakis *et al.*, 2015). In spite of an evidence on the effectiveness of interventions by increasing indoor air humidity to prevent or reduce respiratory diseases at work and in educational settings (Byber *et al.*, 2016), our study indicated that evident dampness or condensation for pregnant mother or young children had consistent positive associations with multiple allergic and infectious effects, and thus early prevention and effective remediation of indoor dampness and condensation are likely to reduce childhood respiratory allergy and infection (Fisk *et al.*, 2010; Mendell *et al.*, 2011).

References

Ahn K. 2014. The role of air pollutants in atopic dermatitis. *J Allergy Clin Immunol* 134:993–999.

Akunne A, Louis V, Sanon M, and Sauerborn R. 2006. Biomass solid fuel and acute respiratory infections: The ventilation factor. *Int J Hyg Environ Health* 209:445–450.

Apte MG, Fisk WJ, and Daisey JM. 2000. Associations between indoor CO_2 concentrations and sick building syndrome symptoms in us office buildings: An analysis of the 1994–1996 base study data. *Indoor Air* 10:246–257.

Atkinson TP. 2013. Is asthma an infectious disease? New evidence. *Curr Allergy Asthma Rep* 13:702–709.

Banerjee S and Annesi-Maesano I. 2012. Spatial variability of indoor air pollutants in schools. A multilevel approach. *Atmos Environ* 61:558–561.

Ben-David T and Waring MS. 2016. Impact of natural versus mechanical ventilation on simulated indoor air quality and energy consumption in offices in fourteen U.S. cities. *Build Environ* 104:320–336.

Biagini JM, LeMasters GK, Ryan PH, *et al.* 2006. Environmental risk factors of rhinitis in early infancy. *Pediatr Allergy Immunol* 17:278–284.

Blencowe H, Cousens S, Oestergaard MZ, *et al.* 2012. National, regional, and worldwide estimates of preterm birth rates in the year 2010 with time trends since 1990 for selected countries: A systematic analysis and implications. *Lancet* 379:2162–2172.

Bornehag C-G, Sundell J, Hägerhed-Engman L, and Sigsgaard T. 2005. Association between ventilation rates in 390 Swedish homes and allergic symptoms in children. *Indoor Air* 15:275–280.

Brauer M, Amann M, Burnett RT, *et al.* 2012. Exposure assessment for estimation of the global burden of disease attributable to outdoor air pollution. *Environ Sci Technol* 46:652–660.

Brundage JF, Scott RM, Lednar WM, Smith DW, and Miller RN. 1988. Building-associated risk of febrile acute respiratory diseases in army trainees. *JAMA* 259:2108–2112.

Bukowski R, Smith GC, Malone FD, *et al.* 2007. Fetal growth in early pregnancy and risk of delivering low birth weight infant: Prospective cohort study. *BMJ* 334:836.

Byber K, Flatz A, Norbäck D, *et al.* 2016. Humidification of indoor air for preventing or reducing dryness symptoms or upper respiratory infections in educational settings and at the workplace. *Cochrane Database Syst Rev.*

Caillaud D, Leynaert B, Keirsbulck M, and Nadif R. 2018. Indoor mould exposure, asthma and rhinitis: Findings from systematic reviews and recent longitudinal studies. *Eur Respir Rev* 27.

Caron F, Guichard R, Robert L, Verriele M, and Thevenet F. 2020. Behaviour of individual VOCs in indoor environments: How ventilation affects emission from materials. *Atmos Environ* 243:117713.

Castro-Rodriguez JA, Forno E, Rodriguez-Martinez CE, and Celedón JC. 2016. Risk and protective factors for childhood asthma: What is the evidence? *J Allergy Clin Immunol Pract* 4:1111–1122.

Cavaleiro Rufo J, Madureira J, Oliveira Fernandes E, and Moreira A. 2016. Volatile organic compounds in asthma diagnosis: A systematic review and meta-analysis. *Allergy* 71:175–188.

Chan P, Chew F, Tan T, Chua K, and Hooi P. 2002. Seasonal variation in respiratory syncytial virus chest infection in the tropics. *Pediatr Pulmonol* 34:47–51.

Chao CY and Wong KK. 2002. Residential indoor PM10 and PM2.5 in Hong Kong and the elemental composition. *Atmos Environ* 36(2):265–277.

Chao CYH and Tung TC. 2001. An empirical model for outdoor contaminant transmission into residential buildings and experimental verification. *Atmos Environ* 35(9):1585–1596.

Chao HJ, Schwartz J, Milton DK, and Burge HA. 2003. The work environment and workers' health in four large office buildings. *Environ Health Perspect* 111:1242–1248.

Chawanpaiboon S, Vogel JP, Moller AB, *et al.* 2019. Global, regional, and national estimates of levels of preterm birth in 2014: A systematic review and modelling analysis. *Lancet Global Health* 7(1):e37–e46.

Chen S and Wu S. 2020. Deeplearning for identifying environmental risk factors of acute respiratory diseases in Beijing, China: Implications for population with different age and gender. *Int J Environ Health Res* 30:435–446.

Chen Y, Li G, Yan R, Zou L, Xin W, and Zhang W. 2013. An epidemiological survey on low birth weight infants in China and analysis of outcomes of full-term low birth weight infants. *BMC Pregnancy Childbirth* 13:242.

Chen Y-C, Tsai C-H, and Lee YL. 2011. Early-life indoor environmental exposures increase the risk of childhood asthma. *Int J Hyg Environ Health* 215:19–25.

Chiesa M, Urgnani R, Marzuoli R, Finco A, and Gerosa G. 2019. Site- and house-specific and meteorological factors influencing exchange of particles between outdoor and indoor domestic environments. *Build Environ* 160:106181.

Choi J, Chun C, Sun Y, *et al.* 2014. Associations between building characteristics and children's allergic symptoms—A cross-sectional study on child's health and home in Seoul, South Korea. *Build Environ* 75:176–181.

Choo CP and Jalaludin J. 2015. An overview of indoor air quality and its impact on respiratory health among Malaysian school-aged children. *Rev Environ Health* 30:9–18.

Cyrys J, Pitz M, Bischof W, Wichmann HE, and Heinrich J. 2004. Relationship between indoor and outdoor levels of fine particle mass, particle number concentrations and black smoke under different ventilation conditions. *J Expo Anal Environ Epidemiol* 14(4):275–283.

D'Amato G, Cecchi L, D'Amato M, and Annesi-Maesano I. 2014. Climate change and respiratory diseases. *Eur Respir J* 23:161–169.

D'Amato G, Cecchi L, D'Amato M, and Liccardi G. 2010. Urban air pollution and climate change as environmental risk factors of respiratory allergy: An update. *J Investig Allergol Clin Immunol* 20:95–102.

D'Amato G, Vitale C, De Martino A, *et al.* 2015. Effects on asthma and respiratory allergy of climate change and air pollution. *Multidiscip Respir Med* 10:1–8.

Dai X, Liu J, Li X, and Zhao L. 2018. Long-term monitoring of indoor CO_2 and PM2.5 in Chinese homes: Concentrations and their relationships with outdoor environments. *Build Environ* 144:238–247.

de Souza A, Fernandes WA, Pavão HG, Lastoria G, and do Amaral Albrez E. 2012. Potential impacts of climate variability on respiratory morbidity in children, infants, and adults. *J Bras Pneumol* 38:708–715.

Deng Q, Lu C, Jiang W, Zhao J, Deng L, and Xiang Y. 2017a. Association of outdoor air pollution and indoor renovation with early childhood ear infection in China. *Chemosphere* 169:288–296.

Deng Q, Lu C, Li Y, *et al.* 2017b. Association between prenatal exposure to industrial air pollution and onset of early childhood ear infection in China. *Atmos Environ* 157:18–26.

Deng Q, Lu C, Li Y, Sundell J, and Norbäck D. 2016a. Exposure to outdoor air pollution during trimesters of pregnancy and childhood asthma, allergic rhinitis, and eczema. *Environ Res* 150:119–127.

Deng Q, Lu C, Norbäck D, *et al.* 2015a. Early life exposure to ambient air pollution and childhood asthma in China. *Environ Res* 143:83–92.

Deng Q, Lu C, Ou C, Chen L, and Yuan H. 2016b. Preconceptional, prenatal and postnatal exposure to outdoor and indoor environmental factors on allergic diseases/symptoms in preschool children. *Chemosphere* 152:459–467.

Deng Q, Lu C, Ou C, and Liu W. 2015b. Effects of early life exposure to outdoor air pollution and indoor renovation on childhood asthma in China. *Build Environ* 93:84–91.

Deng Q, Lu C, Yu Y, Li Y, Sundell J, and Norbäck D. 2016c. Early life exposure to traffic-related air pollution and allergic rhinitis in preschool children. *Respir Med* 121:67–73.

Dick S, Doust E, Cowie H, Ayres JG, and Turner S. 2014b. Associations between environmental exposures and asthma control and exacerbations in young children: A systematic review. *BMJ Open* 4:e003827.

Dick S, Friend A, Dynes K, *et al.* 2014a. A systematic review of associations between environmental exposures and development of asthma in children aged up to 9 years. *BMJ Open* 4:e006554b.

Douwes J, Brooks C, and Pearce N. 2011. Asthma nervosa: Old concept, new insights. *Eur Respir J* 37:986–990.

du Prel J-B, Puppe W, Gröndahl B, *et al.* 2009. Are meteorological parameters associated with acute respiratory tract infections? *Clin Infect Dis* 49: 861–868.

du Prel X, Krämer U, Behrendt H, *et al.* 2006. Preschool children's health and its association with parental education and individual living conditions in East and West Germany. *BMC Public Health* 6:312.

Erdmann CA and Apte MG. 2004. Mucous membrane and lower respiratory building related symptoms in relation to indoor carbon dioxide concentrations in the 100-building base dataset. *Indoor Air* 14(Suppl. 8).

Falagas M, Theocharis G, Spanos A, *et al.* 2008. Effect of meteorological variables on the incidence of respiratory tract infections. *Respir Med* 102:733–737.

Fisk WJ, Eliseeva EA, and Mendell MJ. 2010. Association of residential dampness and mold with respiratory tract infections and bronchitis: A meta-analysis. *Environ Health* 9:72.

Fisk WJ, Lei-Gomez Q, and Mendell MJ. 2007. Meta-analyses of the associations of respiratory health effects with dampness and mold in homes. *Indoor Air* 17:284–296.

Franklin PJ. 2007. Indoor air quality and respiratory health of children. *Paediatr Respir Rev* 8:281–286.

Freitas M, Pacheco A, Verburg T, and Wolterbeek H. 2010. Effect of particulate matter, atmospheric gases, temperature, and humidity on respiratory and circulatory diseases' trends in Lisbon, Portugal. *Environ Monit Assess* 162:113–121.

Gao Y, Zhang Y, Kamijima M, *et al.* 2014. Quantitative assessments of indoor air pollution and the risk of childhood acute leukemia in Shanghai. *Environ Pollut* 187:81–89.

GBD 2015 Risk Factors Collaborators. 2016. Global, regional, and national comparative risk assessment of 79 behavioural, environmental and occupational, and metabolic risks or clusters of risks, 1990–2015: A systematic analysis for the Global Burden of Disease Study 2015. *Lancet* 388:1659–1724.

GBD 2017 Risk Factor Collaborators. 2017. Global, regional, and national comparative risk assessment of 84 behavioural, environmental and occupational, and metabolic risks or clusters of risks for 195 countries and territories, 1990–2017: A systematic analysis for the Global Burden of Disease Study 2017. *Lancet* 392(10159), 1923–1994.

Goldenberg R and Culhane J. 2007. Low birth weight in the united states. *Am J Clin Nutr* 85:584–590.

Goldenberg RL, Culhane JF, Iams JD, and Romero R. 2008. Epidemiology and causes of preterm birth. *Lancet* 371:75–84.

Gordon SB, Bruce NG, Grigg J, *et al.* 2014. Respiratory risks from household air pollution in low and middle income countries. *Lancet Respir Med* 2: 823–860.

Guan WJ, Zheng XY, Chung KF, and Zhong NS. 2016. Impact of air pollution on the burden of chronic respiratory diseases in China: Time for urgent action. *Lancet* 388:1939–1951.

Gunnbjornsdottir M, Franklin K, Norbäck D, *et al.* 2006. Prevalence and incidence of respiratory symptoms in relation to indoor dampness: The Rhine study. *Thorax* 61:221–225.

Gurgel RQ, de Matos Bezerra PG, Duarte MdCMB, *et al.* 2016. Relative frequency, possible risk factors, viral codetection rates, and seasonality of respiratory syncytial virus among children with lower respiratory tract infection in northeastern Brazil. *Medicine* 95.

Hagerhed-Engman L, Bornehag C, Sigsgaard T, *et al.* 2009. Low home ventilation rate in combination with moldy odor from the building structure increase the risk for allergic symptoms in children. *Indoor Air* 19:184–192.

Harville EW and Rabito FA. 2018. Housing conditions and birth outcomes: The national child development study. *Environ Res* 161:153–157.

He C, Morawska L, Hitchins J, and Gilbert D. 2004. Contribution from indoor sources to particle number and mass concentrations in residential houses. *Atmos Environ* 38(21):3405–3415.

Holøs SB, Yang A, Lind M, Thunshelle K, Schild P, and Mysen M. 2019. VOC emission rates in newly built and renovated buildings, and the influence of ventilation – A review and meta-analysis. *Indoor Built Environ* 18(3): 153–166.

Hossain MZ, Bambrick H, Wraith D, *et al.* 2019. Sociodemographic, climatic variability and lower respiratory tract infections: A systematic literature review. *Int J Biometeorol* 63:209–219.

Howard-Reed C, Wallace LA, and Emmerich SJ. 2003. Effect of ventilation systems and air filters on decay rates of particles produced by indoor sources in an occupied townhouse. *Atmos Environ* 37(38):5295–5306.

Hu Y, Xu Z, Jiang F, *et al.* 2020. Relative impact of meteorological factors and air pollutants on childhood allergic diseases in Shanghai, China. *Sci Total Environ* 706:135975.

Huang S, Garshick E, Weschler LB, *et al.* 2020. Home environmental and lifestyle factors associated with asthma, rhinitis and wheeze in children in Beijing, China. *Environ Pollut* 256:113426.

Jendritzky G, de Dear R, and Havenith G. 2012. UTCI—Why another thermal index? *Int J Biometeorol* 56:421–428.

Ji W, Chen Z-R, Guo H-B, *et al.* 2011. Characteristics and the prevalence of respiratory viruses and the correlation with climatic factors of hospitalized children in Suzhou children's hospital. *Zhonghua Yu Fang Yi Xue Za Zhi* 45:205–210.

Jiang W, Lu C, Miao Y, Xiang Y, Chen L, and Deng Q. 2018. Outdoor particulate air pollution and indoor renovation associated with childhood pneumonia in China. *Atmos Environ* 174:76–81.

Kadir M, McClure EM, Goudar SS, *et al.* 2010. Exposure of pregnant women to indoor air pollution: A study from nine low and middle income countries. *Acta Obstet Gynecol Scand* 89:540–548.

Kana MA, Correia S, Peleteiro B, Severo M, and Barros H. 2017. Impact of the global financial crisis on low birth weight in Portugal: A time-trend analysis. *BMJ Global Health* 2:e000147.

Karvonen AM, Hyvärinen A, Roponen M, *et al.* 2009. Confirmed moisture damage at home, respiratory symptoms and atopy in early life: A birth-cohort study. *Pediatrics* 124:e329–e338.

Kilpeläinen M, Terho E, Helenius H, and Koskenvuo M. 2001. Home dampness, current allergic diseases, and respiratory infections among young adults. *Thorax* 56:462–467.

Kim J, Kim J-H, Cheong H-K, *et al.* 2016. Effect of climate factors on the childhood pneumonia in Papua New Guinea: A time-series analysis. *Int J Environ Res Public Health* 13:213.

Kovesi T, Gilbert NL, Stocco C, *et al.* 2007. Indoor air quality and the risk of lower respiratory tract infections in young Canadian Inuit children. *CMAJ* 177: 155–160.

Kukadia V and Palmer J. 1998. The effect of external atmospheric pollution on indoor air quality: A pilot study. *Energy Build* 27:223–230.

Lakshmanan A, Chiu Y-HM, Coull BA, *et al.* 2015. Associations between prenatal traffic-related air pollution exposure and birth weight: Modification by sex and maternal pre-pregnancy body mass index. *Environ Res* 137: 268–277.

Lam H and Chan E. 2019. Short-term association between meteorological factors and childhood pneumonia hospitalization in Hong Kong: A time-series study. *Epidemiology* 30:S107–S114.

Lelieveld J, Evans JS, Fnais M, Giannadaki D, and Pozzer A. 2015. The contribution of outdoor air pollution sources to premature mortality on a global scale. *Nature* 525:367–371.

Lin Z, Zhao Z, Xu H, *et al.* 2015. Home dampness signs in association with asthma and allergic diseases in 4618 preschool children in Urumqi, China—The influence of ventilation/cleaning habits. *PLoS One* 10:e0134359.

Liu L, Oza S, Hogan D, *et al.* 2015a. Global, regional, and national causes of child mortality in 2000-13, with projections to inform post-2015 priorities: An updated systematic analysis. *Lancet* 385:430–440.

Liu W, Huang C, Cai J, Wang X, Zou Z, and Sun C. 2018. Household environmental exposures during gestation and birth outcomes: A cross-sectional study in Shanghai, China. *Sci Total Environ* 615:1110–1118.

Liu W, Huang C, Hu Y, Zou Z, Shen L, and Sundell J. 2015b. Associations of building characteristics and lifestyle behaviors with home dampness-related exposures in Shanghai dwellings. *Build Environ* 88:106–115.

Liu W, Huang C, Hu Y, Zou Z, Zhao Z, and Sundell J. 2014. Association of building characteristics, residential heating and ventilation with asthmatic symptoms of preschool children in Shanghai: A cross-sectional study. *Indoor Built Environ* 23:270–283.

Liu W, Zhang Y, Yao Y, and Li J. 2012. Indoor decorating and refurbishing materials and furniture volatile organic compounds emission labeling systems: A review. *Chin Sci Bull* 57(20):2533–2543.

Liu Y, Liu J, Chen F, *et al.* 2016. Impact of meteorological factors on lower respiratory tract infections in children. *J Int Med Res* 44:30–41.

Long CM, Suh HH, Catalano PJ, and Koutrakis P. 2001. Using time- and size-resolved particulate data to quantify indoor penetration and deposition behavior. *Environ Sci Technol* 35(10):2089–2099.

Lu C, Cao L, Norbäck D, Li Y, Chen J, and Deng Q. 2019. Combined effects of traffic air pollution and home environmental factors on preterm birth in China. *Ecotoxicol Environ Saf* 184:109639.

Lu C, Deng L, Ou C, Yuan H, Chen X, and Deng Q. 2017. Preconceptional and perinatal exposure to traffic-related air pollution and eczema in preschool children. *J Dermatol Sci* 85:85–95.

Lu C, Deng Q, Li Y, Sundell J, and Norbäck D. 2016. Outdoor air pollution, meteorological conditions and indoor factors in dwellings in relation to sick building syndrome (SBS) among adults in China. *Sci Total Environ* 560: 186–196.

Lu C, Miao Y, Zeng J, Jiang W, Shen Y-M, and Deng Q. 2018. Prenatal exposure to ambient temperature variation increases the risk of common cold in children. *Ecotoxicol Environ Saf* 154:221–227.

Lu C, Norbäck D, Li Y, and Deng Q. 2020b. Early-life exposure to air pollution and childhood allergic diseases: An update on the link and its implications. *Expert Rev Clin Immunol* 16(8):813–827.

Lu C, Norbäck D, Zhang Y, *et al.* 2020c. Onset and remission of eczema at preschool age in relation to prenatal and postnatal air pollution and home environment across China. *Sci Total Environ* 755:142467.

Lu C, Zhang W, Zheng X, Sun J, Chen L, and Deng Q. 2020a. Combined effects of ambient air pollution and home environmental factors on low birth weight. *Chemosphere* 240:124836.

Maroni M, Seifert B, and Lindvall T. 1995. *Indoor Air Quality: A Comprehensive Reference Book.* Amsterdam, Oxford: Elsevier.

Matyasovszky I, Makra L, Bálint B, Guba Z, and Sümeghy Z. 2011. Multivariate analysis of respiratory problems and their connection with meteorological parameters and the main biological and chemical air pollutants. *Atmos Environ* 45:4152–4159.

Mendell MJ, Mirer AG, Cheung K, Tong M, and Douwes J. 2011. Respiratory and allergic health effects of dampness, mold, and dampness-related agents: A review of the epidemiologic evidence. *Environ Health Perspect* 119:748.

Meng QY, Turpin BJ, Korn L, *et al.* 2005. Influence of ambient (outdoor) sources on residential indoor and personal PM2.5 concentrations: Analyses of RIOPA data. *J Expo Anal Environ Epidemiol* 15(1):17–28.

Miao Y, Shen Y-M, Lu C, Zeng J, and Deng Q. 2017. Maternal exposure to ambient air temperature during pregnancy and early childhood pneumonia. *J Therm Biol* 69:288–293.

Mirsaeidi M, Motahari H, Taghizadeh Khamesi M, Sharifi A, Campos M, and Schraufnagel DE. 2016. Climate change and respiratory infections. *Ann Am Thorac Soc* 13:1223–1230.

Mohammadyan M and Shabankhani B. 2013. Indoor PM1, PM2.5, PM10 and outdoor PM2.5 concentrations in primary schools in Sari, Iran. *Arh Hig Rada Toksikol* 64(3):371–377.

Moser MR, Bender TR, Margolis HS, Noble GR, Kendal AP, and Ritter DG. 1979. An outbreak of influenza aboard a commercial airliner. *Am J Epidemiol* 110:1–6.

Moster D, Lie RT, and Markestad T. 2008. Long-term medical and social consequences of preterm birth. *N Engl J Med* 359:262–273.

Myatt TA, Johnston SL, Zuo Z, *et al.* 2004. Detection of airborne rhinovirus and its relation to outdoor air supply in office environments. *Am J Respir Crit Care Med* 169:1187–1190.

Nazaroff W and Goldstein A. 2015. Indoor chemistry: Research opportunities and challenges. *Indoor Air* 25:357–361.

Nazaroff WW. 2004. Indoor particle dynamics. *Indoor Air* 7(14 Suppl): 175–183.

Norbäck D, Lu C, Zhang Y, *et al.* 2017. Common cold among pre-school children in China – Associations with ambient PM10 and dampness, mould, cats, dogs, rats and cockroaches in the home environment. *Environ Int* 103:13–22.

Norbäck D, Lu C, Zhang Y, *et al.* 2018. Lifetime-ever pneumonia among pre-school children across China – Associations with pre-natal and post-natal early life environmental factors. *Environ Res* 167:418–427.

Norbäck D, Lu C, Zhang Y, *et al.* 2019. Onset and remission of childhood wheeze and rhinitis across China—Associations with early life indoor and outdoor air pollution. *Environ Int* 123:61–69.

Omonijo AG and Matzarakis A. 2014. Pneumonia occurrence in relation to population and thermal environment in Ondo State, Nigeria. *Afr Rev Phys* 9.

Omrani S, Garcia-Hansen V, Capra B, and Drogemuller R. 2017. Natural ventilation in multi-story buildings: Design process and review of evaluation tools. *Build Environ* 116:182–194.

Othman M, Latif MT, and Matsumi Y. 2019. The exposure of children to PM2.5 and dust in indoor and outdoor school classrooms in Kuala Lumpur City Centre. *Ecotoxicol Environ Saf* 170:739–749.

Pallarés S, Gómez E, Martínez A, and Jordán MM. 2019. The relationship between indoor and outdoor levels of PM10 and its chemical composition at schools in a coastal region in Spain. *Heliyon* 5(8):e02270.

Park JS, Jee NY, and Jeong JW. 2014. Effects of types of ventilation system on indoor particle concentrations in residential buildings. *Indoor Air* 24:629–638.

Peat J, Dickerson J, and Li J. 1998. Effects of damp and mould in the home on respiratory health: A review of the literature. *Allergy* 53:120–128.

Pica N and Bouvier NM. 2012. Environmental factors affecting the transmission of respiratory viruses. *Curr Opin Virol* 2:90–95.

Quang TN, He C, Morawska L, and Knibbs LD. 2013. Influence of ventilation and filtration on indoor particle concentrations in urban office buildings. *Atmos Environ* 79:41–52.

Ritchie H and Roser M. 2013. Indoor Air Pollution. Our World in Data. https://ourworldindata.org/indoor-air-pollution.

Rudnick S and Milton D. 2003. Risk of indoor airborne infection transmission estimated from carbon dioxide concentration. *Indoor Air* 13:237–245.

Salvador CM, Bekö G, Weschler CJ, *et al.* 2019. Indoor ozone/human chemistry and ventilation strategies. *Indoor Air* 29(6):913–925.

Sarigiannis D. 2013. Combined or multiple exposure to health stressors in indoor built environments. In: Proceedings of the Proceedings of the An Evidence-based Review Prepared for the WHO Training Workshop "Multiple Environmental Exposures and Risks", Bonn, Germany 16–18.

Seppänen O and Kurnitski J. 2009. Moisture control and ventilation. In: *Who Guidelines for Indoor Air Quality: Dampness and Mould*. World Health Organization, Geneva.

Shan X, Chen F, Wang W, *et al.* 2014. Secular trends of low birthweight and macrosomia and related maternal factors in Beijing, China: A longitudinal trend analysis. *BMC Pregnancy Childbirth* 14:105.

Shenassa ED, Daskalakis C, Liebhaber A, Braubach M, and Brown M. 2007. Dampness and mold in the home and depression: An examination of mold-related illness and perceived control of one's home as possible depression pathways. *Am J Public Health* 97:1893–1899.

Sørensen M, Andersen A-MN, and Raaschou-Nielsen O. 2010.Non-occupational exposure to paint fumes during pregnancy and fetal growth in a general population. *Environ Res* 110:383–387.

Spengler JD, Jaakkola JJ, Parise H, Katsnelson BA, Privalova LI, and Kosheleva AA. 2004. Housing characteristics and children's respiratory health in the Russian Federation. *Am J Public Health* 94:657–662.

Stark PC, Burge HA, Ryan LM, Milton DK, and Gold DR. 2003. Fungal levels in the home and lower respiratory tract illnesses in the first year of life. *Am J Respir Crit Care Med* 168:232–237.

Stephens B and Siegel JA. 2012. Penetration of ambient submicron particles into single family residences and associations with building characteristics. *Indoor Air* 22:501–513.

Strausz J. 2003. *Asthma bronchial.* Medister GlaxoSmithKline, Budapest.

Sun C, Huang C, Liu W, Zou Z, Hu Y, and Shen L. 2017. Home dampness-related exposures increase the risk of common colds among preschool children in Shanghai, China: Modified by household ventilation. *Build Environ* 124:31–41.

Sun Y, Hou J, Kong X, *et al.* 2018. "Dampness" and "dryness": What is important for children's allergies? A cross-sectional study of 7366 children in northeast Chinese homes. *Build Environ* 139:38–45.

Sun Y and Sundell J. 2011. Life style and home environment are associated with racial disparities of asthma and allergy in northeast Texas children. *Sci Total Environ* 409:4229–4234.

Sun Y, Wang Z, Zhang Y, and Sundell J. 2011. In China, students in crowded dormitories with a low ventilation rate have more common colds: Evidence for airborne transmission. *PLoS One* 6:e27140.

Sundell J, Levin H, Nazaroff WW, *et al.* 2011. Ventilation rates and health: Multidisciplinary review of the scientific literature. *Indoor Air* 21: 191–204.

Takaoka M, Suzuki K, and Norbäck D. 2016. The home environment of junior high school students in Hyogo, Japan—Associations with asthma, respiratory health and reported allergies. *Indoor Built Environ* 25:81–92.

Tasci SS, Kavalci C, and Kayipmaz AE. 2018. Relationship of meteorological and air pollution parameters with pneumonia in elderly patients. *Emerg Med Int* 2018:4183203.

Thatcher TL, Lunden MM, Revzan KL, Sextro RG, and Brown NJ. 2003. A concentration rebound method for measuring particle penetration and deposition in the indoor environment. *Aerosol Sci Technol* 37(11):847–864.

Theodosiou TG and Ordoumpozanis KT. 2008. Energy, comfort and indoor air quality in nursery and elementary school buildings in the cold climatic zone of Greece. *Energy Build* 40:2207–2214.

Tischer C, Chen C, and Heinrich J. 2011. Association between domestic mould and mould components, and asthma and allergy in children: A systematic review. *Eur Respir J* 38:812–814.

Troeger C, Blacker B, Khalil IA, *et al.* 2018. Estimates of the global, regional, and national morbidity, mortality, and aetiologies of lower respiratory infections in 195 countries, 1990–2016: A systematic analysis for the Global Burden of Disease Study 2016. *Lancet Infect Dis* 18(11):1191–1210.

Ucci M, Ridley I, Pretlove S, *et al.* 2004. Ventilation rates and moisture-related allergens in UK dwellings. In: 2nd WHO International Housing and Health Symposium, Vilnius, Lithuania, 2004.

Vardoulakis S, Dimitroulopoulou C, Thornes J, Lai K-M, and Taylor J. 2015. Impact of climate change on the domestic indoor environment and associated health risks in the UK. *Environ Int* 85:299–313.

Wallace LA, Emmerich SJ, and Howard-Reed C. 2004. Effect of central fans and in-duct filters on deposition rates of ultrafine and fine particles in an occupied townhouse. *Atmos Environ* 38(3):405–413.

Walton NA, Poynton MR, Gesteland PH, Maloney C, Staes C, and Facelli JC. 2010. Predicting the start week of respiratory syncytial virus outbreaks using real time weather variables. *BMC Med Inform Decis Mak* 10:68.

Wang F, Meng D, Li X, and Tan J. 2016. Indoor-outdoor relationships of PM2.5 in four residential dwellings in winter in the Yangtze River Delta, China. *Environ Pollut* 215:280–289.

Wang H, Li B, Yu W, Wang J, and Norback D. 2015. Early-life exposure to home dampness associated with health effects among children in Chongqing, China. *Build Environ* 94:327–334.

Wang J, Li B, Yu W, *et al.* 2014. Rhinitis symptoms and asthma among parents of preschool children in relation to the home environment in Chongqing, China. *PLoS One* 9(4):e94731.

Wang Y, Deng B, and Kim CN. 2012. Transient characteristics of VOCs removal by an air cleaner in association with a humidifier combined with different ventilation strategies in an office. *Indoor Built Environ* 21(1):71–78.

Wargocki P and Wyon DP. 2013. Providing better thermal and air quality conditions in school classrooms would be cost-effective. *Build Environ* 59: 581–589.

Wargocki P. 2013. The effects of ventilation in homes on health. *Int J Vent* 12: 101–118.

Warner JA, Frederick JM, Bryant TN, *et al.* 2000. Mechanical ventilation and high-efficiency vacuum cleaning: A combined strategy of mite and mite allergen reduction in the control of mite-sensitive asthma. *J Allergy Clin Immunol* 105:75–82.

WHO. 2011. *Health Co-Benefits of Climate Change Mitigation—Housing Sector.* World Health Organization, Geneva.

WHO. 2014a. Frequently Asked Questions – Ambient and Household Air Pollution and Health. Update 2014.

WHO. 2014b. Fact sheet N°292 – Household Air Pollution and Health. https://www.who.int/en/news-room/fact-sheets/detail/household-air-pollution-and-health.

Wu X, Lu Y, Zhou S, Chen L, and Xu B. 2016. Impact of climate change on human infectious diseases: Empirical evidence and human adaptation. *Environ Int* 86:14–23.

Xu L, Batterman S, Chen F, *et al.* 2017. Spatiotemporal characteristics of PM2.5 and PM10 at urban and corresponding background sites in 23 cities in China. *Sci Total Environ* 599–600:2074–2084.

Xu Y and Zhang J. 2011. Understanding SVOCs. *ASHRAE J* 53(12):121–125.

Xu Z, Etzel RA, Su H, Huang C, Guo Y, and Tong S. 2012. Impact of ambient temperature on children's health: A systematic review. *Environ Res* 117: 120–131.

Yamazaki S, Shima M, Yoda Y, *et al.* 2015. Exposure to air pollution and meteorological factors associated with children's primary care visits at night due to asthma attack: Case-crossover design for 3-year pooled patients. *BMJ Open* 5.

Ye W, Zhang X, Gao J, Cao G, Zhou X, and Su X. 2017. Indoor air pollutants, ventilation rate determinants and potential control strategies in Chinese dwellings: A literature review. *Sci Total Environ* 586:696–729.

Yousif MK and Al Muhyi A-HA. 2019. Impact of weather conditions on childhood admission for wheezy chest and bronchial asthma. *Med J Islam Repub Iran* 33:89.

Zeng J, Lu C, and Deng Q. 2017. Prenatal exposure to diurnal temperature variation and early childhood pneumonia. *J Therm Biol* 65:105–112.

Zhan Z, Zhao Y, Pang S, Zhong X, Wu C, and Ding Z. 2017. Temperature change between neighboring days and mortality in united states: A nationwide study. *Sci Total Environ* 584:1152–1161.

Zhang D, He J, Gao S, Hu B, and Ma S. 2011. Correlation analysis for the attack of respiratory diseases and meteorological factors. *Chin J Integr Med* 17:600.

Zhang J and Smith KR. 2003. Indoor air pollution: A global health concern. *Br Med Bull* 68:209–225.

Zhang X, Shao X, Wang J, and Guo W. 2013a. Temporal characteristics of respiratory syncytial virus infection in children and its correlation with climatic factors at a public pediatric hospital in Suzhou. *J Clin Virol* 58: 666–670.

Zhang Y, Li B, Huang C, *et al.* 2013b. Ten cities cross-sectional questionnaire survey of children asthma and other allergies in China. *Chin Sci Bull* 58:4182–4189.

Zhang Y, Mo J, Li Y, *et al.* 2011. Can commonly-used fan-driven air cleaning technologies improve indoor air quality? A literature review. *Atmos Environ* 45(26):4329–4343.

Zhao B, Zheng H, Wang S, *et al.* 2018b. Change in household fuels dominates the decrease in PM2. 5 exposure and premature mortality in China in 2005–2015. *Proc Natl Acad Sci USA* 115:12401–12406.

Zhao Y, Sun H, and Tu D. 2018a. Effect of mechanical ventilation and natural ventilation on indoor climates in Urumqi residential buildings. *Build Environ* 144:108–118.

Zhou C, Wang Z, Chen Q, Jiang Y, and Pei J. 2014. Design optimization and field demonstration of natural ventilation for high-rise residential buildings. *Energy Build* 82:457–465.

Zhou M, Wang H, Zeng X, *et al.* 2019. Mortality, morbidity, and risk factors in China and its provinces, 1990–2017: A systematic analysis for the Global Burden of Disease Study 2017. *Lancet* 394(10204):1145–1158.

Chapter 14
Ventilation in industry buildings

Yi Wang[1,2] *and Zhixiang Cao*[1,2]

14.1 Introduction

The main task of industrial ventilation is to provide good working conditions for the production process and the workers in it. For a long time, people have been keeping a pursuit and exploration for better industrial ventilation technology. On the one hand, as an old problem, the design of industrial ventilation has accompanied with the industrial revolution. On the other hand, with the constant development of the industry, the expansion of industrial building scale and the improvement of building technology, new challenges have been put forward on the industrial ventilation. What's more, since the main attention was on worker's health and safety; thus, a higher demand was set on indoor thermal conditions and air quality for these industrial occupational environments. In recent decades, numerous research works have been conducted on building environment control; however, these research targets are mainly focused on civil buildings. With respect to the environmental control of industrial buildings, many researchers and designers have devoted a lot to research theoretically and practically, but there are still many problems that have not been fully studied and resolved (Wang and Cao, 2017).

14.1.1 Differences in fundamental research on environmental control between industrial and civil buildings

There are distinct differences between industrial buildings and civil buildings in terms of indoor sources and control demand. Some significant differences exist between industrial buildings and civil buildings.

14.1.1.1 Indoor sources

There are various sources that can generate waste heat, surplus humidity, and contaminants in industrial buildings, of which the contaminants emissions may be

[1]State Key Laboratory of Green Building in Western China, Xi'an University of Architecture and Technology, Xi'an, China
[2]School of Building Services Science and Engineering, Xi'an University of Architecture and Technology, Xi'an, China

ten times higher than found in civil buildings. However, only small differences in contaminants exposure limits are given in industrial standards in comparison to standards given for civil buildings (Howard and Esko, 2001).

In civil buildings, main heat sources include human bodies, electronic devices, and lighting devices while main pollution sources are from human activities. They are to a large extent less harmful than occupational exposure to high-intensity heat sources and pollution in industrial buildings. Due to differences caused by sources, environmental control methods for industrial buildings could vary greatly compared with that for civil buildings. For example, there are high-intensity heat sources in heat treatment plants. The skylight with large opening set on the roof for these kinds of plants would produce the buoyancy-driven ventilation. Thus, natural ventilation would be the most important method for environmental control and energy saving for this type of building (Wang *et al.*, 2016). However, the buoyancy-driven ventilation in most cases cannot be integrated into civil buildings because the ventilation opening with large height difference would not be necessary in these types of buildings. Meanwhile, due to the existence of various pollution sources in industrial buildings, local ventilation would be an important method for environmental control. However, in civil buildings, local ventilation is seldom applied except for fume exhaust in kitchen (Awbi, 2003).

14.1.1.2 Control demand

The indoor environment control of industrial building needs to consider both the production process and workers' health and safety, whereas the purpose of environment control in civil buildings would mainly satisfy the health and well-being requirements of residents. There are many forms of industrial buildings, such as a clean plant with high requirements on temperature and humidity and general plant that produce waste heat and some other pollutants. Some of these are labour-intensive-processing plant and high-level automation plant: totally enclosed plant and open plant. According to different needs of production processes, the demand of industrial building environment control on indoor environmental parameters such as temperature, humidity, cleanliness, and airflow velocity would vary greatly and may even require several orders of magnitude, much higher than that would be required in civil buildings. In addition, there are also great differences between industrial and civil buildings on human demand.

In civil buildings, the main consideration is to create a comfortable environment for occupants, while in industrial buildings, the main consideration is to create an acceptable work environment. Therefore, the difference in environmental control demand for these two kinds of buildings demonstrates not only in their control index, but also in evaluation index. Specifically, due to low labour intensity in civil buildings, the human comfort index such as defined by PMV (Fanger, 1970) or SET (Gagge *et al.*, 1971) would be the main evaluation index. However, since workers in industrial buildings could experience high labour intensity, the evaluation of indoor thermal environment is mainly with aim to reduce impact on human health, such as various kinds of thermal stress indices used for the evaluation of health and safety exposure in workplaces (Wyndham *et al.*, 1967; Belding and Hatch, 1955; Givoni,

1998; Yaglou and Minard, 1957). Thus, there exists a great difference in environmental control between the industrial and civil buildings due to their different control demand requirements.

14.1.2 Efficient ventilation design for industrial buildings

In industrial buildings, ventilation has different particularity problems in different industries. This is shown in two aspects: first, the energy consumption of industrial buildings; and second, the indoor air quality of industrial buildings. The influence of ventilation on building energy consumption is also shown in two aspects: the first is the energy consumption of the fan in ventilation system, and the other one is the air distribution rationality determining air change rate. Air change rate will greatly affect the heating and air-conditioning energy consumption of industrial buildings; therefore, efficient ventilation is very necessary in industrial buildings.

Efficient ventilation is an issue that must be improved and explored continuously; thus energy-efficiency design of ventilation requires specialised technology. The exploration of specialised design scheme refers to different industrial processes; thus, it will face several problems, including increasing of the floor area of industrial buildings, the change in the amount and intensity of pollution, and some problems relevant to the higher environmental control level and energy efficiency request. For the purpose of research, the understanding of pollutant control mechanism should be improved. For the purpose of application, rough design should be turned into refined design.

As shown in Figure 14.1, the relationship between ventilation flow rate and ventilation efficiency is not a simple linear relationship. When the ventilation rate reaches the knee point, the improvement of ventilation efficiency decreases rapidly with the increase of ventilation rate, but the energy consumption increases rapidly. Therefore, in the design of industrial ventilation system, the position of knee point

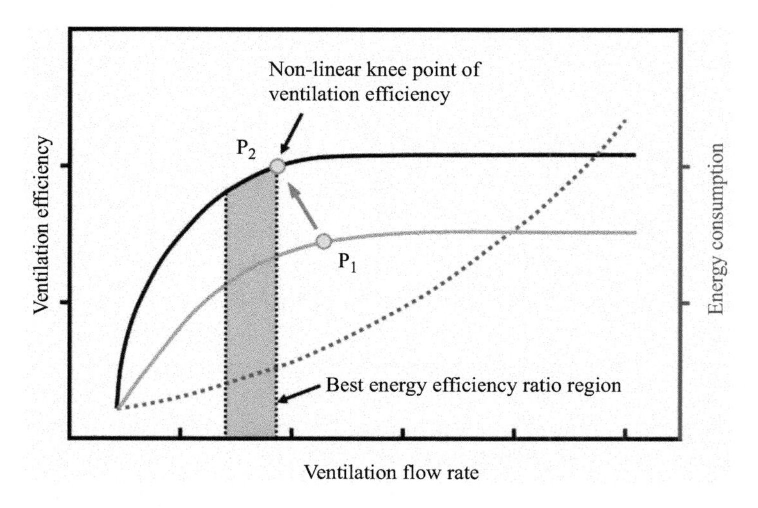

Figure 14.1 Best energy efficiency ratio of industrial ventilation

(from P_1 to P_2) should be improved as much as possible, so that the ventilation system can operate in the best energy efficiency ratio region (Wang *et al.*, 2019).

14.2 Natural ventilation

14.2.1 Conception

Natural ventilation is a type of ventilation mode that uses the heat pressure or wind pressure caused by density differences between air inside and outside a building to form air flows inside and outside a ventilation room. Natural ventilation greatly reduces the building energy consumption relative to mechanical ventilation. The natural ventilation of buildings does not require energy and provides a massive amount of ventilation, making it a very economical ventilation method. Therefore, industrial buildings should make a full use of natural ventilation to eliminate waste heat and humidity in industrial buildings. The advantages of natural ventilation include the following aspects.

1. Energy saving: Natural ventilation does not require fans and/or cold and heat sources; thus, it can effectively reduce energy consumption.
2. Low initial investment cost: Compared with mechanical ventilation, natural ventilation requires less equipment and fewer piping system components.
3. Low maintenance and replacement costs: Buildings with mechanical ventilation and air conditioning require maintenance, and the costs of maintenance and/or replacement of mechanical equipment are often very high. Equipment in natural ventilation usually requires very little maintenance, or none at all.
4. Large ventilation capacity: Compared with a mechanical ventilation system, a natural ventilation system often has large areas for the air inlet and outlet, and ventilation can be continuously conducted as long as there is thermal and wind pressure. Therefore, natural ventilation buildings often have a massive ventilation capacity.

Although natural ventilation is an economic and effective ventilation method in most cases, it is difficult to control the ventilation volume and ventilation effect, owing to the great influences from outdoor meteorological conditions (such as temperature and wind speed). Only by understanding the principles of natural ventilation, can we make rational and efficient use of natural ventilation?

14.2.2 Basic principle

14.2.2.1 Buoyancy-driven natural ventilation

Industrial buildings with natural ventilation often have openings 1 and 2 with different heights. The height difference between them is denoted as h, F_1 is the area of the air inlet, and F_2 is the area of the exhaust port, as shown in Figure 14.2. When the indoor air is heated, the density of the hot air decreases and it rises upwards before being discharged from opening 2 at the upper part of the building. According to the principle of the conservation of mass, the outside air flows into the room from opening 1 at the lower part of the building. The internal pressure of

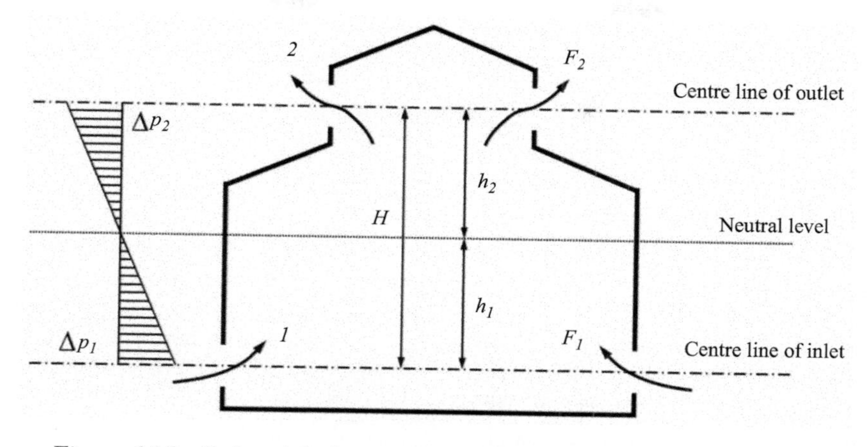

Figure 14.2 Industrial plant with buoyancy-driven natural ventilation

opening 2 in the upper part of the room is higher than the external pressure, and is called the exhaust pressure Δp_2, whereas the internal pressure of the lower opening 1 is lower than the external pressure and is called the inlet pressure Δp_1. According to the principle of hot pressure ventilation, there will be a neutral level between the upper and lower openings of the room.

Neutral level H is the level at which the pressure inside the space equals the ambient pressure at that height. The higher internal pressure at the upper opening h_2 drives the outflow, and the lower internal pressure at the lower opening h_1 drives the inflow. Thus, the neutral level defines the height separating the lower and upper openings: air flows out through openings above the neutral level (upper openings) and flows in through openings below the neutral level (lower openings).

14.2.2.2 Wind-driven natural ventilation

Owing to the obstructions from buildings, the pressure distributions of the outdoor air flows around buildings change. Compared with an undisturbed airflow, the increase or decrease in the static pressure is generally called the wind pressure. As shown in Figure 14.3, the airflow on the windward side of the building is blocked, the dynamic pressure is reduced, the static pressure is increased, and the wind pressure is a positive pressure. Owing to the local vortex generated on the sides and leeward side, the static pressure is reduced, and the wind pressure is a negative pressure. The distribution of the wind pressure is different for buildings with different shapes, under the action of wind in different directions. The wind pressure distribution around a building is related to its geometry and the outdoor wind direction. The indoor airflow caused by the different pressure distributions on the building facade is called the wind-driven natural ventilation.

Owing to the randomness and turbulence characteristics of natural wind, the wind pressure is actually a fluctuation and can be decomposed into the superposition of a time-averaged flow and pulse momentum. It is very difficult to obtain

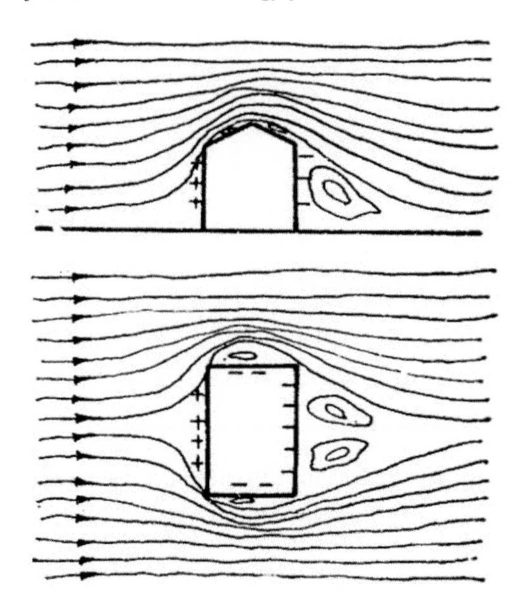

Figure 14.3 Wind pressure distribution around buildings

an accurate average velocity for natural wind; as such, it is usually calculated as a steady flow according to the average flow. There is a certain error between the natural ventilation rate calculated by this method and actual ventilation rate, but it is acceptable for analysing the influences of various building factors.

The wind pressure is affected by the following factors:

1. outdoor wind speed and wind direction (relative angle between building facade and outdoor wind direction);
2. geometry of the building;
3. position of the opening on the building facade;
4. atmospheric boundary layer conditions; and
5. surrounding environmental conditions (including surrounding terrain, building density, and vegetation).

For a rectangular building, if the wind pressure acting on a facade of the building is assumed to be the same, the wind pressure acting on the opening can be calculated as follows:

$$\Delta p_{wind} = p_{wind} - p_{ref} = c_p \cdot \rho \cdot \frac{v_h^2}{2} \tag{14.1}$$

Previously, p_{wind} denotes the wind pressure at the opening, in Pa; p_{ref} denotes the atmospheric pressure, in Pa; ρ is the air density, in kg/m^3; v_h is the incoming wind velocity and generally refers to the wind speed at the height of the building, in m/s; and c_p is the wind pressure coefficient, which reflects the comprehensive effect of the influencing factors of the building shape, opening position, and wind direction (and can be obtained experimentally).

14.2.3 Optimal design principle of natural ventilation

14.2.3.1 Inlet and outlet area

In the process of production, a hot processing workshop emits significant amounts of waste heat, dust, and other polluted gases, deteriorating the internal environment of the factory. Therefore, the quality of the internal environment must be improved by effectively organising the natural ventilation of the plant and quickly removing waste heat and polluted gases. When the height and heat dissipation of the workshop are constant, the key to improving the effect of natural ventilation is to reasonably coordinate the areas of the inlet and outlet.

The principle of natural ventilation design for a plant should be to try to reduce the position of the neutral plane as far as possible. This is because a low position of the neutral plane means that most or all the fresh air entering the plant from the outside will flow through the working area. Evidently, this will play a decisive role in reducing the temperature of the operating area and improving the air quality of the working area, i.e. improving the natural ventilation effect.

According to the principle of natural ventilation, when the intake and exhaust volumes are constant, the key means to lowering the position of the neutral plane is to reasonably coordinate the ratio of the intake and exhaust areas. When the intake area F_1 is larger than the exhaust area F_2, the position of the neutral plane is lower. In contrast, when the exhaust area F_2 is larger than the intake area F_1, the position of the neutral plane is higher.

According to the principle that the air intake volume of the plant is equal to the exhaust volume, equations can be derived as follows:

$$h_1 = \frac{HF_2^2}{F_1^2 + F_2^2} \tag{14.2}$$

$$h_2 = \frac{HF_1^2}{F_1^2 + F_2^2} \tag{14.3}$$

According to the calculations of (14.2), and (14.3), when the inlet and exhaust areas are equal, the position of the neutral plane is in the middle of H, and the distances from the centre line to the inlet and outlet are equal, i.e. $h_1 = h_2$, as shown in Figure 14.2. When the intake area is one half of the exhaust area $F_1 = F_2/2$, the position of the neutral plane is high, that is, $h_1 = 4H/5$ and $h_2 = H/5$, as shown in Figure 14.4.

When the intake area is one-third of the exhaust area ($F_1 = F_2/3$) as shown in Figure 14.5, the position of the neutral plane is higher, that is, $h_1 = 9H/10$ and $h_2 = H/10$. In this case, sometimes even part of the skylight will become part of the air inlet (i.e. part of the height of the skylight will be located in the air intake area). This is not conducive to natural ventilation.

As shown in Figures 14.4 and 14.5, owing to the small h_2 value, the exhaust pressure Δp_2 is also very small, and in part of the skylight area, is even negative.

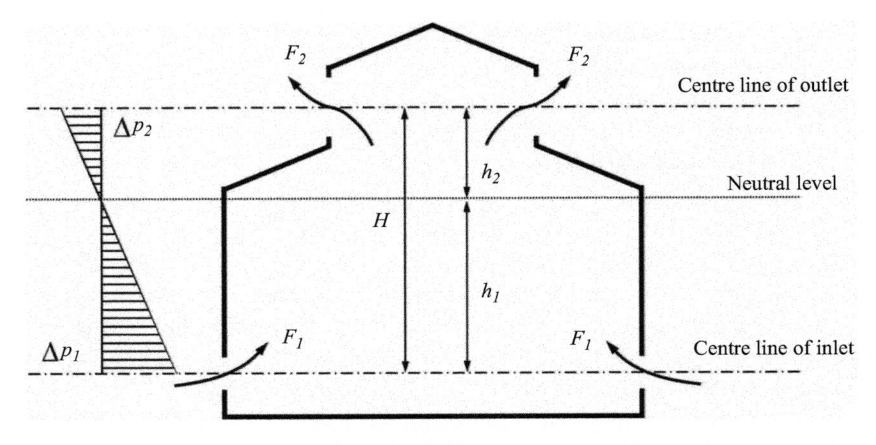

Figure 14.4 Positions of neutral plane ($F_1 = F_2/2$)

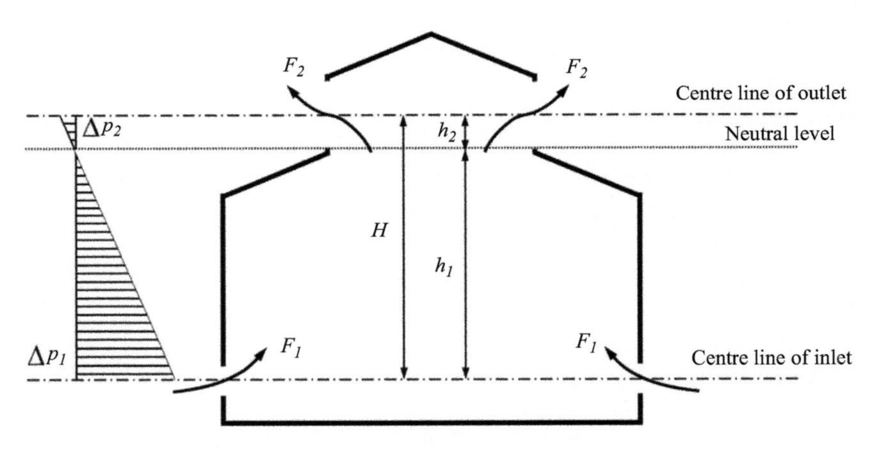

Figure 14.5 Positions of neutral plane ($F_1 = F_2/3$)

In this case, although the exhaust area F_2 of the skylight is large enough because the exhaust pressure is very low, the exhaust function of the skylight is not fully utilised; that is, the ventilation effect is not obtained. In addition, owing to the high position of the neutral plane, the air quality of the operating region is reduced, which is also unacceptable for the natural ventilation design of the plant.

When using external windows as natural ventilation inlets and outlets, the air inlet area and exhaust area should be similar, and the area of the air inlet should be at least the size of the air outlet, if not larger. This should be an extremely important and effective technical measure for improving the natural ventilation effect.

14.2.3.2 Inlet and outlet height

In summer, owing to the small temperature difference between indoors and outdoors, the thermal pressure that is formed is small. To ensure sufficient air intake, eliminate waste heat, and improve the ventilation efficiency, the position of the natural air inlet should be as low as possible, and the position of the air outlet should be as high as possible. Thus, increasing the height difference between the intake and exhaust vents can enhance the effect of the thermal natural ventilation.

14.2.3.3 Position of the inlet and outlet

In addition to the reasonable design of the areas for the air inlet and outlet, it is also necessary to reasonably design the positions of the air inlet and outlet to avoid airflow short circuit. The so-called short circuit of airflow refers to a phenomenon in which fresh air entering a factory building from an air inlet has been heated before it enters the scope of the work area and rises to an air outlet (such as a skylight) to be discharged outside, as shown in Figure 14.6. Therefore, to improve the natural ventilation effect of the plant, air short circuits should be avoided as much as possible.

14.3 Local ventilation

14.3.1 Local exhaust ventilation

In industrial buildings, there is often a regional concentration of waste heat and pollutants. At such times, it is difficult to ensure the indoor environment in these areas. Accordingly, it is necessary to adopt a local exhaust ventilation system to capture pollutants directly near the source and control pollution diffusion in industrial buildings. A well-designed local exhaust ventilation system can remove harmful substances with a small exhaust volume without affecting the production process and operations, while ensuring that the concentrations of pollutants in the working area meet production requirements and sanitary standards.

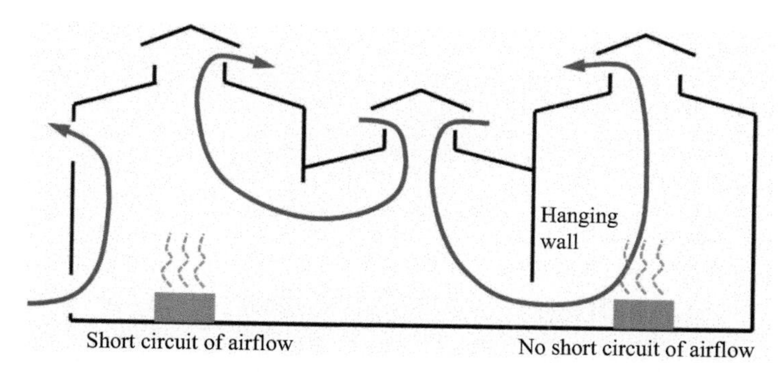

Figure 14.6 Schematic diagram of several airflow short circuit

14.3.1.1 Basic form of local exhaust system

The local exhaust ventilation system is mainly composed of a local exhaust hood, air duct, fan, and gas cleaning device, as shown in Figure 14.7 (Railio *et al.*, 2003).

Local exhaust hood: The local exhaust hood is terminal capture equipment of the local exhaust ventilation system and is used to capture pollutants.

Ductwork: The ductwork transports the captured pollutant gas.

Gas cleaning device: The gas cleaning device removes pollutants from the exhausted air before recycling them or discharging them into the atmosphere.

Fan: The fan provides the power for the airflow of the local exhaust ventilation system.

The efficiency of the local exhaust ventilation system largely depends on the transportation process from the pollutant sources to exhaust outlet. The system can be optimised by adjusting the exhaust outlet structure, exhaust flow volume, momentum distribution of the pollution source, auxiliary air jet(s), distribution of the environmental airflow, and operator behaviour. The performance of the local exhaust ventilation depends on the complex interactions between these factors.

14.3.1.2 Classification of local exhaust ventilation

In the local exhaust ventilation system, the exhaust hood is the terminal capture device. According to its different processes and requirements, exhaust hoods have various shapes, sizes, and setting methods. According to working principles and modes, local exhaust hoods can be divided into the following basic types:

1. enclosures; and
2. external hoods.

Most exhaust hoods can be classified into these two types. Sometimes, an exhaust hood may include the features of both of the above-mentioned types. Different forms of local exhaust hoods should be used according to the production process, as shown in Figure 14.8.

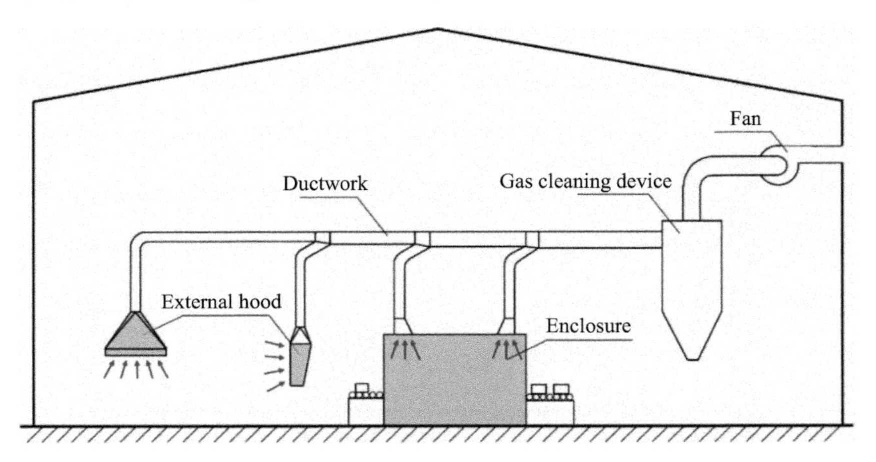

Figure 14.7 Local exhaust ventilation system

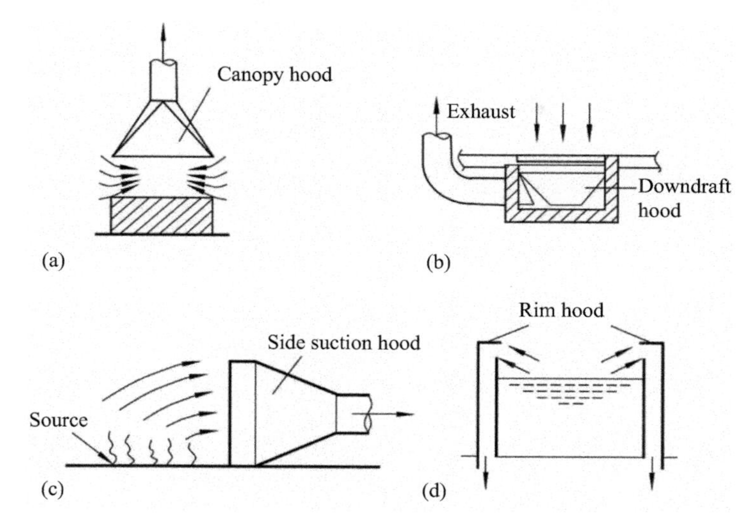

Figure 14.8 Different local exhaust hoods: (a) canopy hood, (b) downdraft hood, (c) side suction hood, and (d) rim hood

14.3.1.3 Optimal design principle for the local exhaust system

Selection of the local exhaust hood

1. Based on the pollutant removal efficiency of the exhaust hood, it should be installed according to the sequence of the enclosure, partial enclosure, and external exhaust hood.
2. The exhaust hood should be selected according to the shape and parameters of the pollution source.

Shape and size of the local exhaust hood

The local exhaust hood should be installed close to the pollution source, and its shape and size should correspond to the pollution source.

For a thermal plume generated in a high-temperature process, at a distance of 1–2 times the heat source diameter or 1–2 times the length of the long side from the heat source surface, the cross-section of the thermal plume will shrink, the width of the airflow coverage will become the smallest, and the velocity will be higher, as shown in Figure 14.9. Therefore, it is easy to obtain higher pollutant removal efficiency when the local exhaust hood is located at this height.

When the exhaust hood is close to the heat pollution source, the shape of the exhaust hood should correspond to the pollution source. This will facilitate the smallest exhaust hood area for capturing pollutants and improve the pollutant removal efficiency of the exhaust hood.

In contrast, when the exhaust hood is far away from the heat pollution source, the shape of the exhaust hood opening should correspond to the cross-sectional shape of the thermal plume. For a thermal plume with a certain aspect ratio, the shape will gradually tend to be circular during the movement, as shown in

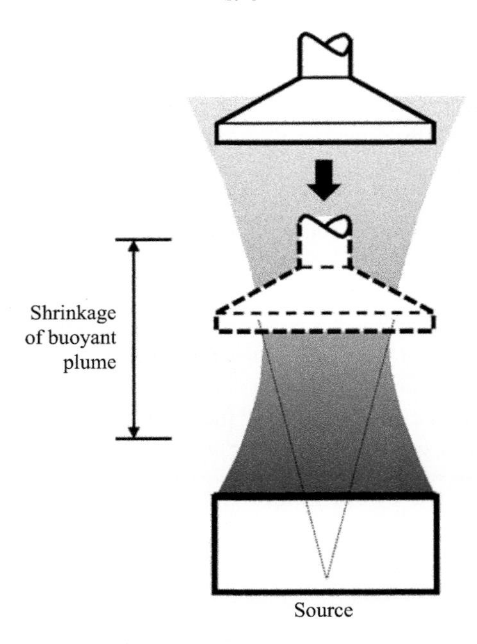

Figure 14.9 Shrinkage of buoyant plume and optimal location of the exhaust hood

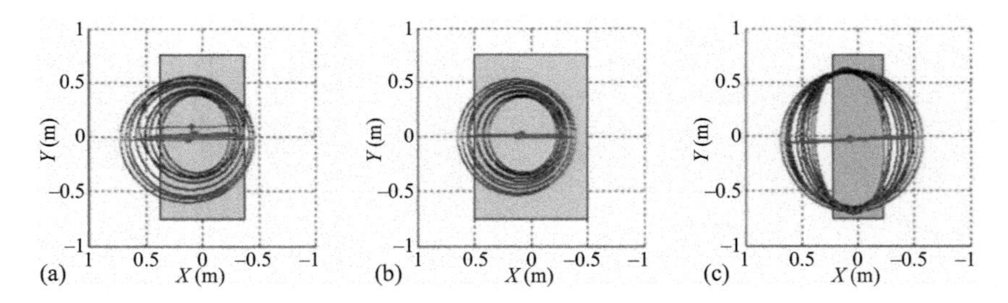

Figure 14.10 Variation of the cross-sectional shape of plumes emitted by rectangular pollution sources with different aspect ratios. (a) aspect ratio = 2, (b) aspect ratio = 1.5, (c) aspect ratio = 3

Figure 14.10 (Devienne and Fontaine, 2012). Therefore, when the exhaust hood is far away from the pollution source (such as the roof exhaust hood in a steel-making plant), the circular exhaust hood should be set according to the development characteristics of the thermal plume.

Direction and position of the local exhaust hood

1. The suction airflow direction of the exhaust hood should be as consistent with the movement direction of the polluted airflow as possible. When the direction of the polluted airflow is the same as that of the exhaust airflow, it can effectively increase the actual suction velocity of the pollutants and reduce their

diffusion, which is conducive to improving the pollutant removal efficiency of the exhaust hood.

2. The suction airflow of the exhaust hood and pollutant flow should not pass through the breathing zone of the operator. As shown in Figure 14.11 (Health and Safety Executive, 2017), the operator should not have his/her face or back oriented towards the exhaust airflow. When facing away from the exhaust airflow, there will be a recirculation flow zone in front of the operator. The airflow in the recirculation flow zone is difficult to diffuse. At this time, the breathing zone of the operator will be in the recirculation flow zone, causing the operator to inhale pollutants. When facing the exhaust airflow, the polluted airflow will pass through the breathing zone of the operator and then enter the exhaust hood, which will also endanger the health of the operator.
3. The design and configuration of the exhaust hood should not affect process operations. For example, in a production process with crane hoisting requirements, a canopy hood or the upper received hood cannot be installed. In this case, a side suction hood or downdraft hood should be considered.
4. When setting the exhaust hood, the influences of the surrounding interferences should be fully considered. The common surrounding interferences are as follows:
   (i) airflow movements generated by other production processes nearby;
   (ii) natural influences from gale weather;
   (iii) influences of cooling air conditioning and fan airflow;
   (iv) influences of opening doors and windows nearby to inlet and exhaust;
   (v) airflows generated by the movement of vehicles and equipment;
   (vi) operators moving nearby; and
   (vii) an unreasonable design of the make-up air.

14.3.2 Local air supply

For industrial workshops with large areas, few workers, and fixed work locations, if general ventilation will cause excessive energy consumption, a local air supply can

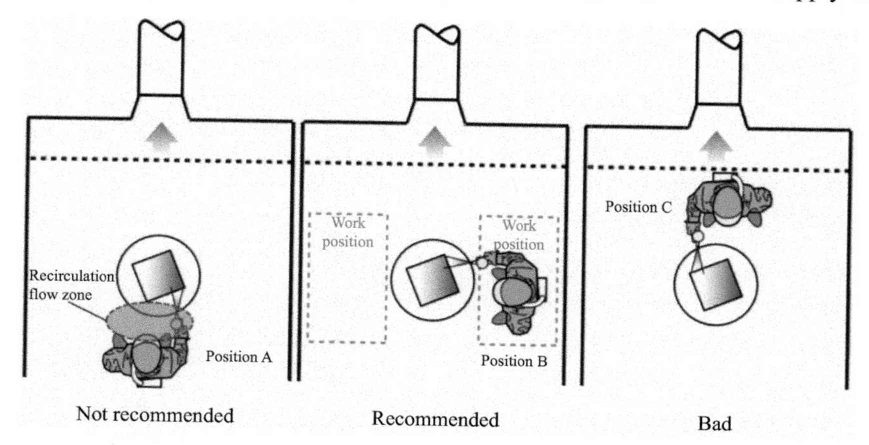

Figure 14.11 Relationships between the exhaust airflow and operator location

be adopted; in this way, it is only necessary to cover the work environment. In addition, when a local air supply is adopted, workers can sometimes be allowed to adjust air supply parameters according to their own needs, so as to realise personalised air supply.

14.3.2.1 Basic form of local air supply

It is not only difficult but also uneconomical to improve the production environment of an entire workshop with a large area and few operators by using general ventilation. For example, some workshops only need to supply air to operators and key locations, which can effectively reduce the local environmental temperature and local pollutant concentration. This ventilation method, i.e. for creating an acceptable local environment, is called a local air supply. In industrial buildings, a local air supply is often used to adjust and protect the local environments of indoor people, important equipment, and products. A local air supply system is generally composed of an air supply outlet, air supply ductwork, air handling unit, and fan, as shown in Figure 14.12.

Air supply outlet: The air supply outlet is the terminal equipment of the local air supply system and is used to deliver fresh air to the designated location.

Air supply ductwork: The air supply ductwork delivers fresh air.

Air-handling unit: The air-handling unit treats outdoor air or indoor circulating air so that its parameters meet air supply requirements.

Fan: The fan provides powers for airflow movement in the local air supply system.

The efficiency of the local air supply system mainly depends on the fresh air transport process from the air outlet to air supply target. The transport process is closely related to the structure of the air supply outlet, air supply volume, air supply parameters, ambient airflow, and air supply target characteristics (such as the operators' behaviour). The performance of the local air supply system depends on the

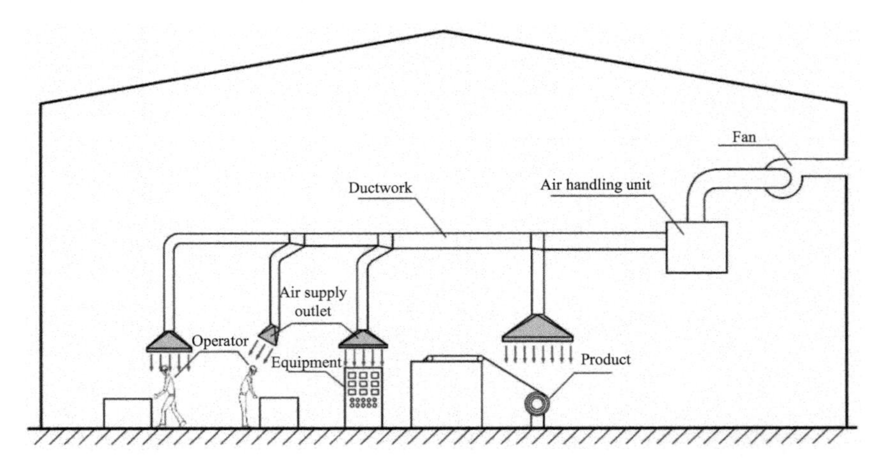

Figure 14.12 Schematic of local air supply

complex interactions between these factors. Therefore, in the design and application of a local air supply system, the influencing factors should be fully considered.

14.3.2.2 Classification of local air supply

The difference in a local air supply system (i.e. from a general system) mainly lies in the structure of the air supply outlet. The flow of the local air supply follows the theory of the supply air jet. The most common type of air supply outlet is a gradually expanding opening, as shown in Figure 14.13(a); this is suitable for fixed workplaces. Figure 14.13(b) shows a rotary air supply outlet, with guide vanes at its outlet. The air outlet and air duct are connected by rotatable movable means, and the direction of the air supply can be adjusted arbitrarily. The rotary air supply outlet is suitable for situations where the working place is not fixed, or where the working place is difficult to determine from the design. Figure 14.13(c) shows a spherical nozzle for arbitrarily adjusting the direction of the air supply airflow; it is widely used for long-distance air supply in workshops. When the working place is fixed and a local air supply is required for the protection of operators, a large air supply outlet can be selected, as shown in Figure 14.13(d). This type of air supply outlet can completely cover the operator, effectively isolate the operator from external pollution, and provide the best protection effect for the operator.

14.3.2.3 Optimal design principle for local air apply

The local air supply system must meet the following rules.

1. Harmful substances should not be blown towards the human body.
2. The air supply flow blows to the breathing zone of the human body from the top or the front of the human body, so that the breathing zone of the human body is surrounded by fresh air, as shown in Figure 14.14 (Health and Safety Executive, 2017). As shown in the figure, part (a) is a narrow air supply jet with a high air supply speed; the air supply effect of this air supply mode is not ideal, the clean area is small, and the pollutants can easily enter the breathing zone by mixing with the supply air. When using this type of air supply outlet, it is necessary to maintain a certain distance between the air supply outlet and human body, so that the air supply jet can diffuse to a certain range and

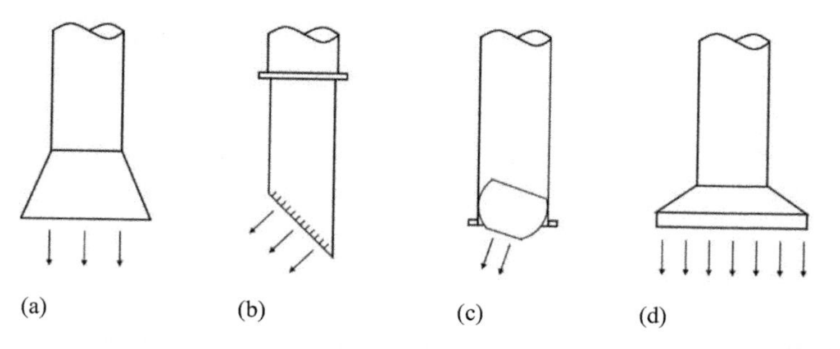

Figure 14.13 Schematic of local air outlets (a) diffuser, (b) rotary air supply outlet, (c) spherical nozzle, (d) large-scale outlet

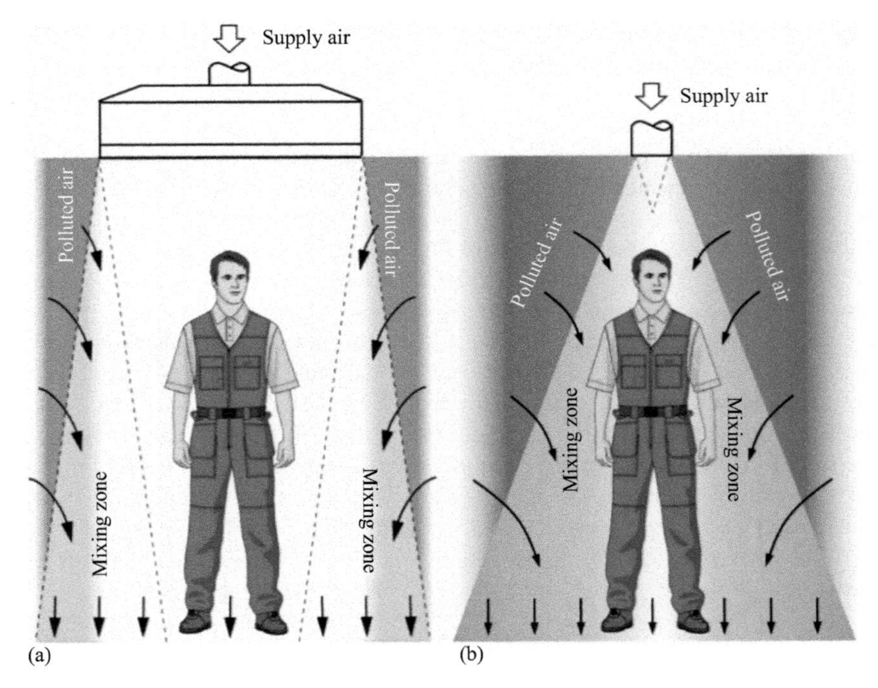

Figure 14.14 Schematic of personalised air supply mode (a) large-scale outlet, (b) narrow outlet

surround the key position of the human body. As also shown in the figure, part (b) is a wide air supply jet with a low air supply velocity; this commonly used air supply mode ensures a clean area, and that the pollutants are far away from the breathing zone. The disadvantage is that with an increase in the air supply area, the required air supply volume is larger.

3. The local air supply system should be provided with a reasonable air supply range, and should not be allowed to interfere with the flows of the polluted air and exhaust ventilation systems.
4. The local air supply system should set the air supply speed and temperature according to the actual situation of the site (such as the work type and work intensity of the operators), so as to reduce the risk to operators as much as possible.
5. When the range of activities of the operators is large, movable or rotatable air supply outlets should be adopted.

14.4 General ventilation

To maintain the production environment, ventilation should be used in the production processes of industrial buildings, e.g. to eliminate fine particulates, toxic gases, other pollutants, and waste heat. These contaminants may threaten the health and safety of operators. In some cases, when the pollutant concentration exceeds a lower explosive limit concentration or lower flammable limit concentration, they may also become flammable or combustible. Therefore, ventilation systems, especially local exhaust ventilation systems, should be used first whenever possible to control heat and pollutants. This is because the local exhaust ventilation system is often set near the heat and pollution sources. Compared with a general ventilation system, a local exhaust ventilation system does not need as high of an airflow rate and can achieve a better ventilation effect. However, when the local exhaust ventilation system cannot meet the indoor environmental requirements for industrial buildings, general ventilation is still needed.

The general ventilation of industrial buildings includes mechanical and natural ventilation. Natural ventilation systems do not consume energy and are only driven by thermal buoyancy or wind pressure. Thus, they are widely used in industrial building environmental control (especially in hot workshops in cold and mild climate areas). However, natural ventilation is highly dependent on external environmental parameters such as outdoor wind speed, air temperature, and cleanliness. Therefore, the effect of natural ventilation is usually uncontrollable in industrial buildings. Therefore, to provide adequate environmental control for industrial buildings, a mechanical ventilation system for general ventilation should still be constructed.

14.4.1 Strategy of general ventilation

Understanding the strategy for indoor ventilation is the basis for controlling the indoor temperature, humidity, pollutant concentration, and air distribution in industrial buildings. Different general ventilation strategies will create different indoor environmental characteristics. The ventilation means (such as the locations of air supply and exhaust outlets and the temperature of the ventilation airflow), production processes, and disturbances will affect the final outcome of the general ventilation.

Table 14.1 lists commonly used general ventilation strategies in industrial buildings (Railio *et al.*, 2003).

14.4.2 Optimal design principle of general ventilation

The general ventilation effect is closely related to air distribution. An industrial building environment is very complex, with a wide variety of pollution sources. The pollutants often have different characteristics, such as high-temperature fume plumes rising vertically and ambient-temperature dust diffusing horizontally. The environmental control requirements for industrial buildings often vary greatly, including workshops with high emissions and high pollution, and clean workshops with high cleanliness requirements. Therefore, in these complex environments, to effectively and efficiently control the indoor environment through general ventilation, it is necessary to fully understand the principles of air distribution. This is necessary for choosing the appropriate general ventilation strategy, which needs to arrange the locations of the air supply inlets and exhaust outlets, distribute the

Table 14.1 Strategy for general ventilation in industrial building

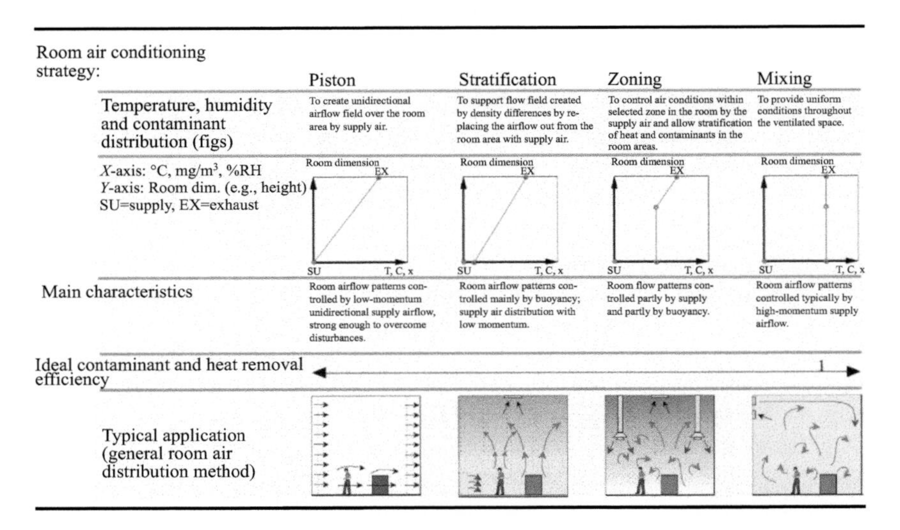

Room air conditioning strategy:	Piston	Stratification	Zoning	Mixing
Temperature, humidity and contaminant distribution (figs)	To create unidirectional airflow field over the room area by supply air.	To support flow field created by density differences by re-placing the airflow out from the room area with supply air.	To control air conditions within selected zone in the room by the supply air and allow stratification of heat and contaminants in the room areas.	To provide uniform conditions throughout the ventilated space.
X-axis: °C, mg/m³, %RH *Y*-axis: Room dim. (e.g., height) SU=supply, EX=exhaust	Room dimension; EX; SU; T, C, x	Room dimension; EX; SU; T, C, x	Room dimension; EX; SU; T, C, x	Room dimension; EX; SU; T, C, x
Main characteristics	Room airflow patterns con-trolled by low-momentum unidirectional supply airflow, strong enough to overcome disturbances.	Room airflow patterns con-trolled mainly by buoyancy; supply air distribution with low momentum.	Room flow patterns con-trolled partly by supply and partly by buoyancy.	Room airflow patterns controlled typically by high-momentum supply airflow.
Ideal contaminant and heat removal efficiency	←			1 →
Typical application (general room air distribution method)				

ventilation volume, and select the appropriate air supply and exhaust outlet forms to achieve the best ventilation effect at the lowest cost. The general principle of airflow distribution is that clean air should pass through the working area and other clean areas first then flow to the polluted area. Moreover, the general ventilation system should not affect the operation of the local ventilation system. Therefore, the room size, locations of the operator and pollutant source, and pollutant characteristics should be fully considered in the design of a general ventilation system.

14.4.2.1 Room size

When designing a general ventilation system, it is necessary to pay attention to the influences of room sizes on the ventilation system performance. For industrial buildings, the room sizes of different production processes vary greatly, so it is necessary to give full consideration to the airflow distribution forms under different room volumes, lengths, widths, and heights.

Room volume

Industrial buildings are often workshops with huge internal spaces, such as assembly workshops for large passenger aircraft. For a large-space workshop, it is difficult to use mixed ventilation, owing to the large amount of ventilation required and difficulty in ensuring uniform mixing. Regarding displacement ventilation, it is also difficult to ensure that the low-temperature fresh air can evenly replace the air in the lower part of the room. Therefore, for large-space workshops, zoning ventilation should be used as much as possible, and the work area should be divided into different horizontal partitions as required for ventilation. In this way, the ventilation effect can be effectively improved, and the ventilation energy consumption can be reduced.

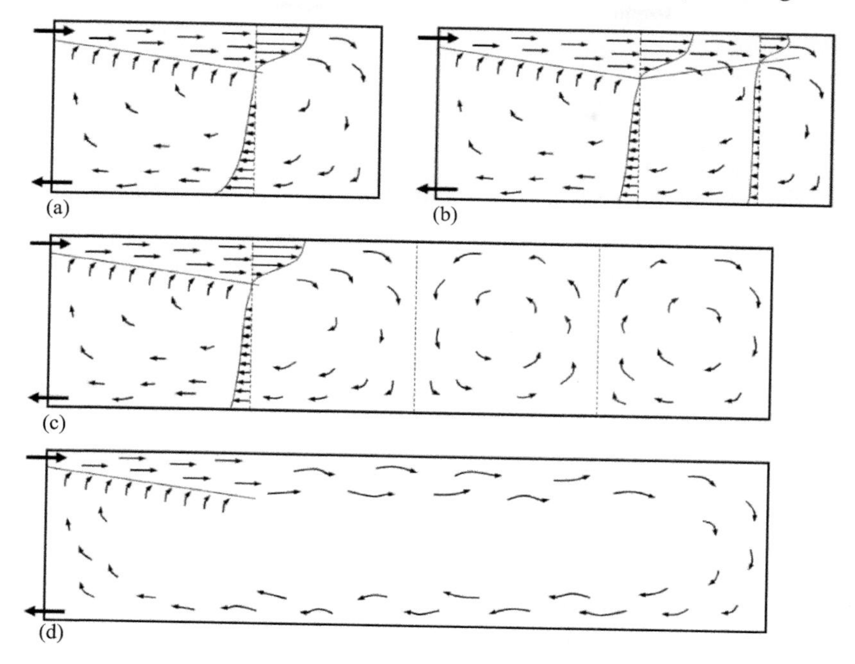

Figure 14.15 Air distribution of side air supply in rooms of different lengths

Room length

For a room with a short length, when supply-side exhaust ventilation is used, the supply air jet will collide with the side wall opposite to the air supply outlet and form a recirculation flow zone in the room, as shown in Figure 14.15(a). If the room length is increased, the velocity of the supply air jet decreases owing to mixing, and the range of the air jet will gradually decrease, as shown in Figure 14.15(b). The critical length of the maximum jet range can be estimated as follows:

$$x_{\max} \approx 0.33\sqrt{BH} \tag{14.4}$$

where B is the width of the room, in m; and H is the height of the room, in m.

Owing to industrial production processes, there are often long rooms in industrial buildings. When the length of the room exceeds the critical value and an indoor temperature gradient is not evident, the side supply air jet will separate from the roof surface after a certain distance, resulting in a local recirculation flow zone. Meanwhile, several secondary recirculation flow zones will be formed in the room outside the main recirculation flow zone, as shown in Figure 14.15(c). When there is an evident buoyancy airflow in the room, thermal stratification will occur. The existence of the thermal stratification will weaken the separation and confluence of the indoor upper air supply airflow, ultimately forming the air distribution shown in Figure 14.15(d) (Howard and Esko, 2001).

Room width

When the room width is large, it is necessary to arrange multiple air supply outlets on one side of the room to achieve a uniform air supply. The layout of the air supply outlets should consider the form of the air supply terminal device and determine the position and spacing of the air supply outlets according to the airflow expansion angle and air supply distance of different air supply terminal devices.

Room height

When there is an evident thermal buoyancy flow in the room or solar radiation on the roof, a thermal stratification phenomenon will appear in the room. When the thermal stratification level is higher than the occupied zone, the use of mixing ventilation will waste part of the ventilation volume, resulting in a waste of ventilation energy. Therefore, it is suitable to use displacement or zoning ventilation in rooms where the uniformity of the temperature and humidity is not particularly high.

14.4.2.2 Position of air supply and exhaust outlet

1. The exhaust outlet should be close to the pollutant source or areas with high concentrations of pollutants as much as possible so as to rapidly eliminate pollutants from the room.
2. The air supply outlet should be as close to the operation site as possible. The clean air sent into the ventilation room should first pass through the operation site and then be discharged outdoors through the polluted area.
3. The locations of the air supply and exhaust outlets should be reasonably set to eliminate stagnation areas of the airflow so as to avoid the accumulation of pollutants.

   Figure 14.16 compares several different forms of air distribution modes. Among these, parts (a) and (b) are typical unacceptable air distribution modes and make it difficult to eliminate indoor pollutants.
4. For a long room with mixing ventilation and side exhaust, if there is no evident thermal stratification, the exhaust outlet should be set at the other end of the room. Otherwise, the ventilation at the far end of the room may be very small, resulting in the accumulation of pollutants as shown in Figure 14.17 (Howard and Esko, 2001).

In a room where the air is completely mixed, the position of the exhaust outlet is not theoretically important. However, it is difficult to achieve adequate mixing in such a room. This may be owing to temperature or density differences in the room. In an industrial building environment, the pollutants released by various pollution sources are often higher temperature than the indoor air temperature, and in some cases, the density of the pollutants themselves is different from the density of the air. Therefore, to quickly eliminate indoor pollutants, the exhaust outlet of the general ventilation system should be set in an area with a high concentration of pollutants as best as possible. When designing a general ventilation system, it is necessary to know the release positions and intensity of pollutants in the workshop. The concentration and distribution of polluted gas in a workshop is not only related to the density of the polluted gas itself, but also to the density of the gas

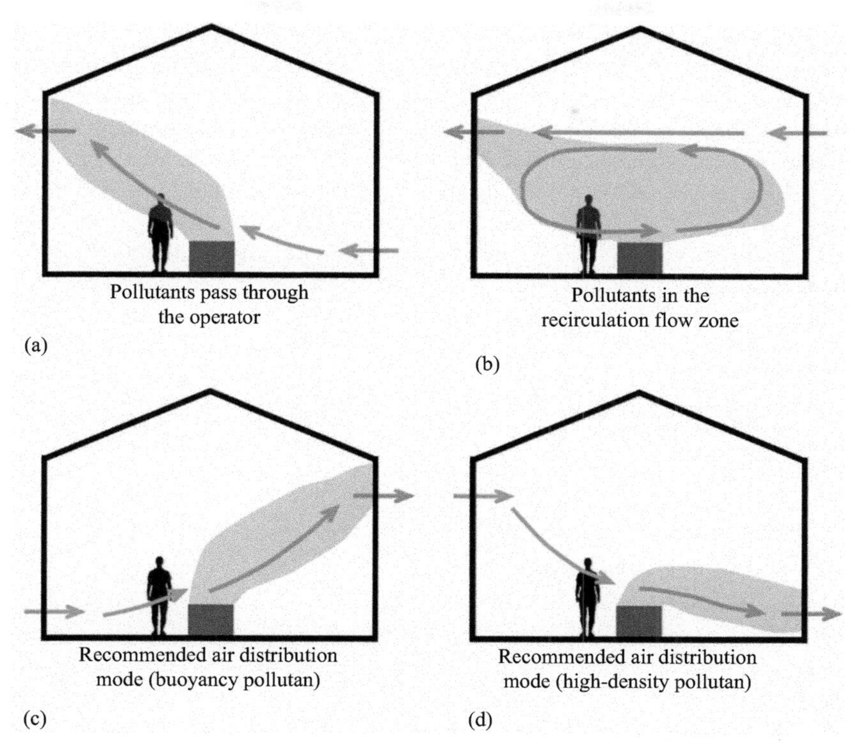

Figure 14.16 Comparison of different air distribution modes

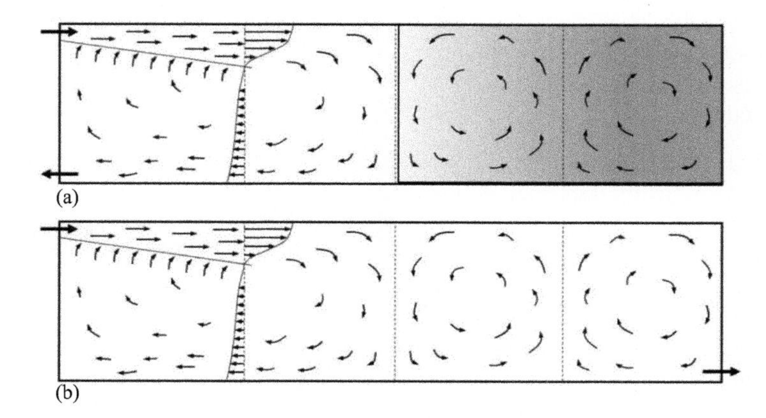

Figure 14.17 Exhaust outlet position of mixing ventilation in long room: (a) accumulation of pollutants at the far end of the room and (b) indoor pollutants are evenly distributed

as mixed with indoor air. It is generally believed that when the density of the pollution gas emitted from a workshop is large, the pollution gas will be deposited in the lower part of the workshop; thus, the exhaust outlet should be set in the lower part of the workshop. Because the concentration of pollution gas in the workshop is generally not too high, the increase in air density caused by this will not exceed 0.30–0.40 g/m^3. In addition, the change in gas density caused by a 1°C air temperature change is 4.0 g/m^3. Therefore, as long as the indoor air temperature is slightly uniform, the pollution gas will move with the indoor airflow; that is, the movement range of pollutants caused by air convection is far greater than that caused by the density difference. When there is no air convection in the room, the polluted gas with a higher surface density will deposit in the lower part of the workshop. In addition, some lighter volatiles (such as gasoline and ether) cool the surrounding air owing to heat absorption by evaporation and are deposited with the surrounding air. Therefore, the specific air supply and exhaust outlet layout should be calculated and set according to the actual situation of the room.

14.4.3 Industrial ventilation based on vortex principle

14.4.3.1 Vortex ventilation

General

As is well known, a tornado is a naturally formed violently rotating column of air with strong suction force. Houses, cars, trees, and dust will be captured and tossed high into the atmosphere when a tornado passes by. In addition, the scope of a tornado is very wide in terms of distance and area. Therefore, the ventilation efficiency can be significantly improved in many cases by making a proper use of the principle of a tornado.

Principle

The vortex flow system is a type of general ventilation flow system based on the column vortex principle. As shown in Figure 14.18, a column vortex flow can be formed combining angular momentum supply airflow and updraft flow, which have been

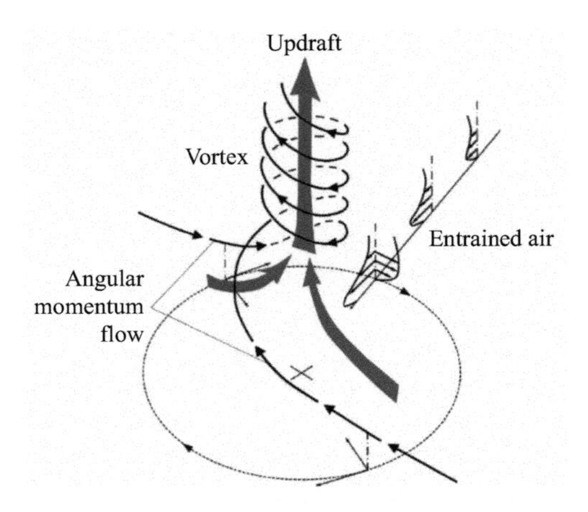

Figure 14.18 Schematic of the column vortex flow

studied through both experimental (Ward, 2010; Hashemi Tari *et al.*, 2010; Refan and Hangan, 2016) and numerical simulation methods (Le *et al.*, 2008; Ishihara *et al.*, 2011; Liu and Ishihara, 2015). The supply airflow converges near the ground and the ambient air/contaminant nearby will be entrained into the vortex since the column vortex formed. A column vortex-like tornado has strong suction force and a long control distance, which has great potential for applications in ventilation. In recent years, preliminary studies have examined the vortex flow for local exhaust ventilation (Cao *et al.*, 2017) and in general ventilation system (Cao *et al.*, 2018a, 2018c).

In a vortex ventilation system, the updraft can be provided by an exhaust outlet. Thus, it is essential to find an appropriate approach to provide angular momentum to form a tornado-like vortex while not disturbing the polluted airflow. To maintain the stability and strength of the vortices, generally, it includes the following methods:

1. Mechanical air supply through polygonal or circle air supply inlets, as shown in Figure 14.19.
2. Setting guide plates or baffles to deflect the mechanical air supply flow, as shown in Figure 14.20(a).
3. Adjust the position of the natural air inlet to make the supplementary air flow with angular momentum, as shown in Figure 14.20(b).

Applicability of sources

According to the flow characteristics of column vortex, the vortex ventilation has a strong negative pressure gradient and axial velocity, which can collect pollutants

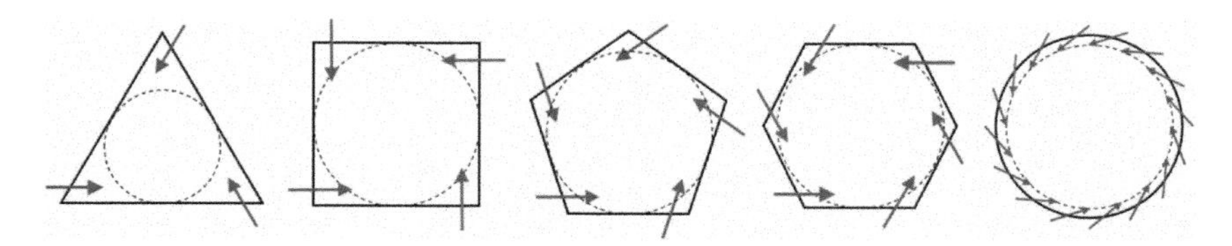

Figure 14.19 Schematic of angular momentum air supply mode with polygonal or circle air supply inlets

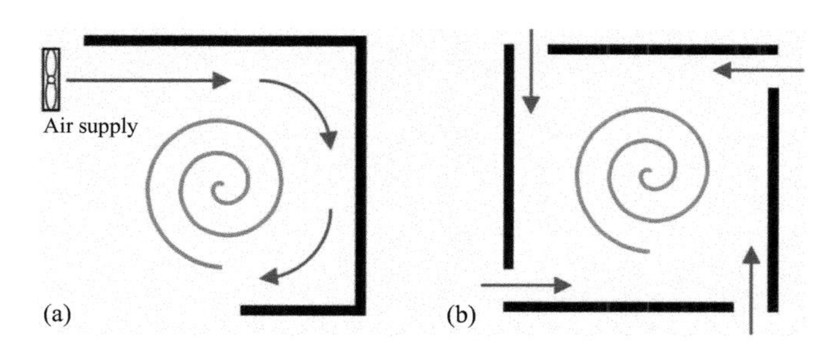

Figure 14.20 Schematic of angular momentum air supply mode with (a) guide plates and (b) natural supply inlets

near the ground and then transport them to the exhaust outlet through the vortex tube. Therefore, the vortex exhaust hood is most suitable for eliminating high density and non-buoyancy pollutants, such as high-density gas, dust, and acid fog at room temperature. Meanwhile, the vortex ventilation can transport pollutants effectively through the vortex tube at a distance from the pollutants.

Different forms

At present, there are many different types of vortex ventilation, including general ventilation system and local exhaust ventilation system. The key of vortex ventilation is the formation of columnar air vortices. Figure 14.21 shows several types of vortex ventilation systems suitable for different industrial environment (Cao *et al.*, 2018a, 2018b, 2018c, 2020).

Specific issues

The vortex exhaust hood is effective but the effect may be drastically reduced if specified design parameters are not followed and maintained. The key factor of vortex ventilation is to form column vortex to capture and transport pollutants. When column vortex cannot be formed, angular momentum air supply may affect the flow of pollutant into the room. At this time, the capture efficiency of the exhaust hood may be lower than that of the basic exhaust hood.

Based on the principle of column vortex, there are many types of vortex ventilation system. Angular momentum air supply can be provided by means of baffle, air curtains, walls, and other methods.

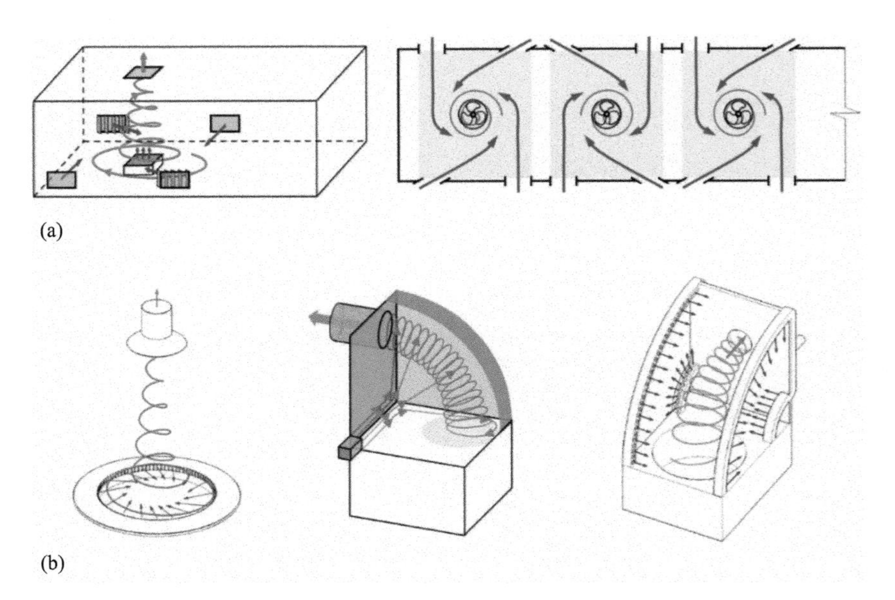

Figure 14.21 Different forms of vortex ventilation system: (a) vortex ventilation in a workshop and (b) vortex local exhaust hood

14.4.3.2 Vortex ring ventilation

General

The vortex ring is a common fluid structure found in nature, such as in the clouds formed from the hot gas, dust, and steam ejected by volcanic eruptions, and in the air bubbles released from a dolphin's mouth in water; now it has been widely used in practical engineering, such as the use of vortex ring structure to achieve pulsed propulsion derived from bionics (Zhang *et al.*, 2020), and the use of synthetic jet under the high-frequency vortex ring to enhance heat transfer and mixing control (Ambole and Chaudhari, 2017; Hasnain *et al.*, 2015; Marchitto *et al.*, 2017). Researches have shown that the vortex ring has excellent characteristics such as less mixing with the ambient fluid (Maxworthy, 2006), strong anti-interference ability (Yagami and Uchiyama, 2011), and low-energy dissipation during the transportation (Akhmetov, 2008). Therefore, the use of vortex ring fluid structure can achieve a long-distance directional efficient air supply.

Principle

Vortex ring ventilation is to use this fluid structure to supply fresh air to working area. Figure 14.22 shows a schematic diagram of the vortex ring ventilation system, which has the following characteristics:

1. Long-distance directional air supply: There are no requirements to set complicated air supply ducts and air supply ports near the breathing zone, improving the flexibility and application range in personalised ventilation.

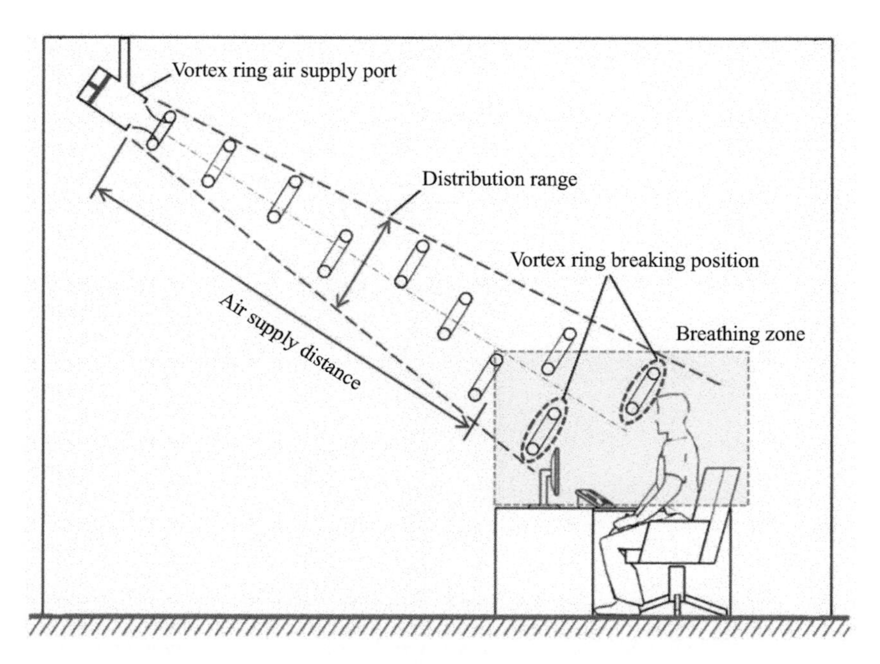

Figure 14.22 Schematic of the vortex ring ventilation system

2. High ventilation efficiency: As vortex rings rarely mix with the ambient air during transportation, the ventilation requirements for breathing zone can be satisfied with a lower ventilation flow rate.

When a vortex ring moves to the vicinity of the breathing zone, as long as the structure of the vortex ring is broken, the fresh air carried by the vortex ring can be released, thereby completing the air supply to the breathing zone. The vortex ring can be broken by, e.g. impacting upon a solid surface; it may also be broken by disturbing the airflow, e.g. by human thermal plumes or local airflow disturbances. The ventilation flow rate of the vortex ring ventilation can be adjusted by adjusting the frequency of the vortex ring.

Applicability of sources

The vortex ring entrains the ambient fluid during its transportation, and the entrained fluid surrounding the vortex ring is called the 'vortex atmosphere' (Gao and Yu, 2015). The 'vortex atmosphere' not only guarantees the isolation of the vortex core from the external ambient fluid, but also the transportation of the isolated vortex ring to be a low-resistance flow (Wang *et al.*, 2020). However, the 'vortex atmosphere' does not need to be considered for ventilation, only the size of the vortex core is calculated. Figure 14.23 shows a schematic diagram of the vortex ring structure. The geometry of the vortex ring can be defined with two parameters: the vortex ring diameter (D_r) and the vortex ring core diameter (d_c).

Vortex rings are commonly produced by pushing fluid out of a tube or through an orifice. And then the fluid is separated at the sharp edge of the orifice to form isolated vortex rings. However, as the duration of the force acting on the fluid varies, the time and its own state at which the vortex ring falls off the orifice are different.

Figure 14.24 shows the three types of formation of the vortex ring due to different duration of the force acting on the fluid. As can be seen, the vortex ring will not entrain fluid from the orifice for the entire formation time but rather experiences a pinch-off volume, i.e. when the airflow volume exceeds the pinch-off volume of the vortex ring, excess air will not be entrained into the vortex ring. Figure 14.24(a)

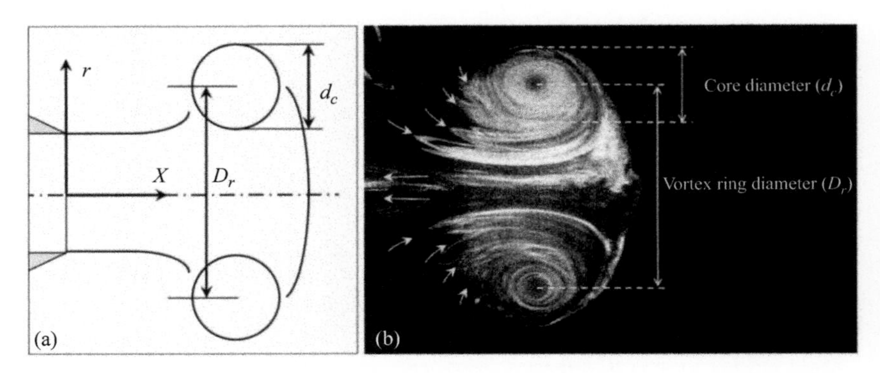

Figure 14.23 Schematic of vortex ring structure (a) from Gao and Yu, (2015) and (b) from Wang et al., (2020)

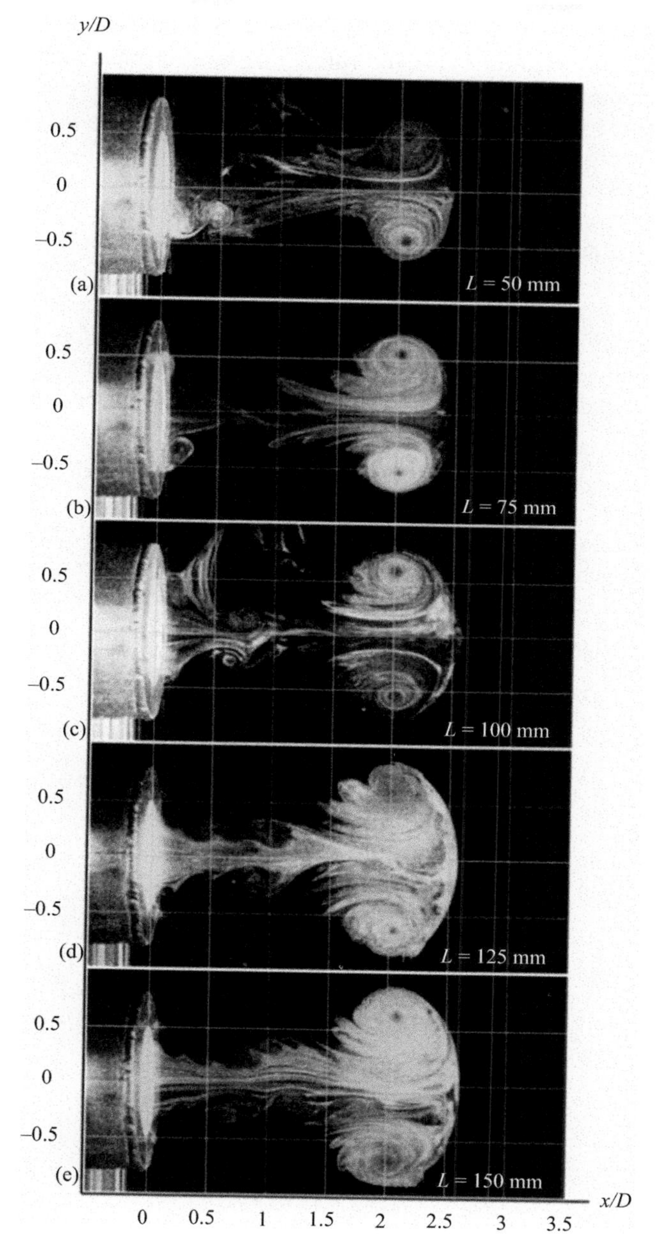

Figure 14.24 Comparison of vortex rings in the same location for different cases (a) L = 50 mm, (b) L = 75 mm, (c) L = 100 mm, (d) L = 125 mm, (e) L = 150 mm

and (b) shows that vortex ring has been completely shed without reaching the pinch-off volume; Figure 14.24(c) shows the vortex ring is shed at the pinch-off volume, and the vortex ring has a higher utilisation rate of the orifice jet; Figure 14.24(d) and (e) shows that the vortex ring has not been shed completely; however, the vortex ring is limited by its maximum size and cannot carry excess fluid, and the vortex ring is squeezed by the orifice jet, showing an unstable trend.

When the vortex ring is used for air supply, it is hoped that fresh air will be delivered to the target area in the form of vortex rings. It is not desirable that the excessive fresh air used to form the vortex ring is not completely entrained, resulting in dissipation and waste. It is even less desirable that the impact of excessive fresh air will affect the stability of the vortex ring, resulting in excessive deviation in the transportation, and finally difficult to achieve directional and precise air supply. Therefore, the vortex ring is designed to fall off as much as possible in a saturated state.

Application scope

In order to illustrate the applicable scope of the vortex ring air supply mode, the fresh air ratio of vortex ring air supply is compared with that of isothermal axi-symmetric circular jet. The variation of the entrainment rate of the isothermal jet with axial distance is provided in Ricou and Spalding (1961) from which the variation of the fresh air ratio with the axial distance for isothermal jets can be

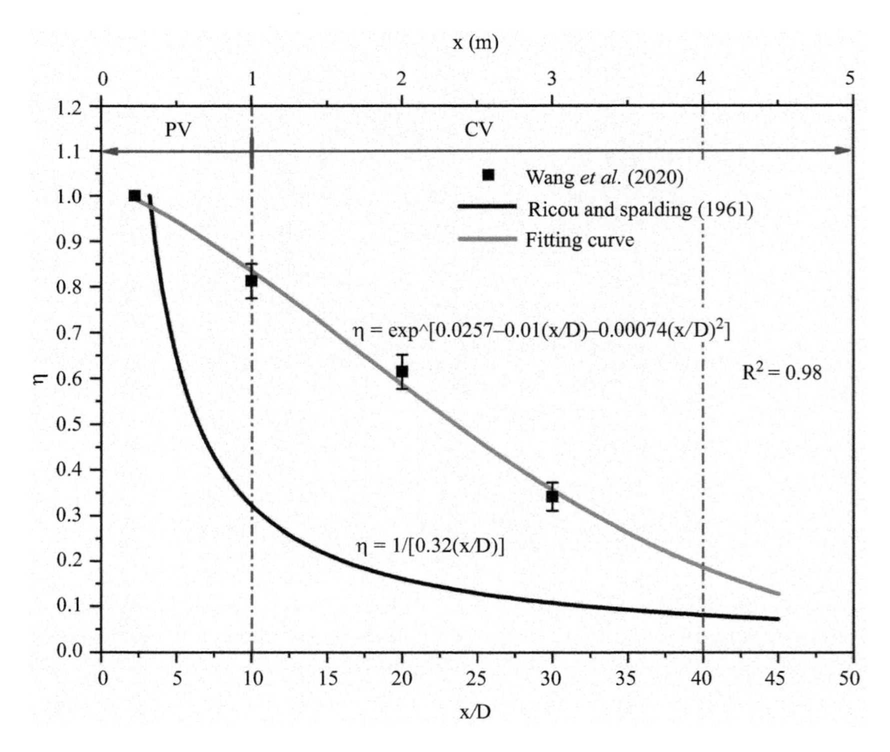

Figure 14.25 Fresh air transportation effectiveness

obtained. And the variation of the fresh air ratio with the axial distance for vortex ring ventilation is provided in Wang *et al.* (2020).

As shown in Figure 14.25, in the commonly used ranges of personalised ventilation (0–1 m) and centralised ventilation (1–4 m), the fresh air ratio of the vortex ring ventilation is significantly higher than that of the traditional jet; it was increased by up to 159.3% (at $x = 0.89$ m), and 37.6% on average (0–4 m). This shows that the vortex ring ventilation has great application potential for long-distance personalised **air supply.**

References

Akhmetov, D. (2008) Loss of energy during the motion of a vortex ring. *Journal of Applied Mechanics & Technical Physics*, 49, 18–22.

Ambole, A. S. and Chaudhari, M. B. (2017) Flow study of synthetic jet: A review. *SSRN Electronic Journal.*

Awbi, H. B. 2003. *Ventilation* of Buildings. London, UK: Taylor and Francis.

Belding, H. S. and Hatch, T. F. (1955) Index for evaluating heat stress in terms of resulting physiological strains. *Heating Piping and Air Conditioning*, 52, 35.

Cao, Z., Wang, Y., Duan, M. and Zhu, H. (2017) Study of the vortex principle for improving the efficiency of an exhaust ventilation system. *Energy & Buildings*, 142, 39–48.

Cao, Z., Wang, Y. and Wang, M. (2018a) Comparison between vortex flow and bottom-supply flow on contaminant removal in a ventilated cavity. *International Journal of Heat and Mass Transfer*, 118, 223–234.

Cao, Z., Wang, Y., Zhai, C. and Wang, M. (2018b) Performance evaluation of different air distribution systems for removal of concentrated emission contaminants by using vortex flow ventilation system. *Building and Environment*, 142, 211–220.

Cao, Z., Zhai, C., Wang, Y., Zhao, T. and Wang, H. (2020) Flow characteristics and pollutant removal effectiveness of multi-vortex ventilation in high pollution emission industrial plant with large aspect ratio. *Sustainable Cities and Society*, 54.

Cao, Z. X., Wang, Y. and Wang, M. (2018c) Numerical study on a novel vortex side hood for high temperature pollutant control. In ROOMVENT & Ventilation Conference 2018. Espoo, Finland.

Devienne, R. and Fontaine, J. R. (2012) Experimental characterisation of a plume above rectangular thermal sources. Effect of aspect ratio. *Building and Environment*, 49, 17–24.

Health and Safety Executive. (2017) Controlling airborne contaminants at work. *A Guide to Local Exhaust Ventilation (LEV)*. Bootle, UK: HSE Books.

Fanger, P. O. (1970) *Thermal Comfort: Analysis and Applications in Environmental Engineering*. New York, USA: Mcgraw-Hill.

Gagge, A. P., Stolwijk, J. A. J. and Nishi, Y. (1971) Effective temperature scale based on a simple model of human physiological regulatory response. ASHRAE Trans., 1, 247–262.

Gao, L. and Yu, S. C. M. (2015) Starting jets and vortex ring pinch-off. In *Vortex Rings and Jets*. Singapore: Springer.

Givoni, B. (1998) *Climate Considerations in Building and Urban Design*. USA: John Wiley & Sons.

Hashemi Tari, P., Gurka, R. and Hangan, H. (2010) Experimental investigation of tornado-like vortex dynamics with swirl ratio: The mean and turbulent flow fields. *Journal of Wind Engineering & Industrial Aerodynamics*, 98, 936–944.

Hasnain, Z., Trollinger, L. N., Hubbard, J. E. and Flatau, A. B. (2015) Dynamic flow control over aerodynamic bodies using phased array vectored synthetic jet actuation. In AIAA Applied Aerodynamics Conference.

Howard, D. G. and Esko, T. (2001) *Industrial Ventilation Design Guidebook*. San Diego, CA: Academic Press.

Ishihara, T., Oh, S. and Tokuyama, Y. (2011) Numerical study on flow fields of tornado-like vortices using the LES turbulence model. *Journal of Wind Engineering & Industrial Aerodynamics*, 99, 239–248.

Le, K., Haan, F. L. and Gallus, W. A. (2008) CFD simulations of the flow field of a laboratory-simulated tornado for parameter sensitivity studies and comparison with field measurements. *Wind and Structures an International Journal*, 11, 75–96.

Liu, Z. and Ishihara, T. (2015) Numerical study of turbulent flow fields and the similarity of tornado vortices using large-eddy simulations. *Journal of Wind Engineering & Industrial Aerodynamics*, 145, 42–60.

Marchitto, L., Valentino, G., Chiatto, M. and De Luca, L. (2017) *Water Spray Flow Characteristics Under Synthetic Jet Driven by a Piezoelectric Actuator. Journal of Physics: Conference Series*, 778, 012005.

Maxworthy, T. (2006) The structure and stability of vortex rings. *Journal of Fluid Mechanics*, 51, 15–32.

Railio, J., Sainio, S. and Hagström, K. (2003) *Industrial Ventilation Primer*. Finland, Suomen Talotekniikan Kehityskeskus Oy – TAKE, Helsinki, Finland.

Refan, M. and Hangan, H. (2016) Characterization of tornado-like flow fields in a new model scale wind testing chamber. *Journal of Wind Engineering & Industrial Aerodynamics*, 151, 107–121.

Ricou, F. P. and Spalding, D. B. (1961) Measurements of entrainment by axisymmetrical turbulent jets. *Journal of Fluid Mechanics*, 11, 21–32.

Wang, Y., Cao, Y. and Meng, X. (2019) Energy efficiency of industrial buildings. *Indoor and Built Environment*, 28, 293–297.

Wang, Y. and Cao, Z. (2017) Industrial building environment: Old problem and new challenge. *Indoor and Built Environment*, 26, 1035–1039.

Wang, Y., Gao, J., Xing, X., Liu, Y. and Meng, X. (2016) Measurement and evaluation of indoor thermal environment in a naturally ventilated industrial building with high temperature heat sources. *Building and Environment*, 96, 35–45.

Wang, Y., Zhai, C., Cao, Z. and Zhao, T. (2020) Potential application of using vortex ring for personalized ventilation. *Indoor Air*, 30, 1296–1307.

Ward, N. B. (2010) The exploration of certain features of tornado dynamics using a laboratory model. *Journal of Atmospheric Sciences*, 29, 1194–1204.

Wyndham, C. H., Allan, A. M., Bredell, G. A. G. and Andrew, R. (1967) Assessing the heat stress and establishing the limits for work in a hot mine. *British Journal of Industrial Medicine*, 24, 255–271.

Yagami, H. and Uchiyama, T. (2011) Numerical simulation for the transport of solid particles with a vortex ring. *Advanced Powder Technology*, 22, 115–123.

Yaglou, C. P. and Minard, D. (1957) Control of heat casualties at military training centers. *AMA Archives of Industrial Health*, 16, 302.

Zhang, X., Wang, J. and Wan, D. (2020) CFD investigations of evolution and propulsion of low speed vortex ring. *Ocean Engineering*, 195.

Chapter 15
Ventilation and fire safety for high-rise buildings

Dahai Qi[1]

This chapter presents the most recent progress in high-rise building design and control methods for achieving energy efficiency and fire safety. First, basic knowledge of high-rise ventilation is introduced. Then, the challenges of designing and controlling high-rise ventilation are pointed out. State-of-the-art energy efficiency and safety research on the modeling, control, and design of high-rise ventilation is also presented. Lastly, two case studies on high-rise fire smoke control and atrium fire smoke control are introduced and discussed. Readers will come away from this chapter with an understanding of the theory of high-rise ventilation and the design challenges related to fire safety concerns, as well as a knowledge about designing safe and energy efficient high-rise ventilation systems.

15.1 Background

The National Fire Protection Association (NFPA) (NFPA, 2018) defines high-rise buildings as buildings higher than 23 m or seven stories. Due to the limited land supply and high population density of cities, high-rise buildings have become increasingly popular, particularly, during periods of economic growth. More than 100 high-rise buildings have been erected in 142 cities and in 2019; among those the city with the most high-rise buildings was New York with 6,034 (SkyscraperPage.com, 2019). Although many cities already have a number of high-rise buildings, the number continues to increase, especially in cities with fast economic growth. For example, at its peak, there were over 150 high-rise buildings under construction in recent years in Toronto (DH Toronto Staff, 2018).

However, the rising demand for space cooling in high-rise buildings is putting pressure on the electrical systems. Air conditioners and fans for space cooling consume around 20% of the total electricity used in buildings around the world (OECD/IEA, 2018). High-rise buildings usually consume more energy due to their higher cooling load than mid- and low-rise buildings. The high cooling load of high-rise buildings is caused by the high internal heat gain (e.g. lights, equipment) and wide use of huge glazed façades (Yuan *et al.*, 2018). Therefore, to mitigate the strain of cooling energy demand, it is essential to reduce the cooling load of high-rise buildings.

[1]Department of Civil and Building Engineering, Université de Sherbrooke, Sherbrooke, Canada

Building ventilation is an effective solution for maintaining indoor air quality, cooling indoor spaces, and reducing building cooling loads, i.e. ventilative cooling (IEA, 2018). Ventilation systems can be categorized as natural ventilation (NV), mechanical ventilation (MV), and hybrid ventilation (HV) (also referred to as mixed-mode ventilation). Previous studies have found that depending on the local climate and building ventilation design, ventilation can reduce cooling-related energy consumption by 56%–86% (Malkawi *et al.*, 2016; Hu and Karava, 2014). There are more than 50 large cities with a significant ventilative cooling potential of more than 2,000 h/year (Chen *et al.*, 2017). In cold climates, such as Canada and Northern Europe, high-rise ventilative cooling can be used for a long time throughout the year, not only during shoulder seasons but also during the summer (Artmann *et al.*, 2007). Furthermore, cold climates have large diurnal temperature variations and relatively low nighttime outdoor temperatures even in the summer. The characteristics of cold climates are more beneficial to high-rise ventilative cooling than other climates. A high-rise building structure can be cooled during the night and becomes a huge heat sink during the daytime to reduce cooling loads and thus reduces peak electricity demands.

However, fire safety concerns, associated with the stack effect in large vertical spaces, are one of the big issues of the use of ventilation in high-rise buildings. In high-rise buildings, cool fresh outdoor air can pass through floors, move upward through the vertical spaces due to the stack effect, arrive at different floors to remove indoor heat, and exit the buildings during NV and HV. The vertical spaces can be atria, stairwells, double-skin façades, and elevator shafts, which are the critical structures of high-rise buildings. During regular operations, many existing features and functions of these large spaces can contribute to the stack effect and high-rise ventilation potentially providing a maximum level of energy savings. However, during a fire outbreak, fire-generated smoke laden with toxic gases can spread far from the origin of the fire deep throughout the building through these large vertical spaces, endangering occupants, damaging property, and creating challenges for firefighters. For example, it was reported that around 145,000 structure fires in high-rise buildings occurred per year from 2009 to 2013 in the United States, causing an average of 40 deaths, 520 injuries, and $154 million in property damage per year (Ahrens, 2016). The problem of smoke control in high-rise buildings can be further complicated by interactions with dynamic weather conditions, including variable winds, temperatures, and building ventilation system operations.

Knowing that energy efficient and safe high-rise buildings are essential for the sustainable development of cities and society, this chapter provides state-of-the-art high-rise ventilation design and fire smoke control methods, which take into consideration the energy efficiency and building safety issues in high-rise ventilation, including NV, MV, and HV.

15.2 Ventilation types

15.2.1 Natural ventilation

NV refers to the intentional introduction of air naturally passing through open windows, doors, grilles, etc. (ASHRAE, 2017). Based on the driving forces, NV can be categorized as buoyancy-driven NV, wind-driven NV, or wind- and buoyancy-driven NV, which are driven by the pressure difference across the building envelope caused by wind, stack effect, or both.

The indoor–outdoor air density difference caused by the indoor–outdoor temperature difference generates the airflow movement of buoyancy (stack effect) ventilation in high-rise buildings (Wood and Salib, 2013). Equation (15.1) can calculate the pressure difference of the stack effect at any vertical location, while neglecting the vertical density gradient.

$$p_s = \rho_0 \left(\frac{T_i - T_0}{T_i} \right) g(H_{NPL} - H) \tag{15.1}$$

where p_s is the stack pressure difference (Pa), ρ_0 is the outdoor air density (kg/m^3), T_i and T_0 are the absolute indoor and outdoor temperature (K), g is the gravitational acceleration (m/s^2), H_{NPL} and H are the height of the neutral pressure level (NPL) and the height above the reference plane respectively (m).

For wind-driven NV, the wind-driven force is dominant, which can be further classified as single-side ventilation or cross ventilation. Single-side ventilation refers to fresh air entering the room and exhausting through the same side. Cross ventilation is the airflow between the windward side and leeward side of a building's envelope. The relationship between wind pressure and wind speed is demonstrated in (15.2) (ASHRAE, 2017):

$$p_w = C_p \rho \frac{U^2}{2} \tag{15.2}$$

where p_w is the windward pressure relative to the outdoor static pressure (Pa), ρ is the outdoor air density (kg/m^3), U is the wind speed (m/s), C_p is the wind surface pressure coefficient, dimensionless. C_p can be decided by the geometry and location of the building, as well as wind direction. The equation to calculate C_p can be found in the 2001 *ASHRAE Handbook—Fundamentals* (Chapter 16). For high-rise buildings, it should be noted that the wind speed increases parabolically as the building height increases (Günel and Ilgin, 2014) and thus creates large wind pressure differences along the building façade according to (15.2). Hence, when designing wind-driven NV in high-rise buildings, the wind pressure variation along the façade must be considered, because the high wind speed at the upper floors may result in uncomfortable indoor air velocity and unacceptable wind-induced noise.

During the winter, the indoor temperature is higher than the outside temperature and the airflow inside the building will rise driven by buoyancy. Due to the air density differences of indoor and outdoor air, there exists an NPL, above which the indoor pressure is higher than the outdoor pressure, so the air tends to exfiltrate to the outside. The NPL can be located in the middle of the building (Figure 15.1(a)), or lower or higher (Figure 15.1(b)), depending on the indoor and outdoor air temperatures, as well as the building structures. During the transient seasons, when indoor and outdoor temperatures are similar, the indoor and outdoor pressure profiles can be parallel, and the outdoor windward pressure can be higher than the indoor pressure as shown in Figure 15.1(c).

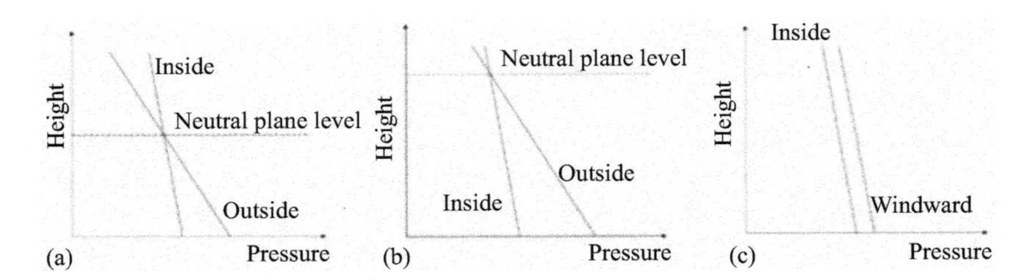

Figure 15.1 Pressure profile of a high-rise building. (a) Neutral plane is in the middle of the building; (b) neutral plane is above the middle of the building; (c) there is no neutral plane

Applying (15.1), it can be found that the pressure difference generated by buoyancy depends on the height above the NPL. Hence, when using buoyancy-driven ventilation for ventilative cooling, its cooling performance is usually highly dependent on the building structure, i.e. the high vertical spaces, such as the ventilation shaft, atrium, solar chimney, and double-skin façade (Sha and Qi, 2020a). To improve buoyancy-driven ventilation, the design of the parameters of these vertical architectural elements, e.g. the cross-sectional size, height, and the location and size of the openings, should be optimized. For example, Moosavi *et al.* (2015) recommended designing a large inlet to outlet opening ratio for achieving high cooling performance in an atrium. Furthermore, to make full use of the vertical space, these architectural elements can be combined together. For example, the atrium, double-skin façade, and solar chimney can be integrated into one building to generate a higher airflow rate (Ding *et al.*, 2005).

When designing wind-induced ventilation, indoor airflow and airflow patterns greatly impact its indoor heat removal performance. The characteristics of the building, including the building positioning, floor planning, and building façade, must be considered to optimize the indoor airflow rate and airflow patterns (Sha and Qi, 2020a). The building positioning refers to the arrangement of buildings on the ground surface according to their locations and dimensions. For example, the location of the building should avoid the shelter effect caused by the surrounding buildings or structures, which can influence the wind pressure distributions on a building and reduce the airflow rate of NV (Chen *et al.*, 2008). Floor planning is the planning of floor partitions and door locations, etc. The reduction of partitions and the creation of wind-path on the floor plan can enhance the airflow rate of NV (Zhou *et al.*, 2014). The elements on the building façade, e.g. balconies and openings, also need to be optimized to achieve effective NV on cooling (Ai *et al.*, 2011; Wang *et al.*, 2007).

15.2.2 Mechanical ventilation

MV refers to the air movement into or out of a building by using mechanical equipment, such as fans, ductwork, and grilles (ASHRAE, 2017). The operation of mechanical equipment affects pressure differences across the building envelope and thus air change rates, which are essential for building energy efficiency and safety (ASHRAE, 2017).

One of the approaches for designing the MV system to reduce the building cooling load is mechanical night ventilation. Medved *et al.* (2014) found that the energy savings of mechanical night ventilation can reach around 50% compared with only using a mechanical cooling (MC) system. Zhang *et al.* (2018) investigated the control of the air exchange rate (ACH) for night MV based on the building energy simulation (BES). The results showed that the optimal ACH of nighttime MV is higher than the ACH for maintaining indoor air quality, which results in a 47% reduction in energy consumption over the entire summer.

However, an appropriate control strategy of the MV system must be used to minimize cooling-energy consumption based on the chiller cooling and mechanical ventilative cooling energy performances. Otherwise, mechanical ventilative cooling may consume more energy than chiller cooling, because the fans in MV systems can consume a considerable amount of energy. This point can be seen in a study conducted by Kolokotroni and Aronis (1999). The MV system is usually designed for maintaining indoor air quality, especially in high-rise buildings. High resistance due to the long ductwork spanning many floors and/or low fan efficiency can generate high-energy consumption of the MV system.

With an appropriate control strategy, the energy efficiency and maximum flow rate of fans are the key factors in determining the cooling performance of mechanical ventilative cooling (Sha and Qi, 2020b). The fans should be energy efficient to provide ventilative cooling, i.e. their specific fan power (SFP) must be low enough. SFP is a parameter for quantifying the energy efficiency of fans, which is defined as the electric power that is needed to transport one unit of air. A low SFP requires low resistance of the ductwork and small energy loss on the belt or motor in the fans. A higher maximum fan flow rate can provide more ventilative cooling, but it is meaningless to design a very high fan flow rate because the cooling demand is not always very high during the day. Therefore, an optimal maximum fan flow rate should exist. In practice, a decentralized system that spans fewer floors than the centralized system in high-rise buildings is recommended in the design, because a decentralized system can more easily satisfy the fan flow rate and energy efficiency requirements than the centralized system.

15.2.3 Hybrid ventilation

An HV system is defined as a system that provides a comfortable internal environment by making use of both NV and MV at different times (Heiselberg, 2002). HV avoids the disadvantages of NV, such as uncertain performance caused by weather conditions, and improves the degree of individual control of the indoor climate. According to Heiselberg (2002) and Li and Heiselberg (2003), the types of HV systems can be classified as an alternate use of NV and MV, fan-assisted NV, and stack and wind-assisted MV. The alternate use of NV and MV means that the NV and MV systems are independent, and a control strategy can switch between the NV and MV systems. The fan-assisted NV system is an NV system combined with fans that are used during periods when the NV is weak. The stack and wind-assisted MV refers to an MV system that uses stack and wind forces to reduce the need for fans.

The main challenge of HV systems is the control strategy and coordination of NV and MV systems. The advanced HV system can save as much energy as possible while maintaining the high indoor environmental performance requirements. Different control methods of HV systems have been previously proposed. For example, Hu and Karava (2014) reported that a model-predictive control strategy to control the opening of the NV system in the HV system can effectively reduce the total cooling energy demand by 85%. Chen *et al.* (2018) proposed a control method based on the reinforcement learning technique, which is a model free. The reinforcement learning control strategy can treat the environment as an unknown black box, and the reinforcement learning algorithm can learn from the interaction with the environment and find the optimal control decision.

15.3 Smoke control for high-rise fires

15.3.1 Pressurization system for stairwells

The pressurization system includes the vertical space and fans. By supplying sufficient air into the vertical space, such as stairwells and elevator shafts, the inside pressure can be high enough to prevent smoke spreading into the protected spaces.

If there is fire, the pressurized stairwells could provide a smoke-free safety route for escape and firefighters. According to the number of air supply locations, the

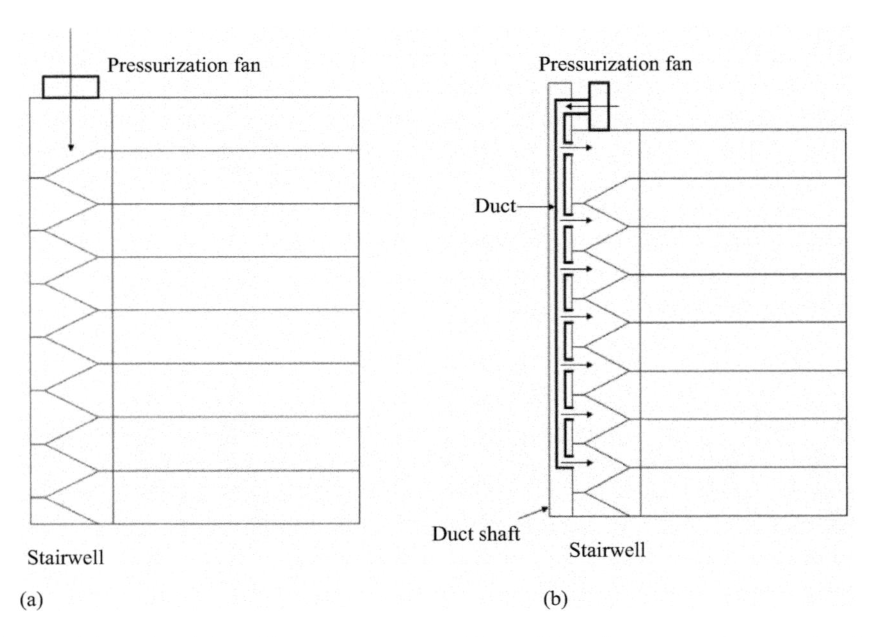

Figure 15.2 (a) Stairwell pressurization with roof fan; (b) stairwell pressurization by multiple injection with roof fan

pressurization can be categorized as a single-injection pressurization system where the air is supplied at one floor (top or bottom of the stairwell, see Figure 15.2(a)), or a multiple injection pressurization system where the air is supplied at multiple floors (see Figure 15.2(b) and (c)). The weakness of the single-injection pressurization system is that it may fail for spaces far from the air supply location, such as in tall buildings. The single-injection pressurization system may also not work when the doors near the air supply location are open, which reduces the pressure. Therefore, in high-rise buildings, to maintain positive pressure in the tall stairwell, the multiple injection pressurization system is often applied, which includes a duct with multiple air supply openings and a fan. The fan can be mounted on the roof of the building or at the ground level.

15.3.1.1 Minimum design pressure difference

To prevent smoke spread to the stairwell, the pressurization system should create an adequate pressure difference across the stairwell door. The required minimum pressure across the closed stairwell door, $\Delta p_{\min}$, can be calculated by using

$$\Delta p_{\min} = \Delta p_{SF} + 3{,}460h\left(\frac{1}{T_o} - \frac{1}{T_F}\right) \tag{15.3}$$

where $\Delta p_{\min}$ is the minimum design pressure difference, Pa; Δp_{SF} is the pressure difference safety factor, Pa; h is the distance above the neutral plane, m; T_o and T_F are the absolute surrounding temperature and hot gas temperature, K.

NFPA requires that stairwells should be pressurized to maintain a 0.10-in. water gauge (25 Pa) across a closed stairwell door for a non-sprinklered building (see Table 15.1), based on the assumption of a hot gas temperature of 927°C next to the stairwell door and a 0.03-in. water gauge (7.5 Pa) pressure difference safety factor. Table 15.1 presents the required minimum design pressure difference across smoke barriers, which depends on the ceiling height and the building type (with or without sprinklers).

15.3.1.2 Maximum design pressure difference

Besides the minimum design pressure difference, there is also a maximum design pressure difference, which is used to avoid pressurization systems creating forces that are too great to open the door. The force required to open a side-hinged

Table 15.1 Minimum design pressure across smoke barriers

Building type	Ceiling height in m (ft)	Design pressure difference in Pa (in.w.g.)
Sprinklered	Any	12.5 (0.05)
Non-sprinklered	2.75 (9)	25 (0.10)
Non-sprinklered	4.56 (15)	35 (0.14)
Non-sprinklered	6.41 (21)	45 (0.18)

Note: The table presents minimum design pressure differences developed for a gas temperature of 1,700°F (927°C) next to the smoke barrier.
Source: NFPA (2012).

swinging door needs to be great enough to overcome the door closer and the pressure difference across the closed door, which can be expressed by (Klote and Milke, 2002)

$$F = F_{dc} + \frac{k_d W A \Delta P}{2(W - d)} \tag{15.4}$$

where F is the total door opening force (N), F_{dc} is the force to overcome the door closer (N), W is the door width (m), A is the door area (m^2), ΔP is the pressure difference across the door (Pa), d is the distance from the doorknob to the edge of the knob side of the door (m), K_d is the coefficient (dimensionless).

The relationship between the pressure difference across the closed door, ΔP, and the resultant force can be expressed as

$$\Delta P = \frac{2(W - d)(F - F_{dc})}{k_d W A} \tag{15.5}$$

Figure 15.3 shows the relationship between the pressure difference and door opening force to overcome the pressure difference for a door 2.13 m in height (H) and a distance from the doorknob to the edge of the knob side of the door, d of 0.06 m. The pressure difference across the closed door, caused by the pressurization system, cannot be too high; otherwise, a person cannot open the stairwell door to escape through the stairwell if there is a fire. NFPA 101 (life safety code) states that the force required to open any door as a means of egress shall not exceed 133 N (30 lb). The force to overcome a door closer is normally greater than 13 N. According to Figure 15.3, for a

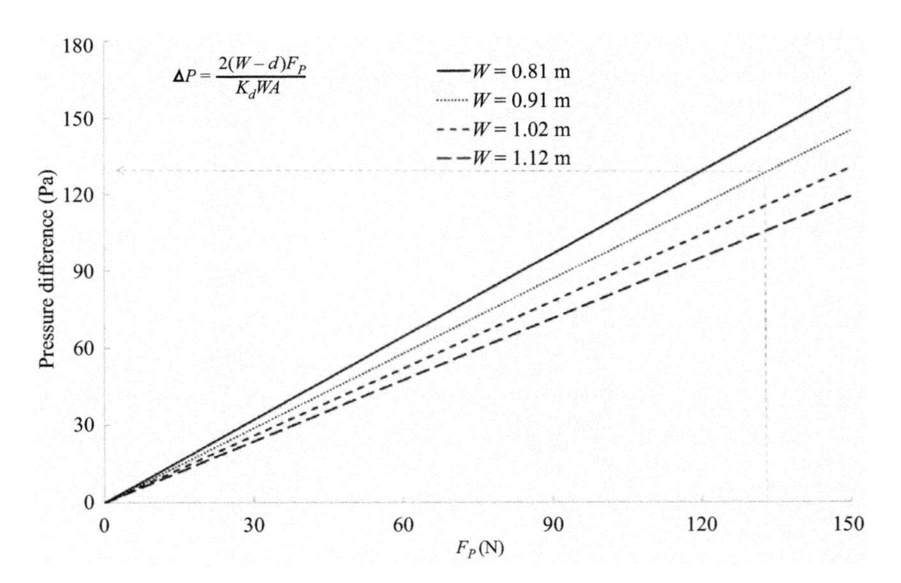

Figure 15.3 Relationship between pressure difference across the door and the force to overcome the pressure difference (H = 2.13 m, d = 0.06 m)

door 2.13 m in height and $d = 0.06$ m, if the force to overcome a door closer is 43 N, to satisfy the NFPA 101 requirement (total door opening force should be less than 133 N), the pressure difference cannot be higher than 88 Pa.

15.3.2 Smoke ventilation

One of the primary concerns for the use of pressurization systems is the possible excessive pressure difference across the closed door, leading to high forces required to open the door and preventing occupants from being able to open them. This issue should be carefully considered especially for tall buildings higher than 30 m (Lay, 2014). One of the alternative solutions is to use a dedicated smoke exhaust system to exhaust smoke by stack effect and/or exhausting fans. This method would reduce the risk of smoke infiltrating into the stairwell so that the pressurization system for the stairwell could create a lower pressure difference across the closed door or not even need to be in operation depending on the fire smoke control performance of the whole system.

One or two dedicated shafts can be installed in the high-rise building to exhaust smoke as indicated in Figure 15.4. Dampers are installed on each floor. If a fire occurs on one of the floors, the damper on the fire floor opens and the smoke is exhausted through the smoke shaft, while the dampers on the non-fire floors remain closed so that the smoke will not infiltrate into the non-fire floors. The shaft can be designed without fans, where the smoke is exhausted due to the stack effect, or with fans to remove the smoke mechanically.

Estimating the smoke temperature inside the smoke shaft is helpful for designing the smoke shaft and smoke exhaust fan at the top of the shaft for safe operation.

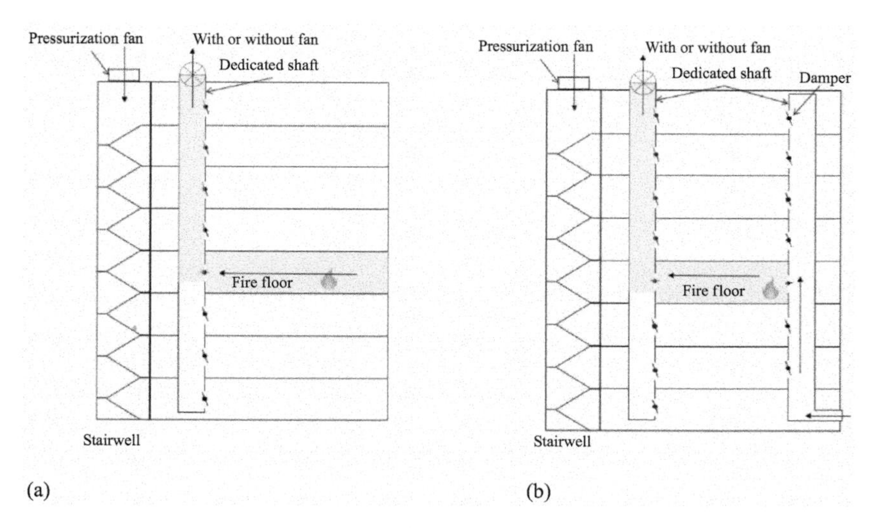

Figure 15.4 (a) Smoke exhaust system with one dedicated shaft; (b) smoke exhaust system with two dedicated shafts

The smoke temperature inside the shaft can be expressed as

$$T_{shx} = T_b + \left(T_f - T_b\right)\exp(-\varphi\alpha) \quad (15.6)$$

where T_{shx} is the temperature profile along the vertical direction of the dedicated shaft (x) where the smoke is exhausted (°C), T_f is the fire room temperature (°C), T_b is the non-fire room temperature (°C).

φ is relative height:

$$\varphi = \frac{x}{H} \quad (15.7)$$

where H is the vertical distance between the fire floor and the top of the shaft (m).

α is the temperature attenuation coefficient:

$$\alpha = \frac{PH}{\dot{m}C_P R_t} \quad (15.8)$$

where P is the perimeter of the shaft (m), $\dot{m}$ is the mass flow rate (kg/s), C_P is the specific heat capacity of the smoke (J/kg K), R_t is the thermal resistance between two sides of the shaft (m^2 K/W).

$$T_{shH} = T_b + \left(T_f - T_b\right)\exp(-\alpha) \quad (15.9)$$

The EN 12101-3 Standard of the European Committee for Standardization classifies smoke exhaust fans by temperature (EN 12101-3 Standard, 2002), e.g. the F200 fan must resist 200°C for at least 2 h. The shaft has a perimeter of $P = 5$ m and a height of $H = 50$ m, and a thermal resistance of $R_t = 0.224$ (m^2 K)/W. If the EN 12101-3 F200 fan is used in the shaft, for an exhaust fan with a given flow rate of 1.6 kg/s, the maximum fire would be $T_f = 374$°C. This means that the fan would withstand a fire smoke of 374°C. A solution could be to change for another fan with a higher temperature rating or to add another F200 fan to share the smoke flow rate. Therefore, (15.9) could be used in practice to evaluate the thermal performance of mechanical exhaust fans installed at the top of the shaft.

15.4 Case studies

15.4.1 Evaluation of high-rise fire smoke control[*]

High-rise buildings comprise many complex structures, like atriums, stairwells, elevator shafts, corridors, and compartments, which also leads to difficulties in designing smoke control systems in these structures. The smoke control systems could interact when they operate together. For example, the required fan speed of the stairwell and elevator shaft-pressurization systems is strongly coupled (Miller and Beasley, 2009).

[*]Section 15.4.1 is modified from the conference paper published by ASHRAE: Qi D, Soubra M, Mashayekh S, Wang L. CFD Modeling of Full-size High-rise Fire Smoke Spread and Smoke Control. ASHRAE 2017 Annual Conference, Long Beach, CA. 2017.

To design a reliable smoke control system, simulation techniques to predict smoke movement and the performance of the smoke control system are often required. To investigate smoke spread and control performance in high-rise buildings, the computational fluid dynamic (CFD) modeling is preferable. In this case study, the fire dynamics simulator (FDS) is used, which is a widely used CFD tool in the prediction of smoke spread inside high-rise buildings (McGrattan *et al.*, 2013). FDS is a CFD model based on large eddy simulation (LES) of turbulent flows, which was developed by the US National Institute of Standards and Technology. The calculation of LES is much more time-consuming than other CFD models based on the Reynolds-averaged Navier–Stokes modeling of turbulences. In recent years, high-performance computing (HPC), using clustered supercomputers, has been more frequently applied in many CFD simulations. HPC can be used in fire dynamics simulation of whole-building fire smoke spreads and controls in full-size high-rise buildings.

In this case study, a full-size 30-story tower with fire smoke spread was studied. Different smoke control strategies were designed and modeled, including sprinklers only, a smoke exhaust system, pressurization system and dedicated smoke exhaust shaft system, and a new approach for high-rise smoke removal (Lay, 2014). To evaluate the smoke control performance, the results of smoke/air temperature distributions, mass flow rates, and pressure distributions in the stairwells were compared among different strategies.

15.4.1.1 Full-size high-rise fire smoke modeling

A 30-story high-rise building, 93 m in height, as shown in Figure 15.5(a), was modeled. The bottom three floors are car parking, which is ignored in the

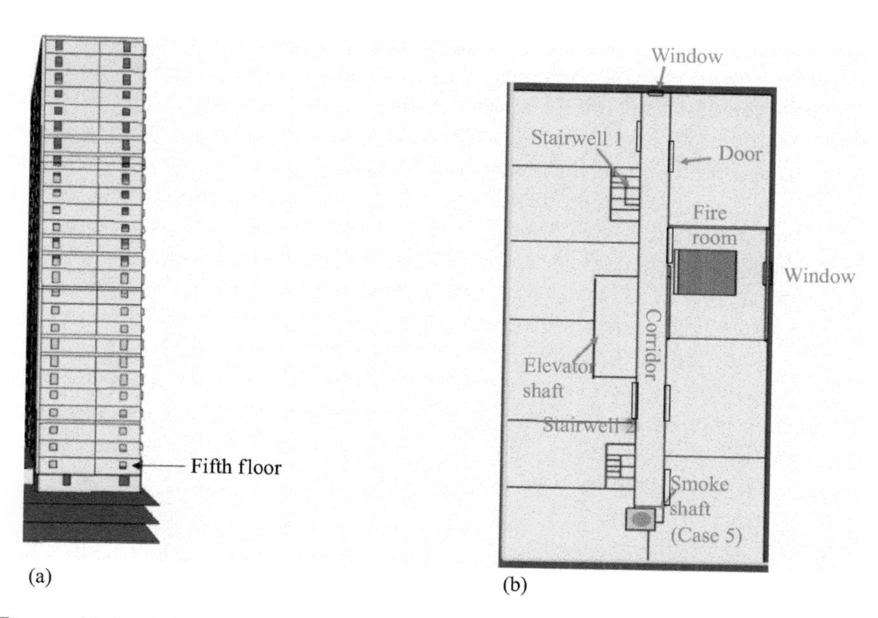

Figure 15.5 Schematic of the building: (a) vertical schematic of the whole building and (b) plan schematic of fifth floor

Table 15.2 Designs of different fire smoke control cases

Cases	Smoke control method	Windows (W (m)×H (m))
Case 1	Sprinkler only, no fan on roof of stairwell	2×0.8
Case 2	Exhaust fan on the roof of Stairwell 2	2×0.8
Case 3	Pressurization fan on roof of Stairwell 1	2×1.6
Case 4	Pressurization fan on roof of two stairwells	2×1.6
Case 5	Dedicated smoke exhaust shaft with roof extraction fan	2×0.8

Note: In Cases 2–5, the volume flow rate of the fan is 20 m^3/s.

simulation. The staircases in each tower do not span to the bottom four floors. The building includes two stairwells, one elevator shaft, corridors, and residential rooms (see Figure 15.5(b)). The door size is 1 m×2 m (*H*).

In this case study, a fire is assumed in a fifth-floor room (see Figure 15.5(b)). The area of the fire room is 70 m^2 and the heat release rate (HRR) is 8,000 kW, which is approximately the total maximum HRR of four chairs (Babrauskas, 1983). The sprinklers were assumed to be installed in the fire room, which are spaced 4 m×4 m. According to the NFPA Standard 13 (NFPA, 2002), the water volumetric flow rate of the sprinklers is set at 60 L/min, and the triggering temperature is 74°C. The ambient temperature is 20°C.

The study focuses on the smoke spread inside the two stairwells that are the evacuation routes for people. To evaluate the performance of different smoke control systems under the worst situation, the window of the fire room and the doors between the room and the stairwells were set as open (assuming the resident opened them due to panic). Five cases with different smoke control systems were designed, and the relevant information is listed in Table 15.2. Case 1 is the base case without any protections in the stairwells. Case 2 includes a smoke exhaust system in Stairwell 2. Cases 3 and 4 use a stairwell pressurization system in the stairwell with different settings. Case 5 has a dedicated smoke exhaust shaft with a cross-sectional area of 2 m×2 m, which is shown in Figue 15.5(b).

The simulation models were created using PyroSim, a graphical user interface for FDS (Thunderbird Engineering, 2015). The total number of cells is around 500,000. The size of each cell is 0.5 m which ensures that there are at least four cells in each opening. Temperature and pressure distributions inside the stairwells were recorded at each floor of the stairwells. Mass flow rates at the stairwell openings were measured by flow measuring devices.

15.4.1.2 Results

Figure 15.6 demonstrates the transient smoke/air temperature in the two stairwells at the 5th (fire floor), 17th, and 29th floors for different cases. In Case 1, the maximum smoke temperature reaches 450°C at around 220 s on the fire floor of Stairwell 2, which is higher than 330°C of the maximum smoke temperature in Stairwell 1 at the

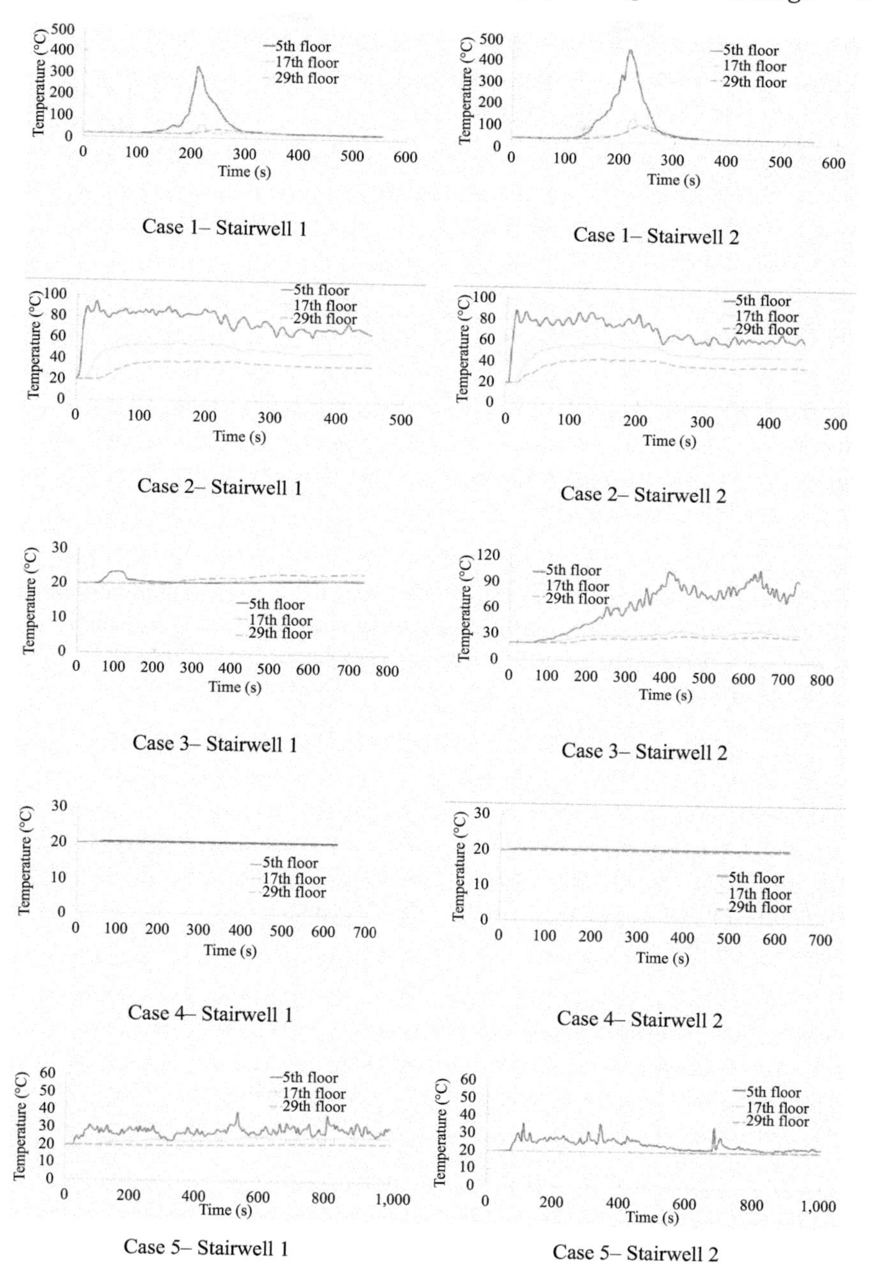

Figure 15.6 Comparison of temperature profiles of smoke/air inside the stairwells for different cases

same time. After reaching the maximum temperature, the smoke temperature decreases to the ambient temperature, which is around 22°C. This is due to the operation of sprinklers in the fire room and the lack of fresh air for combustion since there is no other opening for fresh air flowing into the building except for the open window in the fire room. In Case 2, the stairwell smoke temperature jumps to 90°C in a very short time (around 30 s) and then remains at around 70°C for most of the time since the MV fan on the roof of Stairwell 2 operates and fresh air continuously flows into the fire room through the open window.

For Case 3, the fire floor smoke temperature of Stairwell 1 increases to a maximum of 23°C around 100 s. The temperature then drops to 21°C, because the pressurization fan works well. Since the smoke spreads to Stairwell 2 (no pressurization in Stairwell 2), the smoke temperature increases to around 90°C by 400 s then becomes stable. Both stairwells in Case 4 are smoke free, due to the two pressurization fans on the roof. The air temperatures remain at the ambient temperature, 20°C. For Case 5, most of the smoke is exhausted by the smoke shaft, and only a little smoke infiltrates into the stairwells. The smoke temperature on the fire floor of the stairwells is only a little bit higher than the ambient temperature, which is around 25°C on average.

Figure 15.6 shows that Case 1 (sprinkler system only) is not enough to extinguish the fire, and the smoke can spread throughout the tower. It can also be seen that the smoke spreads through stairwells faster in Cases 1 and 2 than in Cases 3–5. Therefore, for Cases 1 and 2, both stairwells are not safe for evacuation. However, the cases with a pressurization system or smoke shaft (Cases 3–5) could have at least one safe stairwell that smoke cannot flow into. To further investigate the performance of the smoke control systems in Cases 3–5, the mass flow rate and pressure distribution inside the stairwell are discussed later.

According to Figure 15.6, smoke temperatures stabilize after 300 s indicating a stable (or "steady state") smoke flow. The average mass flow rate and pressure distribution inside the stairwells during the time range of 300–400 s are presented in Figures 15.7 and 15.8.

Figure 15.7 compares smoke/air mass flow rates flowing through the open doors of the two stairwells in Cases 3–5. The negative mass flow rate indicates that the smoke/air mass flow exfiltrated from stairwells to the floors whereas positive values mean smoke infiltrated into the stairwells.

For Stairwell 1 in Case 3, pushed by the roof pressurization fan, the smoke leaves the fifth floor, while little airflow infiltrates into the stairwell for the upper floors of the stairwell, so the upper floors are smoke free and the temperature increases a little (see Figure 15.6). The range of stairwell gauge pressure is 9–18 Pa (see Figure 15.8), which is lower than the NFPA 92 requirement of pressure difference across any closed stairwell doors in the pressurized stairwells, which is 25–88 Pa. Considering the stairwell's doors are set as open and smoke cannot be found inside the pressurized stairwell, the design of Stairwell 1 with the roof pressurization fan with a volume flow rate of 20 m^3/s can be acceptable.

For Case 4, most of the air pressurized by the roof fan leaves from the bottom opened doors of the stairwells (exfiltration), because of the large pressure difference

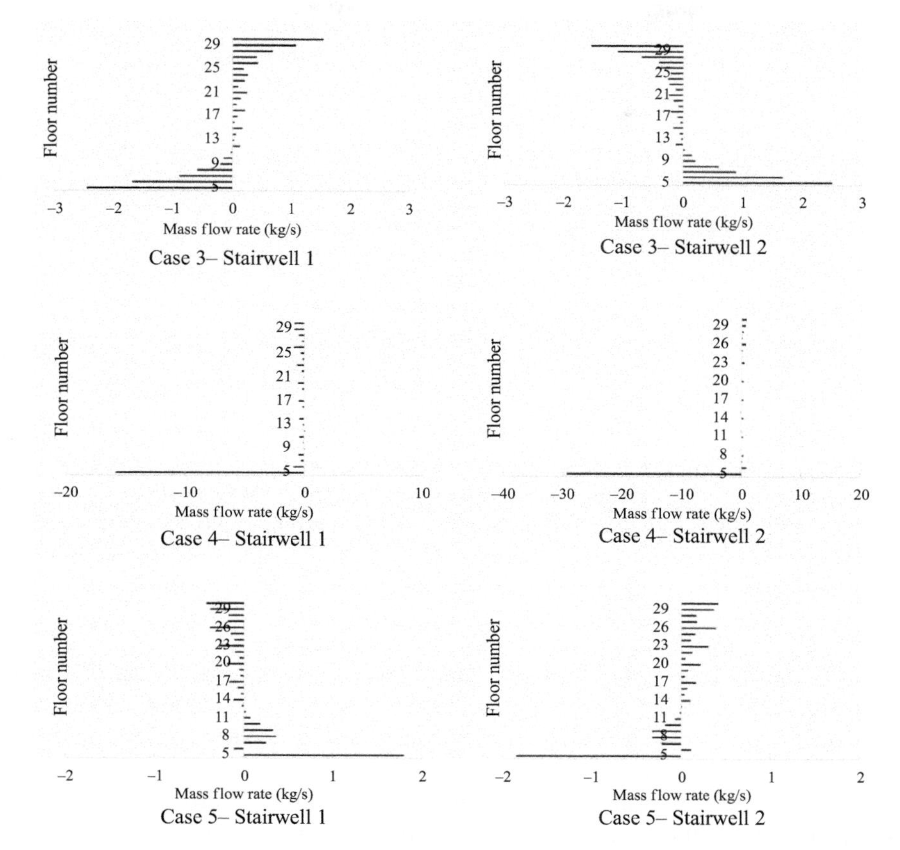

Figure 15.7 Comparison of mass flow rate through stairwell doors for different cases (negative mass flow rate: exfiltration from stairwell; positive mass flow rate: infiltration into stairwell)

across the stairwells' doors. As is shown in Figure 15.8, the largest gauge pressure in both stairwells exceeds 500 Pa, which far exceeds the required range of 25–88 Pa defined by NFPA 92 for stairwell pressurization systems. This high gauge pressure means that the required force to open the door can reach 1,000 N, indicating excessive pressure in the stairwell, which may result in potential problems for people to safely evacuate (e.g. people cannot open the stairwell door easily).

Figure 15.7 shows that Case 5 has smoke infiltrating into Stairwell 1 from the fire floor. However, Stairwell 2 is free of smoke because the smoke is exhausted through the smoke shaft, which is near shaft 2. The gauge pressure of Stairwell 2 at the fire floor is 92 Pa (see Figure 15.8), which is a little bit higher than the NFPA 92

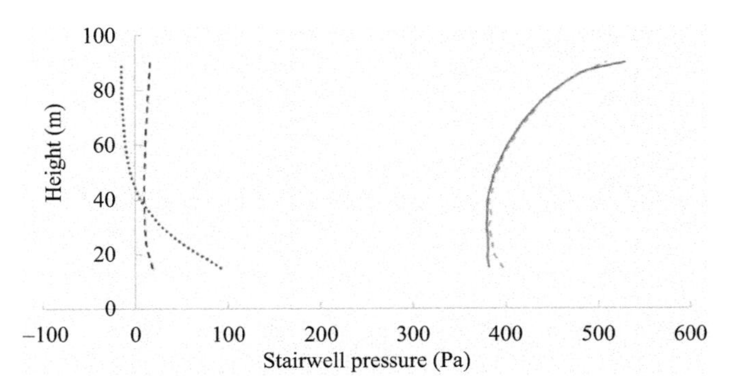

Figure 15.8 Comparison of stairwell pressure distribution for different cases

requirement. In summary, the best smoke control strategy of all five cases is probably Case 3 (pressurization fan on the roof of Stairwell 1) because of its simple implementation and lower working pressures in the evacuation stairwell. It appears that Case 5 (smoke shaft with exhaust fan) may be an alternative approach, but more in-depth analysis needs to be conducted to investigate its performance.

15.4.1.3 Conclusion

This case study uses CFD simulation of fire smoke spread and control in a real high-rise building with two stairwells. Five smoke control methods, including sprinklers only, mechanical exhaust system, two different pressurization systems, and a smoke shaft system, were simulated and compared in the FDS simulations. The comparison of five smoke control methods was shown by using the temperatures, pressures, and mass flow rates of smoke inside the two stairwells. The simulation results show that pressurizing only one stairwell with a fan volumetric flow rate of 20 m^3/s is enough to keep the evacuation stairwell safe for people. However, if pressurization fans are installed in both stairwells, excessive pressure differences across the stairwells could occur leading to a potential failure of the pressurized system (e.g. people cannot open the stairwell door easily). Therefore, the installation of pressurization systems in both stairwells should be carefully analyzed and evaluated. In addition to the pressurization system, an added dedicated smoke exhaust shaft can help to achieve smoke-free conditions in the nearest stairwell. The dedicated smoke exhaust shaft can be an alternative method, especially when the stairwell pressurization system cannot be applied.

15.4.2 Atrium fire smoke control[†]

An atrium,[†] as a popular architecture element of high-rise buildings, can take advantage of its large vertical space to generate the stack effect for NV (Wood and

[†]Section 15.4.2 is modified from the conference paper published by ASHRAE: Sha H, Qi D. A Novel Ventilation Approach of Large Vertical Space for Achieving Fire Safety and Energy Efficiency. 2020 ASHRAE Virtual Conference, 2020.

Salib, 2013). However, an atrium may also become the main route of smoke spread when fire occurs. To avoid smoke spread along the atrium, the height of the atrium is often limited by fire safety regulations. Segmentations are often applied in the atrium, which limits the height of the space and reduces the size of the smoke control zone. For example, the large vertical atrium in the Concordia EV building is divided into five sub-atria by segmentations (Hu and Karava, 2014), and thus each sub-atrium is only one-fifth of the height of the building. However, this solution decreases the NV potential of the atrium, because the added segmentations increase the airflow resistance and reduce the flow rates of NV (Sha and Qi, 2019).

A state-of-the-art ventilation design is proposed to balance the fire safety concerns and potential of NV on the reduction of cooling energy consumption. This ventilation approach consists of two elements: the segmentation slab and the independent ventilation shaft. The segmentation slab is designed to limit the smoke. The ventilation shaft can be used to exhaust smoke when fire occurs and provide an airflow path for NV during daily use. This design was evaluated by using FDS simulations and BESs in a case study of a 30 $(L)\times 12$ $(W)\times 30.6$ (H)-m^3 atrium in Montreal, Canada. With a comparison of the traditional ventilation method (i.e. the atrium without segmentation and ventilation shaft), the smoke layer height predicted by FDS is used to evaluate the fire protection performance. The cooling load in the atrium predicted by BES is used to evaluate the NV energy performance.

15.4.2.1 Fire smoke and energy modeling

FDS was applied for the CFD smoke simulations, and the relevant information is summarized in Table 15.3. The scenarios of the atrium with or without the new ventilation design were considered. The atrium without the new design is called the traditional design, which does not have the segmentation slab and the shaft. In the new ventilation design, a segmentation is added at 16.5 m of the atrium, and a ventilation shaft with a cross-sectional area of 1 $(L)\times 12$ (W) m^2 is added on a sidewall of the atrium (please see Figure 15.9(a)). Apart from the segmentation and shaft, the other settings are the same, i.e. the bottom openings and HRR of the fire source are the same for both scenarios. Two HRRs were considered here: 2 and 5 MW, representing a relatively small fire source and a large fire source. Since the FDS simulation accuracy is highly dependent on the mesh size, the grid independence study was also conducted on grid sizes of 0.125, 0.25, and 0.5 m. The results

Table 15.3 Cases for fire smoke simulations

Cases	Atrium design	Heat release rate (MW)	Exhaust rate (m^3/s)	Bottom opening size (m^2)
SC 1	No new design	2	212	211
SC 2	New design	2	80	78
SC 3	No new design	5	295	298
SC 4	New design	5	115	115

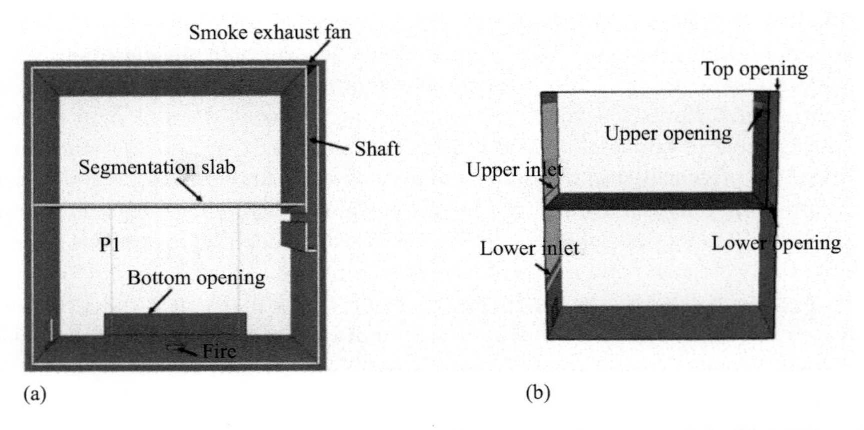

Figure 15.9 Schematic of the atrium: (a) CFD simulation model; (b) BES model

Table 15.4 Energy simulation case design

Cases	**Cooling mode**	**NV design**	**Weather conditions for NV**
Traditional design-MC	MC	All openings closed	/
New design-MC	MC	All openings closed	/
Traditional design-MNV	MC+NV	Inlet opening, 12 m^2 Top outlet opening, 4 m^2	$T_a < T_{in}$ and $15 \leq T_a \leq 24$°C $T_d \leq 13.5$°C; $v_w \leq 7.5$ m/s
New design-MNV	MC+NV	Ventilation shaft and inlet openings, 12 m^2 Top outlet openings, 4 m^2	Same as above

Note: T_a, ambient temperature; T_{in}, indoor temperature; T_d, dew temperature; v_w, wind speed.

of the 0.125 and 0.25 m grids demonstrate little difference in the smoke layer height. Hence, a grid size of 0.25 m was chosen for all the FDS simulations.

EnergyPlus was used to conduct the BES for evaluating the NV energy performance of two scenarios of the atrium. The summer period (from June to September of typical meteorological weather data) was simulated. Four cases were designed and illustrated in Table 15.4; the combination of the atrium with or without the new ventilation design and with or without NV. The design of the atrium with the new ventilation design can be seen in Figure 15.9(b). To calculate the cooling load inside the atrium, the setpoint of MC is 24°C. The atrium is occupied between 8:00 and 22:00. The construction materials and internal heat gain (i.e. people, equipment, and lighting load) are defined in the same way as the prototype building developed by the US

Department of Energy (Goel *et al.*, 2014). The glazing ratios of the west and north facades are set at 0.9 and 0.7, respectively. The NV is calculated by using the airflow network model, which is integrated into EnergyPlus.

15.4.2.2 Results

Figure 15.10 presents the smoke layer height variation at location P1 (P1 is between the entrance door and fire source, as Figure 15.10(a) shows) in all four cases. It can be seen that the cases with the new ventilation design (Cases SC 2 and SC 4) have larger smoke-free spaces than the other two cases (Cases SC 1 and SC 3). For Cases SC 1 and SC 3, the smoke layer height is around 15–20 m, but there is almost no smoke in Case SC 2 and only around a 5-m smoke layer depth in Case SC 4. After adding the segmentation, the new design can keep the entire upper atrium smoke-free and effectively remove the smoke in the lower atrium. Therefore, the new ventilation design can contribute to better smoke control performance. In addition, the cases with the new ventilation design require lower exhaust airflow rates and smaller opening areas than the cases without the new design (see Table 15.3).

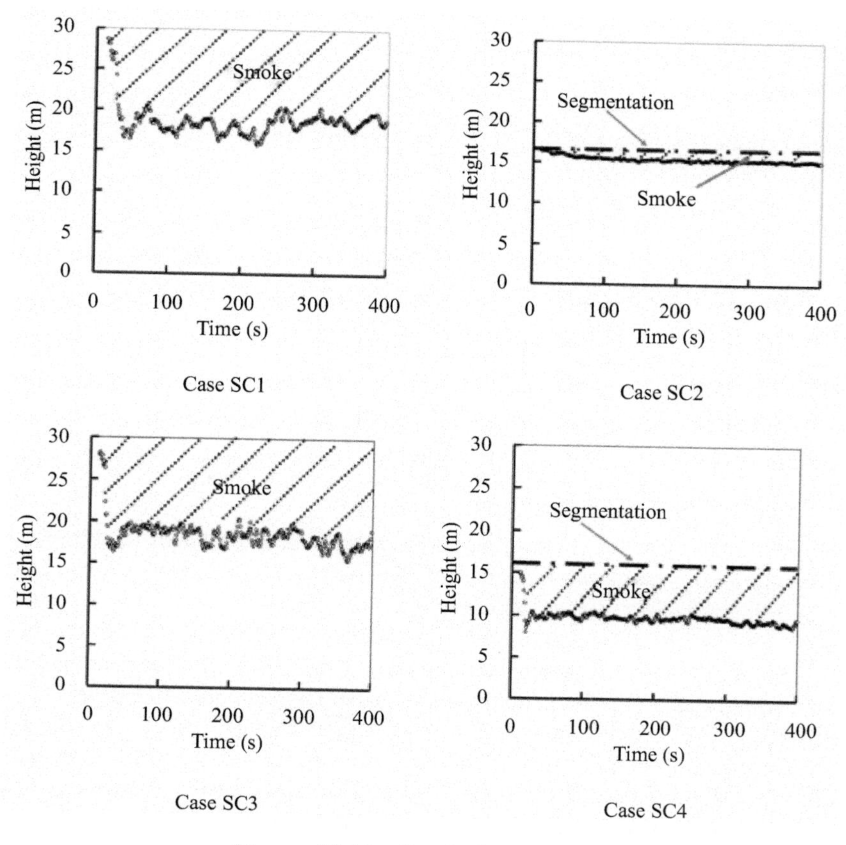

Figure 15.10 Smoke layer height

In BES, the monthly cooling load from June to September was simulated. These two ventilation design methods (the new ventilation design and traditional design) have similar energy performance in terms of reducing the cooling load. Compared with the cases with only MC, both traditional and new design methods with NV (traditional design-MNV and new design-MNV) can reduce about 18% of the total cooling load. The total NV airflow rates of the two design methods are almost the same. However, it should be noted that the total cooling load of the new design-MC case is around 22% higher than that of the traditional design-MC case, which is caused by the increased thermal mass of the segmentation slab and double internal heat gain. Although the new design has a higher total cooling load, the atrium with the new design can accommodate twice as many occupants as the traditional design, because of the segmentation slab.

15.4.2.3 Conclusion

This state-of-the-art novel ventilation design can contribute to better fire safety performance and maintain the NV potential. The results of CFD simulations and BESs proved that the atrium with the novel ventilation design has a larger smoke-free space and almost the same energy savings as NV.

References

Ahrens M. High-rise building fires. 2016. https://doi.org/10.20965/jdr.2007.p0236.

Ai ZT, Mak CM, Niu JL, and Li ZR. The assessment of the performance of balconies using computational fluid dynamics. *Build Serv Eng Res Technol*. 2011;32:229–43. https://doi.org/10.1177/0143624411404646.

Artmann N, Manz H, and Heiselberg P. Climatic potential for passive cooling of buildings by night-time ventilation in Europe. *Appl Energy*. 2007;84:187–201.

ASHRAE. *ASHRAE Handbook of Fundamentals*. Atlanta, GA: American Society of Heating, Refrigerating and Air-Conditioning Engineers; 2017. https://doi.org/10.1017/CBO9781107415324.004.

Babrauskas V. Upholstered furniture heat release rates: Measurements and estimation. *J Fire Sci*. 1983;1:9–32.

Chen H, Liu N, Zhang L, Deng Z, and Huang H. Experimental study on cross-ventilation compartment fire in the wind environment. *Fire Saf Sci*. 2008;9:907–18. https://doi.org/10.3801/iafss.fss.9-907.

Chen Y, Norford LK, Samuelson HW, and Malkawi A. Optimal control of HVAC and window systems for natural ventilation through reinforcement learning. *Energy Build*. 2018;169:195–205. https://doi.org/10.1016/j.enbuild.2018.03.051.

Chen Y, Tong Z, and Malkawi A. Investigating natural ventilation potentials across the globe: Regional and climatic variations. *Build Environ*. 2017;122:386–96.

Ding W, Hasemi Y, and Yamada T. Natural ventilation performance of a double-skin façade with a solar chimney. *Energy Build*. 2005;37:411–8. https://doi.org/10.1016/j.enbuild.2004.08.002.

EN 12101-3 Standard. Smoke and heat control systems—Part 3: Specification for powered smoke and heat exhaust ventilators. Brussels, Belgium. 2002.

Goel S, Rosenberg M, Athalye R, and Xie Y. Enhancements to ASHRAE Standard 90.1 prototype building models. 2014.

Günel MH and Ilgin HE. Tall buildings: Structural systems and aerodynamic form. 2014. https://doi.org/10.4324/9781315776521.

Heiselberg P. *Principles of Hybrid Ventilation*. Aalborg: Aalborg University; 2002.

Hu J and Karava P. Model predictive control strategies for buildings with mixed-mode cooling. *Build Environ*. 2014;71:233–44.

IEA. Annex 62 ventilative cooling design guide. 2018.

Klote JH and Milke JA. Principles of smoke management. 2002.

Kolokotroni M and Aronis A. Cooling-energy reduction in air-conditioned offices by using night ventilation. *Appl Energy*. 1999;63:241–53. https://doi.org/10.1016/S0306-2619(99)00031-8.

Lay S. Pressurization systems do not work & present a risk to life safety. *Case Stud Fire Saf*. 2014;1:13–7. https://doi.org/10.1016/j.csfs.2013.12.001.

Li Y and Heiselberg P. Analysis methods for natural and hybrid ventilation-a critical literature review and recent developments. *Int J Vent*. 2003;1:3–20. https://doi.org/10.1080/14733315.2003.11683640.

Malkawi A, Yan B, Chen Y, and Tong Z. Predicting thermal and energy performance of mixed-mode ventilation using an integrated simulation approach. *Build Simul*. 2016;9:335–46.

McGrattan K, McDermott R, Weinschenk C, and Overholt K. Fire dynamics simulator user's guide. Gaithersburg, MD. 2013.

Medved S, Babnik M, Vidrih B, and Arkar C. Parametric study on the advantages of weather-predicted control algorithm of free cooling ventilation system. *Energy*. 2014;73:80–7. https://doi.org/10.1016/j.energy.2014.05.080.

Miller RS and Beasley D. On stairwell and elevator shaft pressurization for smoke control in tall buildings. *Build Environ*. 2009;44:1306–17. https://doi.org/http://dx.doi.org/10.1016/j.buildenv.2008.09.015.

Moosavi L, Mahyuddin N, and Ghafar N. Atrium cooling performance in a low energy office building in the Tropics, a field study. *Build Environ*. 2015;94:384–94. https://doi.org/10.1016/j.buildenv.2015.06.020.

NFPA. NFPA 101 life safety code. Quincy, MA: NFPA. 2018.

NFPA. NFPA 13: Standard for the installation of sprinkler systems. Quincy, MA: National Fire Protection Association. 2002.

NFPA. NFPA 92: Standard for control smoke systems. Quincy, MA: NFPA. 2012.

OECD/IEA. The future of cooling opportunities for energy-efficient air conditioning. 2018.

Sha H and Qi D. A review of high-rise ventilation for energy efficiency and safety. *Sustain Cities Soc*. 2020a;54:101971. https://doi.org/10.1016/j.scs.2019.101971.

Sha H and Qi D. Energy and fire safety performance of atrium ventilation in high-rise buildings. IBPSA 2019, Rome. 2019.

Sha H and Qi D. Investigation of mechanical ventilation for cooling in high-rise buildings. *Energy Build*. 2020b;228:110440. https://doi.org/10.1016/j.enbuild.2020.110440.

SkyscraperPage.com. Global cities & buildings database. 2019. [cited 2019 Sep 18]. Available from: http://skyscraperpage.com/.

DH Toronto Staff. The 21 tallest buildings under construction in Toronto right now. *Daily Hive*. 2018.

Thunderbird Engineering. PyroSim user manual. 2015.

Wang L, Wong NH, and Li S. Facade design optimization for naturally ventilated residential buildings in Singapore. *Energy Build*. 2007;39:954–61. https://doi.org/10.1016/j.enbuild.2006.10.011.

Wood A and Salib R. Natural ventilation in high-rise office buildings. Chicago. 2013.

Yuan S, Vallianos C, Athienitis A, and Rao J. A study of hybrid ventilation in an institutional building for predictive control. *Build Environ*. 2018;128:1–11.

Zhang Y, Wang X, and Hu E. Optimization of night mechanical ventilation strategy in summer for cooling energy saving based on inverse problem method. *J Power Energy*. 2018;232:1093–102. https://doi.org/10.1177/0957650918766691.

Zhou C, Wang Z, Chen Q, Jiang Y, and Pei J. Design optimization and field demonstration of natural ventilation for high-rise residential buildings. *Energy Build*. 2014;82:457–65. https://doi.org/10.1016/j.enbuild.2014.06.036.

Chapter 16
Urban ventilation and design

Zhengtao Ai[1] *and Cheuk Ming Mak*[2]

16.1 Introduction

Urban ventilation is important for achieving not only a circulated outdoor air, good pedestrian-level air quality, and thermal comfort, but also a good outdoor boundary condition of building indoor ventilation. In the past decades, urban ventilation has been widely investigated from different physical scales, namely, from street canyon, precinct to full city. This chapter presents studies and discussions on the street canyon and precinct scales, which are in the order of approximately 10–1,000 m.

16.2 Street-canyon-scale ventilation

16.2.1 Street canyon microclimates

Microclimate in urban street canyons, including particularly flow, temperature, and pollutants, first represents urban climate and second has a close association with the indoor air quality in their nearby naturally ventilated buildings. Understanding urban microclimate is the basic prerequisite of understanding and thus improving both urban environment and indoor air quality in naturally ventilated urban buildings. Figure 16.1 schematically presents the association between indoor environment quality (IEQ) in naturally ventilated buildings and the microclimate in their nearby street canyon through the breath process at building envelopes.

16.2.1.1 Airflow

Flow patterns inside a street canyon are very distinctive in response to different ambient wind conditions above the canyon (Ghiaus and Allard, 2005; Oke, 1987; Georgakis and Santamouris, 2006; Nakamura and Oke, 1988; Arnfield and Mills, 1994; Longley *et al.*, 2004; Georgakis and Santamouris, 2004; Eliasson *et al.*, 2006), which are particularly associated with the along and the across canyon velocity components above the canyon. Under a parallel ambient flow, if the wind speed above the canyon is high, a strong relationship between the outside and the

[1]College of Civil Engineering, Hunan University, Changsha, China
[2]Department of Building Services Engineering, The Hong Kong Polytechnic University, China

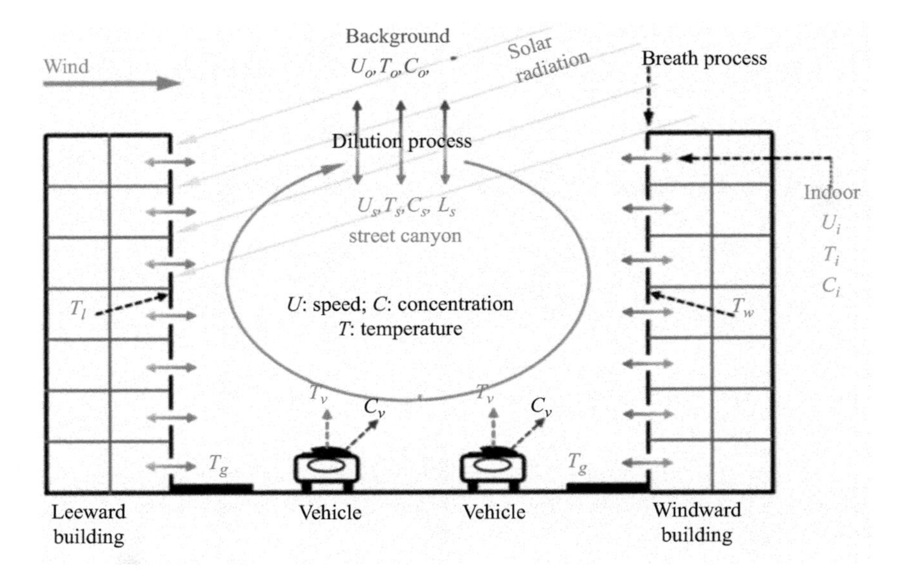

Figure 16.1 Schematic view of the association between the IEQ in naturally ventilated buildings and the microclimate in their nearby street canyon; the subscripts o, s , i , v , w , and g of the U , T , and C indicate outside the canyon (background), street canyon, indoor, vehicle, windward facade, leeward facade, and ground, respectively (Ai and Mak, 2015)

inside flows can be found. Wind inside the canyon flows along the canyon axis with possible uplifts along the building walls, due to the friction of building walls and ground surface. Under a perpendicular ambient flow, if the wind speed above the canyon is high, the canyon flow can be seen as a secondary circulatory flow driven by the imposed flow above the canyon. Under an oblique ambient flow, which occurs in most time, helical vortexes could be developed along street canyons. For all ambient wind directions, if the wind speed above the canyon is low, the coupling between the upper flow and the canyon flow is lost, where thermal flows and mechanical disturbances play a significant role in shaping the wind flow patterns inside the canyon.

The threshold ambient wind speed depends on the canyon AR (ratio of the mean building height H to the street width W). According to DePaul and Sheih (1986), who worked with a street canyon with AR equal to 1.4, the threshold value to establish the coupling and to form circulatory vortexes inside the canyon is between 1.5 and 2.0 m/s. Similar values are reported by Nakamura and Oke (1988), Arnfield and Mills (1994), Yamartino and Wiegand (1986), and Santamouris *et al.* (1999). They worked with canyons having aspect ratios ranging from 1 to 1.52. However, as shown by Santamouris *et al.* (1999), for canyons with a higher AR (that is 2.5 in their study), the coupling is established under a much higher ambient wind speed ranging approximately between 4 and 5 m/s (corresponding to the across canyon wind speed equal to 2–3 m/s). It was also stated that the correlation between the wind speed inside and above the canyon is unclear when the ambient wind speed was below this range (Santamouris *et al.*, 1999). The value 4 m/s was defined as the threshold wind speed in their later studies (Georgakis and Santamouris, 2008; Santamouris *et al.*, 2008) of predicting wind speeds inside street canyons. Comparison of these results may imply that the threshold wind speed increases with the increase of AR value. This should be explained by a faster wind that is required to penetrate into a deeper street canyon. However, the influence of other factors, such as meteorological condition and buoyancy force, on this threshold cannot be revealed from these studies (Nakamura and Oke, 1988; Arnfield and Mills, 1994; Eliasson *et al.*, 2006; DePaul and Sheih, 1986; Yamartino and Wiegand, 1986; Santamouris *et al.*, 1999, 2008; Georgakis and Santamouris, 2008).

16.2.1.2 Air temperature

Air temperature inside a street canyon strongly determines the cooling potential of natural ventilation of its nearby buildings (Georgakis and Santamouris, 2006), while temperature difference between canyon surfaces is an important driving force of flow movements in a canyon, particularly when the ambient wind flow is slow.

Surface temperatures of a canyon are closely associated with local climate, urban morphology, thermal properties of building and street materials, canyon orientation, canyon AR, and sky view factor. Figure 16.2 presents a schematic view of the thermal environment inside a street canyon.

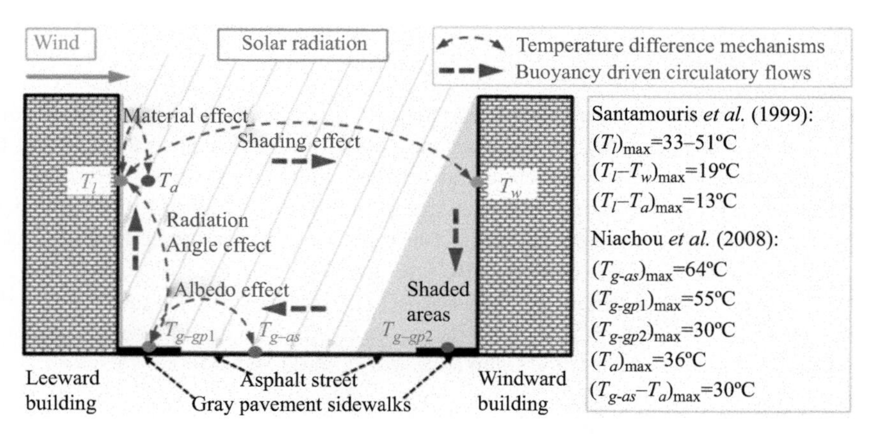

Figure 16.2 A general view of thermal environment inside a street canyon, where the major causes of temperature differences and the maximum temperatures and temperature differences reported in two references (Santamouris et al., 1999; Niachou et al., 2008) are provided; symbols T_a, $T_{g\text{-}gp1}$, $T_{g\text{-}gp2}$, and $T_{g\text{-}as}$ indicate temperature of air, temperature of sunlit ground with gray pavement, temperature of shaded ground with gray pavement, and temperature of sunlit ground with asphalt street, respectively

In general, due to different solar incidence angles, horizontal ground surfaces have much higher temperatures than vertical wall surfaces (Niachou *et al.*, 2008; Pearlmutter *et al.*, 1999; Bourbia and Awbi, 2004; Eliasson, 1990, 1996). With a same solar incidence angle, a surface with a lower albedo can reach a much higher surface temperature than that with a higher albedo (Niachou *et al.*, 2008). The change of solar incidence angle with time could lead to a significant change in surface temperatures inside a canyon, with daily amplitudes up to 35°C, while the maximum temperature between two opposite building walls with and without direct solar radiation was observed to be up to 19°C (Santamouris *et al.*, 1999). In addition, a canyon with a higher AR value has a more restricted sky view factor and thus can produce a stronger shading effect to solar radiation (Qin and Kot, 1993) than a canyon with a lower AR value.

Air temperature inside a street canyon is much lower (up to 13°C) than the temperature on canyon surfaces of direct solar radiation (Santamouris *et al.*, 1999). However, in contrast to surface temperature, canyon configurations and surface materials do not cause an obvious influence in air temperature (Andreou and Axarli, 2012), provided that the AR value is not extremely high (e.g., higher than 7–10 in Givoni (1998)). Slightly lower air temperatures (by 1°C–3°C) were observed in narrower street canyons with AR ranging from 2 to 4 than those in wider street canyons with AR ranging from 0.7 to 0.9 during the daytimes, while such temperature differences were 3°C–4°C during the nighttime (Andreou and Axarli, 2012). It is thus believed that the average air temperature is determined dominantly by large-scale regional factors rather than street-scale factors (Pearlmutter *et al.*, 1999; Barring *et al.*, 1985; Shashua-Bar and Hoffman, 2003). Vertically, no pronounced temperature stratification as a function of canyon's height was observed, which should be attributed to the strong mixing and advective phenomena inside the canyon (Santamouris *et al.*, 1999). However, due to shading effect, the air temperature inside a deep canyon with the AR equal to 3.3 during the daytime was observed to be 3°C–5°C lower than the corresponding air temperature above the canyon (Georgakis and Santamouris, 2006). In addition, the heat island effect, namely, the air temperature difference between a monitoring location inside the canyon and a background station, was mostly close to 2°C (Georgakis and Santamouris, 2006; Andreou and Axarli, 2012). However, the air temperature difference between the monitoring location above the canyon and the background one was between 4°C and 11°C (Mihalakakou *et al.*, 2002, 2004).

16.2.1.3 Traffic pollutants

Due to the increased traffic emissions and/or adverse dispersion conditions (Weber *et al.*, 2006), pollutants can accumulate to reach very high levels in street canyons in comparison with background concentrations.

Under a specific environmental condition, a narrower canyon generally corresponds to a worse street-level air quality. Dispersion and dilution of airborne pollutants in a street canyon rely strongly on the exchange rate of flow between a canyon interior and its upper atmosphere. When the ambient wind speed above a canyon is too low to establish the coupling between the canyon interior and its

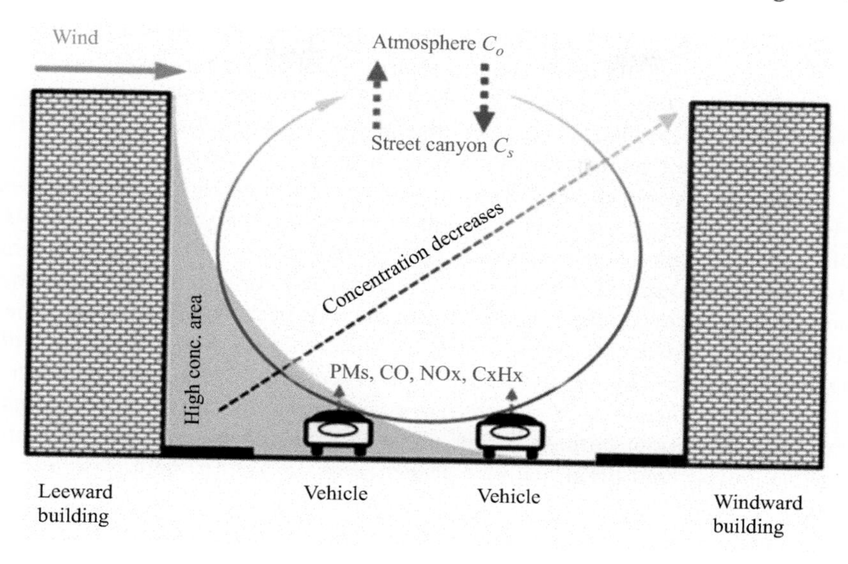

Figure 16.3 Major influencing factors of pollutants level and their distribution inside a street canyon: traffic intensity, mass exchange rate between street canyon and its upper atmosphere, and wind flow pattern

upper atmosphere, pollutants accumulate in the street canyon. On the contrary, when the ambient wind speed is sufficiently high to establish such a coupling, pollutants can be effectively diluted to the upper atmosphere.

Figure 16.3 schematically presents a spatial concentration distribution inside a street canyon. In general, horizontally, traffic-related pollutants maintain relatively high concentrations in a region near the leeward side of a canyon (Qin and Kot, 1993; Tsai and Chen, 2004; DePaul and Sheih, 1985; Vardoulakis *et al.*, 2002), while concentrations decrease vertically along the height above the ground on both sides of the canyon (Weber *et al.*, 2006).

The pollutant concentrations and pedestrian exposure levels in street canyons are closely associated with the traffic intensity (Kaur *et al.*, 2007; Westerdahl *et al.*, 2005; Rakowska *et al.*, 2014). Typically, high concentrations of traffic-related pollutants, such as PM2.5, PM10-2.5, black carbon (BC), and monocyclic aromatic hydrocarbons, in street canyons were reported in rush hours on working days (Cheng *et al.*, 2015; Wang *et al.*, 2002; Chan *et al.*, 2003).

16.2.2 Influence of street canyon configurations on building ventilation

A better building disposition, particularly a wider street, is very important to enhance air penetration and movement inside street canyons and thus potentially to improve natural ventilation performance in buildings (Bady *et al.*, 2011; Cheung and Liu, 2011).

A street canyon model formed by two parallel slab-like buildings is investigated in the previous study (see Figure 16.4(a)). Four aspect ratios (H/B), namely, 1.0, 2.0, 4.0, and 6.0, are considered, which all belong to the skimming flow regime (Ai and Mak, 2015; Oke, 1987). CFD model was used to calculate the flow field distribution and air change per hour (ACH) in the room.

Figure 16.5 presents the ACH values of rooms at both leeward and windward sides of the street canyon under different aspect ratios. It is obvious that the ACH values along height are not uniformly distributed. Figure 16.6 shows the average ACH values of rooms for different aspect ratios. ACH values on both the leeward and windward sides decrease with the increase of aspect ratio. Taking the case of AR = 1.0 as the base case, the percentage decreases of ACH of other cases with a higher AR are calculated. In general, a large decrease of ACH is observed when the aspect ratio is increased. However, such a decrease becomes slow gradually. Obviously, it is more difficult for the atmospheric flow above a street canyon to penetrate deeply into the inside of a deeper street canyon (namely, with a higher aspect ratio).

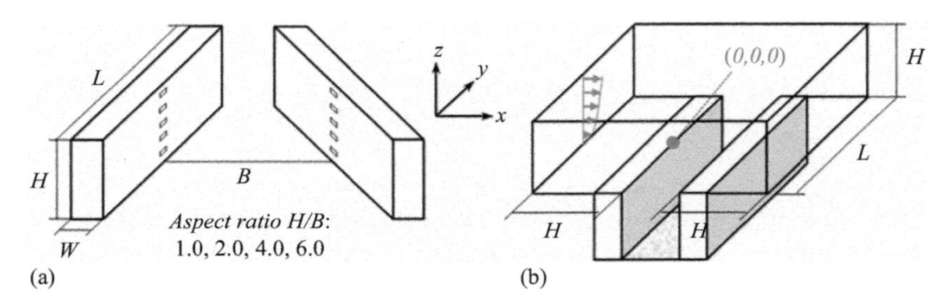

Figure 16.4 Schematic view of the street canyon model (a) and computational domain (b); note that the two buildings are parallel with each other

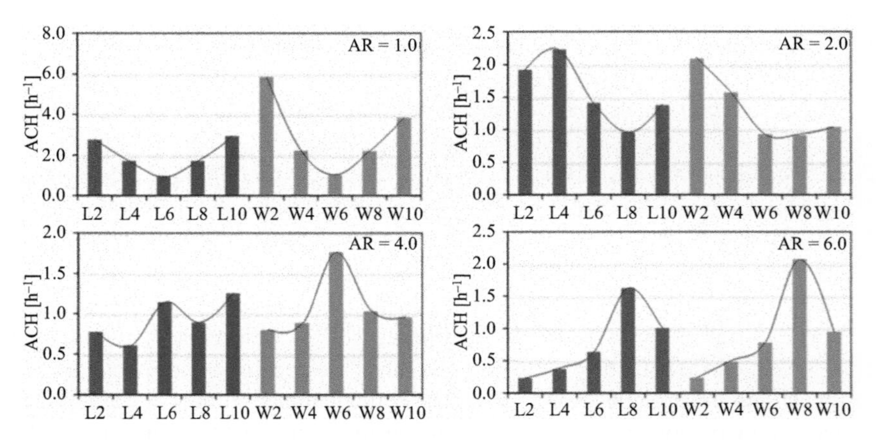

Figure 16.5 ACH values of rooms in buildings near the street canyon under different aspect ratios

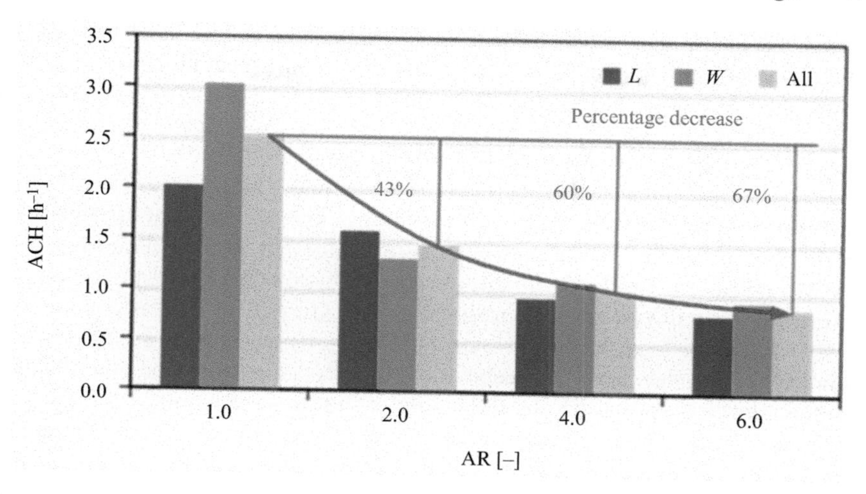

Figure 16.6 Average ACH values of all rooms at leeward facade (L), windward facade (W), and both facades (All), where the percentage decreases of ACH in comparison to the case of AR = 1.0 are also presented

Since the atmospheric flow above a street canyon is more difficult to penetrate deeply into a deeper street canyon, ventilation performance of buildings is decreased with the increase of aspect ratio of a street canyon. Compared to the case of AR = 1.0, the percentage decreases of ACH values are, on average, 43%, 60%, and 67% for the cases of AR = 2.0, 4.0, and 6.0, respectively. Influenced by flow pattern inside a street canyon, ACH values of rooms along the height of a building are not uniformly distributed. Such a distribution varies significantly with the change of aspect ratio. These findings (namely, ACH values and their distributions) suggest that the aspect ratio is an important parameter that should be considered when designing natural ventilation of urban buildings.

16.2.3 CFD simulation of street canyon aerodynamics

Owing to many advantages, CFD simulation has become the most widely used research method to study atmospheric processes in a street canyon. However, the literature review of previous CFD studies indicates arbitrary selections of computational domain, domain dimensions, and inflow boundary conditions. These arbitrary selections should be partially attributed to the fact that they are not included in the general Best Practice Guidelines for CFD simulation of urban aerodynamics. These computational settings have significant influences on the accuracy and reliability of the predicted flow field. It is well known that the accuracy and reliability of CFD simulations are strongly influenced by the computational settings, including the physical geometry, computational domain dimensions, grid quality, boundary conditions, solution methods, and convergence criteria. Therefore, the objective of this study is to evaluate the three computational settings, including the domain configurations, the dimensions of the domain, and the inflow boundary conditions, and then suggest appropriate selections from the viewpoints of both predictive accuracy and computational costs.

16.2.3.1 Evaluation of computational domain configuration

A non-exhaustive literature review of wind tunnel and CFD studies of atmospheric processes in street canyons in recent years shows that six different configurations (see Figure 16.7) of experimental/computational domain were commonly used. In addition to the Domain B, the other three computational domains, namely, Domain A, Domain E, and Domain F, are adopted to predict the flow field inside the same street canyon and the results are compared. The Domains C and D are theoretically excluded because they are first computationally more expensive than other domains and second influenced largely by the length of the street canyon.

Figure 16.8(a) presents the predicted velocity contours on the vertical center-plane of the street canyon using the four computational domains. For Domain A, owing to the effect of the flow impingement in front of the upstream building and the wake flow behind the downstream building, the flow field inside the street canyon is significantly changed when compared to that inside a street canyon with both upstream and downstream buildings (Domain B, the base case). When lifting up the upstream and downstream domain spaces to form a T-shape domain (Domain E), the flow field inside the street canyon is very close to that in Domain B. The flow field predicted using Domain F is still reasonably acceptable, although

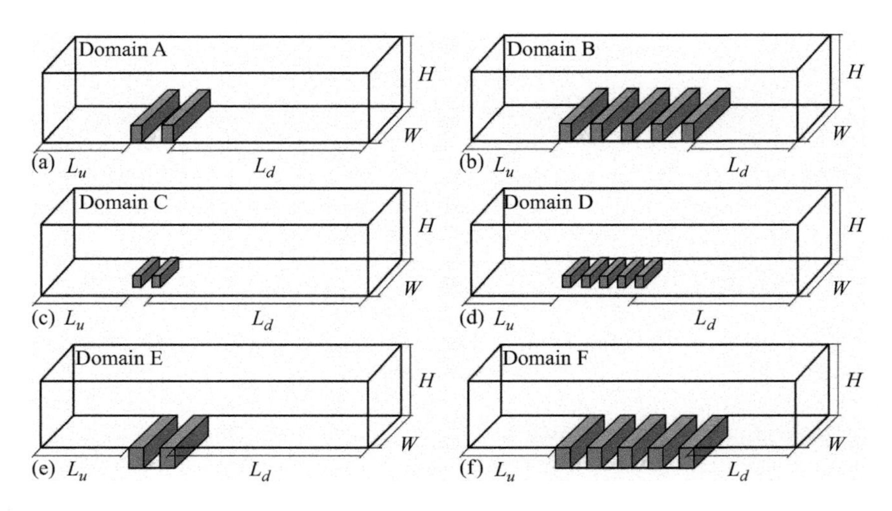

Figure 16.7 Schematic view of computational domains used in previous wind tunnel and CFD studies; in this figure, it should be noted that first (b), (d), and (f) represent those studies using more than two buildings in their street canyon configurations and second (a), (b), (e), and (f) may also represent those performing only two-dimensional CFD simulations

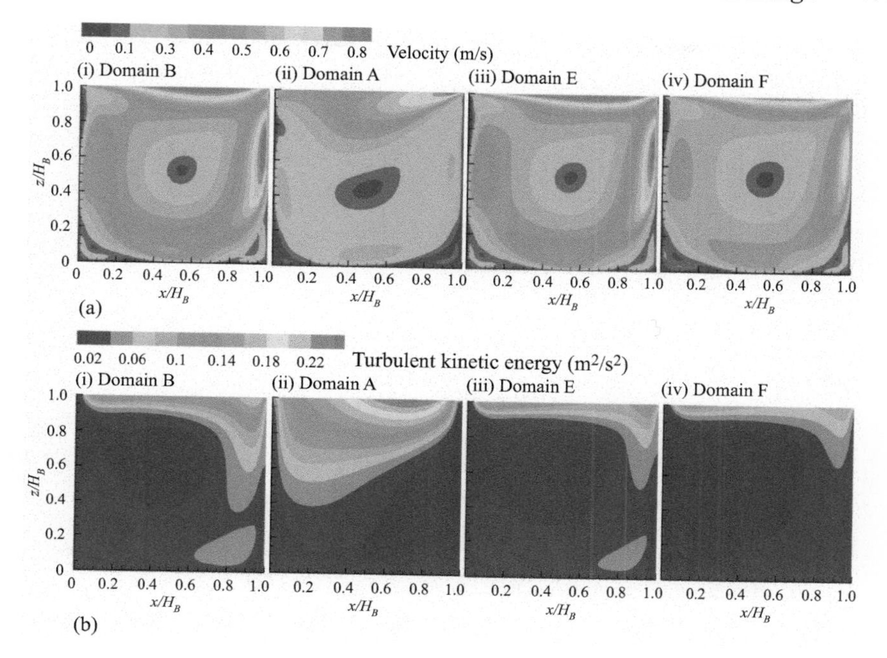

Figure 16.8 Evaluation of domain configurations: (a) velocity contour and (b) turbulent kinetic energy contour on the vertical centerplane ($y = 0$) of the street canyon model when using different street canyon configurations

the flow movements are slightly weaker than those in Domain B. Considering also the lower computational power needed by Domain E, the evaluations would suggest that it is reliable and accurate to use the Domain E (T-shape domain) to predict the flow field in a street canyon in urban areas.

16.2.3.2 Evaluation of inflow boundary conditions

In addition to the uniform inflow boundary conditions, a set of logarithmic law inflow boundary conditions (described by (16.1)–(16.3)) was employed and the predicted flow fields by these two were compared:

$$U = \frac{u^*}{\kappa} \ln\left(\frac{z + z_0}{z_0}\right) \tag{16.1}$$

$$k = \sqrt{M_1 \cdot \ln(z + z_0) + M_2} \tag{16.2}$$

$$\varepsilon = \frac{u^* \sqrt{C_\mu}}{\kappa(z + z_0)} \sqrt{M_1 \cdot \ln(z + z_0) + M_2} \tag{16.3}$$

Figure 16.9 presents the comparison of velocity components along a vertical and a horizontal line on the centerplane of the street canyon. The basic shape of the two lines is

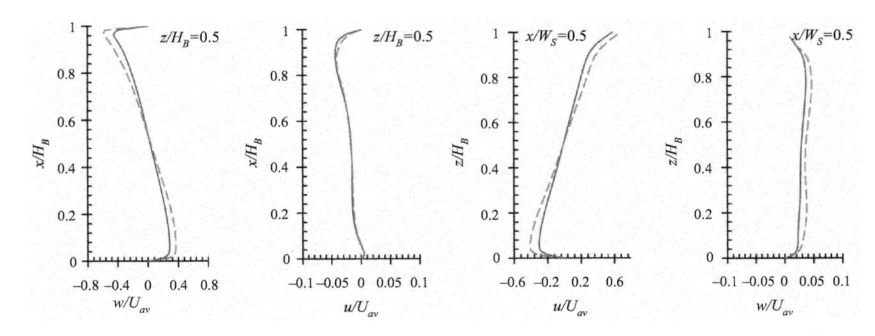

Figure 16.9 Evaluation of inflow boundary conditions: velocity components along the vertical line $x/W_S = 0.5$ and horizontal line $z/H_B = 0.5$ on the centerplane of the street canyon, where the blue lines denote the use of the uniform inflow boundary conditions and the dashed green lines the logarithmic law inflow boundary conditions

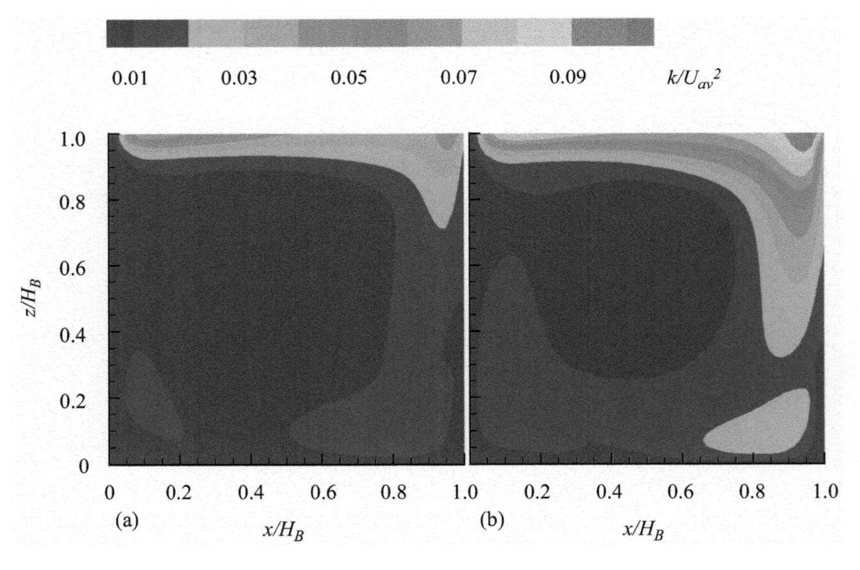

Figure 16.10 Evaluation of inflow boundary conditions: turbulent kinetic energy on the vertical centerplane of the street canyon: part (a) indicates the use of the uniform inflow boundary conditions and (b) the logarithmic law inflow boundary conditions

the same. However, the vortex inside the street canyon predicted using the logarithmic law inflow boundary conditions is stronger than that predicted using the uniform inflow boundary conditions. Figure 16.10 shows the nondimensional turbulent kinetic energy on the vertical centerplane of the street canyon. Despite the difference in values (probably due to different turbulent characteristics at inlet), the basic trend of turbulent kinetic energy predicted by the two types of inflow conditions is the same. In fact, the uniform inflow would change its shape along the computational domain before it reaches the top of the street canyon (Ai and Mak, 2013; Richards and Hoxey, 1993; Blocken *et al.*, 2007). This phenomenon on one hand occurs on building tops in practice and on the other hand would reduce the difference in the approaching flow between the two types of inflow boundary conditions. In addition, the roughness height at the building tops should be much smaller than that above urban ground, where the buildings themselves are roughness elements. Therefore, even if one would use the logarithmic law boundary conditions to represent the free atmospheric boundary layer above building tops, a very thin boundary layer (namely, with a relatively low roughness height) should be applied. In general, the comparison of the velocity and turbulent kinetic energy fields suggest that it is reasonable and appropriate to use the uniform inflow boundary conditions to predict the flow field inside a street canyon within a T-shape domain.

16.3 Precinct-scale ventilation

16.3.1 Pedestrian wind comfort for weak wind condition

16.3.1.1 New comfort criteria for weak wind condition

Outdoor human comfort is an important parameter to be considered in urban planning since it can not only be beneficial to the physical and mental health of residents but also can help alleviate energy consumption for residential buildings. Urban heat island and global warming have become a serious issue, especially in high-density cities at low- or mid-latitudes, such as Hong Kong and Singapore. There is a rising trend in average temperature in Hong Kong and the hot period has become significantly longer over the last century based on the finding of Chan *et al.* (2012). Thus, it is very useful to create a local cooling spot against the background of global warming. Successful achievement of the acceptable pedestrian level wind environment is seriously compromised by congested airflow in megacities. Moreover, existing wind comfort criteria aim at strong wind conditions, which cannot represent weak wind environment. Thus, in this chapter, a wind comfort criterion for weak wind condition (as shown in Table 16.1) was created. Note that this wind comfort criterion is proposed based on extensive literature review and scientific comparison. For more detailed information, please refer to Du *et al.* (2017b). The attributes of this wind comfort criterion are as follows: (i) the criteria are proposed seasonally; (ii) a low wind speed bound is used for summer criterion; (iii) the parameters are selected to adapt weak wind conditions.

Threshold wind velocity parameters

The mean wind velocity is used here as the threshold wind velocity parameter as it can readily combine with local meteorological data. The overall mean wind velocity (OMV) is utilized, which is calculated as

$$OMV = \sum_{i=1}^{16} F_i \times U_i \qquad (16.4)$$

Table 16.1 New wind comfort criteria for Hong Kong

Wind comfort criteria for summer (Jun–Aug)				
Category	**Threshold velocity**	**Exceedance probability (P_{exc}) (%)**	**Activity description**	**Remark**
Unfavorable	$OMV < 1.5$	50	N/A	No noticeable wind
Acceptable	$OMV < 1.8$	2	Sitting long	Light breeze
	$OMV < 3.6$	2	Sitting short	Gentle breeze
	$OMV < 5.3$	2	Strolling	Moderate breeze
Tolerable	$OMV < 7.6$	2	Walking fast	Fresh breeze
Intolerable	$OMV > 7.6$	2	Not suitable for activities	Strong breeze
Danger	$OMV > 15$	0.05	Dangerous	Gale
Wind comfort criteria for winter (Dec–Feb)				
Category	**Threshold velocity**	**Exceedance probability (P_{exc}) (%)**	**Activity description**	**Remark**
Acceptable	$OMV < 1.8$	2	Sitting long	Light breeze
	$OMV < 3.6$	2	Sitting short	Gentle breeze
	$OMV < 5.3$	2	Strolling	Moderate breeze
Tolerable	$OMV < 7.6$	2	Walking fast	Fresh breeze
Intolerable	$OMV > 7.6$	2	Not suitable for activities	Strong breeze
Danger	$OMV > 15$	0.05	Dangerous	Gale

where F_i is the probability of the approaching wind coming from i direction and U_i denotes the mean wind velocity for i direction.

Exceedance probability

The statistical data of the environmental wind follows the two-parameter Weibull distribution. The Weibull probability density function $p(U_{ri})$ for direction i is as follows:

$$p(U_{ri}) = F_i\left\{\frac{k_i}{c_i}\left(\frac{U_{ri}}{C_i}\right)^{k_{i-1}}\exp\left[-\left(\frac{U_{ri}}{c_i}\right)^{k_i}\right]\right\} \tag{16.5}$$

where k_i is the shape parameter, c_i is the scale parameter. They are correlated in the following way:

$$\bar{U}_{ri} = c_i\Gamma\left(1+\frac{1}{k_i}\right) \tag{16.6}$$

$$\sigma_i^2 = c_i^2\left[\Gamma\left(1+\frac{2}{k_i}\right) - \Gamma^2\left(1+\frac{1}{k_i}\right)\right] \tag{16.7}$$

where $\overline{U}_{ri}$ is the mean wind velocity, Γ is the gamma function, and σ_i^2 stands for the corresponding standard deviation.

The Weibull cumulative distribution:

$$P(U < U_{ri}) = F_i \left\{ 1 - \exp\left[-\left(\frac{U_{ri}}{c_i} \right)^{k_i} \right] \right\} \tag{16.8}$$

The overall probability of exceeding U_{ri} for 16 wind directions:

$$P_{exc}(U > U_{ri}) = \sum_{i=1}^{16} F_i \left\{ \exp\left[-\left(\frac{U_{ri}}{c_i} \right)^{k_i} \right] \right\} \tag{16.9}$$

16.3.1.2 Case study

The wind environment of the Hong Kong Polytechnic University (HKPolyU) campus is chosen for the case study. The previously proposed wind comfort criteria are used to evaluate its pedestrian level wind environment. To obtain pedestrian level wind environment, the wind tunnel tests were conducted in the low-speed section of CLP Power Wind/Wave Tunnel Facility of Hong Kong University of Science and Technology, see Figure 16.11. Sixteen wind directions were measured, and two wind profiles were used. Profile A for the incident wind directions of 0°, 45°, 90°, 112.5°, 135°, 180°, 202.5°, 225°, 292.5°, and Profile B for the rest, see Figure 16.11(b). Seventy Kanomax velocity sensors were used. The test scale was 1:200, and pedestrian level was 0.01 m above ground. The studied region was 10 km in prototype.

According to the proposed new wind comfort criteria for Hong Kong, there are three and two levels of wind comfort in summer and in winter, respectively. In hot and humid summer, some regions have weak wind and unfavorable for any activities, see Figure 16.12(a). Some places are favorable for sitting long in summer, like regions in red-dashed boxes in Figure 16.12(a). Some places have relatively higher velocity than other (green), which can only sit for short time. In temperate winter, the majority of regions are favorable for sitting long, see Figure 16.12(b). The region in green can only sit for short time because of high wind velocity.

To make a comparison, wind comfort at the HKPolyU campus assessed by the widely used standard NEN 8100 is shown here, see Figure 16.13. The whole campus is classified as grade A. This means that based on NEN 8100, the wind environment at the HKPolyU is good for all activities. It fails to reveal the unfavorable weak wind conditions in Hong Kong. This therefore proves that the proposed new comfort criteria are more accurate and useful to assess weak wind conditions.

16.3.2 Lift-up design

With rapid urbanization, there are more high-rise buildings and narrow streets, which causes serious heat island effect in densely populated urban areas. Heat island effect increases urban air temperature and decreases people's outdoor

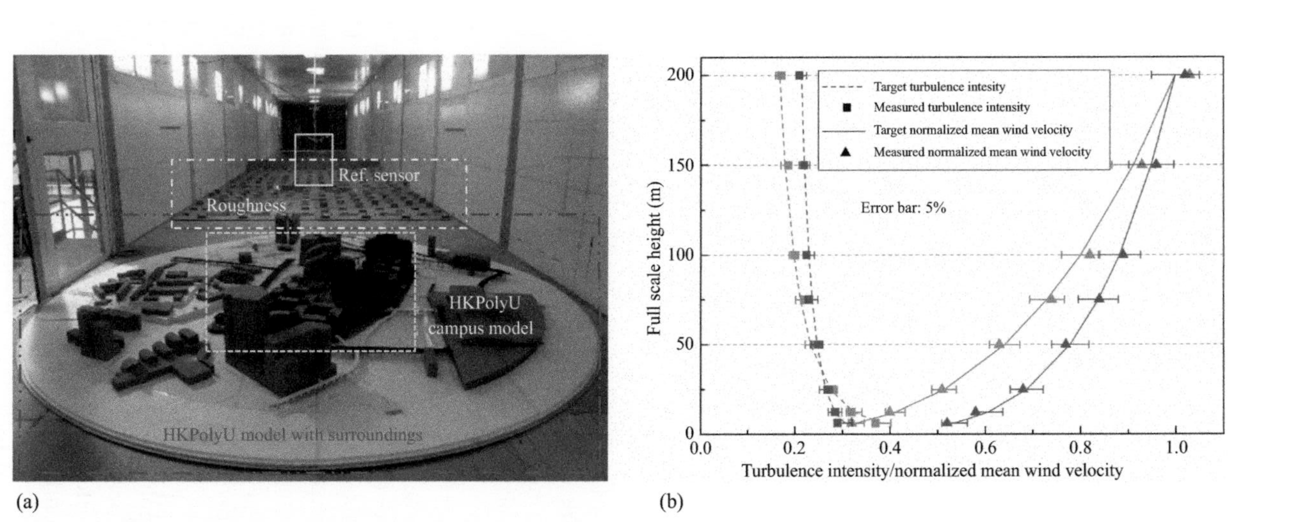

Figure 16.11 (a) Wind tunnel test photo; (b) wind profiles: red for Profile A, blue for Profile B (Du et al., 2017b)

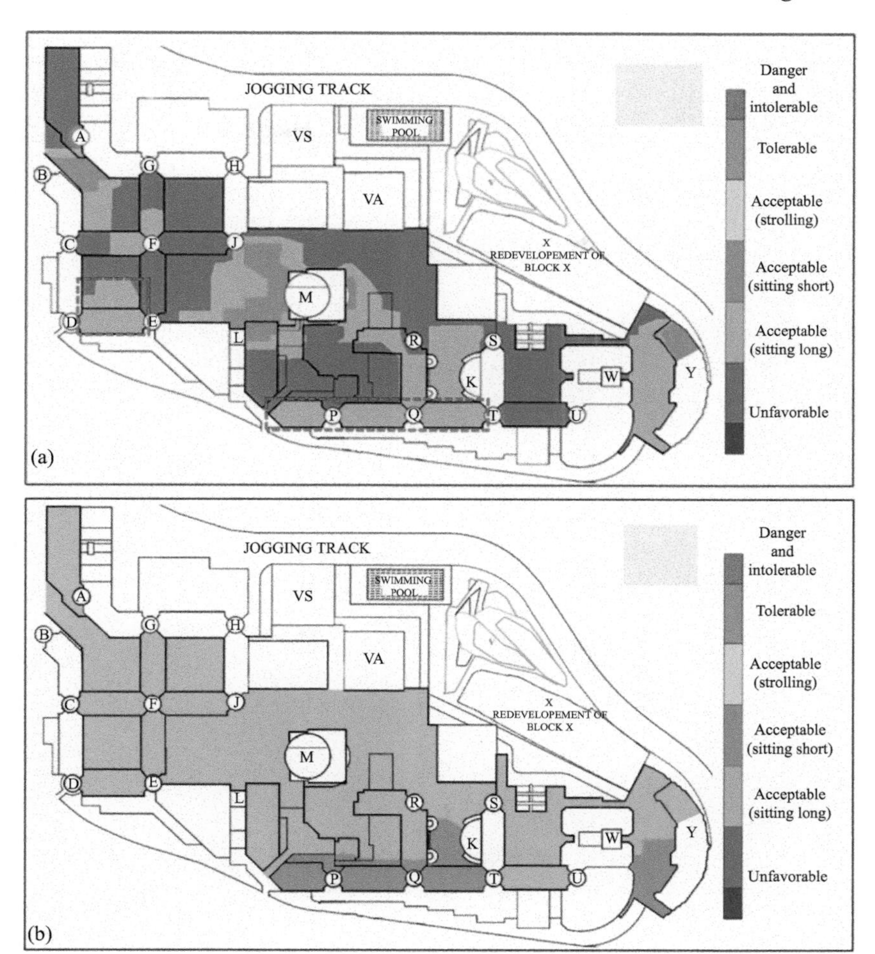

Figure 16.12 Assessment results of wind comfort: (a) summer and (b) winter (Du et al., 2017b)

activities. Methods that can create cooling spots would encourage people going for outdoor activities. Since the geometric characteristics of buildings can modify airflow patterns at pedestrian level, the lift-up building designs were proposed to create such a cooling place, where direct solar radiation is shaded and wind could blow though. Consequently, the lift-up design (see Figure 16.14), where the main structure is elevated by pillars or a combination of columns and shear walls, is a potential solution for improving low wind environment at pedestrian level.

16.3.2.1 Isolated building with lift-up design

In order to have a generic study of the effect of lift-up design on wind environment at pedestrian level, an isolated building with lift-up design was examined. The

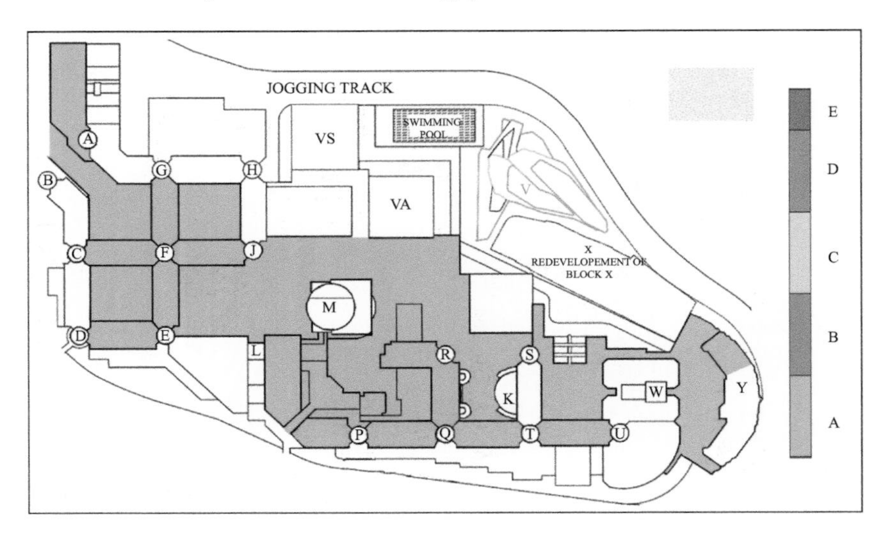

Figure 16.13 Assessment results of wind comfort by NEN 8100 (Du et al., 2017b)

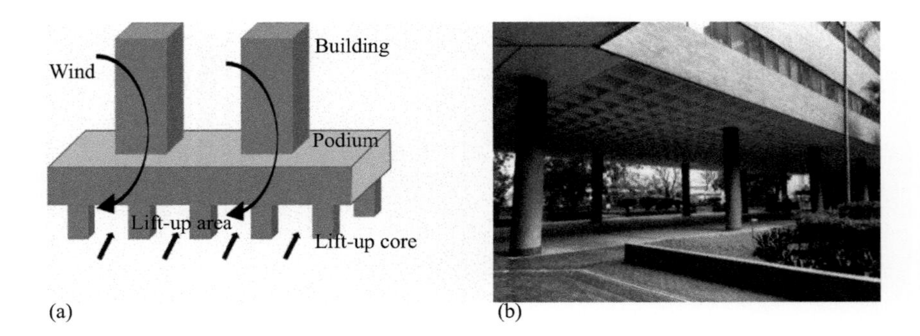

Figure 16.14 (a) The lift-up design (schematic figure); (b) photo of the lift-up design in a university campus (Du, 2018)

mean velocity ratio (*MVR*) was usually used as an index to evaluate pedestrian level wind environment, which is defined as

$$MVR = \frac{U}{U_r} \tag{16.10}$$

where U denotes mean wind velocity and U_r is the reference mean wind velocity of the approaching flow. Note, $U_r = 5$ m/s is used in this chapter.

There are three zones in Figure 16.15 both around the building with and without lift-up design: upstream low wind velocity (ULWV) zone, lateral high wind velocity (LHWV) zone, and downstream far-filed low wind velocity (DFLWV) zone. Two different zones around these two building configurations: the

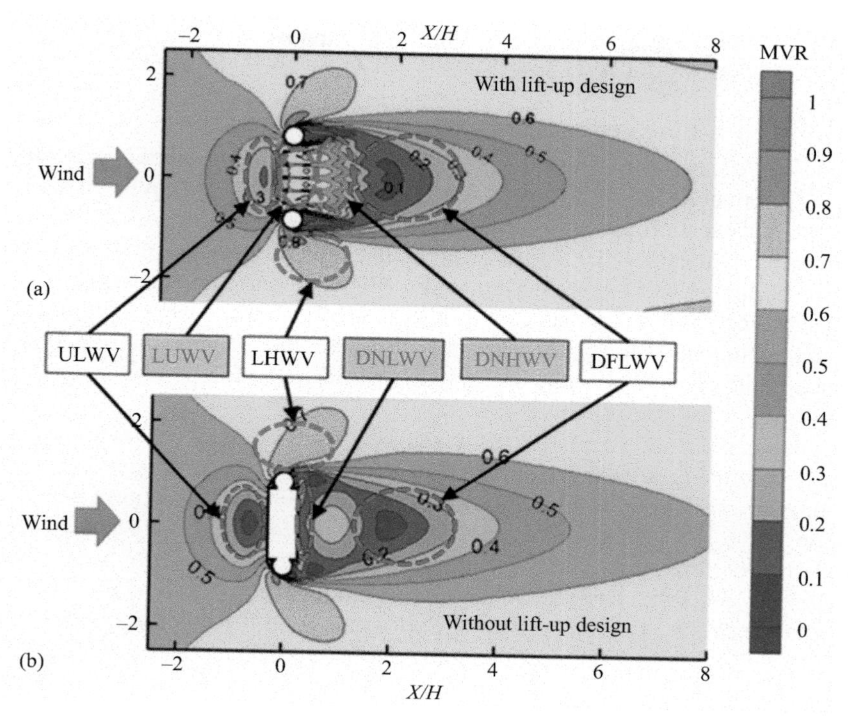

Figure 16.15 The MVR distributions at pedestrian level: (a) with lift-up design; (b) without lift-up design (Du et al., 2017c)

downstream near-filed low wind velocity (DNLWV) zone is replaced by near-field high wind velocity(DNHWV) zone for building with lift-up design. This is because airflow can pass through the lift-up area directly. Another one is a local wind amplification zone underneath the elevated building: lift-up wind velocity (LUWV) zone, which can be explained by the Venturi effect.

16.3.2.2 Different shaped building with lift-up design

Followed by the isolated building, the effects of lift-up design in more complicated and practical buildings with different shapes were further studied (Figure 16.16). In total, four building shapes were considered, including "-," "L," "U," and "□." For "-" shaped building, the values of *MVR* for LHWV, LUWV, and DNLWV zones are over 0.3, which is over 1.5 m/s according to (16.10). According to wind comfort established in Section 16.3.1, these zones can be considered wind comfort in low wind environment. For "L" shaped building, it is wind comfort in LHWV, LUWV and DNLWV zones for the upstream building. However, the LUWV zone has low wind condition for the lateral building due to the blockage of upstream column. These conclusions also apply for the "U" shaped building because of its symmetrical configuration. For the "□" shaped building, the wind environment of LUWV zone is generally comfortable. The values of *MVR* are overall higher in the LUWV zone for the upstream building than that of the downstream building.

16.3.2.3 A university campus with lift-up design

The results obtained from using isolated buildings provide important and fundamental information on the effect of lift-up design on pedestrian level wind environment. This section presents the study and results based on a real-life building community, which is the HKPolyU campus. The light-dashed areas in Figure 16.17 are the buildings with lift-up design in the HKPolyU campus, and the letters are representations of building names. For instance, “A” means A Core building. Two different colors of monitor points are indicated here: the orange monitor points are located in lift-up regions, while blue monitor points located in podium regions (not in lift-up regions).

To clearly understand the influence of lift-up design in a complex university campus (HKPolyU), the results for campus without lift-up design (the lift-up part was blocked for this case) are also presented in Figure 16.18(b). Figure 16.18 can provide an overall and generalized picture of wind environment in the campus for 16 wind directions. The results of *MVR* with lift-up design are generally larger than the case without lift-up design: the range of *MVR* is mostly from 0.1 to 0.46 for the condition with lift-up design while from 0.1 to 0.41 without lift-up design. Thus, the lift-up design can have an evident influence on the wind environment at pedestrian level even for a complex precinct like the HKPolyU campus.

16.3.3 CFD simulation of precinct-scale ventilation

CFD simulation, among others, is an important research and design method of precinct-scale ventilation, which has many advantages and is widely used in both academic and industrial fields. However, quality and efficiency in CFD simulation of pedestrian level wind environment in a complex urban area are always concerned by scholars and urban planners. Works that can improve this still remain of great significance. This section introduces briefly some key parts that influence largely quality and efficiency of CFD simulation, which include mesh generation and model validation.

16.3.3.1 Mesh generation method

The resolution and quality of mesh system is essentially important for an accurate reproduction of wind flow in a complicated urban environment, especially when the near-ground pedestrian level wind environment is the concern. However, generating high-quality and high-resolution grid for a domain with complicated structures is certainly not straightforward. An efficient and systematic mesh generation method is therefore developed to have a full control over local grids while generating high-quality grid in the computational domain, see Figure 16.19. This contains three parts: mesh generation preparation, mesh generation technique, and near-wall mesh generation technique. Each step is explained in Figure 16.19.

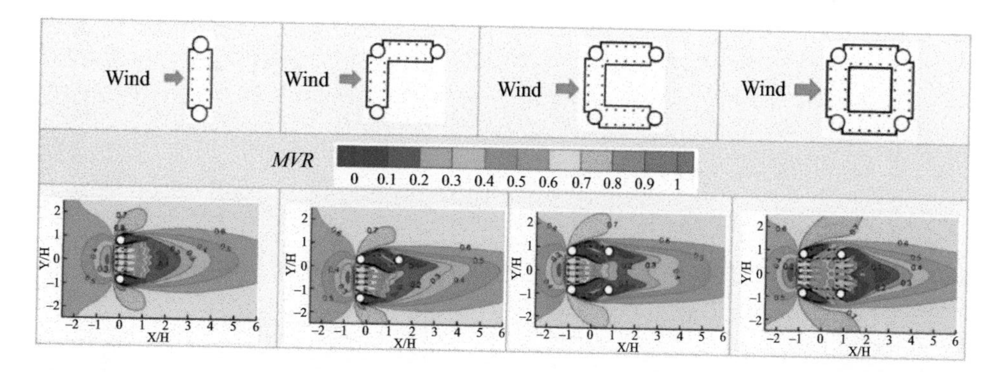

Figure 16.16 The MVR distributions at pedestrian level around different shaped buildings with lift-up design (Du et al., 2017c)

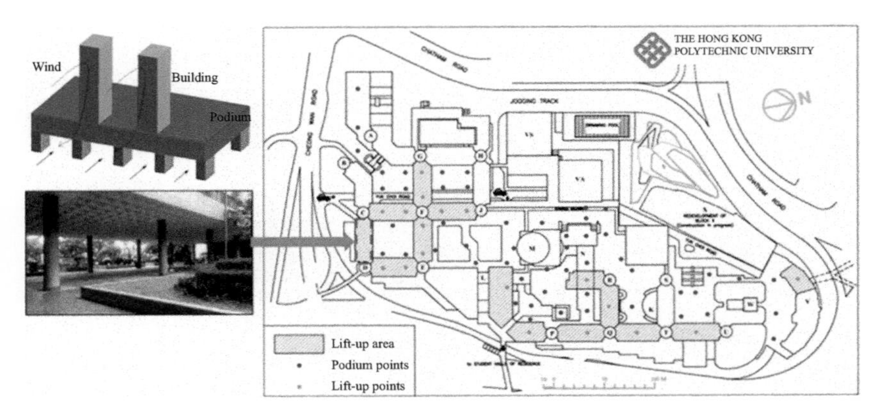

Figure 16.17 Lift-up design in the Hong Kong Polytechnic University (HKPolyU) campus

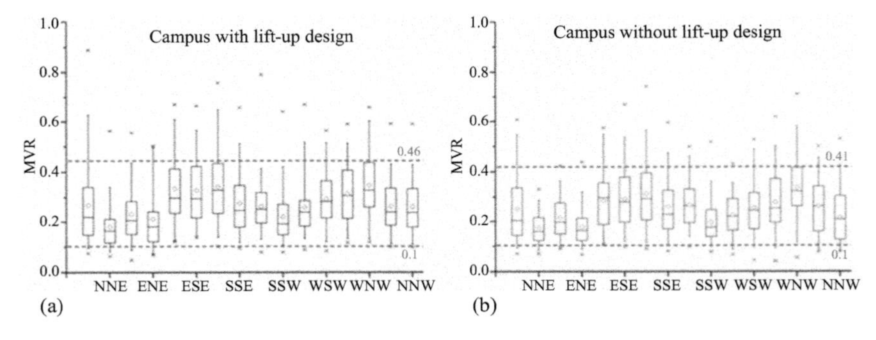

Figure 16.18 Box plots of MVR results at pedestrian level for 16 incident wind directions: the box edges represent the 25th and 75th percentiles, the whiskers for the 5th and 95th percentiles, the lines in the boxes for median values, and the symbols (◊) for mean values: (a) with lift-up design; (b) without lift-up design (Du et al., 2017a)

16.3.3.2 Computational model and grid

The HKPolyU campus refers in the previous chapters is used as the computational model. This model is constructed in a 1:200 scale, which is the same as in the wind tunnel tests. The geometrical complexity of the campus is modeled in great detail in the computational model, and any configurations more than 1 m in prototype were reproduced. Obviously, the campus is a very complicated urban area and the proposed mesh generation method is therefore adopted to construct the mesh of the campus model. Overviews of the mesh and its corresponding image from map are presented in Figure 16.20(a) and (b). The specific view of the computational cells for lift-up design and its corresponding photos are presented in Figure 16.20(c) and (d).

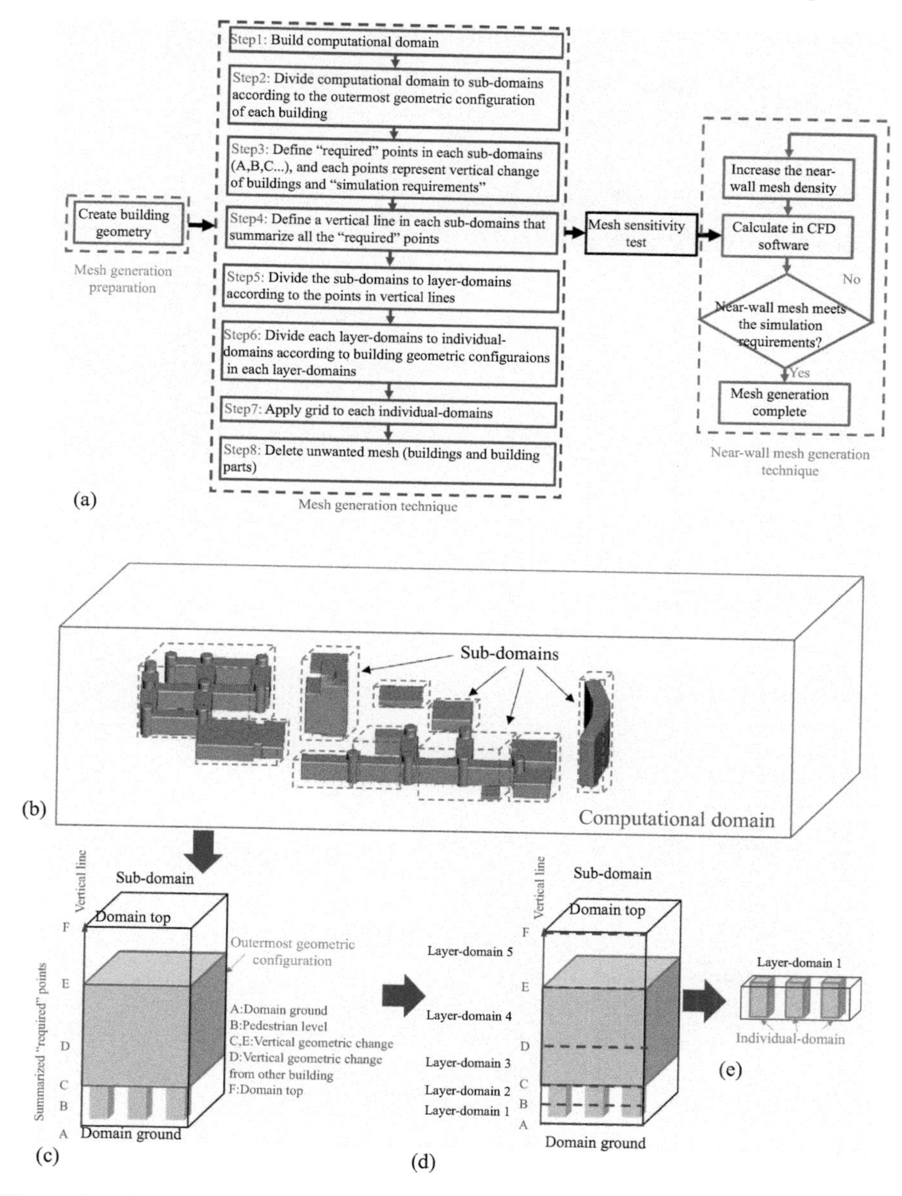

Figure 16.19 (a) Workflow of the mesh generation method; (b) schematic illustration of Step 2; (c) Steps 3 and 4; (d) Step 5; and (e) Step 6 (Du et al., 2018)

At least ten cells have been applied over the height of the lift-up, which suggests that there are over five cells at pedestrian level. The computational grid has high-quality and high-resolution all over the computational campus terrain, which

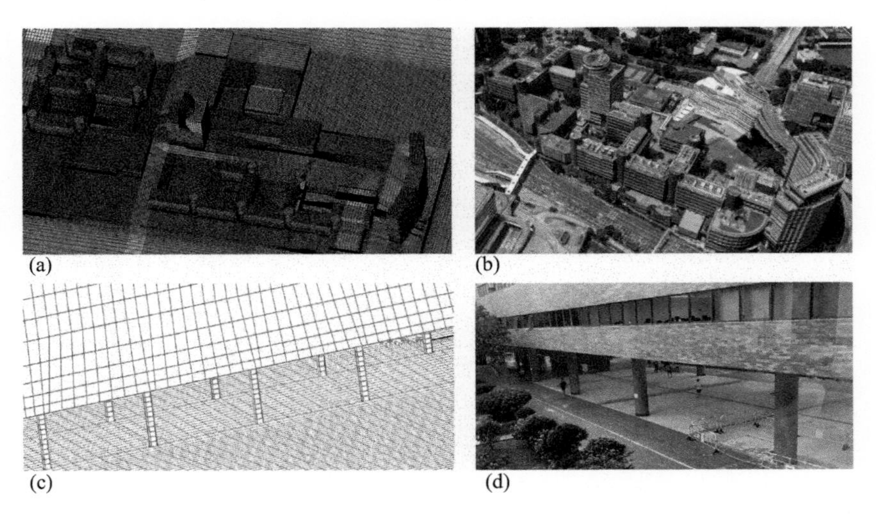

Figure 16.20 (a) Overview of high-quality computational mesh (8.9 million cells); (b) satellite image from the Google Map (Google Map, 2020); ((c) and (d)) grids for lift-up design

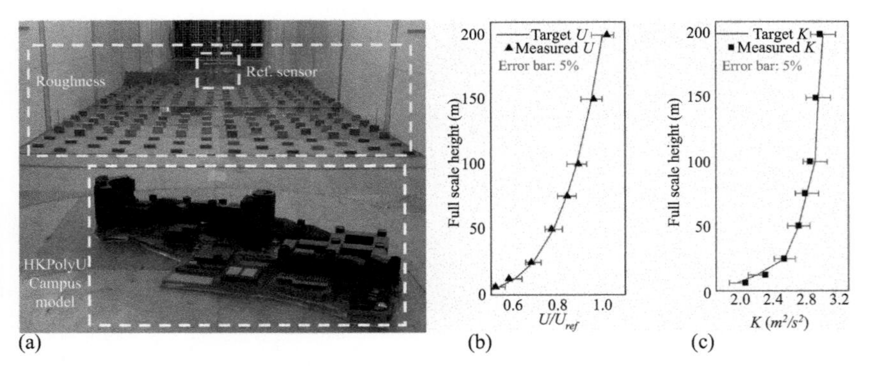

Figure 16.21 (a) Wind tunnel test photo; (b) approaching profile of mean wind velocity; and (c) turbulent kinetic energy (Du et al., 2018)

provide the prerequisite for accurately reproducing wind flow at pedestrian level in the lift-up area.

16.3.3.3 Wind tunnel test and validation

To validate the CFD simulation, the wind tunnel tests of corresponding campus model were carried out, see Figure 16.21. Fifty measurement points at the pedestrian level (0.01 m) were used for this validation.

The inlet boundary profiles for CFD simulation were interpreted from the wind tunnel tests. Zero normal gradients were utilized on the ceiling and lateral domain.

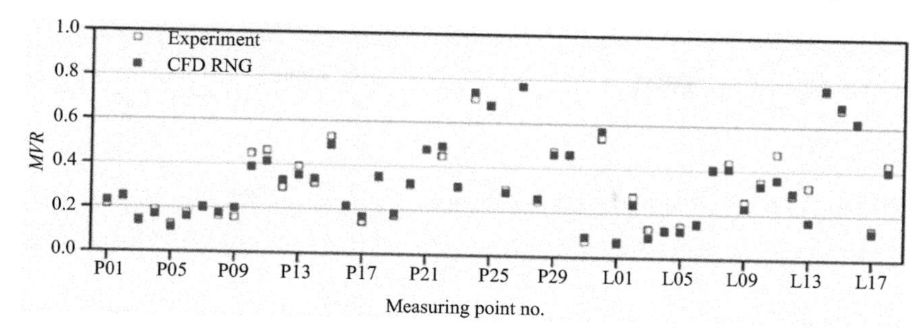

Figure 16.22 Comparison between wind tunnel test and CFD simulation (Du et al., 2018)

Zero statistic pressure was used on the outlet. The two-layer model was employed for ground. The RNG $k - \varepsilon$ model was used for turbulence modeling. The momentum and pressure were coupled by SIMPLEC algorithm with second-order upwind scheme. The convergence was achieved when residuals were less than 10^{-5}. After mesh sensitivity test, 8.9 million cells were used here.

The normalized mean wind parameter (MVR) is used to evaluate wind environment at pedestrian level. The comparison between wind tunnel test and CFD simulation is shown in Figure 16.22. It can be seen that a good agreement is achieved and the deviations are almost within 20% between CFD and wind tunnel test results, see Figure 16.22. Over 60% measuring points were within the deviation of 10%. The validation justifies the CFD method, including mesh system described earlier.

16.4 Summary

This chapter deals with urban ventilation and design, with a focus on the street canyon and precinct scales, which are in the order of approximately 10–1,000 m. For the street canyon scale, the focus is more on the physics, building aerodynamics and internal microclimates. An ideal long street canyon flanking with two parallel buildings was investigated. The characteristics of microclimates (air speed, air temperature, air pollution, and noise) inside the street canyon and the influence of aspect ratio of the street canyon were analyzed. The influence of the microclimates, especially air speed, on the building indoor ventilation performance was then examined. The implications to the design of both urban ventilation and building ventilation were discussed. For the precinct scale, the focus is more on the pedestrian microclimates and urban design. A specific urban precinct containing a university campus was investigated. The pedestrian-level wind environment and comfort at different wind speeds and directions were analyzed. The influence of a lift-up design on the local wind environment was particularly examined. The potential of using such a lift-up design to create a local cooling spot to increase urban residents' outdoor activities was discussed. Based on the generic and applied studies using these two urban configurations, an overall discussion and a perspective on urban ventilation and design were made.

References

Ai, ZT and Mak, CM. CFD simulation of flow and dispersion around an isolated building: Effect of inhomogeneous ABL and near-wall treatment. *Atmospheric Environment*. 2013, 77, 568–78.

Ai, ZT and Mak, CM. From street canyon microclimate to indoor environmental quality in naturally ventilated urban buildings: Issues and possibilities for improvement. *Building and Environment*. 2015, 94, 489–503.

Andreou, E and Axarli, K. Investigation of urban canyon microclimate in traditional and contemporary environment: Experimental investigation and parametric analysis. *Renewable Energy*. 2012, 43, 354–63.

Arnfield, AJ and Mills, G. An analysis of the circulation characteristics and energy budget of a dry, asymmetric, east, west urban canyon, 1 circulation characteristics. *International Journal of Climatology*. 1994, 14, 119–34.

Bady, M; Kato, S; Takahashi, T; and Huang, H. Experimental investigations of the indoor natural ventilation for different building configurations and incidences. *Building and Environment*. 2011, 46, 65–74.

Blocken, B; Stathopoulos, T; and Carmeliet, J. CFD simulation of the atmospheric boundary layer: Wall function problems. *Atmospheric Environment*. 2007, 41, 238–52.

Bourbia, F and Awbi, HB. Building cluster and shading in urban canyon for hot dry climate Part 1: Air and surface temperature measurements. *Renewable Energy*. 2004, 29, 249–62.

Barring, L; Mattson, JO; and Lindovist, S. Canyon geometry, street temperatures and urban heat island in Malmo Sweden. *International Journal of Climatology*. 1985, 5, 433–44.

Cheng, Y; Lee, SC; Gao, Y; *et al.* Real-time measurements of PM2.5, PM10-2.5, and BC in an urban street canyon. *Particuology*. 2015, 20, 134–40.

Chan, H; Kok, M; and Lee, T. Temperature trends in Hong Kong from a seasonal perspective. *Climate Research*. 2012, 55, 53–63.

Cheung, JOP and Liu, CH. CFD simulations of natural ventilation behavior in high-rise buildings in regular and staggered arrangements at various spacings. *Energy and Buildings*. 2011, 43, 1149–58.

Chan, LY; Lau, WL; Wang, XM; and Tang, JH. Preliminary measurements of aromatic VOCs in public transportation modes in Guangzhou, China. *Environment International*. 2003, 29, 429–35.

Du, Y. Investigation of wind and thermal comfort in high-density urban environment. The Hong Kong Polytechnic University (PhD thesis). 2018.

Du, Y; Mak, CM; and Ai, Z. Modelling of pedestrian level wind environment on a high-quality mesh: A case study for the HKPolyU campus. *Environmental Modelling & Software*. 2018, 103, 105–19.

Du, Y; Mak, CM; Huang, T; and Niu, J. Towards an integrated method to assess effects of lift-up design on outdoor thermal comfort in Hong Kong. *Building and Environment*. 2017a, 125, 261–72.

Du, Y; Mak, CM; Kwok, K; *et al.* New criteria for assessing low wind environment at pedestrian level in Hong Kong. *Building and Environment*. 2017b, 123, 23–36.

Du, Y; Mak, CM; Liu, J; Xia, Q; Niu, J; and Kwok, KCS. Effects of lift-up design on pedestrian level wind comfort in different building configurations under three wind directions. *Building and Environment*. 2017c, 117, 84–99.

DePaul, FT and Sheih, CM. Measurements of wind velocities in a street canyon. *Atmospheric Environment*. 1986, 20(3), 455–9.

DePaul, FT and Sheih, CM. A tracer study of dispersion in an urban street canyon. *Atmospheric Environment*. 1985, 19(4), 555–9.

Eliasson, I; Offerle, B; Grimmond, CSB; and Lindqvist, S. Wind fields and turbulence statistics in an urban street canyon. *Atmospheric Environment*. 2006, 40, 1–16.

Eliasson, I. Urban nocturnal temperatures, street geometry and land use. *Atmospheric Environment Part B*. 1996, 30(3), 379–92.

Georgakis, C and Santamouris, M. On the estimation of wind speed in urban canyons for ventilation purposes – Part 1: Coupling between the undisturbed wind speed and the canyon wind. *Building and Environment*. 2008, 43, 1404–10.

Eliasson, I. Urban geometry, surface temperature and air temperature. *Energy and Buildings*. 1990, 15–16, 141–5.

Georgakis, C and Santamouris, M. Experimental investigation of air flow and temperature distribution in deep urban canyons for natural ventilation purposes. *Energy and Buildings*. 2006, 38, 367–76.

Ghiaus, C and Allard, F. *Natural Ventilation in the Urban Environment: Assessment and Design*. 1st ed. New York, NY: Routledge. 2005, https://doi.org/10.4324/9781849772068.

Georgakis, C and Santamouris, M. On the airflow in urban canyons for ventilation purposes. *International Journal of Ventilation*. 2004, 3(1), 53–65.

Givoni, B. *Climate Considerations in Building and Urban Design*. New York, NY: Johan Wiley and Sons. 1998.

Google Map. The Hong Kong Polytechnic University. https://www.google.com.hk/maps/place/The+Hong+Kong+Polytechnic+University.html. Accessed in July. 2020.

Kaur, S; Nieuwenhuijsen, M; and Colvile, R. Fine particulate matter and carbon monoxide exposure concentrations in urban street transport microenvironments. *Atmospheric Environment*. 2007, 41, 4781–810.

Longley, ID; Gallagher, MW; Dorsey, JR; Flynn, M; and Barlow, JF. Short-term measurements of airflow and turbulence in two street canyons in Manchester. *Atmospheric Environment*. 2004, 38, 69–79.

Mihalakakou, G; Santamouris, M; Papanikolaou, N; and Cartalis, C. Simulation of the urban heat island phenomenon in Mediterranean climates. *Pure and Applied Geophysics*. 2004, 161, 429–51.

Mihalakakou, G; Flocas, H; Santamouris, M; and Helmis, C. The impact of synoptic scale atmospheric circulation on the urban heat island effect over Athens, Greece. *Journal of Applied Meteorology*. 2002, 41(5), 519–27.

Niachou, K; Livada, I; and Santamouris, M. Experimental study of temperature and airflow distribution inside an urban street canyon during hot summer weather conditions – Part 1: Air and surface temperatures. *Building and Environment.* 2008, 43, 1383–92.

Nakamura, Y and Oke, TR. Wind, temperature and stability conditions in an east-west oriented urban canyon. *Atmospheric Environment.* 1988, 22(12), 2691–700.

Oke, TR. *Boundary Layer Climates*. 2nd ed. New York, NY: Routledge. 1987.

Pearlmutter, D; Bitan, A; and Berliner, P. Microclimatic analysis of 'compact' urban canyons in an arid zone. *Atmospheric Environment.* 1999, 33, 4143–50.

Richards, PJ. and Hoxey, R. Appropriate boundary conditions for computational wind engineering models using the turbulence model. *Journal of Wind Engineering and Industrial Aerodynamics.* 1993, 46–47, 145–53.

Qin, Y and Kot, SC. Dispersion of vehicular emission in street canyons, Guangzhou city, South China (P.R.C.). *Atmospheric Environment.* 1993, 27B, 283–91.

Rakowska, A; Wong, KC; Townsend, T; *et al.* Impact of traffic volume and composition on the air quality and pedestrian exposure in urban street canyon. *Atmospheric Environment.* 2014, 98, 260–70.

Santamouris, M; Georgakis, C; and Niachou, A. On the estimation of wind speed in urban canyons for ventilation purposes – Part 2: Using of data driven techniques to calculate the more probable wind speed in urban canyons for low ambient wind speeds. *Building and Environment.* 2008, 43, 1411–8.

Shashua-Bar, L and Hoffman, ME. Geometry and orientation aspects in passive cooling of canyon streets with trees. *Energy and Buildings.* 2003, 35, 61–8.

Santamouris, M; Papanikolaou, N; Koronakis, I; Livada, I; and Asimakopoulos, D. Thermal and air flow characteristics in a deep pedestrian canyon under hot weather conditions. *Atmospheric Environment.* 1999, 33, 4503–21.

Tsai, MY and Chen, KS. Measurements and three-dimensional modeling of air pollutant dispersion in an urban street canyon. *Atmospheric Environment.* 2004, 38, 5911–24.

Vardoulakis, S; Gonzalez-Flesca, N; and Fisher, BEA. Assessment of traffic-related air pollution in two street canyons in Paris: Implications for exposure studies. *Atmospheric Environment.* 2002, 36, 1025–39.

Weber, S; Kuttler, W; and Weber, K. Flow characteristics and particle mass and number concentration variability within a busy urban street canyon. *Atmospheric Environment.* 2006, 40, 7565–78.

Westerdahl, D; Fruin, S; Sax, T; Fine, PM; and Sioutas, C. Mobile platform measurements of ultrafine particles and associated pollutant concentrations on freeways and residential streets in Los Angeles. *Atmospheric Environment.* 2005, 39, 3597–610.

Wang, XM; Sheng, GY; Fu, JM; *et al.* Urban roadside aromatic hydrocarbons in three cities of the Pearl River Delta, People's Republic of China. *Atmospheric Environment.* 2002, 36, 5141–8.

Yamartino, RJ and Wiegand, G. Development and evaluation of simple models for the flow, turbulence and pollution concentration fields within an urban street canyon. *Atmospheric Environment.* 1986, 20, 2137–56.

Chapter 17
Conclusion and future perspectives

Shi-Jie Cao[1] and Zhuangbo Feng[1]

17.1 Basic concept and knowledge of building ventilation

In Chapters 2 and 3, basic knowledge of building ventilation was summarized and analyzed, including definition, purpose, types and requirement of ventilation technologies. The state-of-the-art ventilation technologies and future development of ventilation were also discussed. Finally, the theory and design principles of typical building ventilation systems were summarized.

17.2 Basic components and patterns of building ventilation

17.2.1 Chapter 4: Ventilation system: duct network and fluid machinery

This chapter mainly described the basic theory, mechanism, process, strategy and detailed quantitative method of ventilation duct network design/control. Basic requirements of duct network design were summarized, including duct strength, installation/construction, air leakage rate, duct air velocity, thermal insulation and anti-fire requirements. The hydraulic calculation method for duct network was also described in detail, including three main methods: assumed velocity, average pressure loss and static regain. The design strategies of some essential devices/components in ventilation duct network were also introduced, such as air dampers and air volume adjustment, connection of duct fittings and fan (elbow, tee), materials and specification of ventilation ducts, and reinforcement and sealing of ducts. The systematic design strategy of ventilation duct network could provide powerful tool and reference for engineering design. The quantitative design strategy was mainly based on semiempirical formulas/equations. In future, the numerical method could be applied in duct network design or control in order to improve accuracy level and operation/adjustment efficiency.

[1]School of Architecture, Southeast University, Nanjing, China

17.2.2 Chapter 5: Ventilation system: air filtration technologies

This chapter introduces the electrostatic-enhanced air filtration method, including electrostatic precipitator, air cleaning technologies, ion generator, hybrid electrostatic filtration system (HEFS) and electret filters. Except for particle removal, the electrostatic-enhanced air filtration system could effectively disinfect biological aerosols, such as bacteria and virus. The available researches indicated that the newly proposed design strategy of HEFS could achieve three goals: high efficiency (100%), lower ozone generation (<10 ppb) and 90% decrease of energy consumed by HEPA. The properly designed HEFS is much more promising in future building ventilation design.

17.2.3 Chapter 6: Air distribution in mechanical ventilation system

This chapter mainly introduced the quantitative design methods of two typical mechanical ventilation patterns: mixing and displacement. In many buildings, the main aim of mechanical systems was to transport and supply clean and cooled/heated air to occupied spaces, creating safe, comfortable, healthy and energy-efficient indoor environment. The physical mechanisms of two widely adopted indoor air distribution patterns were described: mixing ventilation and displacement ventilation. For these two ventilation types, key devices (e.g. diffuser and air jet) and design strategies (including the selection of air diffuse, air jet and air distribution method) were summarized and described in detail. The essential formulas were listed.

17.2.4 Chapter 7: Air ventilation system: air distribution in a natural ventilation system

The basic physical mechanisms of natural ventilation were introduced: wind driven and buoyancy driven. Typical types of natural ventilation (e.g. single-sided, cross and stack ventilation) and quantitative design methods (e.g. envelope flow models, multi-zone models and CFD) were also briefly described. In reality, buildings combine different natural ventilation types leading to a very complicated airflow so that CFD techniques must be applied to analyze the detailed air distribution of natural ventilation.

Before a detailed design of natural ventilation, a potential evaluation of natural ventilation should be applied to determine whether natural ventilation can be used or not and provide support in detailed design period. To comprehensively evaluate the natural ventilation potential, a systematic method concerning various environmental factors, including climate characteristics, building characteristics, surrounding wind fields and outdoor air pollution, was developed. It was helpful both in the early design stages and for the control of building operations. The climate characteristics and building thermal characteristics determined indoor thermal condition, and urban text determined surrounding wind environment and natural ventilation rate. Based on outdoor air pollution, indoor thermal condition and natural ventilation rate, the potential of natural ventilation can be evaluated.

17.3 Advanced building ventilation system with a function of heating/cooling

17.3.1 Chapter 8: Ventilation systems and heating and cooling

One of the main aims of building ventilation is removing waste heat or cooling in indoor environment, and creating comfortable thermal environment. In order to overcome the disadvantages of traditional ventilation heating/cooling system, a hybrid "radiation+ventilation" system was proposed, which combined radiant heating/cooling panels with decentralized displacement ventilation units together.

Although natural ventilation and mechanical ventilation (without heating/cooling source) were energy-saving, these were quite limited to use in extreme weather. The hybrid "radiation+ventilation" system was reliable in different climate conditions. Compared with the all-air mechanical ventilation system, radiant panel systems could save around 25%–33% of cooling energy and more than 93% of fan energy, while it consumes around three times more hydronic pump energy. And decentralized ventilation system could save about 6%–7% additional total cooling energy. Compared with the typical radiant cooling system, the hybrid radiant+ventilation system can reduce around 9%–10% of the cooling energy consumption due to the highest cooling supply water temperature. Besides, hybrid radiant heating and cooling system reduced moisture condensation risk on the surface of the radiant cooling panels.

The hybrid "radiation+ventilation" system could have lower thermal energy consumption, higher thermal comfort, less construction cost, time and materials compared with the centralized HVAC system. The main disadvantage of this hybrid system is its complexity. How to efficiently control the performance of hybrid system and avoid system failure is quite challenging and warrants further investigations in future study.

17.4 Prediction and design of different ventilation types

17.4.1 Chapter 9: Natural ventilation system design: predictive methods

This chapter introduced different research methods in characterization of natural ventilation, including on-site measurement, wind tunnel experiment, empirical models and CFD simulation. Overall, taking the advantages of each method and combining them together would be the best way to investigate natural ventilation. Due to the complexity of natural ventilation, CFD is most effective and reliable tool in engineering design.

Regarding the complexity of natural ventilation that involves multiple length scales, i.e. from urban area, building, room, down to ventilation opening, any inappropriate settings of computational parameters in CFD simulation would result in incorrect solutions. More studies are required to support the establishment of guideline for CFD simulations of natural ventilation, including outdoor flow around a building, the indoor flow and the interaction of indoor and outdoor flows at a ventilation opening. The general issues in natural ventilation simulation include (1) the effect of envelope features; (2) single-story versus multistory building model; (3) isolated buildings versus urban buildings; and (4) thermal comfort criteria for naturally ventilated spaces.

Generally, ignoring the envelope features could result in large deviations between the predicted and the real-life natural ventilation performance in buildings. It is possible to use these envelope features to improve natural ventilation performance through careful designs to ingeniously utilize the building aerodynamics and to adapt to the local wind environment. In natural ventilation modeling, using single-story model could save lots of computing cost, while its simulation error was relatively large. How to utilize single-story model in practical multistory building ventilation design needs further investigations. The surrounding buildings significantly influence the natural ventilation performance of a certain building. The simplified building configurations without surrounding buildings may generate some errors. With the improvement of computational power, coupling urban environment and natural ventilation of buildings is promising. The key issue is how to ensure the necessary urban scope. Finally, thermal sensation in naturally ventilated environment was quite different from that of mechanically ventilated rooms. Thermal adaptation and tolerance should be considered thermal comfort criteria for naturally ventilated spaces.

17.4.2 Chapter 10: Ventilation system design: numerical method

The traditional design of ventilation system by CFD uses a trial-and-error process. The trial-and-error process is robust and easy to use. However, the design process can take days and the obtained air distributions may not be optimal. The CFD-based adjoint methods were developed for indoor environment optimization. For an indoor mechanical ventilation case, both the CFD-based GA and trial-and-error approaches require the calculation of about 200 cases to obtain the converged solutions, and the computing effort was about 20 times greater than that of the adjoint method. Therefore, CFD-based adjoint method has been proven to be an effective tool in built environment design. Coupling of different optimization methods (adjoint, POD and machine learning) is more promising in future research.

17.4.3 Chapter 11: Ventilation system design: fast prediction

The mixing and nodal models were very fast. However, a nodal model was difficult to use for most designers as the detailed prior knowledge of the airflow pattern was required to specify mass flow in the thermal network. Zonal models introduce more flow dynamics into the prediction of mean airflows compared to nodal models. However, the existing zonal models are limited to the prediction of mean temperature and mass flow rate in each zone. More specifically, zonal models employ two main assumptions: (1) the primary driving flows can be pre-predicted and (2) users have good knowledge of the entire flow structures. Although the nodal and zonal models are very fast in simulation, the prior knowledge, limited information and accuracy limit the application of these methods in practical prediction processes. Compared with nodal models, CFD could provide much more informative prediction of indoor environment. However, CFD models were computationally expensive to be solved due to the strong nonlinearity of the flow-governing equations.

The voice for fast and reliable simulation has been leading to various means and efforts to make CFD models less computationally expensive. Some of these efforts include developing simplified turbulence models such as zero-equation models; utilization of coarse grid; reforming solution algorithms for pressure–velocity decoupling such as Pressure-Implicit with Splitting of Operators (PISO) and semi-Lagrangian-based PISO (SLPISO) algorithm; creating reduced-order models to quantitatively describe the original dynamics of flow systems with simplified numerical models. For displacement ventilation case, the optimized coarse grid strategy could save 95% of computing time consumed by fine grid strategy. For convection case, PISO and SLPISO could, respectively, save 50% and 31% of the computing time consumed by SIMPLE algorithm. However, more efforts are needed to achieve fast-than-real time simulations in practical engineering design.

17.5 Intelligent control of building ventilation

17.5.1 Chapter 12: HVAC online monitoring and control strategy

One of the main aims of HVAC system is to satisfy the dynamics of indoor environment, including thermal environment and airborne pollutant field. Traditional static design and manual adjusting of ventilation system could not achieve accurate control, resulting in worse indoor environment and large building energy consumption. This chapter introduces a newly proposed online monitoring and control strategy, including three key components: online monitoring, "faster-than-real-time" prediction and optimal evaluation and control. The bottleneck problems are how to achieve the "faster-than-real-time" prediction of indoor environment, and how to combine limited monitoring and optimal air supply parameter evaluation/control. The coupling "faster-than-real-time" model was proposed, including low-dimensional linear ventilation model, artificial neural network and contribution ratio of indoor environment. Based on limited monitoring, fast prediction and multiple objectives, real-time control strategy was developed. By importing control algorithms into intelligent hardware, an indoor environment monitoring and control module was established.

17.6 Special topics of building ventilation

17.6.1 Chapter 13: Ventilation and health

The main function of building ventilation is controlling airborne pollutant concentration and indoor climate. Climatic factors, such as temperature, relative humidity, wind velocity and precipitation, were indicated to be related with the epidemics of allergic diseases and respiratory infections. The relationship between ventilation and human health involves air pollution and climate factors. The overall effects of early-life exposure to ventilation on childhood allergy/infection were investigated and analyzed.

The associations between natural ventilation/mechanical ventilation effects and allergies/infections in indoor environment were investigated. Poor ventilation indicators (e.g. mold/damp stains, window condensation) were positively correlated with allergy/infection (e.g. asthma, pneumonia and frequent common cold). Sensitivities of boys and girls were quite different. For example, boys were more susceptible to the protective effect of frequent opening windows especially in winter and spring on allergy/infection. The influences of wind speed on allergies/ infections of children were quite complex.

The quantitative relationship between allergy/infection and ventilation is very essential for building ventilation design, from the perspectives of human health. However, the current researches and data in literature were very limited and insufficient for practical engineering design. In future, the popular Big Data technology may be a powerful tool in revealing deep relationship between building ventilation and allergy/infection.

17.6.2 Chapter 14: Ventilation in industry buildings

This chapter summarizes the design strategy of industry ventilation. First, distinct differences between industrial buildings and civil buildings in terms of indoor sources and control demand were analyzed. Second, the design strategies of natural ventilation, local ventilation and general ventilation were introduced, including concept, basic principle and optimal design principle. Especially, highly efficient ventilation patterns were proposed: vortex and vortex ring. Vortex ring ventilation could high-efficiently achieve long-distance directional air supply. The design strategy of vortex ring ventilation was also summarized.

17.6.3 Chapter 15: Ventilation and fire safety for high-rise buildings

This chapter presents the most recent progress in high-rise building design and control methods for achieving energy efficiency and fire safety. Basic theories of high-rise ventilation were introduced, including natural ventilation, mechanical ventilation and smoke control for high-rise fires. The challenges of designing and controlling high-rise ventilation were summarized. Strategies of how to model, control and design high-rise ventilation were presented. Two case studies on high-rise fire smoke control and atrium fire smoke control are introduced and discussed. The basic knowledge, design strategies and typical cases could provide essential reference for a practical engineering design of ventilation system of high-rise buildings.

17.6.4 Chapter 16: Urban ventilation and design

This chapter introduces how to utilize numerical simulations to design urban ventilation pattern in street canyon and precinct scales with the order of approximately

10–1,000 m. For street canyon ventilation, the microclimates (air speed, air temperature, air pollution and noise) inside the street canyon were characterized, and the influences of aspect ratio of the street canyon on ventilation performance were analyzed. Besides, the influence of the microclimates, especially air speed, on building indoor ventilation performance was also evaluated. For the precinct ventilation, the pedestrian-level wind environment and comfort were evaluated. The influences of wind speed, wind direction and lift-up design on local environment were characterized. The potential of using lift-up design to create a local cooling spot was analyzed. The design strategy could provide important tools for urban planner and designer in engineering applications.

17.7 Future perspectives

In future, the fast prediction and intelligent control of building ventilation are becoming the mainstreams for future building ventilation systems. Fast prediction will involve more physical fields, including urban climate, coupling of indoor/outdoor environments, network of ventilation ducts and air cleaning system. The accuracy and computing speed will be further improved by using advanced mathematical algorithm (e.g. topology) and intelligent hardware devices (e.g. quantum computer). Incorporating fast prediction models into optimization design method (e.g. adjoint, AI) can achieve fast design, which is very promising in engineering applications. Intelligent building ventilation control will couple model-predictive control and general industrial control technology to improve control accuracy and reduce response time. Besides, advanced ventilation components will be developed to improve building ventilation performance, such as high-efficiency air cleaning and disinfection devices, low-resistance tee and damper, flexible air duct and intelligent ventilation terminals. This book mainly describes and discusses the design and control of building ventilation system. In future, the intelligent construction of ventilation system will be investigated, and some powerful tools (building information modeling) in construction management will be analyzed.

Index